The Conceptual Development of Quantum Mechanics

INTERNATIONAL SERIES IN PURE AND APPLIED PHYSICS

Leonard I. Schiff, Consulting Editor

ADLER, BAZIN, AND SCHIFFER Introduction to General Relativity
ALLIS AND HERLIN Thermodynamics and Statistical Mechanics
BECKER Introduction to Theoretical Mechanics
BJORKEN AND DRELL Relativistic Quantum Fields
BJORKEN AND DRELL Relativistic Quantum Mechanics
CHODOROW AND SUSSKIND Fundamentals of Microwave Electronics
CLARK Applied X-rays
COLLIN Field Theory of Guided Waves
EVANS The Atomic Nucleus
FEYNMAN AND HIBBS Quantum Mechanics and Path Integrals
GINZTON Microwave Measurements
GREEN Nuclear Physics
GURNEY Introduction to Statistical Mechanics
HALL Introduction to Electron Microscopy
HARDY AND PERRIN The Principles of Optics
HARNWELL Electricity and Electromagnetism
HARNWELL AND LIVINGOOD Experimental Atomic Physics
HENLEY AND THIRRING Elementary Quantum Field Theory
HOUSTON Principles of Mathematical Physics
JAMMER The Conceptual Development of Quantum Mechanics
KENNARD Kinetic Theory of Gases
LANE Superfluid Physics
LEIGHTON Principles of Modern Physics
LINDSAY Mechanical Radiation
LIVINGSTON AND BLEWETT Particle Accelerators
MIDDLETON An Introduction to Statistical Communication Theory
MORSE Vibration and Sound
MORSE AND FESHBACH Methods of Theoretical Physics
MUSKAT Physical Principles of Oil Production
NEWTON Scattering Theory of Waves and Particles
PRESENT Kinetic Theory of Gases
READ Dislocations in Crystals
RICHTMYER, KENNARD, AND LAURITSEN Introduction to Modern Physics
SCHIFF Quantum Mechanics
SEITZ The Modern Theory of Solids
SLATER Introduction to Chemical Physics
SLATER Quantum Theory of Atomic Structure, Vol. I
SLATER Quantum Theory of Atomic Structure, Vol. II
SLATER Quantum Theory of Matter
SLATER Quantum Theory of Molecules and Solids, Vol. 1
SLATER Quantum Theory of Molecules and Solids, Vol. 2
SLATER AND FRANK Electromagnetism
SLATER AND FRANK Introduction to Theoretical Physics
SLATER AND FRANK Mechanics
SMYTHE Static and Dynamic Electricity
STRATTON Electromagnetic Theory
TINKHAM Group Theory and Quantum Mechanics
TOWNES AND SCHAWLOW Microwave Spectroscopy
WHITE Introduction to Atomic Spectra

The late F. K. Richtmyer was Consulting Editor of the series from its inception in 1929 to his death in 1939. Lee A. DuBridge was Consulting Editor from 1939 to 1946; and G. P. Harnwell from 1947 to 1954.

The CONCEPTUAL DEVELOPMENT *of* QUANTUM MECHANICS

MAX JAMMER

Bar-Ilan University, Israel

McGRAW-HILL BOOK COMPANY

New York, St. Louis, San Francisco, Toronto, London, Sydney

The Conceptual Development of Quantum Mechanics

Library of Congress Catalog Card Number 66-17914

1 2 3 4 5 6 7 8 9 0 MP 7 3 2 1 0 6 9 8 7 6

Preface

It is the purpose of this study to trace the conceptual development of quantum mechanics from its inception to its formulation as a full-fledged theory of atomic physics, from its status as a rather doubtful *ad hoc* hypothesis to that of an imposing intellectual structure of great beauty.

All the great theories in the history of physics—from Aristotelian mechanics and its medieval elaborations, through Newtonian dynamics with its Lagrangian or Hamiltonian modifications, to Maxwellian electrodynamics and Einsteinian relativity—have been subjected repeatedly to historico-critical investigations, and their conceptual foundations have been thoroughly analyzed. But no comprehensive scholarly study of the conceptual development of quantum mechanics has heretofore appeared. The popular or semiscientific publications available hardly skim the surface of the subject. And the few, though important, essays on the topic written by the originators of the theory themselves are mostly confined to a particular aspect or to the defense of a specific philosophical position. The publication of a comprehensive and coherent analysis of the conceptual development of quantum mechanics, the only consistent theory of atomic processes, and hence a foundation of modern science, seems therefore to fill an important lacuna in the literature on the history and philosophy of physics.

Such a study, however, should not be regarded as an end in itself. Never before has a theory been treated in so many excellent texts within so short a period of time. Since every additional text, as is natural in a developing branch of science, attempts to construct the theory on the basis of an ever more concise logical structure, the traditional texts presenting the material in the order of its historical development are gradually losing ground. Students, it is rightfully claimed, may be saved much trouble "if they are not led through all the historical pitfalls, and instead acquainted from the very beginning with concepts, such as the spin, that cannot be grasped except by quantum mechanical means." [1] Admittedly, a good working knowledge, even a high standard of efficiency and competence in applying the theory to physical problems may be obtained by this method of instruction. But as long as it is true that "however far the phenomena transcend the scope of classical physical explanation, the account of all evidence must be expressed in classical terms," [2] a profound comprehension of quantum mechanics requires more than a study of the subject in its shortest possible logical formulation. "It is impossible to understand the methods of modern quantum mechanics without a knowledge of the way in which the theory has been developing." [3] Indeed, some knowledge of the dramatic struggle of ideas that preceded the formation of quantum conceptions and of the intricate ways of reasoning that led to the generally accepted formulation of the theory is indispensable for a profound comprehension of its physical significance, for an intelligent appreciation of its philosophical implications, and even for an enlightened understanding of its logical structure. Duhem's statement "faire l'histoire d'un principe physique, c'est en même temps en faire l'analyse logique" is particularly meaningful for the science of quantum mechanics.

The subject of our discussion may be of interest not only from the historical and methodological, but also from the purely scientific, point of view. If the fundamental conceptions of theoretical physics cannot be extracted from experience alone but are, as Einstein once said, "free inventions of the human intellect" [4] a study of their formation and its attendant empirical, mathematical, philosophical, and psychological factors may well throw some light on the questions of whether the choice of the actually accepted theoretical scheme was unique and of whether a selection of other factors, under discussion at the time, could have led to an alternative formulation of the theory.

Such conceptual and circumstantial inquiries, of course, do not exclude the possibility of an irrational residue in the process of scientific theory construction. If reductions of such irrationalities to their bare minima form the historical and methodological components of the contents of a study such as the present, considerations of the kind mentioned before

[1] F. A. Kaempffer, *Concepts in Quantum Mechanics* (Academic Press, New York and London, 1965), p. v.

[2] Niels Bohr, "Discussion with Einstein on epistemological problems in atomic physics," in *Albert Einstein: Philosopher-Scientist* (The Library of Living Philosophers, Evanston, Illinois, 1949), p. 209.

[3] A. March, *Quantum Mechanics of Particles and Wave Fields* (J. Wiley & Sons, New York: Chapman and Hall, London, 1951), p. vi.

[4] A. Einstein, *The World As I See It*, translated from *Mein Weltbild* by A. Harris (John Lane, The Bodley Head Limited, London, 1935), p. 134.

which bring this study into relation with current research constitute its scientific component.[5]

By now, the reader will have understood that the present volume, which confines itself to nonrelativistic quantum mechanics of a finite number of degrees of freedom and which, it is hoped, will be followed by a similar treatment of relativistic quantum mechanics and quantum field theory, is neither a textbook nor a collection of biographical notes nor even a study of priority questions. In fact, even the strict chronological order of presenting the material has often been violated in favor of the logical coherence of the discussion. The primary objective of this book is the study of the problems of how empirical phenomena led to a clash with established principles of classical physics, how experimental research, combined with mathematical reasoning and philosophical thought, opened entirely novel perspectives, and how in the process of constructing the conceptual edifice of quantum mechanics each stage depended on those preceding it without necessarily following from them as a logical consequence. In comparison to these, all other considerations are deemed of secondary importance.

Although various issues in the development of quantum mechanics have been pursued, as the reader will see, to the very front line of current research, the main theme of the study concludes with a discussion of the conceptual situation as brought about by the establishment of the so-called Copenhagen interpretation. As is well known, this interpretation is still espoused today by the majority of theoreticians and practicing physicists. Though not necessarily the only logically possible interpretation of quantum phenomena, it is *de facto* the only existing fully articulated consistent scheme of conceptions that brings order into an otherwise chaotic cluster of facts and makes it comprehensible. The exclusion of alternative interpretative attempts—which, it is hoped, will be discussed in another volume—should therefore not impair the logical completeness and current significance of the present study.

The sources I have been using fall into three different categories.

1. The main sources are, of course, the papers and monographs written by the originators of the theory. Full bibliographical references have been supplied in the present work so that the interested reader can easily check or pursue any issue involved. Since science consists of *recorded* verifiable

[5] Thus, for example, P. A. M. Dirac, preparing himself for an interview (conducted by Professor T. S. Kuhn in Cambridge, England, in May 1963, in connection with the project "Sources for History of Quantum Physics") reexamined some of his early unpublished notes and found that certain previously rejected ideas could now be instrumental in overcoming current convergence difficulties in quantum electrodynamics. P. A. M. Dirac, *Lectures on Quantum Electrodynamics*, delivered at Bar-Ilan University, December 1965 (oral communication).

or falsifiable general statements, these sources will have to serve as the ultimate basis for our discussion as far as the subject matter is concerned.

2. Einstein once said: "If you want to find out anything from the theoretical physicists about the methods they use, I advise you to stick closely to one principle: don't listen to their words, fix your attention on their deeds." For, he continued, "to the discoverer in this field the products of his imagination appear so necessary and natural that he regards them, and would have them regarded by others, not as creations of thought but as given realities."[6] Generally speaking, Einstein was certainly right. With respect to our present subject, however, which deals with conceptual constructs that on the whole defy all visualization or picturability and hence can hardly be called "necessary and natural," the danger referred to cannot be serious. I felt therefore entitled to ignore this warning and discussed the subject with quite a number of prominent physicists who contributed decisively to the development of the theory.

Appointments at the *Institut Henri Poincaré* in Paris, at the *Federal Institute of Technology* (*E.T.H.*) in Zurich, at the *University of Göttingen*, and at the *Catholic University of America* in Washington, D.C., gave me the opportunity to discuss various issues of the subject with Prince Louis de Broglie, Professors Max Born, Werner Heisenberg, Paul A. M. Dirac, Friedrich Hund, Walter Heitler, B. L. Van der Waerden, Marcus Fierz, Res Jost, and to receive valuable suggestions from Sir George Thomson, Professors E. N. da C. Andrade, John C. Slater, Erwin Fues, and Franz Tank as well as from Professors Gerald Holton, Paul K. Feyerabend, and Thomas S. Kuhn. I am grateful to these distinguished physicists and philosophers of science for their stimulating comments.

3. I wish to thank also the U.S. National Science Foundation for having granted me the opportunity to study thoroughly the *Archive for the History of Quantum Physics* deposited at the Library of the *American Philosophical Society* in Philadelphia, at the Library of the *University of California* in Berkeley, and at the *Universitets Institut for Teoretisk Fysik* in Copenhagen. The material gathered in this *Archive* contains numerous interviews, recorded in tape and transcript, with many physicists who contributed to the development of quantum mechanics. These interviews proved highly informative and reliable. However, since there was always the danger that the subject interviewed retrojected later conceptions or results into his recollections of earlier developments, the conclusions drawn from such interviews were critically compared with other sources of information whenever available.

I am also indebted to Mrs. Wolfgang Pauli for her courtesy in arranging for me to peruse the *Pauli Collection* in Zollikon-Zurich.

[6] REF. 4, p. 131.

Having soon realized that in quantum mechanics, though less ostensibly so than in relativity, almost every phase of the development bears the mark of Einstein's ideas, I studied the documentary material of the *Einstein Estate* at the *Institute for Advanced Study* in Princeton, N.J. Einstein's correspondence with the leading originators of quantum mechanics proved indeed most revealing.

I am greatly indebted to Professor P. A. M. Dirac, who read parts of the manuscript when visiting Bar-Ilan University in the winter of 1965–1966.

Finally, I wish to express my deep gratitude to Professor Werner Heisenberg for the privilege he afforded me of discussing with him, at the *Max-Planck-Institut für Physik* in Munich, various aspects of the work and for his kindness in critically reading the entire typescript of this book. Needless to say, the full responsibility for any errors or misinterpretations rests only upon me.

Max Jammer

Contents

The Conceptual Development of Quantum Mechanics

1 The Formation of Quantum Conceptions

1.1 Unsolved Problems in Classical Physics

Quantum theory, in its earliest formulation, had its origin in the inability of classical physics to account for the experimentally observed energy distribution in the continuous spectrum of black-body radiation.

It was unfortunate that the first problem dealt with in the systematic development of quantum theory was that of the energy quantization of harmonic electromagnetic vibrations. There can be no doubt that much intellectual effort could have been saved on the part of the early quantum theoretists—as well as on the part of the reader of the present text—if a conceptually less involved issue had initiated the development of the theory.

This raises the question: Was it possible that other problems at issue at the time, if worked out consistently, could have brought about the same conceptual reorientation as did the problem of black-body radiation? And perhaps in a logically less complicated way? Consider, for example, the well-known irreconcilability with classical physics of the specific heat of solids at low temperatures, a problem whose solution was obtained *de facto* in terms of concepts formed in solving the black-body problem. One may conjecture how an independent and consistent solution of this problem on specific heat would have influenced the progress of theoretical physics. It seems highly probable that in this hypothetical case energy quantization of material systems (atoms or molecules) would have preceded that of waves and the approach to quantum theory would have been conceptually less difficult.

It may even be argued that the energy quantization of atoms, at least so far as their ground state is concerned, could have been inferred without great difficulties from the empirical foundations of the classical kinetic theory.[1] Did not the fundamental observation that the specific heat per mole of a monatomic gas at constant temperature is $\frac{3}{2}R$ clearly indicate, if explained as it was in terms of the average kinetic energy $\frac{3}{2}kT$, that any energy supplied to the gas can increase only the kinetic energy of the atoms and not the energy of the internal motions of their constituent parts, whatever they were thought to be? Was it not common knowledge that in the process of atomic or molecular collisions the kinetic energy is conserved and none of it transformed into internally bound energy? (Today this fact is explained by saying that the kinetic energy of particles, which at room temperature is of the order of $\frac{1}{40}$ eV, is much too small to excite the atom, its excitation energy being of the order of a few electron volts.) In retrospect it seems quite possible that a profound critical study of these and similar facts could have led to anticipating the idea of energy quantization of atoms many decades before Bohr and thus to a more convenient approach to quantum-theoretic conceptions. For, as it has once been said, "research is to see what everybody has seen and to think what nobody has thought." But *post iacturam quis non sapit?*

There were, of course, reasons that quantum theory developed just the way it did. In order to understand fully how and why just the phenomenon of black-body radiation was to play such a decisive role in the process of physical concept formation, our analysis of the conceptual development of quantum mechanics will begin with a discussion on the foundations of the classical theory of thermal radiation.

"A few weeks ago I had the honor of addressing the Academy with a memoir on some observations which seemed to me most interesting as they allow drawing conclusions on the chemical constitution of the solar atmosphere. Starting from these observations I have now derived, on the basis of rather simple theoretical considerations, a general theorem which, in view of its great importance, I allow myself to present to the Academy. It deals with a property of all bodies and refers to the emission and absorption of heat and light." With these words Kirchhoff began his classic paper "On the relation between emission and absorption of light and heat"[2] which he read before the Berlin Academy in 1859 and in which he showed that "for rays of the same wavelength at the same temperature the ratio of the emissive power to the absorptivity is the same for all bodies."[3] To prove this theorem, which was later called "Kirchhoff's law," Kirchhoff

[1] For the didactical implications of this statement see Sir Nevill Mott, "On teaching quantum phenomena," *Contemporary Physics 5*, 401–418 (1964).

[2] Gustav Robert Kirchhoff, "Über den Zusammenhang zwischen Emission und Absorption von Licht und Wärme," *Monatsberichte der Akademie der Wissenschaften zu Berlin 1859* (December), pp. 783–787; reprinted in G. Kirchhoff, *Gesammelte Abhandlungen* (Leipzig, 1882), pp. 566–571, and in *Ostwald's Klassiker der exakten Wissenschaften No. 100* (Akademische Verlagsgesellschaft, Leipzig, 1889), pp. 6–10.

[3] "Für Strahlen derselben Wellenlänge bei derselben Temperatur ist das Verhältnis des Emissionsvermögens zum Absorptionsvermögen bei allen Körpern dasselbe." *Ibid.* (1859), p. 783.

calculated the equilibrium conditions for the radiation exchange between two emitting and absorbing infinite parallel plates facing each other and covered on the outside surfaces by ideal reflectors. He assumed that one of the plates emits and absorbs radiation only of wavelength λ whereas the other does so for all wavelengths. Since all radiation of wavelength different from λ emitted by the second plate is finally reabsorbed by the emitter after repeated reflections, the calculation could be restricted to the exchange of radiation of wavelength λ. If E is the emissive power (i.e., the energy emitted at wavelength λ per unit time) and A the absorptivity (i.e., the fraction of the incident radiation at wavelength λ absorbed) of one plate and e and a those of the other, Kirchhoff showed that the latter absorbs the amount $aE/(1 - k)$ from E, where $k = (1 - A)(1 - a)$, and the amount $a(1 - A)e/(1 - k)$ from e. Since at equilibrium the energy emitted equals the energy absorbed,[4]

$$e = \frac{aE}{1-k} + \frac{a(1-A)e}{1-k}$$

and hence

$$\frac{e}{a} = \frac{E}{A}$$

which proved the theorem. A year later Kirchhoff published a second paper[5] in which he treated the same subject in greater detail and supplied a more rigorous analytical proof of the theorem.

From the historical point of view it is interesting to note that it was the study of the Fraunhofer lines in the solar spectrum which led Kirchhoff to his investigations on the relation between emission and absorption. This is shown not only by his own introductory remarks to his first paper,

[4] With this statement Kirchhoff implicitly adopted Pierre Prévost's "theory of exchange" according to which "the equilibrium of heat between two neighboring free spaces consists in the equality of the exchanges" (P. Prévost, *Du Calorique Rayonnant*, p. 258). Prévost, a disciple of LeSage, conceived heat as a caloric fluid which radiates constantly from every point of the surface of all bodies in straight lines, the intensity of the radiation being proportional to the quantity of heat contained in the body; a constant exchange of heat takes place among neighboring bodies, each body radiating as if the other were not present; under conditions of equilibrium each body absorbs just as much as it emits. Prévost published his theory in his *Du Calorique Rayonnant* (Paris, Geneva, 1809) and in his *Exposition Élémentaire des Principes qui Servent de Base à la Théorie de la Chaleur Rayonnante* (Geneva, Paris, 1832). Some passages of the former treatise have been translated into English by D. B. Brace in *The Laws of Radiation and Absorption—Memoirs by Prévost, Stewart, Kirchhoff, and Kirchhoff and Bunsen* (Harper's Scientific Memoirs, edited by J. S. Ames; American Book Company, New York, 1901).

[5] "Über das Verhältnis zwischen dem Emissionsvermögen und dem Absorptionsvermögen der Körper für Wärme und Licht," *Poggendorffs Annalen der Physik 109*, 275–301 (1860); translated into English by F. Guthrie, "On the relation between the radiating and absorbing powers of different bodies for light and heat," *Philosophical Magazine 20*, 1–21 (1860); *Gesammelte Abhandlungen* (REF. 2), pp. 571–598; *Ostwald's Klassiker* (REF. 2), pp. 11–36. Cf. also the translation in Brace (REF. 4), pp. 73–97.

as quoted above, but also by the fact that his second paper was part of a study entitled *Investigations on the Solar Spectrum and the Spectra of Chemical Elements.*[6] Because of its fundamental importance for the physics of radiation and for the subsequent development of theoretical physics, Kirchhoff's law was repeatedly subjected to critical examinations and was given increasingly rigorous proofs.[7]

Simultaneously with Kirchhoff's publications, Stewart[8] established independently the validity of the law on the basis of experimental measurements. Stewart's paper, like Kirchhoff's, was based on Prévost's theory. The primary concern of the Scottish physicist, however, was an experimental verification[9] of radiation relations, in contrast to that of Kirchhoff, who stressed the theoretical aspects involved. In fact, Kirchhoff seems to have fully recognized the theoretical significance of the ratio E/A for he emphasized the importance of an exact experimental determination of this function of temperature and expressed the hope that this would raise no special difficulties since "all functions encountered so far which are independent of the nature of bodies are of simple structure."

In the beginning of his second paper Kirchhoff introduced the notion of what he called "a *perfectly black*, or more briefly *black body*,"[10] defining it—just as it is defined today—as a body which absorbs all the radiation incident upon it. Thus the absorptivity A of a black body being unity, its emissive power E becomes identical with the aforementioned universal function. On the basis of simple thermodynamic considerations Kirchhoff then proceeded to show that cavity radiation, i.e., the radiation inside an enclosure of adiathermanous walls of equal temperature T, is of the same

[6] *Untersuchungen über das Sonnenspectrum und die Spectren der chemischen Elemente* (Berlin, Dümmler, 2d ed., 1862); English translation by H. E. Roscoe, *Researches on the Solar Spectrum, and the Spectra of the Chemical Elements* (Macmillan, Cambridge, London, 1862).

[7] A derivation of Kirchhoff's law on less restrictive assumptions (i.e., without assuming the existence of perfectly black bodies, perfect reflectors, etc.) was given by E. Pringsheim in his paper "Einfache Herleitung des Kirchhoff'schen Gesetzes," *Verhandlungen der Deutschen Physikalischen Gesellschaft 3*, 81–84 (1901); *Zeitschrift für wissenschaftliche Photographie 1*, 360–364 (1903). Rigorous derivations were given by G. C. Evans, "Note on Kirchhoff's law," *Proceedings of the American Academy 46*, 95–106 (1910), and by D. Hilbert, on the basis of integral equations, in his "Begründung der elementaren Strahlungstheorie," *Physikalische Zeitschrift 13*, 1056–1064 (1912), *14*, 592–595 (1914); *Göttinger Nachrichten 1912*, pp. 773–789, *1913*, pp. 409–416; *Jahresbericht der Deutschen Mathematiker Vereinigung 22*, 1–16 (1913); reprinted in David Hilbert, *Gesammelte Abhandlungen* (J. Springer, Berlin, 1935), vol. 3, pp. 217–230, 231–237.

[8] Balfour Stewart, "Account of some experiments on radiant heat," *Transactions of the Royal Society of Edinburgh 22*, 1–3, 95–97, 426–429 (1858).

[9] Cf. his "An account of some experiments on radiant heat," *Proceedings of the Royal Society of Edinburgh 1858–1859*, pp. 203–204.

[10] "Ich will solche Körper *vollkommen schwarze*, oder kürzer *schwarze* nennen," REF. 5, p. 277; p. 573. The term "black body" had already been used, though of course not yet as a *terminus technicus*, by Newton in *Query 6* of his *Opticks* (1704), where he asked: "Do not black bodies conceive heat more easily from light than those of other colours do . . . ?" (Dover edition, New York, 1952), p. 339.

"quality and intensity" as the emissive power E of a black body[11] of the same temperature T. Kirchhoff's cavity theorem, in modern terminology, summarized the following situation. If $B_\nu = B(\nu,T)$ denotes the energy emitted per unit frequency interval at ν, per unit area, per unit solid angle, and per unit time in a direction normal to the surface of a black body[12] of temperature T so that for a surface obeying Lambert's law the total emissive power over a hemisphere is $E = \pi\int_0^\infty B_\nu\, d\nu$, and if $u_\nu = u(\nu,T)$ is the radiation density inside an enclosure of temperature T per unit frequency interval at ν so that the total energy density is $u = \int_0^\infty u_\nu\, d\nu$, then

$$u_\nu = \frac{4\pi}{c} B_\nu \qquad \text{or} \qquad u = \frac{4}{c} E \tag{1.1}$$

a relation which can be proved by simple geometric considerations.[13] The second law of thermodynamics implies that u_ν and u are independent of the nature of the walls of the enclosure.

Kirchhoff's law, the fundamental theorem for the *theoretical* study of thermal radiation, was, as we have seen, historically viewed, the outcome of astrophysical or, more precisely, solar-physical investigations. The most important single device for the *experimental* study of thermal radiation owes its existence likewise to solar-physical investigations. For it was in the course of his study of the atmospheric absorption of solar radiation that S. P. Langley invented in 1880 the bolometer or, as he called it at first tentatively, the "actinic balance." Prior to his appointment in 1887 as Secretary of the Smithsonian Institution, Langley was professor at the Western University of Pennsylvania (today the University of Pittsburgh) and in charge of the Allegheny Observatory at Allegheny City, which specialized in solar research. Dissatisfied with the available equipment and, in particular, with the thermopile, invented by Nobili and improved by Melloni and Tyndall, Langley, supported by a grant from the American Academy of Arts and Sciences, had been, during some time, working experiments on the device and construction of an instrument more delicate and more prompt than the thermopile.[14] His bolometer,[15] which consisted

[11] In view of the identity of cavity and black-body radiation M. Thiesen suggested, forty years later, in a paper "Über das Gesetz der schwarzen Strahlung," *Verhandlungen der Deutschen Physikalischen Gesellschaft 2*, 65–70 (1900), that this radiation be called "black radiation" ("schwarze Strahlung"), a term which gained some popularity mainly through its adoption by M. Planck in his *Vorlesungen über die Theorie der Wärmestrahlung* (Barth, Leipzig, 1st ed. 1906, 2d ed. 1913; *Theory of Heat Radiation*, Dover, New York, 1959), which presents his lectures on this subject at the Berlin University.

[12] The brightness B_ν is often expressed as $2K_\nu$, where the factor 2 is introduced because of the two planes of polarization.

[13] The velocity of light c appears because of the energy flux density which has to be introduced into the calculation.

[14] Samuel Pierpont Langley, "The actinic balance," *American Journal of Science 21*, 187–198 (1881).

[15] S. P. Langley, "The bolometer and radiant energy," *Proceedings of the American Academy 16*, 342–359 (1881). The term "actinic balance" referred to the bridge or balance for measurement of radiation (ἀκτίς = ray); "bolometer" (the term is introduced on p. 349) is derived from βολή = beam of light.

of two blackened platinum strips, 0.5 mm wide and 0.002 mm thick, placed in opposite arms of a Wheatstone bridge, improved the sensitivity of radiation measurements by at least a factor of 10. The instrument was soon employed in spectroscopic research, but its importance for the study of black-body radiation was not realized until a convenient source for such radiation was made available.

Although the idea that isothermal enclosures are black-body radiators had been suggested, at least implicitly, as early as 1884 by Christiansen,[16] who stated that "holes" ("Löcher") act like "small completely black spots" ("kleine vollkommen schwarze Flecken"), and although Boltzmann, with reference to Christiansen, applied this idea in the same year in a theoretical analysis of the relation between thermal radiation and thermodynamics,[17] it was only in 1895, on the basis of Kirchhoff's cavity theorem, that Lummer and Wien constructed the first isothermal enclosure to serve as a source of radiation.[18] This device, combined with Langley's bolometer, facilitated considerably the systematic study of black-body radiation. This study was carried out toward the turn of the century at various centers of research and particularly at the Imperial Physico-Technical Institute ("Physikalische-Technische Reichsanstalt"), founded in 1884 in Berlin, Charlottenburg, where Lummer, Pringsheim, Holborn, Rubens, and others worked on the problem.

Meanwhile the problem began to be tackled also by theoretical physicists. Having read in Wüllner's textbook on heat[19] about John Tyndall's experiments,[20] according to which the total emission of a heated platinum wire at 1200°C (1473°K) was 11.7 times that at 525°C (798°K) and realizing that $(\frac{1473}{798})^4$ is approximately 11.7, Stefan[21] concluded in 1879 that the total radiation E is proportional to T^4. Stefan claimed that he found additional support for the validity of this relation in the measurements performed by de la Provostayes, Desains, Draper, and Ericsson. Today we know that Stefan's inference, based after all on relatively meager experimental evidence, was to a high degree fortuitous. A modern repetition of Tyndall's experiment—Tyndall measured a radiation which was far from that of a black body—would yield a ratio of 18.6 [22] rather than 11.7.

[16] C. Christiansen, "Über die Emission der Wärme von unebenen Oberflächen," *Wiedemannsche Annalen der Physik 21*, 364–369 (1884).

[17] Ludwig Boltzmann, "Über eine von Hrn. Bartoli entdeckte Beziehung der Wärmestrahlung zum zweiten Hauptsatz," *Wiedemannsche Annalen der Physik 22*, 31–39 (1884).

[18] Otto Lummer and Willy Wien, "Methode zur Prüfung des Strahlungsgesetzes absolut schwarzer Körper," *Wiedemannsche Annalen der Physik 56*, 451–456 (1895).

[19] A. Wüllner, *Die Lehre von der Wärme vom Standpunkte der mechanischen Wärmetheorie* (2d ed., Teubner, Leipzig, 1875), p. 215.

[20] J. Tyndall, "Ueber leuchtende und dunkle Strahlung," *Poggendorff's Annalen der Physik 124*, 36–53 (1865).

[21] Josef Stefan, "Über die Beziehung zwischen der Wärmestrahlung und der Temperatur," *Wiener Berichte 79*, 391–428 (1879).

[22] See, for example, A. G. Worthing and D. Halliday, *Heat* (Wiley, New York, 1948), p. 438.

Stefan's "fourth-power law" for black-body radiation

$$E = \sigma T^4 \tag{1.2}$$

was put on firm experimental foundations only in 1897 by Paschen,[23] by Lummer and Pringsheim,[24] and by Mendenhall and Saunders.[25]

A theoretical proof, however, of Stefan's law had already been provided by Boltzmann in 1884. The principal idea of this proof occurred to Boltzmann in the course of a critical study of a publication by Bartoli[26] to which Wiedemann had drawn his attention. The Italian physicist described in a *Gedankenexperiment* a cyclic process by which, with the help of moving mirrors, heat in the form of radiation could be transferred from a cold body to a hot one; since according to the second law of thermodynamics, as formulated by Clausius, such a heat transfer requires the performance of work, he concluded that heat radiation exerts pressure. Boltzmann[27] recognized immediately the importance of these qualitative conclusions and attempted to obtain a quantitative elaboration in accordance with Maxwell's electromagnetic theory of light. In a second paper, published shortly afterward, Boltzmann[28] showed, in accordance with Maxwell's theory[29] and Krönig's statistical averaging method,[30] that this pressure is equal to one-third of the energy density. Having thus associated with the "radiation gas" two thermodynamic variables, the temperature T and the pressure p, Boltzmann proved by thermodynamic reasoning that $u = aT^4$ and consequently $E = \sigma T^4$. The modern derivation of Stefan's law subjects the radiation gas to a Carnot cycle or, equivalently, computes the thermal energy dQ necessary for an isothermal expansion and inte-

[23] F. Paschen, "Über Gesetzmäßigkeiten in den Spektren fester Körper," *Wiedemannsche Annalen der Physik 60*, 662–723 (1897).

[24] O. Lummer and E. Pringsheim, "Die Strahlung eines 'schwarzen Körpers' zwischen 100°C und 1300°C," *Wiedemannsche Annalen der Physik 63*, 395–410 (1897); "Die Vertheilung der Energie im Spectrum des schwarzen Körpers," *Verhandlungen der Deutschen Physikalischen Gesellschaft 1*, 23–41 (1899); "(1) Die Vertheilung der Energie im Spectrum des schwarzen Körpers und des blanken Platins, (2) Temperaturbestimmung fester glühender Körper," *ibid.*, 215–235.

[25] C. E. Mendenhall and F. A. Saunders, "Das Energiespectrum eines absolut schwarzen Körpers," *Naturwissenschaftliche Rundschau 13*, 457–460 (1898); "The energy spectrum of an absolutely black body," *Johns Hopkins University Circulars 17*, 55 (1898).

[26] A. Bartoli, *Sopra i Movimenti Produtti dalla Luce e dal Calore e sopra il Radiometro di Crookes* (Florence, 1876); cf. also "Il calorico raggiante e il secondo principio di termodinamica," *Il Nuovo Cimento 15*, 193–202 (1884).

[27] See REF. 17.

[28] L. Boltzmann, "Ableitung des Stefan'schen Gesetzes betreffend die Abhängigkeit der Wärmestrahlung von der Temperatur aus der electromagnetischen Lichttheorie," *Wiedemannsche Annalen der Physik 22*, 291–294 (1884).

[29] Boltzmann referred explicitly to article 792 of Maxwell's *Treatise of Electricity and Magnetism* (1873), where in the chapter on "Energy and Stress of Radiation" it is stated that "in a medium in which waves are propagated there is a pressure in the direction normal to the waves and numerically equal to the energy in unit of volume." The first experimental verification of radiation pressure was obtained in 1899 by P. Lebedew.

[30] A. Krönig, "Grundzüge einer Theorie der Gase," *Poggendorffs Annalen der Physik 99*, 315–322 (1856).

grates the integrability conditions of the total differential dQ/T. This is but a slight modification of a second proof which Boltzmann supplied at the end of his paper. The exact determination of the constant a or, equivalently, of the "Stefan constant" $\sigma = ac/4$ was the object of numerous measurements at the turn of the century. The value of σ, averaged over the results obtained by Kurlbaum, Féry, Bauer and Moulin, Valentiner, Féry and Drecq, Shakespear, and Gerlach, was found to be $\sigma = 5.672 \times 10^{-5}$ erg/(sec)(cm²)(deg⁴).

Neither Stefan's law nor Boltzmann's derivation of it paid any attention to the spectral distribution of the radiation. Boltzmann's thought experiment which involved the reflection of radiation from a moving piston clearly implied a redistribution with respect to frequency in accordance with the Doppler effect. In order to calculate the change of u_ν or B_ν due to the Doppler shift, Wien[31] studied the adiabatic contraction of a perfectly reflecting sphere and showed that B_ν must satisfy the equation $B_\nu = \nu^3 F(\nu/T)$, where $F(\nu/T)$ is an as yet unknown function of the argument ν/T. This equation or its equivalent form[32]

$$E_\lambda = \lambda^{-5}\varphi(\lambda T) \tag{1.3}$$

if referred not to the frequency ν but, as is usually done in experimental physics, to the wavelength λ, was called "Wien's displacement law."[33] Among its numerous experimental confirmations one of the most painstaking was that by Paschen.[34] Wien's displacement law was rightfully regarded as an important achievement, for it showed that from the spectral distribution of black-body radiation for any given temperature the distribution for any other temperature could be deduced. Another immediate conclusion from the displacement law was the relation

$$\lambda_{\max} T = b \tag{1.4}$$

stating that the wavelength $\lambda_{\max}$ at which E_λ has its maximum value is

[31] Willy Wien, "Temperatur und Entropie der Strahlung," *Wiedemannsche Annalen der Physik 52*, 132–165 (1894).

[32] E_λ is the emissive power per unit interval of wavelength so that $E_\lambda\, d\lambda = -B_\nu\, d\nu$.

[33] For a derivation of the law in modern notation but closely akin to the spirit of Wien's reasoning, see F. K. Richtmyer, E. H. Kennard, T. Lauritsen, *Introduction to Modern Physics* (McGraw-Hill, New York, 5th ed., 1955), pp. 113–118. See also E. Buckingham's paper "On the deduction of Wien's displacement law," *Philosophical Magazine 23*, 920–931 (1912), which studies the necessary conditions for a rigorous proof. For a particularly lucid derivation of the law cf. Max Born, *Atomic Physics* (Blackie & Son, London, 6th ed., 1957), pp. 410–412. For a derivation from the Stefan-Boltzmann law by dimensional analysis cf. L. Hopf, "Über Modellregeln und Dimensionsbetrachtungen," *Die Naturwissenschaften 8*, 110 (1920). That the displacement law implies the Stefan-Boltzmann law can be shown easily by simple integration.

[34] F. Paschen, "Über die Vertheilung der Energie im Spectrum des schwarzen Körpers bei niederen Temperaturen," *Berliner Berichte 1899*, pp. 405–420; "Über die Vertheilung der Energie im Spectrum des schwarzen Körpers bei höheren Temperaturen," *ibid.*, pp. 959–976.

inversely proportional to the absolute temperature T. Lummer and Pringsheim confirmed[35] this relation and found for the constant b the numerical value 0.294 cm deg. The problem of explaining the empirical distribution curves of E_λ or, equivalently, of B_ν or u_ν seemed almost to have found its solution. It was reduced to the explanation of a single function $\varphi(\lambda T)$. It was clearly understood, however, that—contrary to the numerous characteristic physical constants and functions encountered in thermodynamics, electrodynamics, etc., whose particular values seemed merely to express more or less accidental properties of matter—the function $\varphi(\lambda T)$ must play a much more fundamental role. In view of the universal character of black-body radiation, its independence of the particular properties of the material substances, and thus, ultimately, in view of the general validity of Kirchhoff's law and the second law of thermodynamics, a theoretical explanation of this function was rightfully regarded as a problem involving fundamental issues of far-reaching importance. In fact, as we shall see in the sequel, the inability of nineteenth-century physics to account for the experimentally established function $\varphi(\lambda T)$ signalized the breakdown of classical mechanics. For the time being, however, desperate attempts were made to explain the empirical data.

In his search for a theoretical derivation of this function Wien was convinced that all previous attempts[36] to solve this problem not only were unsuccessful in attaining their aim but attacked the problem from a wrong point of departure. The only exception, in his opinion, was in this respect Michelson's "Theoretical essay on the distribution of energy in the spectra of solids,"[37] which explained the continuity of these specta on the basis of the irregularity in the vibrations of the atoms. As a consequence of this assumption Michelson based his explanation on statistical considerations and employed the theory of probability. It was, of course, the classical conception of probability as used in statistical mechanics. In fact, Michelson assumed that Maxwell's classic formula for the distribution of velocities of gas molecules holds also for the molecules of the solid radiating black body.

In Wien's view Michelson's approach was a "fortunate idea."[38] Following these suggestions, Wien assumed that the wavelength λ of the radiation emitted by any molecule and the corresponding intensity are functions of the molecule's velocity v alone and that, conversely, v^2 is

[35] See REF. 24.

[36] One of the earliest of these attempts was E. Lommel's "Theorie der Absorption und Fluorescenz," *Wiedemannsche Annalen der Physik 3*, 251–283 (1877).

[37] V. A. Michelson, "Essai théorique sur la distribution de l'énergie dans les spectra des solides," *Journal de Physique 6*, 467–480 (1887); the paper was originally published, under the same title, in the *Journal de la Société Physico-Chimique Russe 19*, 79–92 (1887); an English version appeared in *Philosophical Magazine 25*, 425–435 (1888).

[38] W. Wien, "Über die Energievertheilung im Emissionsspectrum eines schwarzen Körpers," *Wiedemannsche Annalen der Physik 58*, 662–669 (1896). On p. 663 Wien says, "der glückliche Gedanke Michelsons."

a function of λ. From the Maxwell-Boltzmann distribution function $v^2 \exp(-v^2/aT)\, dv$, which determines the number of molecules whose velocity lies between v and $v + dv$, where a is a constant for a given gas, he inferred that $E_\lambda = g(\lambda) \exp[-f(\lambda)/T]$, where $g(\lambda)$ and $f(\lambda)$ denote unknown functions. Since, however, this expression for E_λ had to be consistent with his previous result (1.3), Wien concluded that[39]

$$E_\lambda = c_1\lambda^{-5} \exp\left(-\frac{c_2}{\lambda T}\right) \qquad \text{or} \qquad u_\nu = \alpha\nu^3 \exp\left(-\frac{\beta\nu}{T}\right) \tag{1.5}$$

where c_1, c_2, α, and β are constants.

1.2 The Concept of Quanta of Energy

Although based on rather questionable arguments, Wien's law of radiation, as the preceding formula (1.5) was called, seemed to give an adequate account of all experimental data available at that time. Paschen,[40] in a series of painstaking measurements, and Wanner,[41] who used for this purpose a photometric method, confirmed the law, at least for the visible region, for temperatures up to 4000°C. Furthermore, a more rigorous derivation of Wien's law of radiation was proposed by Max Planck in a series of papers[42] presented to the Berlin Academy of Sciences during the years 1897 to 1899.

Planck, it is known, attended Kirchhoff's lectures in Berlin but derived his main inspiration from Clausius' writing on thermodynamics, a subject which early became his field of specialization. The measurements performed by Lummer and Pringsheim at the Imperial Physico-Technical Institute in Berlin attracted his attention to the problem of black-body radiation. This problem, as he later admitted in his autobiography,[43] had for him a special fascination for, in view of its universal character, it represented something absolute and he "had always regarded the search for the absolute as the loftiest goal of all scientific activity."

Taking advantage of the fact that according to Kirchhoff the radiation

[39] *Ibid.*

[40] See REF. 23.

[41] F. Paschen and H. Wanner, "Eine photometrische Methode zur Bestimmung der Exponentialconstanten der Emissionsfunction," *Berliner Berichte 1899*, pp. 5–11.

[42] M. Planck, "Über irreversible Strahlungsvorgänge," *Berliner Berichte*: 1st communication, Feb. 4, 1897, pp. 57–68; 2d communication, July 8, 1897, pp. 715–717; 3d communication, Dec. 16, 1897, pp. 1121–1145; 4th communication, July 7, 1898, pp. 449–476; 5th communication, May 18, 1899, pp. 440–480. These papers are summarized in his article "Über irreversible Strahlungsvorgänge," *Annalen der Physik 1*, 69–122 (1900), and reprinted in Planck's *Physikalische Abhandlungen und Vorträge* (Friedr. Vieweg & Sohn, Braunschweig, 1958), vol. 1, pp. 493–600.

[43] Wissenschaftliche Selbstbiographie, *Physikalische Abhandlungen und Vorträge*, (REF. 42), vol. 3 (1958), pp. 389–390; *Scientific Autobiography* (Philosophical Library, New York, 1949), pp. 34–35.

distribution at equilibrium is independent of the nature of the radiators, Planck started with the simplest possible assumption that the radiators are linear harmonic oscillators of frequency ν. The theory of electric-dipole radiation had been well known since 1889, when Hertz published his classic paper[44] on this subject.

Planck most probably based his program of research on the following considerations. Wien's derivation of his radiation law, in spite of its untenable theoretical foundations, seemed to produce a correct result. Planck therefore assumed that Wien's approach was probably not altogether wrong. The basic element in Wien's derivation was the Maxwell-Boltzmann velocity distribution, which, according to the kinetic theory of gases, characterizes the equilibrium reached by irreversible processes from arbitrary initial conditions.

Now, since cavity radiation is a process related to electromagnetism rather than to the kinetic theory of gases, Planck decided that he could do the same for Maxwell's theory of the electromagnetic field as Boltzmann had done for mechanics with his famous, though still disputed, H theorem. Planck thus attempted to prove that the Maxwell-Hertz equations, if applied to the resonators with arbitrary initial conditions, would lead to irreversible processes converging toward a stationary state whose energy distribution is that of cavity radiation and which therefore determines the energy spectrum of black-body radiation. In short, what Planck seems to have had in mind was a translation of the reasoning that led to the Maxwell-Boltzmann velocity distribution on the basis of the kinetic theory into the conceptual structure of electromagnetic theory.

In his study[45] of the absorption and emission of electrical waves by resonance he thought he had found such an irreversible process in the interaction between absorbing and emitting resonators.[46] When this result was disproved by Boltzmann,[47] however, Planck proposed—in analogy to Boltzmann's statistical hypothesis of the so-called "molecular chaos" in the kinetic theory—a theory of "natural radiation" according to which the harmonic partial vibrations composing a wave of thermal radiation are completely incoherent.[48] Equating the emission and absorption rates of the resonators, Planck[49] obtained as equilibrium conditions, using only principles of classical electrodynamics, the equation

$$u_\nu = a\nu^2 U \tag{1.6}$$

[44] H. Hertz, "Die Kräfte electrischer Schwingungen, behandelt nach der Maxwell'schen Theorie," *Wiedemannsche Annalen der Physik 36*, 1–22 (1889).

[45] "Absorption und Emission electrischer Wellen durch Resonanz," *Berliner Berichte 1895, 15* and *16*, 289–301 (Session on Mar. 21, 1895); *Wiedemannsche Annalen der Physik 57*, 1–14 (1896); *Physikalische Abhandlungen und Vorträge* (REF. 42), vol. 1, pp. 445–458.

[46] REF. 42, 1st communication.

[47] "Über irreversible Strahlungsvorgänge," *Berliner Berichte 1897*, pp. 660–662, 1016–1018 (Session on June 17, 1897).

[48] REF. 42, 4th communication.

[49] See Appendix A.

where $U = \bar{U} = U(\nu,T)$ denotes the average energy of a harmonic oscillator at temperature T, a a constant $(8\pi/c^3)$, and u_ν the energy density as defined previously. U could, of course, be easily determined from the equipartition theorem of statistical mechanics according to which, loosely speaking, the total kinetic energy, under equilibrium conditions, is equally divided among the degrees of freedom, or more rigorously stated, the same amount of energy $(\frac{1}{2}kT)$ has to be allotted, on the average, to every Hamiltonian coordinate (q or p) in terms of which the Hamiltonian can be expressed as a quadratic function.

In view of the precarious status of the equipartition theorem within the subsequent development of quantum physics, it seems appropriate to include a brief digression on the doctrine of energy equipartition.

The earliest enunciation of the equipartition theorem, though only with reference to translational motion and based on hardly defendable arguments, was in a memoir "On the physics of media that are composed of free and perfectly elastic molecules in a state of motion" by J. J. Waterston. In this paper, which was later called "a foundation-stone of a new branch of scientific knowledge"[50] since it anticipated a great deal of the work of Krönig, Clausius, and Maxwell in the kinetic theory of gases, Waterston contended that "in mixed media the mean square molecular velocity is inversely proportional to the specific weight of the molecules" (the last six words meaning "the mass of the molecules"). A former student of John Leslie and since 1839 a naval instructor at the East India Company in Bombay, Waterston sent this paper to the Royal Society in 1845. The Society, however, rejected its publication after one of the two referees described it as "nothing but nonsense, unfit even for reading before the Society." In accordance with the rules of the Society the manuscript became its property and was buried in the archives. Only a very brief report appeared in the *Proceedings*.[51] Convinced of the importance of his paper, Waterston circulated an abstract privately.[52] Five years later, at the 21st meeting of the British Association, in Ipswich, a short extract[53] was read in which it was stated that "equilibrium of pressure and heat between two gases takes place when the number of atoms in unity of volume is equal, and the *vis viva* of each atom is equal." But, again, nobody took notice of the statement or its potential implications. "It is probable

[50] J. S. Haldane (p. lxv) in the introduction to *The Collected Scientific Papers of John James Waterston* (Oliver & Boyd, Edinburgh, London, 1928).

[51] *Proceedings of the Royal Society of London* (*A*) *5*, 604 (1846). The report was communicated by Captain Beaufort, the head of the Hydrographer's Department in the Admiralty, where Waterston had worked prior to his departure to the Far East.

[52] "An account of a mathematical theory of gases—being an outline of the demonstrations contained in a paper 'On the physics of media that consist of perfectly elastic molecules in a state of motion' submitted to the Royal Society in October 1845." Reprinted in *The Collected Scientific Papers* (REF. 50), pp. 320–331.

[53] J. J. Waterston, "On a general theory of gases," *British Association Reports, Ipswich*, *21*, p. 79 (1851).

that in the long and honourable history of the Royal Society no mistake more disastrous in its actual consequences for the progress of science and the reputation of British science than the rejection of Waterston's papers was ever made. . . . There is every reason for believing that had the papers been published physical chemistry and thermodynamics would have developed mainly in this country and along much simpler, more correct, and more intelligible lines than those of their actual development."[54] More than forty years had to pass until the 1845 memoir, under its original title, was finally published in the *Philosophical Transactions.*[55] With reference to Waterston's enunciation of the equipartition theorem, Lord Rayleigh, through whose efforts the paper was published, declared in the introduction: "The omission to publish it at the time was a misfortune which probably retarded the development of the subject by ten or fifteen years. It is singular that Waterston appears to have advanced no claim for subsequent publication, whether in the *Transactions* of the Society, or through some other channel. At any time since 1860 reference would naturally have been made to Maxwell, and it cannot be doubted that he would have at once recommended that everything possible should be done to atone for the original failure of appreciation."

Rayleigh's reference to Maxwell alludes to his paper "Illustrations of the dynamical theory of gases,"[56] in which Maxwell elaborated some conclusions submitted one year earlier at the Aberdeen meeting of the British Association.[57] There he gave his first formulation of the equipartition theorem as follows: "Two different sets of particles will distribute their velocities, so that their *vires vivae* will be equal." At first he considered only the case of "smooth spherical particles" but later, in a corollary, extended the theorem to the case of a mixture of particles of any form and included rotation.

In 1868 the theorem was further generalized by Boltzmann, who proved[58] its validity also for particles which are not necessarily rigid but have a number of internal degrees of freedom. Finally, Maxwell[59] removed certain restrictions on the interaction among particles and showed, using generalized Lagrangian coordinates for systems with an arbitrary number of degrees of freedom, that the equipartition of energy holds even if "the

[54] REF. 50, p. lxv.

[55] *Philosophical Transactions of the Royal Society of London, 183,* 1–79 (1892); reprinted in *The Collected Scientific Papers* (REF. 50), pp. 207–319.

[56] *Philosophical Magazine 20,* 21–37 (1860); reprinted in *The Scientific Papers of James Clerk Maxwell,* edited by W. D. Niven (Cambridge University Press, 1890; republished by Dover, New York), vol. 1, pp. 378–409.

[57] J. C. Maxwell, "On the dynamical theory of gases," *British Association Reports, Aberdeen, 29,* 9 (1859).

[58] L. Boltzmann, "Studien über das Gleichgewicht der lebendigen Kraft zwischen bewegten materiellen Punkten," *Wiener Berichte 58,* 517–560 (1868).

[59] J. C. Maxwell, "On Boltzmann's theorem on the average distribution of energy in a system of material points," *Transactions of the Cambridge Philosophical Society 12,* 547–570 (1878); *Scientific Papers* (REF. 56), vol. 2, pp. 713–741.

material points may act on each other at all distances, and according to any law which is consistent with the conservation of energy . . . the only assumption which is necessary for the direct proof is that the system, if left to itself in its actual state of motion, will, sooner or later, pass through every phase which is consistent with the equation of energy."

Yet the very generalizations of the theorem jeopardized its validity. For, as Tait[60] put it in his critical examination of Boltzmann's approach and in his search for an unassailable proof of the theorem: "There can be no doubt that each individual particle of a gas has a very great number of degrees of freedom besides the six which it would have if it were rigid; the examination of its spectrum while incandescent proves this at once. But if all these degrees of freedom are to share the whole energy (on the average) equally among them, the results of theory will no longer be consistent with our experimental knowledge of the two specific heats of a gas, and the relations between them." A still more drastic description of the problems raised by the theorem was given by Lord Kelvin[61] in a lecture at the Royal Institution on April 27, 1900, when he said: "The beauty and clearness of the dynamical theory, which asserts heat and light to be modes of motion, is at present obscured by two clouds. The first . . . involved the question, How could the earth move through an elastic solid, such as essentially is the luminiferous ether? The second is the Maxwell-Boltzmann doctrine of partition of energy."

It should be obvious from these remarks that the scientific literature at the end of the nineteenth century, both in England and on the continent, contained numerous articles dealing with the doctrine of energy equipartition, and there cannot be any doubt that Planck must have had knowledge of the equipartition theorem.

However, fortunately for the future development of physics, Planck did not make use of the theorem. It is hard to say whether it was in view of these difficulties or because of his unfamiliarity with the Boltzmann-Gibbs methods of statistical mechanics[62] or his profound aversion to the molecular approach[63] or, finally, because of his strong conviction in the power of thermodynamic reasoning based on the concept of entropy. One thing is certain: had he used the equipartition theorem at this stage of

[60] P. G. Tait, "On the foundations of the kinetic theory of gases," *Transactions of the Royal Society of Edinburgh, 33,* 65–95, 251–277 (1886; all four parts of this paper, published between 1886 and 1891, are reprinted in *Scientific Papers by Peter Guthrie Tait* (Cambridge University Press, 1900), vol. 2, pp. 124–208.

[61] Lord Kelvin, "Nineteenth century clouds over the dynamical theory of heat and light," *Philosophical Magazine 2,* 1–40 (1901); *Baltimore Lectures on Molecular Dynamics and the Wave Theory of Light* (London, Baltimore, 1904), pp. 486–527.

[62] This reason is given by Werner Heisenberg, *Das Plancksche Wirkungsquantum* (Walter de Gruyter, Berlin, 1945); reprinted in *Max Planck, Erinnerungen* (W. Keiper, Berlin, 1948), pp. 69–82.

[63] Arguments for the plausibility of this reason are adduced by Martin J. Klein, "Max Planck and the beginning of the quantum theory," *Archives for History of Exact Sciences 1,* 459–479 (1962).

his work, he would necessarily have arrived at the Rayleigh-Jeans law of radiation, which is incompatible with experience, and would probably have given up further research on this problem. Instead, adopting what he later called the "thermodynamic approach," Planck[64] defined the entropy S of an oscillator by the equation

$$S = \frac{U}{a\nu} \log \left(\frac{U}{eb\nu}\right) \tag{1.7}$$

where a and b are constants and e is the base of natural logarithms. The quantity $\partial^2 S/\partial U^2$, which proved important in connection with the principle of the increase of entropy, thus satisfied the equation

$$\frac{\partial^2 S}{\partial U^2} = \frac{\text{const}}{U} \tag{1.8}$$

The reasoning which led Planck to this apparently arbitrary definition of S in terms of U and ν may be reconstructed on the basis of certain remarks found at the end of his fifth communication.[65] From (1.6) and guided by the form of (1.5), Planck obtained $U = C\nu \exp(-\beta\nu/T)$, where C is a constant. Solving this equation for T^{-1}, which according to thermodynamics equals, at constant volume, $\partial S/\partial U$, Planck obtained (1.7) by simple integration. Consistent with his assumption concerning the irreversibility associated with "natural radiation," he then showed that the "total electric entropy" $S_t = \sum S + \int s\, d\tau$, where the summation extends over all oscillators and the integration over all volume elements $d\tau$ of the radiation field with entropy density s, is a function of state which increases in time and reaches a maximum at equilibrium. Planck now assumed that a small amount of energy passes from one oscillator of frequency ν, entropy S, and energy U, to another of frequency ν', entropy S', and energy U'. The entropy and energy principles require that $\delta S_t = \delta S + \delta S' = 0$ and $\delta U + \delta U' = 0$, which in view of (1.7) lead to the equation $-(a\nu)^{-1} \log (U/b\nu) = -(a\nu')^{-1} \log (U'/b\nu')$. Hence the expression on the left-hand side of this equation is a constant for all oscillators considered and therefore, in virtue of (1.6), a common parameter of u_ν for all ν. Setting this expression—which has just been shown to be a function only of T—equal to T^{-1}, Planck obtained $U = b\nu \exp(-a\nu/T)$, which, in combination with (1.6), yielded Wien's radiation law (1.5). Being fully aware that the result was determined by the particular choice of (1.7), Planck contended that an equation only of the form of (1.5) does lead to an expression for S which satisfies the entropy principle.

[64] REF. 42, 5th communication.

[65] In section 23 of this paper Planck refers to a "retrogressive computation" of the entropy from the energy distribution law: ". . . berechnet man daraus rückwärts den Ausdruck der Entropie . . . ." *Physikalische Abhandlungen und Vorträge* (REF. 42), vol. 1, p. 596.

Before the turn of the century, however, the unrestricted validity of Wien's radiation law (1.5) was seriously challenged. Lummer and Pringsheim[66] recorded systematic deviations for smaller frequencies; when additional measurements[67] in the range from 12 to 18 μ confirmed their suspicion, they had the courage to declare: "It has been demonstrated that black-body radiation is not represented, in the range of wavelengths measured by us, by the Wien-Planck spectral equation."[68] In addition to these objections based on experiment also Planck's theoretical procedure in deriving equation (1.5) became the target of severe criticisms.[69] No wonder that Thiesen,[70] Lummer and Jahnke,[71] and Lummer and Pringsheim[72] proposed new distribution laws to fit also the experimental data obtained for longer waves.

Meanwhile Lord Rayleigh, in a two-page paper "Remarks upon the law of complete radiation,"[73] published in June, 1900, showed that the equipartition theorem of statistical mechanics, if applied to the electromagnetic vibrations of cavity radiation, led necessarily to a formula radically different from (1.5). Rayleigh, an expert in the mathematical treatment of standing waves, as he had already shown in his *Theory of Sound*,[74] computed the number N_λ of modes of free electromagnetic vibrations per unit volume in an enclosure and per unit range of wavelength at λ and found[75]—if we consider Jean's subsequent correction[76] of Rayleigh's result—that N_λ is equal to $8\pi/\lambda^4$. Assuming that the average energy of each mode at temperature T, according to the equipartition theorem, is

[66] REF. 24 (1899).

[67] O. Lummer and E. Pringsheim, "Über die Strahlung des schwarzen Körpers für lange Wellen," *Verhandlungen der Deutschen Physikalischen Gesellschaft 2*, 163–180 (1900).

[68] *Ibid.*, p. 171.

[69] W. Wien, "Les lois théoriques du rayonnement," in *Rapports Présentés au Congrès International de Physique* (Gauthier-Villars, Paris, 1900), vol. 2, pp. 23–40. Cf. also REF. 67, p. 166.

[70] REF. 11.

[71] O. Lummer and E. Jahnke, "Über die Spectralgleichung des schwarzen Körpers und des blanken Platins," *Annalen der Physik 3*, 283–297 (1900).

[72] O. Lummer and E. Pringsheim, "Über die Strahlung des schwarzen Körpers für lange Wellen," *Verhandlungen der Deutschen Physikalischen Gesellschaft 2*, 163–180 (1900).

[73] *Philosophical Magazine 49*, 539–540 (1900); *Scientific Papers of Lord Rayleigh* (Cambridge University Press, 1902–1910; Dover, New York, 1964), vol. 4, pp. 483–485.

[74] John William Strutt, Baron Rayleigh, *The Theory of Sound* (Macmillan, London, 1877–1878; 2d ed. 1929; Dover, New York, 1945).

[75] In the second volume of *The Theory of Sound* (*ibid.*), sec. 267, Rayleigh had solved the same problem for acoustical vibrations. See chap. 13, which begins with the words: "We will now inquire what vibrations are possible within a closed rectangular box . . ." (1878 ed., p. 65; 1929 ed., p. 67; 1945 ed., p. 69).

[76] J. H. Jeans, "On the partition of energy between matter and ether," *Philosophical Magazine 10*, 91–98 (1905). Jeans pointed out that the original Rayleigh formula had to be divided by 8 since only the octant of positive integers, and not the whole sphere, has to be taken into account. Thus, Jeans's contribution to the "Rayleigh-Jeans" law was only the statement: "It seems to me that Lord Rayleigh has introduced an unnecessary factor 8 by counting negative as well as positive values of his integers." *Ibid.*, p. 98, Postscript, added June 7.

$(R/N)T$, where R is the universal gas constant and N Avogadro's number, or kT, where k is Boltzmann's constant, introduced at that time by Planck, Lord Rayleigh obtained for the energy density per unit interval of wavelength $u_\lambda = 8\pi kT/\lambda^4$ or equivalently

$$u_\nu = \frac{8\pi\nu^2 kT}{c^3} \tag{1.9}$$

This formula, the "Rayleigh-Jeans radiation law," agreed, of course, with Wien's displacement law (1.3). It also agreed with all experimental data in the region of extremely low frequencies, just where Wien's radiation law failed. On the other hand, it was immediately clear that (1.9) must be wrong for high frequencies. It assigned no maximum to u_ν or B_ν, contrary to experience, and led—in view of the unlimited increase for higher frequencies—to a divergent integral for the total energy density u, a situation which was later, following Ehrenfest,[77] referred to as the "ultraviolet catastrophe."

However, that for low frequencies and high temperatures u was proportional to T, as required by (1.9) in contrast to (1.5), had meanwhile been established irrefutably in a series of measurements carried out by Rubens and Kurlbaum.[78] The importance of these measurements for the future development of quantum theory is best characterized by Planck himself, who admitted that "without the intervention of Rubens the formulation of the radiation law and consequently the foundation of quantum theory would have perhaps taken place in a totally different manner and perhaps even not at all in Germany."[79] A few days before presenting their results to the Berlin Academy,[80] which was to convene on October 25, 1900, Rubens and Kurlbaum reported their observations to Planck. Convinced by their report of the inadequacy of Wien's radiation law, Planck realized that his reasoning leading to the Wien formula had to be revised so that it would lead to a new formula which for large ν and small T agrees with Wien's expression but for small ν and large T reduces to a proportionality of u_ν with T. Planck's point of departure was, of course, definition (1.7).

[77] P. Ehrenfest, "Welche Züge der Lichtquantenhypothese spielen in der Theorie der Wärmestrahlung eine wesentliche Rolle," *Annalen der Physik 36*, 91–118 (1911). The fourth chapter of this paper is entitled "Die Vermeidung der Rayleigh-Jeans-Katastrophe im Ultravioletten," where the term "ultraviolet catastrophe" appeared for the first time. Reprinted in P. Ehrenfest, *Collected Scientific Papers*, edited by Martin J. Klein (North-Holland Publishing Co., Amsterdam; Interscience, New York, 1959), pp. 185–212.

[78] H. Rubens and F. Kurlbaum, "Anwendung der Methode der Reststrahlen zur Prüfung des Strahlungsgesetzes," *Annalen der Physik 4*, 649–666 (1901).

[79] "Gedächtnisrede des Hrn Planck auf Heinrich Rubens," *Berliner Berichte 1923* (June 28), p. cxi.

[80] "Über die Emission langwelliger Wärmestrahlen durch den schwarzen Körper bei verschiedenen Temperaturen," *Berliner Berichte 1900*, pp. 929–941.

Since under the latter conditions with u_ν, according to (1.6), also U has to be proportional to T and since $\partial S/\partial U = T^{-1}$, Planck inferred that S is proportional to log U or

$$\frac{\partial^2 S}{\partial U^2} = \frac{\text{const}}{U^2} \tag{1.10}$$

whereas for the former conditions (1.8) has to remain valid. Compromising, therefore, between (1.8) and (1.10), Planck assumed[81]

$$\frac{\partial^2 S}{\partial U^2} = \frac{a}{U(U+b)} \tag{1.11}$$

which, in fact, reduces for small values of U to (1.8) and hence to Wien's law, and for large values of U to (1.10) and hence to the Rubens-Kurlbaum results.

This interpolation, though mathematically a mere trifle, was one of the most significant and momentous contributions ever made in the history of physics. Not only did it lead Planck, in his search for its logical corroboration, to the proposal of his elementary quantum of action and thus initiate the early development of quantum theory, as we shall see presently; it also contained certain implications which, once recognized by Einstein, affected decisively the very foundations of physics as well as their epistemological presuppositions. Never in the history of physics was there such an inconspicuous mathematical interpolation with such far-reaching physical and philosophical consequences.

Now, from this interpolation, Eq. (1.11), Planck deduced that

$$\frac{1}{T} = \frac{\partial S}{\partial U} = a' \log \frac{U+b}{U}$$

or

$$U = \frac{b}{\exp(1/a'T) - 1}$$

where $a' = -a/b$ and b are, of course, still functions of ν. To find their dependence on ν, Planck referred to (1.5) and (1.6) and obtained

$$U = \nu \Phi\left(\frac{\nu}{T}\right) \tag{1.12}$$

where $\Phi(\nu/T)$ is a function of ν/T. Hence, he concluded,

$$U = \frac{\text{const } \nu}{\exp(c'\nu/T) - 1}$$

[81] a and b are constants.

and finally

$$E_\lambda = \frac{C\lambda^{-5}}{\exp\,(c/\lambda T) - 1}$$

or

$$u_\nu = \frac{A\nu^3}{\exp\,(B\nu/T) - 1} \tag{1.13}$$

where c', C, A, and B are constants.

Planck obtained this result just in time to prepare an extended "comment" ("Diskussionsbemerkung") to follow Kurlbaum's report to the German Physical Society, which met on October 19, 1900.

In this "comment," published under the title "On an improvement of Wien's radiation law,"[82] Planck announced formula (1.13), the formulation of what was later called "Planck's law of radiation." For the time being, it was an empirical formula since its basic assumption (1.11) had no rigorous theoretical justification. But it seemed to be a correct formula. Rubens,[83] who through the night following the Academy session checked it against his experimental results, reported complete agreement, as did Lummer and Pringsheim after correcting their own errors of calculation a short time afterward.

To change the status of (1.11) from that of a "lucky guess" ("eine glücklich erratene Interpolationsformel")[84] to that of a "statement of real physical significance," Planck ultimately found it necessary to abandon his "thermodynamic approach" and to turn to Boltzmann's probabilistic conception of entropy.[85] Writing S_N for the entropy of a system of N oscillators of frequency ν, Planck, apparently following Boltzmann, posited $S_N = k \log W$, where W is the number of distributions compatible with the energy of the system. In order to determine W, Planck had to assume that the total energy $U_N = NU$ consists of an integral number P of "energy elements" ϵ ("Energie-elemente") so that $U_N = P\epsilon$, for the traditional conception of U_N as a continuous magnitude would not have admitted a combinatorial procedure for the determination of W. Since it is exactly at this point that the methodological requirement for a combinatorial procedure motivated Planck's introduction of the quantum of

[82] "Über eine Verbesserung der Wienschen Spektralgleichung," *Verhandlungen der Deutschen Physikalischen Gesellschaft 2*, 202–204 (1900); the paper was read at the meeting of the German Physical Society on Oct. 19, 1900: *Physikalische Abhandlungen und Vorträge* (REF. 42), vol. 1, pp. 687–689.

[83] *Ibid.*, vol. 3, p. 263.

[84] *Ibid.*, p. 125.

[85] L. Boltzmann, "Über die Beziehung zwischen dem zweiten Hauptsatz der mechanischen Wärmetheorie und der Wahrscheinlichkeitsrechnung respective den Sätzen über das Wärmegleichgewicht," *Wiener Berichte 76*, 373–435 (1877). Reprinted in L. Boltzmann, *Wissenschaftliche Abhandlungen* (Barth, Leipzig, 1909), vol. 2, p. 164.

action which, in its turn, led eventually to the development of quantum theory and its departure from the principles of classical physics, it is appropriate to quote Planck's first explicit reference to h: "Now we have to consider the distribution of the energy U_N among the N resonators of frequency ν. If U_N were regarded as an infinitely divisible quantity, the distribution could be performed in an infinite number of ways. We consider, however—and this is the cardinal point of the whole computation—U_N as composed of a finite number of discrete equal parts and employ for this purpose the natural constant $h = 6.55 \times 10^{-27}$ erg sec. This constant multiplied by the common frequency ν of the resonators gives the energy element ϵ in ergs, and by dividing U_N by ϵ we obtain the number P of energy elements which are distributed among the N resonators."[86]

Interpreting W in the equation $S_N = k \log W$ as the number of possible ways of distributing P energy elements ϵ among N oscillators, Planck[87] obtained

$$W = \frac{(N + P - 1)!}{(N - 1)!P!}$$

[86] "... Nun ist noch die Verteilung der Energie U_N auf die N Resonatoren mit der Schwingungszahl ν vorzunehmen. Wenn U_N als unbeschränkt teilbare Größe angesehen wird, ist die Verteilung auf unendlich viele Arten möglich. Wir betrachten aber—und dies ist der wesentliche Punkt der ganzen Berechnung—U_N als zusammengesetzt aus einer ganz bestimmten Anzahl endlicher gleicher Teile und bedienen uns dazu der Naturconstante $h = 6.55 \times 10^{-27}$ erg sec. Diese Constante mit der gemeinsamen Schwingungszahl ν der Resonatoren multipliziert ergiebt das Energieelement ε in erg, und durch Division von U_N durch ε erhalten wir die Anzahl P der Energieelemente, welche unter die N Resonatoren zu verteilen sind." *Berliner Berichte* (Dec. 14, 1900); *Physikalische Abhandlungen und Vorträge* (REF. 42), vol. 1, pp. 700–701. Planck used the letter E instead of our U_N.

[87] A very simple proof of the combinatorial formula was given by P. Ehrenfest and H. Kamerlingh Onnes in their paper "Vereenvoudigde afleiding van de formule uit de combinatieleer, welke Planck aan zijne theorie der straling ten groundslag heeft gelegd," *Verslag van de Gewone Vergaderingen der Wis- en Natuurkundige Afdeeling, Koninklijke Akademie van Wetenschappen te Amsterdam 23*, 789–790 (1914); the paper had an appendix: "De tegenstelling tusschen de hypothese der energietrappen van Planck en de hypothese der energiequanta van Einstein," *ibid.*, 791–792. The English version, "Simplified deduction of the formula from the theory of combinations which Planck uses as the basis of his radiation theory," appeared in the *Proceedings of the Amsterdam Academy 17*, 870–872 (1914), and the appendix "The contrast between Planck's hypothesis of the energy-grades and Einstein's hypothesis of the energy quanta," in *ibid.*, 872–873; both are reprinted in P. Ehrenfest, *Collected Scientific Papers* (REF. 77), pp. 353–356. Cf. also "Vereinfachte Ableitung der kombinatorischen Formel, welche der Planckschen Strahlungstheorie zugrunde liegt," *Annalen der Physik 46*, 1021–1022 (1915), with the appendix "Der Gegensatz zwischen der Energiestufenhypothese von Planck und der Energiequantenhypothese von Einstein," *ibid.*, 1022–1024. The possible distributions are represented by a set of P identical symbols for the P energy elements and by a different set of $N - 1$ identical symbols for "partitions." The $(N + P - 1)!$ possible permutations of all symbols, divided by the $P!$ permutations of the energy elements and the $(N - 1)!$ permutations of the partitions, represent all possible modes of distribution. It is tacitly assumed, as we see, that the energy elements are indistinguishable or, in other words, that the exchange of any two energy elements, even if they belong to different resonators, does not produce a new mode of distribution.

and by the use of Stirling's formula

$$W = \frac{(N + P)^{N+P}}{N^N P^P}$$

so that

$$S_N = k[(N + P) \log (N + P) - N \log N - P \log P]$$

or finally

$$S_N = kN\left[\left(1 + \frac{U}{\varepsilon}\right) \log \left(1 + \frac{U}{\varepsilon}\right) - \frac{U}{\varepsilon} \log \frac{U}{\varepsilon}\right] \tag{1.14}$$

Since the entropy $S = S_N/N$ of a single oscillator did indeed satisfy Eq. (1.11), Planck felt sure he was on the right track. From $\partial S/\partial U = 1/T$ he now obtained for the average energy U of the oscillators of frequency ν

$$U = \frac{\epsilon}{\exp (\epsilon/kT) - 1} \tag{1.15}$$

which is compatible with his previous result $U = \nu\Phi(\nu/T)$ only if $\epsilon = h\nu$, where h is a constant independent of ν. Finally, in view of Eq. (1.6) Planck arrived at his famous radiation law

$$u_\nu = \frac{8\pi\nu^2}{c^3} \frac{h\nu}{\exp (h\nu/kT) - 1} \tag{1.16}$$

in agreement with Eq. (1.13). Integrating Eq. (1.16) over all frequencies, Planck obtained the Stefan-Boltzmann law and established a relation between k^4/h^3 and σ; calculating the frequency at which u_ν reaches a maximum, he confirmed Eq. (1.4) and related h/k to b. From the known values of σ and b Planck computed the numerical value of the constant of action and found $h = 6.55 \times 10^{-27}$ erg sec. In addition he computed $k(1.346 \times 10^{-16}$ erg deg$^{-1})$ and, with the help of the gas constant R, Avogadro's number $(6.175 \times 10^{23}$ mole$^{-1})$. Finally, from Faraday's constant he determined the elementary unit charge $e(4.69 \times 10^{-10}$ esu).

These results were obtained within a period of about eight weeks which Planck described two decades later: "After a few weeks of the most strenuous work of my life, the darkness lifted and an unexpected vista began to appear."[88] At the meeting of the German Physical Society on December 14, 1900, a date which is often regarded as the "birthday of

[88] "Die Entstehung und bisherige Entwicklung der Quantentheorie," *Nobel Prize Lecture*, delivered to the Royal Swedish Academy, Stockholm, on June 2, 1920. *Physikalische Abhandlungen und Vorträge* (REF. 42), vol. 3 pp. 121–134; English translation in Planck, *A Survey of Physics* (Methuen, London, 1922; Dover, New York, 1960).

quantum theory,"[89] Planck read his historic paper "On the theory of the energy distribution law of the normal spectrum,"[90] in which he presented these results and introduced the "universal constant h," destined to change the course of theoretical physics.

It should be noted that Planck's combinatorial approach differed from Boltzmann's probabilistic method[91] in so far as Planck associated W with S_N at the equilibrium state without maximizing it. For Planck W was merely the total number of possible complexions and not, as for Boltzmann, the number of possible complexions corresponding to the macro state which can be realized by the largest number of complexions. The reason for this deviation was probably the fact, as already pointed out by Rosenfeld,[92] that Planck's actual point of departure was the expression (1.14) for S, to fit conjecture (1.11), and that therefore W, to satisfy the equation $S_N = k \log W$, necessarily had to be of the form $(N + P)^{N+P}/N^N P^P$, which, because of its similarity to the well-known combinatorial formula[93] $(N + P - 1)!/P!(N - 1)!$, prompted him to adopt the combinatorial procedure the way he did.

It is also interesting to note that nowhere in this paper, nor in any other of his early writings, did Planck bring into prominence the fundamental fact that U is an integral multiple of $h\nu$. At that time Planck apparently was not yet quite sure whether his introduction of h was merely a mathematical device or whether it expressed a fundamental innovation of profound physical significance. In an unpublished letter[94] (1931), addressed to R. W. Wood, Planck described in detail the psychological motives which led him to the postulate of energy quanta: he called it "an act of desperation," done because "a theoretical explanation *had* to be supplied at all cost, whatever the price." As he admitted later in his *Autobiography*,[95] he was dissatisfied with his own approach and attempted repeatedly, though unsuccessfully, to fit the introduction of h somehow ("irgendwie") into the framework of classical physics. On the other hand, his son reported how his father, on long walks through the Grunewald, a forest in the suburbs of

[89] E.g., by Max von Laue in his Memorial Address, delivered at Planck's funeral in the Albani Church, Göttingen, on Oct. 7, 1947. Cf. *Physikalische Abhandlungen und Vorträge* (REF. 42), vol. 3, p. 419; *Scientific Autobiography* (Philosophical Library, New York, 1949), p. 10.

[90] "Zur Theorie des Gesetzes der Energieverteilung im Normalspektrum," *Verhandlungen der Deutschen Physikalischen Gesellschaft 2*, 237–245 (1900); "Über das Gesetz der Energieverteilung im Normalspektrum," *Annalen der Physik 4*, 553–563 (1901); *Physikalische Abhandlungen und Vorträge* (REF. 42), vol. 1, pp. 717–727.

[91] Boltzmann's method would also have led to (1.14).

[92] L. Rosenfeld, "La première phase de l'évolution de la Théorie des Quanta," *Osiris 2*, 149–196 (1936).

[93] This formula had already appeared in Boltzmann's paper referred to in REF. 85.

[94] "Kurz zusammengefasst kann ich die ganze Tat als einen Akt der Verzweiflung bezeichnen." The letter (Oct. 7, 1931) is deposited at the Center for History and Philosophy of Physics, American Institute of Physics, New York.

[95] *Physikalische Abhandlungen und Vorträge*, vol. 3, p. 267.

Berlin, intimated to him his feelings of having made a discovery comparable perhaps only to the discoveries of Newton.[96]

For the time being, at least until 1905, nobody in fact seems to have realized that Planck's was indeed "a discovery comparable perhaps only to the discoveries of Newton." Germany's official *Physical Abstracts* of that time, the *Fortschritte der Physik*, edited and published by the German Physical Society, mentioned Planck's contribution only in outline ("in äußersten Umrißen").[97] Outside Germany it seems to have attracted still less attention. An exception was Arthur L. Day's report on Planck's work to the 547th meeting of the Philosophical Society of Washington[98] in 1902. J. H. Jeans's first edition of his *Dynamical Theory of Gases*,[99] published in 1904, contained no reference whatever to Planck's law. In short, Planck's introduction of h seems to have been regarded at that time as an expedient methodological device of no deeper physical significance, although his radiation law was repeatedly subjected to experimental test. It was confirmed by Holborn and Valentiner,[100] by Coblentz,[101] and by Warburg and his collaborators.[102] On the other hand, as late as 1919 Nernst and Wulf[103] thought they had found deviations from Planck's law. Subsequent research, however, fully vindicated his result from both the experimental and the theoretical points of view.[104] With the increasing number of experimental

[96] E.g., cf. W. Heisenberg, *Physics and Philosophy* (G. Allen & Unwin, London, 1959), p. 35.

[97] 56. *Jahrgang*, for 1900 (1901), p. 338.

[98] A. L. Day, "Measurement of high temperature," *Science* (n.s.) *15*, 429–433 (1902).

[99] Cambridge University Press, 1904.

[100] L. Holborn and S. Valentiner, "Eine Vergleichung der optischen Temperaturskala mit dem Stickstoffthermometer bis 1600°," *Annalen der Physik 22*, 1–48 (1907).

[101] W. W. Coblentz, "A characteristic of spectral energy curves," *Physical Review 31*, 314–319 (1910). Cf. also E. Baisch, "Versuche zur Prüfung des Wien-Planckschen Strahlungsgesetzes im Bereich kurzer Wellenlängen," *Annalen der Physik 35*, 543–590 (1911).

[102] E. Warburg, G. Leithäuser, E. Hupka, and C. Müller, "Über die Konstante c des Wien-Planckschen Strahlungsgesetzes," *Annalen der Physik 40*, 609–634 (1913); E. Warburg and C. Müller, "Über die Konstante c des Wien-Planckschen Strahlungsgesetzes," *ibid. 48*, 410–432 (1915).

[103] W. Nernst and T. Wulf, "Über eine Modifikation der Planckschen Strahlungsformel auf experimenteller Grundlage," *Verhandlungen der Deutschen Physikalischen Gesellschaft 21*, 294–337 (1919). They proposed to add on the right-hand side of (1.16) a factor $(1 + \alpha)$, where α is a function of ν.

[104] H. Rubens and G. Michel, "Prüfung der Planckschen Strahlungsformel," *Physikalische Zeitschrift 22*, 569–577 (1921), confirmed Planck's formula by precision measurements and showed that the results obtained by Nernst and Wulf (see REF. 103) were erroneous. For subsequent theoretical derivations of Planck's formula cf. J. Weiss, "Über das Plancksche Strahlungsgesetz," *Physikalische Zeitschrift 10*, 193–195 (1909); P. Debye, "Der Wahrscheinlichkeitsbegriff in der Theorie der Strahlung," *Annalen der Physik 33*, 1427–1434 (1910); J. Larmor, "On the statistical and thermodynamical relations of radiant energy," *Proceedings of the Royal Society of London* (*A*), *83*, 82–95 (1910); W. Nernst, "Zur Theorie der spezifischen Wärme und über die Anwendung der Lehre von den Energiequanten auf physikalisch-chemische Fragen überhaupt," *Zeitschrift für Elektrochemie 17*, 265–275 (1911); P. Franck, "Zur Ableitung der Planckschen Strahlungsformel," *Physikalische Zeitschrift 13*, 506–507 (1912); A. Einstein and O. Stern, "Einige Argumente für die Annahme einer molekularen Agitation beim absoluten Nullpunkt," *Annalen der Physik 40*, 551–

confirmations of Planck's law, numerous attempts were made to evade Rayleigh's conclusion (1.9) without abandoning classical statistical mechanics and, in particular, the equipartition theorem.[105] The reason, as Lorentz put it, was undoubtedly that "we cannot say that the mechanism of the phenomena has been unveiled [by Planck's theory], and it must be admitted that it is difficult to see the reason for this partition of energy by finite portions, which are not even equal to each other, but vary from one resonator to the other."[106]

Another conceptual difficulty, which prevented the general acceptance of Planck's introduction of h, was undoubtedly the following fact. As shown by its dimension, this quantity represented an invariable unit of "action" (energy $\times$ time) or an "elementary quantum of action" ("elementares Wirkungsquantum"), as it was subsequently called. But it was clear that no principle of conservation of action exists in physics. It is therefore not surprising that the attempt to reconcile Planck's law with classical statistical mechanics was not abandoned even after Lorentz had shown that classical physics, that is, the equipartition theorem and Hamilton's principle, leads necessarily to Rayleigh's radiation law and its empirically untenable implications. As Lorentz put it, the ether is a system of infinitely many degrees of freedom, and the temperature of a ponderable body, in thermal equilibrium with it, on this assumption must necessarily be absolute zero, a result contrary to experience.

Lorentz made these statements in a series of lectures which he delivered in 1910 at the University of Göttingen. In this context the following historical comments are not without interest.

In 1908 the mathematician Paul Wolfskehl[107] of Darmstadt bequeathed the sum of 100,000 marks to the Academy of Sciences in Göttingen as an award for the first person to publish a complete proof of Fermat's famous *Last Theorem* (1637). In this theorem, it will be recalled, Fermat denied

---

560 (1913); M. Wolfke, "Zur Quantentheorie," *Verhandlungen der Deutschen Physikalischen Gesellschaft 15*, 1123, 1215 (1913); M. Wolfke, "Welche Strahlungsformel folgt aus der Annahme der Lichtatome?" *Physikalische Zeitschrift 15*, 308–310, 463 (1914); A. Einstein, "Zur Quantentheorie der Strahlung," *Mitteilungen der Physikalischen Gesellschaft, Zürich, 18*, 47–62 (1916), *Physikalische Zeitschrift 18*, 121–128 (1917); A. Rubinowicz, "Zur Quantelung der Hohlraumstrahlung," *ibid.*, 96–99 (1917); C. G. Darwin and R. H. Fowler, "On the partition of energy," *Philosophical Magazine 44*, 450–479, 823–842 (1922); C. G. Darwin and R. H. Fowler, "Partition functions for temperature radiation and the internal energy of a crystalline solid," *Proceedings of the Cambridge Philosophical Society 21*, 262–273 (1922); S. N. Bose, "Plancks Gesetz und Lichtquantenhypothese," *Zeitschrift für Physik 26*, 178–181 (1924); S. N. Bose, "Wärmegleichgewicht im Strahlungsfeld bei Anwesenheit von Materie, *ibid. 27*, 384–392 (1924); A. S. Eddington, "On the derivation of Planck's law from Einstein's equation," *Philosophical Magazine 50*, 803–808 (1925).

[105] H. A. Lorentz, "On the emission and absorption by metals of rays of heat of great wave-length," *Amsterdam Proceedings 1902–1903*, p. 666.

[106] H. A. Lorentz, *The Theory of Electrons* (1st ed. 1909, 2d ed. 1915; quoted from the Dover edition, New York, 1952), p. 80.

[107] "Bekanntmachung," *Göttinger Nachrichten 1908*, p. 103.

the existence of integers x, y, z, and n which satisfy $xyz \neq 0$, $n > 2$, $x^n + y^n = z^n$. It is also well known that the theorem has not yet been proved but has gained the unique distinction of being the problem for which the greatest number of incorrect "proofs" has ever been published.

What is not so well known, however, is the wise decision of the Wolfskehl committee to use the interest of the amount for the purpose of inviting prominent scientists as guest speakers to Göttingen. Such an invitation brought Poincaré there at the end of April, 1909. He gave six lectures on problems in pure and applied mathematics.[108] In his first talk (April 22) he spoke of Fredholm's equations in connection with the work of Hill and Helge von Koch, a subject whose relevance to quantum theory was not recognized until 1925; in his last lecture "La Mécanique Nouvelle" (April 28)—the only one he gave in French—he discussed the theory of relativity—incidentally, without mentioning the name of Einstein.

In the following year Lorentz was invited. From October 24 to 29, 1910, he delivered six lectures on "Old and New Problems in Physics"[109] which were subsequently edited by Born and published in the *Physikalische Zeitschrift.* The last three of these lectures dealt with the problem of black-body radiation. Three years later Sommerfeld spoke on problems of mathematical physics, and in the summer semester of 1914 Debye gave a series of lectures.

The last scientist to be invited on this program was Niels Bohr. His Göttingen lectures, delivered on June 12 to 22, 1922, had, as we shall see in due course, a decisive influence upon Pauli and Heisenberg. Bohr's "Seven Lectures on the Theory of Atomic Structure"[110] began with a general survey of atomic theory (first lecture), dealt with the correspondence principle and the adiabatic principle (second lecture), their applications (third lecture), discussed polyelectronic systems (fourth lecture), the periodic system (fifth lecture), x-rays and atomic structure (sixth lecture), and concluded with remarks on problems still to be solved. The subjects covered in these lectures were essentially the same as contained in Bohr's paper on the structure of atoms, published at that time.[111]

Mathematical research so far seems to have profited very little from Wolfkehl's incitement, and since the inflation in Germany depreciated the prize and in view of the historic impact of Bohr's lectures upon Pauli and Heisenberg it is perhaps no exaggeration to say that quantum theory

[108] H. Poincaré, *Sechs Vorträge aus der Reinen Mathematik und Mathematischen Physik* (Teubner, Leipzig, Berlin, 1910).

[109] H. A. Lorentz, "Alte und neue Fragen der Physik," *Physikalische Zeitschrift 11,* 1234–1257 (1910).

[110] A manuscript with notes on these lectures which Bohr delivered in German under the title "Sieben Vorträge über die Theorie des Atombaus" is found in the *Bohr Archive* under the title "Optegnelser til Forelaesningerne i Göttingen," Bohr Mss. No. 10.

[111] N. Bohr, "Der Bau der Atome und die physikalischen und chemischen Eigenschaften der Elemente," *Zeitschrift für Physik 9,* 1–67 (1922).

was the main beneficiary of the Wolfskehl Prize. Whether this statement will have to be modified in view of the recently proposed revival of the prize remains to be seen.

In concluding our discussion on the early development of the conception of energy quanta, in which Planck's derivation of the radiation law played the dominant role, we think it necessary to stress the following critical remarks.

As we have pointed out, Planck's derivation consisted of two separate parts: (1) a derivation of the relation (1.6) between the radiative energy density u_ν and the oscillator energy U,

$$u_\nu = \frac{8\pi}{c^3} \nu^2 U \tag{1.6}$$

a formula which Planck obtained by using exclusively the principles of classical electrodynamics (as shown in Appendix A); (2) a statistical treatment of the interaction among oscillators of different proper frequencies which resulted in the formula (1.15),

$$U = \frac{h\nu}{\exp{(h\nu/kT)} - 1} \tag{1.15}$$

By combining (1.6) and (1.15) Planck obtained his radiation law (1.16). We have also emphasized that these conclusions were adduced by Planck in order to provide a logical justification of his far-reaching interpolation mentioned above.

Planck's reasoning was inconsistent, however, as Einstein, in 1906, was the first to recognize.[112] For although either part of Planck's derivation of (1.16) was in itself consistent, their combination was logically incompatible. The reason was this: in the electrodynamical part (1) formula (1.6) is based on Maxwell's theory (see Appendix A) and the assumption that the oscillator energy is a continuously variable quantity, whereas in the statistical part (2) this same energy is treated as a discrete quantity, capable of assuming only values which are multiples of $h\nu$.

Referring to this inconsistency, Einstein remarked that "if the energy of a resonator can change only discontinuously, the usual theory of electricity cannot be applied for the calculation of the average energy of such a resonator in a radiation field. Planck's theory has, therefore, to assume that, although Maxwell's theory of elementary resonators is not applicable, the average energy of such a resonator, surrounded by radiation, is equal to that which would result from the calculation on the basis of Maxwell's theory of electricity."

"Such an assumption," continued Einstein, "would be plausible

[112] A. Einstein, "Zur Theorie der Lichterzeugung und Lichtabsorption," *Annalen der Physik 20*, 199–206 (1906).

provided $\epsilon = h\nu$ were small throughout the observable spectrum compared to the average energy U of the resonator; but this is not the case."[113]

Three and a half years later, at the 81st meeting of the German Association of Scientists, held at Salzburg in September, 1909, Einstein[114] spoke on the development of our ideas on the nature and constitution of radiation. Repeating on this occasion his challenge to Planck's reasoning and resuming the question of whether the two parts cannot be reconciled with each other, Einstein pointed out that the previously mentioned condition, namely, that the energy quantum $\epsilon = h\nu$ be small in comparison with U, is certainly not satisfied. "A simple calculation shows," he declared, "that ϵ/U for $\nu = 0.5\ \mu$ and $T = 1700°K$ is not only not small compared to unity, but very large. It is approximately 6.5×10^7."

For Einstein this inconsistency was no reason to reject Planck's quantum theory as such. Having meanwhile proposed his ideas concerning light quanta, as we shall see in the next paragraph, Einstein saw in this inconsistency an indication that the foundations of the traditional radiation theory, based on Maxwell's electromagnetic theory, had to be revised.

The logical incompatibility of the two parts in Planck's derivation of his radiation law was a matter of great concern also for Peter Debye.[115] But contrary to Einstein, who hoped to overcome the difficulty by modifying the Maxwellian interaction between resonator and field, Debye attempted to resolve the inconsistency by eliminating altogether the role of ponderable resonators in Planck's derivation. Referring to the Jeans-Rayleigh computation of the number $N\,d\nu$ of vibrations in an enclosure of unit volume and frequency interval $d\nu$,

$$N\,d\nu = \frac{8\pi\nu^2}{c^3}\,d\nu$$

Debye assumed that the $N\,d\nu$ vibrations consist of $f(\nu)$ quanta of energy content $h\nu$ each, so that

$$u_\nu\,d\nu = \frac{8\pi h\nu^3}{c^3}\,f(\nu)\,d\nu$$

Defining "black radiation" as the most "probable radiation," that is, as the state with the greatest possible number of distributions of the $f(\nu)$ quanta among the $N\,d\nu$ receptors, Debye proved by using Planck's combinatorial formula that in this case $f(\nu) = [\exp\,(h\nu/kT) - 1]^{-1}$, a result which in combination with the preceding formula implied Planck's law of radiation. Debye thus showed that Planck's law and its implications

[113] *Ibid.*, p. 203.

[114] A. Einstein, "Über die Entwicklung unserer Anschauungen über das Wesen und die Konstitution der Strahlung," *Physikalische Zeitschrift 10*, 817–825 (1909).

[115] P. Debye, "Der Wahrscheinlichkeitsbegriff in der Theorie der Strahlung," *Annalen der Physik 33*, 1427–1434 (1910).

follow from the assumption alone that the energy as such is quantized in units of $h\nu$ and no knowledge concerning the properties of resonators or their mechanism is needed for this purpose. Debye's[116] assumption may be referred to as the "weak quantum postulate," in contrast to Planck's "quantum postulate," according to which also the energy content of an oscillator is always a multiple of $h\nu$.

1.3 *The Concept of Quanta of Radiation*

In the development of quantum theory discussed so far, the concept of energy elements or quanta had been regarded as applicable only to the mechanism regulating the interaction between matter and radiation: it was the material oscillator of frequency ν which could emit or absorb energy only in multiples of $h\nu$.

Meanwhile, however, an important conceptual development took place which led to a certain generalization of the conception of quanta. It began in 1905, when the general validity of the electromagnetic theory of light was seriously called into question by Einstein's article "On a heuristic viewpoint concerning the production and transformation of light."[117] In its importance for the future development of theoretical physics this essay may be compared with Einstein's classic paper on special relativity with which it appeared—together also with his famous study on Brownian motion—in the same volume of the *Annalen der Physik*. Although commonly referred to as Einstein's paper on the photoelectric effect, it discussed a problem of much wider significance and contained a suggestion which challenged classical physics perhaps to the same extent as did Planck's historic paper of 1900.

Einstein considered monochromatic radiation of frequency ν and of small density within the range of ν/T where Wien's radiation law (1.5) is valid. If v is the volume of the enclosure and $u(\nu)$ the spectral distribution function, the entropy could be expressed by the equation $S = v\int_0^\infty \varphi(u,\nu)\, d\nu$, where φ is a function of u and of ν. In order to find the explicit dependence of φ on u and ν, Einstein had two equations to start with: $\delta\int\varphi\, d\nu = 0$, expressing the fact that the entropy for the equi-

[116] Strictly speaking, to attain a stationary state of radiation (i.e., of maximum entropy) Debye needed the property of ponderable bodies to exchange radiation from one wavelength to another. He therefore defined his postulate of elementary quanta as follows: "Schwingungsenergie kann von ponderabelen Körpern aufgenommen werden und eventuell in Energie von anderer Schwingungszahl übergeführt werden nur in Form von Quanten von der Größe $h\nu$." *Ibid.*, p. 1430.

[117] Albert Einstein, "Über einen die Erzeugung und Verwandlung des Lichtes betreffenden heuristischen Gesichtspunkt," *Annalen der Physik 17*, 132–148 (1905). Recently translated into English by A. B. Arons and M. B. Peppard, *American Journal of Physics 33*, 367–374 (1965).

librium state of the cavity radiation is a maximum, and $\delta \int u \, d\nu = 0$, expressing the conservation of energy. Introducing an undetermined multiplier, he obtained, for every choice of δu as a function of ν, the equation

$$\int \left(\frac{\partial \varphi}{\partial u} - \lambda \right) \delta u \, d\nu = 0$$

where λ and consequently also $\partial \varphi / \partial u$ are independent of ν. Taking $v = 1$, Einstein calculated the increase of entropy for dT as

$$dS = \int_{\nu=0}^{\infty} \frac{\partial \varphi}{\partial u} \, du \, d\nu$$

or, in view of the independence just proved, $dS = (\partial \varphi / \partial u) \, dE$, where dE is the heat added reversibly and hence subject to $dS = dE/T$. Comparison of the last two equations showed that $\partial \varphi / \partial u = 1/T$ irrespective of the particular form of $u(\nu)$. Solving Eq. (1.5) for T^{-1}, Einstein obtained the differential equation

$$\frac{\partial \varphi}{\partial u} = -(\beta \nu)^{-1} \log \frac{u}{\alpha \nu^3}$$

and after integration

$$\varphi(u,\nu) = -\frac{u}{\beta \nu} \left(\log \frac{u}{\alpha \nu^3} - 1 \right)$$

The entropy of the radiation within the interval from ν to $\nu + d\nu$ and with the energy $E_\nu = v u_\nu \, d\nu$ was therefore given by the expression

$$S = -\frac{E_\nu}{\beta \nu} \left[\log \left(\frac{E_\nu}{\alpha v \nu^3 \, d\nu} \right) - 1 \right]$$

If the radiation, originally in volume v_0, is assumed to occupy volume v, the last equation shows that the change in entropy is

$$S - S_0 = \frac{E_\nu}{\beta \nu} \log \frac{v}{v_0}$$

or equivalently

$$S - S_0 = \frac{R}{N} \log \left(\frac{v}{v_0} \right)^{NE_\nu / \beta \nu R} \tag{1.17}$$

where N is Avogadro's number and R the gas constant. On the other hand, from the kinetic theory of gases, Einstein argued, it is well known, that the probability of finding n particles at an arbitrary instant of time within a partial volume v of the volume v_0 in which they were originally moving is given by $(v/v_0)^n$. Hence

$$S - S_0 = \frac{R}{N} \log \left(\frac{v}{v_0}\right)^n \tag{1.18}$$

Guided by the identity of the mathematical structure of (1.17) and (1.18), Einstein concluded that $E_\nu = n(R\beta\nu/N)$ and declared that with respect to the theory of heat, "monochromatic radiation of small density (within the range of validity of Wien's radiation law) behaves as if it consisted of independent energy quanta of magnitude $R\beta\nu/N$."[118] Wien's exponential coefficient β, expressed by Planck's constants, was of course h/k, as Planck's law readily showed for $h\nu \gg kT$ and $R/N = k$. In effect, therefore, Einstein stated that radiation behaved as if it were composed of a finite number of localized energy quanta $h\nu$ or "photons," as they were later called after G. N. Lewis[119] introduced this term in 1926.

The idea of a discontinuous distribution of radiant energy in space was, of course, completely at variance with the prevailing undulatory electromagnetic theory of light. Furthermore Einstein's suggestion of a granular structure of radiation seemed to counter one of the most well-founded and indisputable results of physical research. Was not the discovery of diffraction, first reported by Leonardo da Vinci,[120] rediscovered and investigated by Grimaldi,[121] and accountable only in terms of the wave theory of Huygens[122] and Young,[123] as Fresnel[124] has so masterly

[118] "Monochromatische Strahlung von geringer Dichte (innerhalb des Gültigkeitsbereiches der Wienschen Strahlungsformel) verhält sich in wärmetheoretischer Beziehung so, wie wenn sie aus voneinander unabhängigen Energiequanten von der Größe $R\beta\nu/N$ bestünde." *Ibid.*, p. 143.

[119] Lewis thought it inappropriate to speak of a "quantum of light," "if we are to assume that it spends only a minute fraction of its existence as a carrier of radiant energy, while the rest of the time it remains an important structural element within the atom. . . . I therefore take the liberty of proposing for this hypothetical new atom, which is not light but plays an essential part in every process of radiation, the name photon." G. N. Lewis, "The conservation of photons," *Nature 118*, 874–875 (1926).

[120] G. Libri, *Histoire des Sciences Mathématiques en Italie* (J. Renouard, Paris, 1838–1841), vol. 3, p. 54.

[121] Francesco Maria Grimaldi, *Physicomathesis de lumine, coloribis et iride aliisque adnexis libri duo* (Benatii, Bologna, 1665).

[122] Christiaan Huygens, *Traité de la Lumière* (P. Van der Aa, Leiden, 1960); *Treatise on Light*, translated by S. P. Thompson (Macmillan, London, 1912).

[123] Thomas Young, "On the theory of light and colour," *Philosophical Transactions of the Royal Society of London 92*, 12–24 (1802); *A Course of Lectures on Natural Philosophy and the Mechanical Arts* (J. Johnson, London, 1807), especially Lecture 39, vol. 1, pp. 457–471.

[124] Augustin Jean Fresnel, "Sur la diffraction de la lumière, où l'on examine particulièrement le phénomène des fringes colorées que présentent les ombres des corps éclairés par un point lumineux," *Annales de Chimie et de Physique 1*, 239–281 (1816); *Oeuvres Complètes d'Augustin Fresnel* (Imprimerie Impériale, Paris, 1866), vol. 1, pp. 89–122, 129–170; "Mémoire sur la diffraction de la lumière," *Mémoires de l'Académie des Sciences 1819; Annales de Chimie et de Physique 11*, 246–296, 377–378 (1819); *Oeuvres*, vol. 1, pp. 247–384; English translation in *The Wave Theory of Light*, edited by H. Crew (American Book Company, New York, 1900), pp. 79–144; German translation in *Abhandlungen über die Beugung des Lichts*, translated by F. Ritter, *Ostwald's Klassiker der exakten Wissenschaften No. 215*, (Akademische Verlagsgesellschaft, Leipzig, 1926).

shown, an incontestable deathblow to any corpuscular conception of light such as proposed by Newton,[125] Laplace,[126] or Biot?[127] Have not Fizeau,[128] Foucault,[129] and Breguet,[130] following a suggestion by Arago,[131] established without doubt that the velocity of light is less in water than in air and thus brought forth a crucial[132] decision of the "particle-versus-wave" issue in favor[133] of the latter? As Hanson[134] recently pointed out, these experiments proved the undulatory nature of light but certainly did not prove that light cannot also be corpuscular. For classical physics has only gradually shaped the notions of particle and wave to logical contraries or opposites. In fact, a thorough scholarly study of the history of the logical relationship between these two notions, so important for modern physics, and on their development to the status of fundamentally incompatible and mutually exclusive conceptions still remains a project for future research. In any case, when Young declared, "It is allowed on all sides, that light either consists in the emission of very minute particles from luminous substances, which are actually projected, and continue to move, with the velocity commonly attributed to light, or in the excitation of an undulatory motion, analogous to that which constitutes sound, in a highly light and elastic medium pervading the universe; but the judgments of philosophers of all ages have been much divided with respect to the preference of one or the other of these opinions,"[135] he obviously used the connective "or" in the dis-

[125] Sir Isaac Newton, *Philosophiae Naturalis Principia Mathematica* (jussu Societatis regiae, London, 1687), book 1, section 14, propositions 94–98; *Opticks, or a Treatise of the Reflexions, Refractions, Inflexions and Colours of Light* (S. Smith, London, 1704), book 1, part 1, proposition 6, theorem 5.

[126] Pierre-Simon Laplace, *Traité de Mécanique Céleste* (Duprat, Paris, 1808), vol. 4, p. 241; *Exposition du Système du Monde* (Courcier, Paris, 1813), 4th ed., p. 327.

[127] Jean Baptiste Biot, *Traité de Physique Expérimentale et Mathématique* (Deterville, Paris, 1816), vols. 3 and 4.

[128] Armand Hippolyte Fizeau, "Sur une expérience relative à la vitesse de propagation de la lumière," *Comptes Rendus 29*, 90–92 (1849); "Versuch, die Fortpflanzungsgeschwindigkeit des Lichts zu bestimmen," *Poggendorff's Annalen der Physik 79*, 167–169 (1850).

[129] Jean Bernard Léon Foucault, "Méthode générale pour mesurer la vitesse de la lumière dans l'air et les milieux transparents," *Comptes Rendus 30*, 551–560 (1850); "Allgemeine Methode zur Messung der Geschwindigkeit des Lichts in Luft und durchsichtigen Mitteln," *Poggendorff's Annalen der Physik 81*, 434–442 (1850).

[130] H. Fizeau and L. Breguet, "Note sur l'expérience relative à la vitesse comparative de la lumière dans l'air et dans l'eau," *Comptes Rendus 30*, 562–563, 771–774 (1850); "Notiz in Betreff eines Versuchs über die comparative Geschwindigkeit des Lichts in Luft und in Wasser," *Poggendorff's Annalen der Physik 81*, 442–444 (1850), *82*, 124–127 (1851).

[131] Dominique François Jean Arago, "Sur un système d'expérience à l'aide duquel la théorie de l'émission et celle des ondes seront soumises à des épreuves décisives," *Comptes Rendus 7*, 954–965 (1838); "Über ein System von Versuchen, mit Hülfe dessen die Emissions- und die Undulationstheorie auf entscheidende Proben gestellt werden können," *Poggendorff's Annalen der Physik 46*, 28–41 (1839).

[132] On the "cruciality" of such experiments see N. R. Hanson, *The Concept of the Positron* (Cambridge University Press, 1963), pp. 18–24.

[133] "Hiermit haben sich Fizeau und Foucault das Verdienst erworben, die Emissionstheorie endgiltig widerlegt zu haben." A. Winkelmann, *Handbuch der Physik* (E. Trewendt, Breslau, 1894), p. 10.

[134] REF. 132, p. 12.

[135] REF. 123, Lecture 39, p. 457.

junctive sense of the Latin *aut* (and not *vel*). Arago[136] even considered the issue as a mathematically or logically unequivocal dichotomy. Yet, for the physics of the later nineteenth century the spatial distribution of energy was either discrete, as in the corpuscular-kinetic theory of Newtonian mechanics, or continuous, as in Maxwell's electromagnetic theory, but never both discrete and continuous for one and the same category of physical phenomena.

Strictly speaking, as Einstein[137] once pointed out, the success of the wave theory of light was the first breach in Newtonian physics, for corpuscular-kinetic conceptions were replaced by field-theoretic notions. But throughout the later nineteenth century these two schemes of conceptions enjoyed a rather peaceful coexistence.

Even Einstein, in the beginning of the paper under discussion,[138] admitted that the classical theory of light based on continuous space functions was so firmly established that it would probably never be replaced by another theory. But, he continued to say, optical observations take account only of time averages, and it is quite conceivable that such a theory of light, in spite of its convincing verifications by experiments on interference and diffraction, may prove itself insufficient whenever instantaneous values of those functions have to be considered or whenever interactions of matter with radiation, as in the processes of emission and absorption, are involved. Einstein, it seems, did not know that similar doubts had previously been raised by J. J. Thomson. Faced by difficulties in explaining quantitatively the ionization caused by Röntgen rays, as x-rays were still called at that time, Thomson declared in 1903: "If, for example, we consider a plane at right angles to the direction of propagation of the rays the energy is not distributed uniformly over this plane, but the distribution of energy has as it were a structure, although an exceedingly fine one, places where the energy is large alternating with places where it is small, like mortar and bricks in a wall."[139] The effect which led Thomson to his conjecture of "patches of energy," namely, photoionization, was one of the instances, together with Stokes' law and the photoelectric effect, for

[136] "Je me propose de montrer dans cette Note, comment il est possible de décider, sans équivoque, si la lumière se compose de petites particules *émanant* des corps rayonnants, ainsi que le voulait *Newton*, ainsi que l'ont admis la pluspart des géomètres modernes; ou bien si elle est simplement le résultat des *ondulations* d'un milieu très rare et très élastique, que les physiciens sont convenus d'appeler l'Ether. Le système d'expériences que je vais décrire, ne permettra plus, ce me semble, d'hésiter entre les deux théories rivales. Il tranchera *mathématiquement* (j'emploie à dessein cette expression); il tranchera mathématiquement une des questions les plus débattues de la philosophie naturelle." REF. 131, p. 954.

[137] "The theoretical system, built up by Newton with his powerful and logical intellect, should have been overthrown precisely by a theory of light . . . by the Huygens-Young-Fresnel wave theory of light." A. Einstein, "The new field theory," *The Times* (London), No. 45,118, Feb. 4, 19 9, pp. 13–14.

[138] REF. 117.

[139] J. J. Thomson, *Conduction of Electricity through Gases* (Cambridge University Press, 1903), p. 258.

which Einstein suggested the application of his new conception of quanta of light.

Owing to Einstein's paper of 1905, it was primarily the photoelectric effect to which physicists referred as an irrefutable demonstration of the existence of photons and which thus played an important part in the conceptual development of quantum mechanics. It seems appropriate, therefore, to discuss it in greater detail.

It will be recalled that in a series of ingenious experiments begun in 1886 Heinrich Hertz demonstrated the existence of electromagnetic waves and thereby confirmed the electromagnetic theory of light in a way which Maxwell himself had feared would never be achieved. The same series of experiments, however, which so brilliantly confirmed the electromagnetic theory of light—paradoxical as it may sound—also produced the first evidence toward its refutation. For it was in the course of these investigations that Hertz discovered as early as 1887 that the length of the spark induced in the secondary circuit was greatly reduced when the spark gap was shielded from the light of the spark in the primary circuit and "that ultraviolet light has the power to increase the spark gap of the discharge of an inductor and of related discharges."[140]

Although this statement may rightfully be regarded as the earliest description of the photoelectric effect which eventually served to disprove the universal validity of the electromagnetic theory of light, in the interest of historical accuracy, it should be remarked that Schuster[141] and Arrhenius,[142] in the course of their investigations of related subjects, independently of Hertz and at about the same time, called attention to essentially the same effect, although they were not fully aware of the immediate cause of the phenomenon. Hertz refrained from formulating any theory as to the mechanism by which radiation facilitates electrical discharge, recognizing that much more experimental research was needed for such a task. In fact, his discovery initiated a long series of investigations. First, Wiedemann and Ebert[143] discovered in 1888 that irradiation from an electric arc discharges the negative electrode without affecting the positive one. Hallwachs,[144] also in 1888, found on the basis of his experiments with freshly polished zinc plates that a rise in temperature does not significantly

[140] "Nach den Resultaten unserer Versuche hat das ultraviolette Licht die Fähigkeit, die Schlagweite der Entladungen eines Inductoriums und verwandter Entladungen zu vergrößern." H. Hertz, "Über einen Einfluß des ultravioletten Lichtes auf die electrische Entladung," *Wiedemannsche Annalen der Physik 31*, 982–1000 (1887).

[141] A. Schuster, "Experiments on the discharge of electricity through gases," *Proceedings of the Royal Society of London* (*A*), *42*, 371–379 (1887).

[142] S. Arrhenius, "Über das Leitungsvermögen der phosphorescirenden Luft," *Wiedemannsche Annalen der Physik 32*, 545–572 (1887); *Proceedings of the Swedish Academy 44*, 405–429 (1887).

[143] E. Wiedemann and H. Ebert, "Über den Einfluß des Lichtes auf die electrischen Entladungen," *Wiedemannsche Annalen der Physik 33*, 241–264 (1888).

[144] W. Hallwachs, "Über den Einfluß des Lichtes auf die electrostatisch geladene Körper," *ibid.*, 301–312.

increase the effect, red and infrared radiation being ineffective, and that negatively charged particles seem to be emitted. He also demonstrated[145] that even an electrically neutral insulated plate, if irradiated, required a small positive potential. In the following year Stoletow showed that for a certain metal, wavelengths above 2950 Å were ineffective, and he devised the first cell for the production of a photoelectric current. Stoletow found[146] that this current is strictly proportional to the intensity of the light absorbed, and he demonstrated that the time lapse between the incidence of radiation and the discharge is less than a millisecond. At the same time Elster and Geitel, in a long series of measurements with the help of an Exner electroscope, systematically examined metals and alloys for their photoelectric response to sunlight, to arc light, to the light of a petroleum lamp, and finally to that from a glass rod just heated to redness.[147] They found that the more electropositive a metal is, the greater is its photoelectric sensitivity and that rubidium, potassium, sodium, and their alloys are sensitive even to visible light.

Once these experimental features were determined, research concentrated on the identification of the carriers of the photoelectric current and on their physical characteristics. That it was not the molecules of the gas surrounding the cathode which were the carriers of this current followed from the fact that the effect persisted independently of the gas pressure as soon as a certain low pressure had been reached, and continued to persist at the highest vacuum attainable. In 1898 Rutherford measured[148] the velocity of these carriers of charge and in the same year J. J. Thomson,[149] on the basis of his famous measurements of the velocity and of the charge-to-mass ratio of the electron, suggested identifying those carriers with

[145] "Über die Electrisierung von Metallplatten durch Bestrahlung mit electrischem Licht," *ibid. 34*, 731–734 (1888). In German scientific literature the photoelectric effect was often called "Hallwachs effect."

[146] A. G. Stoletow, "Sur une sorte de courants électriques provoqués par les rayons ultraviolets," *Comptes Rendus 106*, 1149–1152 (1888); "Sur les courants actino-électriques au travers de l'air," *ibid.*, 1593–1595; "Sur des recherches actino-électriques," *ibid. 107*, 91–92 (1888); "Sur les phénomènes actino-électriques," *ibid. 108*, 1241–1243 (1889); "Sur les courants actino-électriques dans l'air raréfié," *Journal de Physique 9*, 468–473 (1889).

[147] J. Elster and H. Geitel, "Notiz über die Zerstreuung der negativen Electricität durch das Sonnen- und Tageslicht," *Wiedemannsche Annalen der Physik 38*, 40–41 (1889); "Über den Einfluß negativ electrischer Körper durch das Sonnen- und Tageslicht," *ibid.*, 497–514; "Über einen hemmenden Einfluß der Belichtung auf electrische Funken- und Büschelentladungen," *ibid. 39*, 332–335 (1890); "Über die Verwendung des Natriumamalgams zu lichtelectrischen Versuchen," *ibid. 41*, 161–165 (1890); "Über den hemmenden Einfluß des Magnetismus auf lichtelectrische Entladungen in verdünnten Gasen," *ibid.*, 166–176; "Weitere lichtelectrische Versuche," *ibid. 52*, 433–454 (1894); "Lichtelectrische Untersuchungen an polarisiertem Licht," *ibid. 55*, 684–700 (1895).

[148] E. Rutherford, "The discharge of electrification by ultraviolet light," *Proceedings of the Cambridge Philosophical Society 9*, 401–416 (1898); *The Collected Papers of Lord Rutherford of Nelson* (Interscience, New York, 1962), vol. 1, pp. 149–162.

[149] J. J. Thomson, "On the masses of the ions in gases at low pressure," *Philosophical Magazine 48*, 547–567 (1899).

the cathode rays in Geissler tubes. This suggestion found additional support in a series of experiments performed by Lenard[150] in which a sodium amalgam served as a cathode and a chemically pure platinum wire as an anode in an atmosphere of hydrogen while a photoelectric current carrying about 3 μcoul was allowed to flow through the circuit. Had sodium atoms been the carriers of the charge, the current would have deposited almost 10^{-6} mg of sodium on the platinum wire, an amount sufficient to be detectable by spectroscopic means. But no trace of sodium was found. Having demonstrated that the carriers are electrons, Lenard showed that their emission occurs only at a frequency that exceeds a certain minimum value ν_0, confirming Stoletow's result, that the energy of the ejected electrons increased with increasing difference $\nu - \nu_0$, irrespective of the distance of the source from the irradiated surface. The existence of such a threshold frequency and the fact, also irrefutably verified by Ladenburg,[151] that the energy of the photoelectrons is independent of the light intensity but proportional to the frequency of the incident light proved irreconcilable with Maxwell's electromagnetic theory of light, but it received a most natural explanation by the hypothesis of light quanta as proposed by Einstein in 1905. For in view of the supposed corpuscular localization or concentration of their energy $R\beta\nu/N = h\nu$, each quantum could be thought of as being capable of interacting *in toto* with one electron only.

It is well known that all quantitative features of the photoelectric effect could be accounted for by Einstein's photoelectric law, according to which the maximum kinetic energy of the photoelectrons is given by $(R/N)\beta\nu - P$ or, in modern notation, by $h\nu - P$, where P, a constant for a given metal, is the work necessary to remove an electron from the metal. Einstein's photoelectric equation was verified in 1912 by Hughes,[152] who measured the maximum velocity of photoelectrons ejected from a number of elements (K, Ca, Mg, Cd, Zn, Pb, Sb, Bi, As) which he carefully prepared by distillation in a vacuum, and by Richardson and Compton,[153] who improved the measuring method by using a modern Hilger monochromator and making allowance for the contact-potential differences of the electrodes. Einstein's remark that the stopping potential "should be a linear function of the frequency of the incident light, when plotted in cartesian coordinates, and its slope should be independent of the nature of the substance investigated"[154] found its most striking verification in

[150] P. Lenard, "Erzeugung von Kathodenstrahlen durch ultraviolettes Licht," *Wiener Berichte 108*, 1649–1666 (1899); *Annalen der Physik 2*, 359–375 (1900); Über die lichtelectrische Wirkung," *Annalen der Physik 8*, 149–198 (1902).

[151] E. Ladenburg, "Untersuchungen über die entladende Wirkung des ultravioletten Lichtes auf negativ geladene Metallplatten im Vakuum," *Annalen der Physik 12*, 558–578 (1903).

[152] A. L. Hughes, "On the emission velocities of photoelectrons," *Philosophical Transactions of the Royal Society of London 212*, 205–226 (1912).

[153] O. W. Richardson and K. T. Compton, "The photoelectric effect," *Philosophical Magazine 24*, 575–594 (1912).

[154] REF. 117, p. 146.

the brilliant experimental work performed by Millikan at the Ryerson Laboratory of the University of Chicago. In his Nobel Prize Address, delivered in Stockholm in 1923,[155] Millikan summarized his work in these words: "After ten years of testing and changing and learning and sometimes blundering, all efforts being directed from the first toward the accurate experimental measurement of the energies of emission of photoelectrons, now as a function of the temperature, now of wavelength, now of material (contact E.M.F. relations), this work resulted, contrary to my own expectations, in the first direct experimental proof in 1914 of the exact validity, within narrow limits of experimental errors, of the Einstein equation, and the first direct photoelectric determination of Planck's constant *h*."

Millikan's work consisted of a long series of painstaking measurements. As early as 1908, at the Boston meeting of the American Physical Society, he was in a position to report some striking results.[156] Essentially what he intended to do was measure the kinetic energy of the liberated electrons by a determination of the stopping potential, i.e., the potential just sufficient to prevent them from reaching a second metallic plate. The actual carrying out of these measurements, however, was complicated by the necessity of a precise determination of the energy that had to be supplied to the electrons to overcome retaining forces, such as that of the image charge, and others which, as functions of the nature of the surface, were most difficult to control and reproduce. Millikan worked with the alkali metals sodium, potassium, and lithium. With monochromatic beams of mercury light he illuminated their surfaces, which he obtained clean by cutting shavings in a high vacuum with the aid of a specially designed and magnetically operated knife. In 1916 Millikan finally provided incontrovertible proof of the direct proportionality between the kinetic energy of the photoelectrons and the frequency of the absorbed light.[157] He showed that the constant of proportionality is independent of the nature of the surface and that its numerical value is 6.57×10^{-27} erg sec, in close agreement with the value obtained by Planck in 1900.[158] As a result of Millikan's confirmation of Einstein's photoelectric equation, the quantum of action became a physical reality, accessible directly to experi-

[155] *The Autobiography of Robert A. Millikan* (Prentice-Hall, Englewood Cliffs, N.J., 1950), pp. 102–103.

[156] R. A. Millikan, "Some new values of the positive potentials assumed by metals in a high vacuum under the influence of ultra-violet light," *Physical Review 30*, 287–288 (1910).

[157] R. A. Millikan, "A direct photoelectric determination of Planck's '*h*,'" *Physical Review 7*, 355–388 (1916). Cf. also his preceding papers: "A direct determination of *h*," *ibid. 4*, 73–75 (1914); "New tests of Einstein's photoelectric equation," *ibid. 6*, 55 (1915); and his article "Quantenbeziehungen beim photoelektrischen Effekt," *Physikalische Zeitschrift 17*, 217–221 (1916).

[158] See p. 21.

ment, and Einstein's conjecture of light quanta was endowed with physical significance and an experimental foundation.[159]

If a particle is defined as a carrier of energy and momentum, Planck's clarification of the physical content of the energy-momentum equations in special relativity and electromagnetic theory[160] showed that the quanta of light which Einstein always conceived as directed in space were indeed particles. For according to Planck every flux of energy is accompanied by a transfer of momentum. Hence each quantum of light, moving in empty space with the velocity c, carries with it a momentum $h\nu/c$. Stark, one of the earliest proponents of the hypothesis of light quanta, suggested as early as 1909 a possibility of experimentally measuring these momenta. In fact, in a paper published in 1909, Stark[161] outlined essentially the theory of an experiment by which Compton in 1922 strikingly confirmed Einstein's hypothesis of light quanta and their transfer of momentum, as we shall see in due course.

Although the full comprehension of the dual nature of light belongs to a later stage in the development of quantum theory, its roots may be found in one of Einstein's early papers on black-body radiation. In an article "On the present state of the problem of radiation"[162] Einstein calculated the fluctuations of radiative energy in a partial volume v of an isothermal enclosure of temperature T. Denoting by E the instantaneous energy within the frequency interval between ν and $\nu + d\nu$, with $\bar{E} = \int Ee^{-E/kT}\,dp\,dq/\int e^{-E/kT}\,dp\,dq$ its average energy and with $\varepsilon = E - \bar{E}$ the energy fluctuation, he could easily show that the mean-square fluctuation $\overline{\varepsilon^2}$, which equals $\overline{E^2} - \bar{E}^2$, where $\overline{E^2} = \int E^2 e^{-E/kT}\,dp\,dq/\int e^{-E/kT}\,dp\,dq$, is given by $\overline{\varepsilon^2} = kT^2\,d\bar{E}/dT$, a result which also could be obtained on the basis of the quantum-theoretic energy distribution according to which E assumed only the discrete values $E_n = nh\nu$. Substituting in $\bar{E} = vu_\nu\,d\nu$

[159] For further details on the history of the photoelectric effect cf. R. Ladenburg, *Jahrbuch der Radioaktivität 6,* 428–433 (1909); A. L. Hughes, *Photo-Electricity* (Cambridge University Press, 1910); A. L. Hughes, *Die Lichtelektrizität* (Barth, Leipzig, 1914); R. Pohl and E. Pringsheim, *Die Lichtelektrischen Erscheinungen* (Friedrich Vieweg & Sohn, Braunschweig, 1913); W. Hallwachs, "Die Lichtelektrizität," *Handbuch der Radiologie* (Akademische Verlagsgesellschaft, Leipzig, 1916), vol. 3, pp. 245–395; H. S. Allen, *Photoelectricity* (Longmans, Green & Co., London, 1913, 2d. ed. 1925); A. L. Hughes and L. A. DuBridge, *Photoelectric Phenomena* (McGraw-Hill, New York, 1932); V. K. Zworykin and E. G. Ramberg, *Photoelectricity and Its Applications* (Wiley, New York, 1949, 1956).

[160] M. Planck, "Bemerkungen zum Prinzip der Aktion und Reaktion in der allgemeinen Dynamik," *Physikalische Zeitschrift 9,* 828–830 (1908); *Verhandlungen der Deutschen Physikalischen Gesellschaft 10,* 728–732 (1908); *Physikalische Abhandlungen und Vorträge* (REF. 42), vol. 2, pp. 215–219.

[161] J. Stark, "Zur experimentellen Entscheidung zwischen Aetherwellen- und Lichtquantenhypothese," *Physikalische Zeitschrift 10,* 902–913 (1909).

[162] A. Einstein, "Zum gegenwärtigen Stand des Strahlungsproblems," *Physikalische Zeitschrift 10,* 185–193 (1909); *Verhandlungen der Deutschen Physikalischen Gesellschaft 11,* 482–500 (1909).

for u_ν, in conformance with Planck's law, the right-hand member of Eq. (1.16), Einstein obtained

$$\overline{\varepsilon^2} = kT^2 v\, d\nu \frac{du_\nu}{dT} = \frac{8\pi h^2 \nu^4 v\, d\nu\, e^{h\nu/kT}}{c^3 (e^{h\nu/kT} - 1)^2}$$

and after eliminating T, again by Eq. (1.16), the equation

$$\overline{\varepsilon^2} = \bar{E} h\nu + \frac{c^3 \bar{E}^2}{8\pi \nu^2 v\, d\nu} \tag{1.19}$$

The second term of the sum in Eq. (1.19) is the mean-square energy fluctuation due to interferences between partial waves, as Einstein showed in his paper by a simple dimensional analysis and as Lorentz[163] subsequently demonstrated in rigorous detail. It is therefore exactly the term that had to be expected on the basis of the undulatory or Maxwellian theory of light. The first term, on the other hand, while inexplicable from this point of view, could easily be accounted for, as Einstein pointed out, on the basis of his hypothesis of light quanta, according to which the latter are subject to the same statistical laws as are particles or molecules in the kinetic theory of the ideal gas. In fact, if $\bar{n}$ denotes the average of the instantaneous number n of light quanta in volume v and $\delta = n - \bar{n}$ the fluctuation of n, then $\overline{\varepsilon^2} = \overline{(\delta h\nu)^2} = \bar{n} h^2 \nu^2 = \bar{E} h\nu$ since $\overline{\delta^2} = \bar{n}$ and $\bar{E} = \bar{n} h\nu$. Einstein corroborated his result with a discussion of the Brownian motion of a mirror which he assumed to be a perfect reflector for radiation within the frequency interval $d\nu$ and transparent for all other frequencies. For the mean square of the pressure fluctuation he obtained again a sum of two terms, one term corresponding to the fluctuation of waves and the other to that of particles. Einstein, it seems, did not realize at that time the momentous importance of his conclusions. His point was primarily to show that the wave theory of light, as conceived at the time, was incompatible with Planck's experimentally well-established law of radiation.

However, at the conclusion of an address on "The Development of Our Conceptions on the Nature and Constitution of Radiation,"[164] which he delivered before the 81st assembly of the German Association of Scientists, held at Salzburg in September, 1909, Einstein proposed, though only in outline, a model for his theory of light quanta and their statistical behavior. He described each quantum of light as a singularity in space surrounded by a field of forces whose magnitudes decrease with the distance from the singularity at the center. A superposition of these fields, he

[163] H. A. Lorentz, *Les théories statistiques en thermodynamique*, Conférences faites au Collège de France en Novembre 1912 (Teubner, Leipzig, Berlin, 1916), Appendix IX, pp. 114–120.

[164] A. Einstein, "Über die Entwicklung unserer Anschauungen über das Wesen und die Konstitution der Strahlung," *Physikalische Zeitschrift 10*, 817–825 (1909); *Verhandlungen der Deutschen Physikalischen Gesellschaft 11*, 482–500 (1909).

contended, would produce an undulatory field of structure similar to that of the electromagnetic field in Maxwell's theory of light. The purpose of this suggestion, as Einstein explicitly stated, was not to construct such a model but merely to show "that the two structure qualities, the undulatory structure and the quantum structure, both of which Planck's formula ascribes to radiation, are not incompatible with each other."[165] Concerning the details of a model for light quanta or the details of their theory, Einstein pointed out that the first term of the sum in Eq. (1.19) does not supply sufficient information, just as Maxwell's equations cannot be deduced from its second term.

Nevertheless, the search for an acceptable mechanical model to explain the mechanism of quanta of light and, more generally, the intimate connection between the corpuscular concept of energy E and the undulatory notion of frequency ν, a relation nowhere encountered in classical physics and yet indisputably assured, was the subject of much discussion at that time. The construction of such models, it was hoped, would be particularly useful for the clarification of the real meaning of Planck's constant h. In his reminiscences of the early years of quantum theory Max Born[166] mentions the following example of such a model which, by the way, Planck is said to have used himself in one of his lectures. Consider, says Born, an apple tree and assume that the length l of the stems of its apples, regarded as pendula, is inversely proportional to the square of their height H above ground, so that $\nu \sim l^{-1/2} \sim H$. If now the tree is shaken with a certain frequency ν, the apples at the corresponding height H will resonate and fall to earth with a kinetic energy E proportional to H and consequently also proportional to ν. Needless to say, models of this kind and even more refined macroscopic models did not contribute very much toward an understanding of Planck's constant.

Among the microscopic models, based on the contemporary knowledge concerning the structure of the atom, those proposed by Schidlof and by Haas were widely—though not always favorably—discussed. Atomic theory, at that time, was still based on J. J. Thomson's model of the atom, which described it as consisting of a positively charged sphere of uniform density inside which a compensating number of negatively charged electrons revolve on circular paths. Schidlof[167] modified Thomson's model by

[165] "Ich wollte durch dasselbe nur kurz veranschaulichen, daß die beiden Struktureigenschaften (Undulationsstruktur und Quantenstruktur), welche gemäß der Planckschen Formel beide der Strahlung zu kommen sollten, nicht als miteinander unvereinbar anzusehen sind." *Ibid.*, p. 825.

[166] M. Born, *Physik im Wandel meiner Zeit* (Vieweg, Braunschweig, 1957), p. 224; "Albert Einstein und das Lichtquantum," *Die Naturwissenschaften 42*, 425–431 (1955).

[167] A. Schidlof, "Zur Aufklärung der universellen elektrodynamischen Bedeutung der Planckschen Strahlungskonstante h," *Annalen der Physik 35*, 90–100 (1911); "Sur quelques problèmes récents de la théorie du rayonnement—(II) La signification électromagnétique de l'élément de l'action," *Comptes Rendus de la Société de Physique de Genève, Archives des Sciences Physiques et Naturelles 31*, 385–387 (1911).

assuming that at the center of the positive sphere a negatively charged sphere is located whose density equals that of the electrons. On the basis of this model Schidlof showed that Planck's constant is given by the relation $h = 2\pi e(am/N^{1/3})^{1/2}$, where N is the number of electrons, m their mass, e their charge, and a the radius of the atom.

Of greater importance, though likewise based on J. J. Thomson's assumptions, was a model of the hydrogen atom proposed in 1910 by Arthur Erich Haas.[168] Since it made it possible to deduce for the first time the Rydberg constant, though only with a wrong numerical factor, in terms of the elementary charge of the electron, its mass, and Planck's constant and since it thus led, albeit on the basis of erroneous assumptions, to a result which preceded[169] Bohr's famous derivation of this spectroscopic constant by three years, it is appropriate to discuss Haas's model and the peculiar circumstances which led to its conception in greater detail.

In December, 1908, Haas completed a scholarly historicocritical analysis of the principle of energy conservation[170] which he submitted to the philosophical faculty of the University of Vienna for his habilitation as Privatdocent. The experimental physicists Franz Exner and Victor Lang, who were appointed as referees and who knew that Haas had obtained his Ph.D. only two years earlier, turned it down, but asked him to renew his application after some time with additional research of a more technical nature in physics. Haas complied with this request, and within almost a year's time he established a theory of the hydrogen atom, which he later regarded as "the greatest achievement"[171] of his life, and submitted it to the Viennese Academy of Science, where it was read on March 10, 1910.

In the first part of the paper Haas showed that an electron of mass m and charge $-e$, moving along a circle of radius r, concentric with a uniformly distributed positive spherical charge e of radius a, attains the

[168] A. E. Haas, "Über die elektrodynamische Bedeutung des Planck'schen Strahlungsgesetzes und über eine neue Bestimmung des elektrischen Elementarquantums und der Dimensionen des Wasserstoffatoms," *Wiener Berichte 119*, 119–144 (1910); "Uber eine neue theoretische Bestimmung des elektrischen Elementarquantums und des Halbmessers des Wasserstoffatoms," *Physikalische Zeitschrift 11*, 537–538 (1910); *Jahrbuch der Radioaktivität 7*, 261–268 (1910).

[169] Cf. A. Haas, *Das Naturbild der neuen Physik* (Walter de Gruyter, Berlin, Leipzig, 1920), p. 78: "Im Jahre 1910 fand der Verfasser dieser Schrift, in dem er zuerst das Quantenprinzip auf die Theorie des Atoms und die Theorie der Spektren anwandte, eine Beziehung, die die Rydbergsche Konstante mit den Grundgrößen der Quantentheorie und der Elektronentheorie verknüpft. . . ." Cf. also A. Haas, *Einführung in die Theoretische Physik* (Walter de Gruyter, Berlin, Leipzig, 1921), vol. 2, p. 14, footnote 2; A. Haas, *Atomic Theory*, translated by T. Verschoyle (Van Nostrand, New York, 1926), p. 32: "This relation, which connects the fundamental constant of spectroscopy with the fundamental quantities of the electron theory and the elementary quantum of action, was first derived by the author of this work in 1910, and more accurately later by Bohr from his theory." Cf. also A. Haas, *The World of Atoms*, translated by H. S. Uhler (Van Nostrand, New York, 1928), pp. 130–131.

[170] A. Haas, *Die Entwicklungsgeschichte des Satzes von der Erhaltung der Kraft* (A. Hölder, Vienna, 1909).

[171] Haas's unpublished autobiography. Cf. A. Hermann, "A. E. Haas und der erste Quantenansatz für das Atom," *Sudhoffs Archiv 49*, 255–268 (1965).

maximum of its total energy, namely, e^2/a, if $r = a$. Assuming also that in this case the energy is given by $h\nu_\infty$, where ν_∞ ($= 8.23 \times 10^{14}$ sec^{-1}) is the frequency c/λ_∞, λ_∞ being the limit of the Balmer expression

$$\lambda = \frac{n^2}{n^2 - 4} \times 3.647 \times 10^{-5} \text{ cm}$$

for $n = \infty$ (or, in modern notation, $\nu_\infty = Rc/4$, R being the Rydberg constant), and that ν_∞ is the frequency of the electronic revolution, Haas deduced from the equality between the Coulombian attraction e^2/a^2 and the centripetal force $ma\omega^2 = 4\pi^2 ma\nu_\infty{}^2$ that $\nu_\infty = e/(2\pi a\sqrt{am})$. Thus, on the assumption that the electron revolves on the surface of Thomson's positive sphere, he derived the relations

$$h\nu_\infty = \frac{e^2}{a} \qquad \text{and} \qquad \nu_\infty = \frac{e}{2\pi a\sqrt{am}}$$

which, after elimination of a, implied that

$$\nu_\infty = \frac{4\pi^2 me^4}{h^3}$$

(or, in modern notation, $R = 16\pi^2 me^4/h^3c$). Had Haas taken $h\nu_\infty = \frac{1}{2}e^2/a$ (which would have agreed with Bohr's result derived from the quantization of the angular momentum), he would have obtained the eight times smaller correct Bohr formula $R = 2\pi^2 me^4/h^3c$. In view of the relatively great uncertainty in the knowledge of the size of elementary constants—it was only in 1911 that Millikan introduced the technique of oil droplets, which considerably improved the accuracy in the determination of e—Haas's result and his deductions of the values of e ($= 3.18 \times 10^{-10}$ esu), of m ($= 5.68 \times 10^{-28}$ g), and of a ($= 1.88 \times 10^{-8}$ cm) did not bring him into conflict with experience.

Haas's work, however, was rejected and even ridiculed. In fact, as reported in his unpublished autobiography, when Haas lectured on his ideas at a meeting of the Vienna Chemical-Physical Society, Ernst Lecher, the director of the Institute of Physics at the Vienna University, called it a joke ("ein Faschingsscherz") and Hasenöhrl, when asked for his opinion on this matter, remarked that Haas's jumbling together of two completely unrelated subjects such as quantum theory, which was still regarded as a part of the theory of heat, and spectroscopy, a branch of optics, was "naïve" and should not be taken seriously. It was therefore only in 1912, after the publication of another technical paper on the equilibrium distribution of electron groups in the Thomson model of the atom,[172] that Haas obtained his appointment as Privatdocent.

[172] A. Haas, "Über Gleichgewichtslagen von Elektronengruppen in einer äquivalenten Kugel von homogener positiver Elektrizität," *Wiener Berichte 120*, pp. 1111–1171 (1911) (presented on June 28, 1911).

And yet, although generally rejected, Haas's work was not completely ignored. Explicit reference to it and to his article in the *Physikalische Zeitschrift*[173] was made by Arnold Sommerfeld in an extremely important paper which Sommerfeld presented to the first Solvay Congress[174] in 1911 and on which we shall have to comment in another context later on. Furthermore, Niels Bohr, who, as stated in his Rutherford Memorial Lecture,[175] learned of Haas's work only after having accomplished his own famous derivation of the Rydberg constant in 1913, got "a vivid account of the discussions at the first Solvay meeting"[176] from Rutherford and thus probably also of Sommerfeld's paper, which was to some extent influenced by Haas's conceptions, as we shall see in due course. Could such an influence, circuitous and scanty as it may have been, be fully established between Haas and Bohr, Haas's work, although fortuitous and based on untenable assumptions, would prove to have been of great historical importance for the conceptual development of quantum theory.

As we have seen, Sommerfeld did not reject the idea of connecting Planck's constant with atomic constants, but he objected to the philosophy on the basis of which such attempts were carried out. He opposed the idea that a dynamic interpretation of h had to be found. In an address to the 83d meeting of the German Association of Scientists, held in 1911 in Karlsruhe, he declared that he thought it futile to explain h on the basis of molecular dimensions; the existence of molecules should rather be educed as a "function and consequence of the existence of an elementary quantum of action. . . . An electromagnetic or mechanical 'explanation' of h," he stated, "seems to me just as worthless and unpromising as a mechanical 'explanation' of Maxwell's equations. It would be much more profitable to science if the numerous consequences of the h hypothesis were investigated and other phenomena were retraced to it."[177] Modern quantum mechanics, which views Planck's constant h as a kind of connecting link between the concept of waves and that of particles, fully justified Sommerfeld's contention. Something which delimits and restricts the validity of models

[173] See REF. 168.

[174] *La Théorie du Rayonnement et les Quanta—Rapports et Discussions de la Réunion Tenue à Bruxelles*, 1911, edited by P. Langevin and M. de Broglie (Gauthier-Villars, Paris, 1912), p. 362.

[175] "Thus, as I later learned, A. Haas had in 1910 attempted, on the basis of Thomson's atomic model, to fix dimensions and period of electronic motions by means of Planck's relation between the energy and the frequency of a harmonic oscillator" (p. 1086) in N. Bohr, "The Rutherford Memorial Lecture 1958," *Proceedings of the Physical Society of London 78*, 1083–1115 (1961); reprinted in N. Bohr, *Essays 1958–1962 on Atomic Physics and Human Knowledge* (Interscience, New York; Bungay, Suffolk, 1963), pp. 30–73; and in J. B. Birks (ed.), *Rutherford at Manchester* (Heywood, London, 1962; Benjamin, New York, 1963), pp. 114–167. Yet, in his classic paper (REF. 56 OF CHAP. 2) Bohr explicitly mentioned Haas's work.

[176] N. Bohr, "The Solvay Meetings and the Development of Quantum Physics," in *La Théorie Quantique des Champs* (Interscience, New York, 1962), p. 17; reprinted in N. Bohr, *Essays 1958–1962*.

[177] A. Sommerfeld, "Das Plancksche Wirkungsquantum und seine allgemeine Bedeutung für die Molekularphysik," *Physikalische Zeitschrift 12*, 1057–1066 (1911).

based either on the particle conception or on the undulatory conception cannot itself be represented by a model and certainly cannot be explained in the terms of classical physics.

This argument, however, had not yet been proposed in the early years of the century. In fact, it was that very belief in the unrestricted validity of a model that obstructed the general acceptance of Einstein's idea of quanta of light. For in spite of the force of its argument and the cogent reasoning behind its statistical conclusions, Einstein's hypothesis of localized quanta of light was still regarded as unacceptable. Even Lorentz rejected these ideas. Experiments had been performed by Lummer and Gehrcke[178] in which two rays of the green mercury line with a phase difference of about two million wavelengths were shown to be still capable of being brought into interference, and these experiments clearly indicated that the spatial extension of quanta of light in the direction of their propagation, if such quanta existed as presumably incoherent entities, was at least a meter. Similarly, the improvement in resolution due to the increase of large-scale apertures of telescopes, for example, was thought to be inexplicable unless the lateral dimensions of these quanta were of the same order of magnitude. How then, asked Lorentz, for example,[179] was it possible for the human eye to see? Its pupil could admit only a small fraction of a quantum whereas, according to the basic tenets of the hypothesis, a whole quantum would be necessary to affect the retina. Does not the experiment of Newton's rings, in which a ray of light divides by reflection into two parts which, after proceeding over different paths, reunite for interference, clearly demonstrate the divisibility of quanta? Inconsistencies like those raised by Lorentz and others were the first examples of the difficulties originating from an indiscriminate application and simultaneous use of the models of wave and particle. It was mainly because of these difficulties that Einstein's corpuscular conception of light quanta met such strong opposition.

Incidentally, there exists an interesting document which ostensibly puts on record the reaction toward Einstein's notion of light quanta, prior to Millikan's final empirical confirmation of the photoelectric equation, on the part of Germany's most prominent physicists. To ensure the election of Einstein on June 12, 1913, to membership in the Prussian Academy of Science, in replacement of the recently deceased J. H. van't Hoff, the noted Dutch physical chemist and first Nobel laureate in chemistry, four of the most eminent German physicists, Planck, Warburg, Nernst, and Rubens, submitted to the Prussian Ministry of Education a petition in which they recommended Einstein for this vacancy. In this signed document

[178] O. Lummer and E. Gehrcke, "Über die Interferenz des Lichtes bei mehr als zwei Millionen Wellenlängen Gangunterschied," *Verhandlungen der Deutschen Physikalischen Gesellschaft 4*, 337–346 (1902).

[179] REF. 109, p. 1250.

they described Einstein's work on the special theory of relativity, his contributions to the quantum theory of specific heats—which we shall discuss in the next section—and his treatment of the photoelectric and photochemical effects. In concluding their recommendation, they declared: "Summing up, we may say that there is hardly one among the great problems, in which modern physics is so rich, to which Einstein has not made an important contribution. That he may sometimes have missed the target in his speculations, as, for example, in his hypothesis of light quanta, cannot really be held too much against him, for it is not possible to introduce fundamentally new ideas, even in the most exact sciences, without occasionally taking a risk."[180] It is instructive to compare this assessment of Einstein's work on light quanta with its evaluation at a later period as, for example, with K. T. Compton's description of it as "a contribution to physical theory certainly comparable in importance, and thus far more useful in its applications than his more impressive and wider publicized general theory of relativity."[181]

It should be understood that the ultimate reason for Einstein's statistical discovery of the duality of light as wave and particle was Planck's law of radiation. In fact, when Planck, in order to fit his theory with experience, interpolated between

$$\frac{\partial^2 S}{\partial U^2} = \frac{\text{const}}{U} \tag{1.8}$$

which led to Wien's law of radiation, and

$$\frac{\partial^2 S}{\partial U^2} = \frac{\text{const}}{U^2} \tag{1.10}$$

which led to the Rayleigh-Jeans formula, the resulting equation (1.11),

$$\frac{\partial^2 S}{\partial U^2} = \frac{a}{U(b + U)} \tag{1.11}$$

was precisely an amalgamation of the wave and particle aspects of radia-

[180] "Zusammenfassend kann man sagen, daß es unter den großen Problemen, an denen die moderne Physik so reich ist, kaum eines gibt, zu dem nicht Einstein in bemerkenswerter Weise Stellung genommen hätte. Daß er in seinen Spekulationen auch einmal über das Ziel hinausgeschossen haben mag, wie zum Beispiel in seiner Hypothese der Lichtquanta, wird man ihm nicht allzusehr anrechnen dürfen. Denn ohne einmal ein Risiko zu wagen, läßt sich auch in der exaktesten Naturwissenschaft keine wirkliche Neuerung einführen." Quoted after T. Kahan, "Un document historique de l'Académie des Sciences de Berlin sur l'activité d'Albert Einstein (1913)," *Archive Internationale d'Historie des Sciences 15*, 337 342 (1962). The original text of this document is also found in C. Seelig, *Albert Einstein—Eine dokumentarische Biographie* (Europa Verlag, Zürich, Stuttgart, Wien, 1954), pp. 174–175, and an English version in C. Seelig, *Albert Einstein* (Staples Press Ltd., London, 1956), p. 145.

[181] K. T. Compton, "The electron; its intellectual and social significance," *Nature 139*, 229–240 (1937).

tion, the first of its kind in the development of quantum theory, though, of course, not recognized as such by its author.

This may be understood from the following analysis. If radiation were treated solely according to Eq. (1.10) or, equivalently, according to the Rayleigh-Jeans formula (1.9), a straightforward calculation of the energy fluctuation would yield only the second term in the sum of Eq. (1.19), corresponding to the wave picture of light. If, however, Eq. (1.8) or Wien's law, Eq. (1.5), were taken as the point of departure, the fluctuation formula would be reduced to the first term of the sum in Eq. (1.19), corresponding to the particle picture of light. Alternatively, and without calculation, this conclusion also follows from the fact that for low frequencies, the range of validity of Eq. (1.10), the first term of the sum under discussion can be neglected in comparison with the second, while for high frequencies, where Eq. (1.8) is valid, Eq. (1.19) reduces to $\varepsilon^2 = \bar{E}h\nu$, the fluctuation formula for the particles of an ideal gas.

The reader will now understand our reference to Planck's interpolation formula (1.11), announced on October 19, 1900, as containing "certain implications which, once recognized by Einstein, affected decisively the very foundations of physics."[182]

It makes little sense, of course, to regard a certain day as "the birthday" of a development conceptually so complicated as that of quantum theory. However, an inquiry concerning the earliest formulation of a statement or formula which characterizes the fundamental idea of a theory seems to have a place in a conceptual analysis such as ours. From this point of view we would prefer to consider October 19, 1900, as the birthday of our theory and not, as Max von Laue[183] and others suggested, December 14, 1900. The following considerations will explain our choice.

December 14, 1900, has usually been regarded as this date because on that day Planck introduced his constant h and in terms of it the discrete energy spectrum of a harmonic oscillator of given frequency. Now, it can be shown[184] that discontinuous change is not necessarily an essential characteristic of quantum theory. It ought also to be noted that as early as May 18, 1899, a year and a half before December 14, 1900, Planck had read a paper[185] before the Berlin Academy in which he calculated on the basis of Wien's radiation law and certain experimental data obtained by Kurlbaum[186] and Paschen[187] the numerical value of a constant which he

[182] See p. 18.
[183] REF. 89.
[184] Cf. for example, P. Jordan, *Anschauliche Quantentheorie* (Springer Verlag, Berlin, 1936), p. 85.
[185] M. Planck, "Über irreversible Strahlungsvorgänge" (see REF. 42, 5th communication).
[186] F. Kurlbaum, "Ueber eine Methode zur Bestimmung der Strahlung in absolutem Maass und die Strahlung des schwarzen Körpers zwischen 0 und 100 Grad," *Wiedemannsche Annalen der Physik 65*, 746–760 (1898).
[187] F. Paschen, "Über die Vertheilung der Energie im Spectrum des schwarzen Körpers bei niederen Temperaturen," *Berliner Berichte 1899*, pp. 405–420.

called b and found that $b = 6.885 \times 10^{-27}$ cm^2 g/sec. He also recognized it as a fundamental natural constant. For, together with the velocity of light $c = 3.00 \times 10^{10}$ cm/sec, the gravitational constant $f = 6.685 \times 10^{-8}$ cm^3/g sec^2, and another radiation constant[188] $a = 0.4818 \times 10^{-10}$ sec deg, it made it possible to introduce a system of natural units ("natürliche Maßeinheiten") "which in view of their independence of particular bodies or substances are of everlasting significance also beyond the limits of our global human civilization."[189] Planck's introduction of h, though under a different name and without reference to the physical dimension of action,[190] as well as its recognition as a fundamental natural constant, though not yet with respect to the discreteness of energy values, preceded considerably, as we see, the date of December 14, 1900. Since, however, the wave-particle duality may be regarded, in our view, as the essential characteristic of quantum theory, since this duality was implicitly contained, as we have shown following Einstein, in Planck's paper of October 19, 1900, and since it was ultimately the interpolation formula (1.11) for the logical corroboration of which Planck turned to Boltzmann's statistics and quantized the energy of the harmonic oscillator—or, in modern terminology, the energy of the field of electromagnetic harmonic vibrations—our choice of October 19, 1900, as the birthday of quantum theory seems to be justified.

1.4 *Elaborations of the Concept of Quanta*

In the early development of quantum theory, as we have seen, two conceptions of quanta had been proposed: Planck's energy quanta, associated with the interaction between radiation and matter, and Einstein's light quanta, associated with radiation in transit. Both were described by the same equation $\epsilon = h\nu$. Yet in spite of this identity in formula, Planck's conception was universally accepted as indispensable for further research while Einstein's notion was generally rejected for reasons previously explained.

Also Planck refused to accept the idea of light quanta. Einstein, it will be recalled, drew attention to the logical inconsistency in Planck's derivation of the radiation law as being based on a simultaneous espousal and rejection of classical electrodynamics;[191] but, viewing Planck's law as an empirically established fact, he saw in this inconsistency an argument against the unrestricted applicability of Maxwell's electrodynamics and

[188] The constant a is essentially h/k.

[189] Planck's proposed system of units was $(bf/c^3)^{1/2} = 4.13 \times 10^{-33}$ cm as unit of length, $(bc/f)^{1/2} = 5.56 \times 10^{-5}$ g as unit of mass, $(bf/c^5)^{1/2} = 1.38 \times 10^{-43}$ sec as unit of time, and $a(c^5/bf)^{1/2} = 3.50 \times 10^{32}$ deg as unit of temperature.

[190] In his *Annalen* paper "Über irreversible Strahlungsvorgänge" (see REF. 42), written also before Dec. 14, 1900, Planck has already put $b = 6.885 \times 10^{-27}$ erg sec (p. 120).

[191] REF. 114.

thus an argument in favor of his own ideas on light quanta. Now, in striking contrast to Einstein, Planck attempted to resolve this inconsistency by a stricter adherence to Maxwell's electrodynamics.

When Einstein raised his objections at the Salzburg meeting, Planck's reply, in which he described Einstein's "atomistic interpretation" of radiation and his "rejection of Maxwell's equations" as "an as yet unnecessary step,"[192] already contained the germ of his subsequent modification in deriving his radiation law. Referring to Einstein's thought experiment on radiation fluctuations and their effect on a movable mirror,[193] Planck commented that Einstein's reasoning would be conclusive provided one really understood the interaction between radiation and matter; "but if this is not the case, the bridge which is needed to connect the motion of the mirror with the intensity of the incident radiation is missing. Indeed," Planck continued, "the interaction between free electric energy in the vacuum and the motion of ponderable atoms, I think, is little understood. It is essentially the process of emission and absorption light. . . . But just emission and absorption are an obscure point on which we know so little. Perhaps we know something of absorption, but how about emission? One pictures it as produced by accelerations of electrons. But this is the weakest point of the whole electron theory. One assumes that the electron has a definite volume and a definite finite charge density . . . in violation of the atomistic conception of electricity. These are not impossibilities but difficulties, and I almost wonder that no stronger objection has been voiced. It is at this very point, I believe, where the quantum theory can be useful. . . ."[194]

It was precisely in connection with this issue and in this spirit that at the end of 1911 Planck completed his so-called "second theory," which he presented to the German Physical Society on January 12, 1912. In it[195] he proposed to derive the radiation law in a way which he thought "is free of inner inconsistencies" and which "does not depart from the core of the classical electrodynamics and electron theory more than is absolutely necessary in view of the undeniably irreconcilable differences with the quantum hypothesis."

Contrary to his first theory and in closer agreement with Maxwell's theory Planck now suggested that the absorption of electromagnetic energy by oscillators is a continuous process. On the other hand, he contended, emission occurs only in integral energy quanta ϵ and is a discontinuous process regulated by a law of chance; not every oscillator

[192] "... ein Schritt, der in meiner Auffassung noch nicht als notwendig geboten ist." *Ibid.*, p. 825.

[193] See p. 38.

[194] REF. 114, p. 825.

[195] M. Planck, "Eine neue Strahlungshypothese," *Verhandlungen der Deutschen Physikalischen Gesellschaft 13*, 138–148 (1911); *Physikalische Abhandlungen und Vorträge* (REF. 42), vol. 2, pp. 249–259. "Über die Begründung des Gesetzes der schwarzen Strahlung," *Annalen der Physik 37*, 642–656 (1912); *Physikalische Abhandlungen und Vorträge*, vol. 2, pp. 287–301.

after having accumulated the amount ϵ during the absorption necessarily emits this quantum, but only a fraction η of such oscillators does so. Planck now argued as follows. Of a total of N oscillators Nw_n have at a given instant of time t energies between $n\epsilon$ and $(n + 1)\epsilon$, where $n = 0$, 1, 2, ... If dV/dt is the rate of increase of energy content per oscillator during the time interval dt, the energy threshold $(n + 1)\epsilon$ is passed by all those oscillators whose energy, at the beginning of dt, was between $(n + 1)\epsilon$ and $(n + 1)\epsilon - dV$. On the assumption of a uniform distribution, $Nw_n\, dV/\epsilon$ is the number of oscillators within dV, and of these $\eta N w_n\, dV/\epsilon$ emit their total energy while $(1 - \eta)Nw_n\, dV/\epsilon$ pass the aforementioned energy threshold to replace the $Nw_{n+1}\, dV/\epsilon$ oscillators which at the beginning of dt were in the energy interval from $(n + 1)\epsilon$ to $(n + 1)\epsilon + dV$. Under conditions of equilibrium $(1 - \eta)Nw_n\, dV/\epsilon = Nw_{n+1}\, dV/\epsilon$ or $w_{n+1} = w_n(1 - \eta)$. Thus $w_n = w_0(1 - \eta)^n$ and, since $\sum_0^\infty Nw_n = N$, $w_0 = \eta$. The average energy of the oscillators between $n\epsilon$ and $(n + 1)\epsilon$ being $(n + \frac{1}{2})\epsilon$, the total energy E is

$$E = \sum Nw_n(n + \tfrac{1}{2})\epsilon = \frac{N\epsilon(1 - \eta)}{\eta} + \frac{N\epsilon}{2}$$

Hence the average energy per oscillator is

$$U = \frac{\epsilon}{2} + \epsilon\,\frac{1 - \eta}{\eta}$$

Planck now determined the dependence of U on the temperature T. The thermodynamic probability of the distribution is $W = N!/\prod_0^\infty (Nw_n)!$ and hence the entropy is

$$S = -kN \log \eta - \frac{kN}{\eta}(1 - \eta) \log (1 - \eta)$$

Calculating $dS/dE = (dS/d\eta)(d\eta/dE) = T^{-1}$, Planck obtained $T^{-1} = -(k/\epsilon) \log (1 - \eta)$ or $1 - \eta = \exp(-\epsilon/kT)$, so that finally

$$U = \frac{\epsilon}{2} + \epsilon\,\frac{1 - \eta}{\eta} = \frac{h\nu}{2} + \frac{h\nu}{\exp(h\nu/kT) - 1}$$

an expression which differed from Planck's previous value of U (cf. p. 21) by the additive constant $h\nu/2$. In order to obtain the radiation law Planck had to introduce an additional hypothesis. He postulated that the ratio between the probabilities of emission and absence of emission is proportional to the reciprocal rate of the absorption: $\eta/(1 - \eta) = \alpha/(dV/dt)$. Thus the more slowly an oscillator passes the critical threshold $n\epsilon$, the greater is the probability of emission. Since at equilibrium the brightness K_ν

(cf. note 12, p. 5) and consequently also the energy density $u_\nu = (8\pi/c)K_\nu$ are proportional to dV/dt, Planck obtained

$$u_\nu = \beta \frac{1-\eta}{\eta} = \beta \left[\exp \frac{h\nu}{kT} - 1 \right]^{-1}$$

where the constant β by the use of the Rayleigh-Jeans formula $[\lim(\nu/T) = 0]$ turned out to be $8\pi h\nu^3/c^3$, so that Planck's original law of radiation (1.16) has been recovered.

Although soon rejected by its author and replaced by a "third theory," which regarded emission as well as absorption as continuous phenomena,[196] Planck's "second theory" deserves our attention for at least three important reasons. In the first place, from the new expression of the average energy U of the harmonic oscillator Planck deduced that at absolute zero the energy is not zero, as implied by his original formula for U, but equals $\frac{1}{2}h\nu$. The notion of "zero-point energy" thus made its first appearance in modern physics; though forged by an untenable theory, it was eventually, as is well known, vindicated by quantum mechanics. The other point of interest is that Planck's "second theory" contained probably the earliest suggestion, as far as quantum theory is concerned, that elementary processes are subject to a law of chance. For Planck assumed that the oscillator, while absorbing energy continuously from the surrounding field of radiation, accumulates this energy and that its energy content may therefore have any value whatever; as soon, however, as its energy content reaches the amount $h\nu$, it has a definite probability of emitting the quantum $h\nu$. It should be emphasized that Planck's conception of probability was the classical conception as used in the kinetic theory of gases. For he stated explicitly: "Not that causality is denied for emission; but the factors which determine emission causally seem to be so deeply hidden that for the time being their laws can be only statistically ascertained."[197] Planck obviously regarded the use of the law of chance, which, as he said, "always signifies a renunciation of the completeness of causal interconnection,"[198] merely as a provisional device. In fact, this was the main reason why in his own opinion his new theory was but a "hypothetical attempt" to reconcile the law of radiation with the foundations of Maxwell's doctrine, and not a final solution of the problem.

[196] M. Planck, "Eine veränderte Formulierung der Quantenhypothese," *Berliner Berichte 1914*, pp. 918–923; *Physikalische Abhandlungen und Vorträge* (REF. 42), vol. 2, pp. 330–335. The main idea of this "third theory," namely, that quantum effects are due only to collisions, was shown to be untenable by A. D. Fokker, "Die mittlere Energie rotierender Dipole im Strahlungsfeld," *Annalen der Physik 43*, 810–820 (1914).

[197] "Nicht als ob für die Emission keine Kausalität angenommen würde; aber die Vorgänge, welche die Emission kausal bedingen, sollen so verborgener Natur sein, daß ihre Gesetze einstweilen nicht anders als auf statistischem Wege zu ermitteln sind." REF. 195, p. 644.

[198] *Ibid.*, p. 148; p. 259.

The third point of interest is his method of determining this probability. Having postulated that the ratio between the probabilities of emission and absence of emission is proportional to the reciprocal rate of the absorption or, equivalently, that the ratio of the probability of no emission to that of emission is proportional to the intensity of the exciting vibration,[199] Planck determined the constant of proportionality by recourse to the limiting case of extremely strong radiations where Rayleigh's formula and classical dynamics become valid.[200] Planck's derivation of the constant of proportionality as proposed presently was probably the earliest instance in quantum theory of applying what more than ten years later became known as the "correspondence principle" and what proved to be immensely fruitful not only for the determination of constants through such limiting considerations but also for the inference of conclusions of a much wider scope.

Those who rejected Einstein's light quanta had, of course, to refute the statistical evidence he had adduced in favor of this conception. As long as only the experimental evidence of the existence of interference and diffraction served as the basis for this rejection, no clear-cut decision could be made. Thus, various attempts to disprove Einstein's conclusions were made at the turn of the first decade of the century. For example, in his lectures delivered at Columbia University in 1913 on "Recent Problems in Theoretical Physics,"[201] in which he rejected Einstein's conception of light quanta for reasons similar to those adduced by Lorentz,[202] Wien attempted to invalidate Einstein's conclusion regarding Eq. (1.19) as follows. He pointed out that in virtue of Eq. (1.16), with the abbreviation $x = h\nu/kT$,

$$\frac{\bar{E}h\nu}{e^x - 1} = \frac{c^3\bar{E}^2}{8\pi\nu^2 v\, d\nu}$$

an expression which he called B. Denoting $\bar{E}h\nu$ briefly by A, so that $A = (e^x - 1)B$, Wien obtained $\overline{\epsilon^2} = A + B = Be^x$. "This formula," he concluded, "shows that there is no special reason to attribute the fluctuations to two causes which are independent of each other."[203]

Those who rallied to support Einstein's idea of light quanta thought it imperative to investigate their relation to Planck's conception of quanta.

[199] "Das Verhältnis der Wahrscheinlichkeit, daß keine Emission stattfindet, zu der Wahrscheinlichkeit, daß Emission stattfindet, ist proportional der Intensität der den Oszillator erregenden Schwingung." *Ibid.*, p. 645.

[200] "Den Wert der Proportionalitätskonstanten bestimmen wir nachher durch die Anwendung auf den speziellen Fall, daß die Strahlungsintensität sehr groß ist." *Ibid.*, p. 645.

[201] W. Wien, *Neuere Probleme der theoretischen Physik*, six lectures delivered at Columbia University in April, 1913 (Teubner, Leipzig, Berlin, 1913).

[202] See p. 43.

[203] REF. 201, p. 51.

An important advance in this direction was made when two simultaneous yet independent papers were read in 1911; Ehrenfest, in an address to the Physical Society of Petersburg,[204] and Natanson in a paper read before the Academy of Sciences in Cracow,[205] showed that Einstein's hypothesis of noninteracting light quanta does not lead to Planck's radiation law (1.16) but only to Wien's law, Eq. (1.5), which was known to be a special case of the former.

In his analysis of the precise assumptions underlying Planck's combinatorial procedure in which P energy elements ϵ were distributed among N "receptacles of energy" so that N_j receptacles each contain j energy elements, subject to the restrictions $\sum N_j = N$ and $\sum jN_j = P$, Natanson called the correlation of $j\epsilon$ with N_j, that is, sorting the receptacles according to their energy content and specifying the number of receptacles which have equal energy content, a "mode of distribution." In setting up a "mode of distribution," Natanson emphasized, no account is taken of a possible "identifiability" or distinguishability, as we would say today, of receptacles or of energy elements. However, as soon as the former are regarded as individually identifiable, every given "mode of distribution" ramifies into a number of "modes of collocation" which specifies the number of energy elements in each individual receptacle. Finally, if the energy elements are also considered as identifiable, each "mode of collocation" splits into a number of "modes of association" which associates individual energy elements with individual receptacles. Natanson then pointed out that the thermodynamic probability of a given "mode of distribution" depends notably on whether all "modes of association" or all "modes of collocation" are regarded as equally probable, and he showed that Planck, in contrast to Einstein, adopted the latter alternative. It was in Natanson's analysis of Planck's statistical procedure that the problem of the distinguishability of elementary entities was raised for the first time.

In his article of 1911 and later in his appendix to the paper quoted in REF. 87, Ehrenfest distinguished between "identical and discrete quanta" ("gleichartige, von einander losgelöste Quanten") such as Einstein's quanta of light, and "nondisjointed quanta" ("nicht von einander losgelöste Quanten") such as Planck's. He based this distinction on their different statistical behavior. For, if A_1 is the number of ways of distributing P energy elements of the former kind among receptacles or cells in space and A_2 the corresponding number for N_2 receptacles, Einstein's reasoning, as Eq. (1.18) shows, postulates that $A_1 : A_2 = N_1^P : N_2^P$ since $v = N_1$ units of volume and $v_0 = N_2$ such units. If, however, A_1 and A_2

[204] REF. 77.

[205] L. Natanson, "On the statistical theory of radiation," *Bulletin de l'Académie des Sciences de Cracovie* (*A*) *1911*, pp. 134–148; "Über die statistische Theorie der Strahlung," *Physikalische Zeitschrift 12*, 659–666 (1911).

refer to energy quanta as conceived by Planck, Ehrenfest continued, then

$$A_1 : A_2 = \frac{(N_1 + P - 1)!}{(N_1 - 1)!P!} : \frac{(N_2 + P - 1)!}{(N_2 - 1)!P!}$$

Hence, Ehrenfest concluded, "nondisjointed quanta" correspond to Planck's law of radiation whereas "disjoint quanta" as conceived by Einstein lead to Wien's law. It is clear that Ehrenfest's distinction is essentially that between the number of ways in which P distinguishable objects may be placed into N receptacles and the number of ways in which P indistinguishable objects may be placed into N receptacles. An attempt to modify the Einsteinian conception of quanta so as to reconcile it with Planck's law of radiation was made, also in 1911, by Joffé.[206] Similarly, a few years later, Krutkow showed that the two assumptions that any resonator of frequency ν possesses only energy values $0, h\nu, 2h\nu, \ldots$ and that these energy values result from "appositions" ("Aneinanderlagerungen") of discrete energy elements $h\nu$ lead to Wien's and not to Planck's law of radiation; to obtain Planck's law, the second assumption has to be abandoned. The question of what conditions have to be imposed upon quanta was the subject of a prolonged dispute between Krutkow and Wolfke.[207]

For the time being, none of these considerations succeeded in clarifying a still obscure situation. No wonder that Lorentz, in the opening statement of his address to the Solvay Congress held in Brussels from October 30 to November 3, 1911, still called black-body radiation a "most mysterious phenomenon and a most difficult one to unveil."[208] The Solvay Congress of 1911, attended among others by Marcel Brillouin, Maurice de Broglie, Einstein, Jeans, Kamerlingh Onnes, Nernst, Planck, Rubens, Rutherford, Sommerfeld, Warburg, and Wien, was convened with the declared purpose of attempting to achieve some clarification of the complicated conception of quanta. On one point everybody was agreed: that, as Nernst put it at the opening of the Congress, "the fundamental and fruitful ideas of Planck

[206] A. Joffé, "Zur Theorie der Strahlungserscheinungen," *Annalen der Physik 36*, 534–552 (1911); part of this paper was the substance of an address given earlier, in September, 1910, before the Physical Society of Petersburg.

[207] M. Wolfke, "Welche Strahlungsformel folgt aus der Annahme der Lichtatome?" *Physikalische Zeitschrift 15*, 308–310 (1914). G. Krutkow, "Bemerkung zu Herrn Wolfkes Note: 'Welche Strahlungsformel folgt aus der Annahme der Lichtatome?' " *ibid.*, 363–364. M. Wolfke, "Antwort auf die Bemerkung Herrn Krutkows zu meiner Note: 'Welche Strahlungsformel folgt aus der Annahme der Lichtatome?' " *ibid.*, 463–464.

[208] H. A. Lorentz, "Sur l'application au rayonnement du théorème de l'équipartition de l'énergie," in *La Théorie du Rayonnement et les Quanta—Rapports et Discussions de la Réunion Tenue à Bruxelles*, 1911, edited by P. Langevin and M. de Broglie (Gauthier-Villars, Paris, 1912), p. 12. A German edition, under the title *Die Theorie der Strahlung und der Quanten*, of the Proceedings of this Congress was published in the *Abhandlungen der Deutschen Bunsen-Gesellschaft*, 7 (Halle, 1914), under the editorship of A. Eucken.

and Einstein must serve as the basis for our discussions; they may be modified and elaborated, but they cannot be ignored."[209] In fact, it was only a few weeks after the Congress that Poincaré published his paper "On the theory of quanta"[210] in which he conclusively demonstrated that "the hypothesis of quanta is the only one which leads to Planck's law of radiation" and that the existence of a discontinuity in the probability function of the energy distribution is a necessary consequence of any assumed radiation law which leads to a finite total radiation.

Planck[211] contributed an important paper to the Congress in which he showed how classical statistical mechanics had to be modified in order to yield his radiation law (1.16) instead of the Rayleigh-Jeans law (1.9). According to the classic theory of Gibbs[212] the probability of finding the representative point for the state of a system in the element $dq\,dp$ of the phase space is given by $e^{-E/kT}\,dq\,dp/\iint e^{-E/kT}\,dq\,dp$, where the energy E is a function of the coordinate q and the generalized momentum p. For the linear harmonic oscillator $E = p^2/2m + 2\pi^2\nu^2 mq^2$. Classical mechanics, as a simple integration shows, yields for the average energy $U = \bar{U} = \iint Ee^{-E/kT}\,dq\,dp/\iint e^{-E/kT}\,dq\,dp = kT$ and hence leads to Eq. (1.9). If, however, it is postulated that E can take on only the discrete values $E_n = n\epsilon = nh\nu$ with $n = 0, 1, 2, \ldots$, then[213]

$$U = \frac{\sum E_n e^{-E_n/kT}}{\sum e^{-E_n/kT}} = \frac{\epsilon}{e^{\epsilon/kT} - 1}$$

in agreement with Eq. (1.16). It was in view of this elementary derivation of the radiation formula that Planck made for the first time the explicit statement that the energy, and not only the average energy, of an oscillator of frequency ν is an integral multiple of the energy element $h\nu$. On the basis of these considerations Planck interpreted h as the finite extension of the elementary area in phase space. The motions of linear harmonic oscillators of energy $E = p^2/2m + \frac{1}{2}\beta q^2$ are described in phase space by a family of concentric ellipses with semiaxes $(2E/\beta)^{1/2}$ and $(2mE)^{1/2}$. If it is now assumed, continued Planck, that these ellipses, instead of forming a continuum, are separated from each other so that the intervening areas of

[209] *La Théorie du Rayonnement et les Quanta* (REF. 208), p. 11.

[210] H. Poincaré, "Sur la théorie des quanta," *Comptes Rendus 153*, 1103–1108 (1911); *Journal de Physique 2*, 1–34 (1912). Cf. also M. Planck, "Henri Poincaré und die Quantentheorie," *Acta Mathematica 38*, 387–397 (1915).

[211] M. Planck, "La loi du rayonnement noir et l'hypothèse des quantités élémentaires d'action" (REF. 208), pp. 93–114; "Die Gesetze der Wärmestrahlung und die Hypothese der elementaren Wirkungsquanten," *Abhandlungen der Bunsengesellschaft 3*, 77–94 (1913); *Physikalische Abhandlungen und Vorträge* (REF. 42), vol. 2, pp. 269–286.

[212] Josiah Willard Gibbs, *Elementary Principles in Statistical Mechanics* (Scribner, New York, 1902); *Collected Works* (Longmans, Green & Co., New York, 1928), vol. 2.

[213] Put $\epsilon/kT = x$ and use $\Sigma e^{-nx} = (1 - e^{-x})^{-1}$ and $\Sigma ne^{-nx} = e^{-x}(1 - e^{-x})^{-2}$.

phase space are of magnitude h or, in other words, that the nth ellipse encloses an area $\iint dq\,dp = nh$, then the energies of the oscillators are multiples of $h\nu$.[214]

These considerations, which *prima facie* might be regarded as merely graphical representations, constituted an important turning point in the development of quantum theory. They led Planck to the far-reaching idea that the appearance of energy quanta is the result of a much more profound and general principle. Planck consequently declared that the concept of energy quanta, which until then had been at the forefront of theoretical research, was in fact only of secondary importance since it is the consequence or a more basic condition

$$\iint_{E}^{E+\epsilon} dq\,dp = h$$

"One should therefore confine oneself to the principle that the elementary region of probability h has an ascertainable finite value and avoid all further speculation about the physical significance of this remarkable constant."[215]

Planck's address, in fact, was an appeal to abandon all attempts at finding a classical interpretation of h and to search, instead, for a clarification of the basic principles which underlie the manifestation of energy quanta. It was generally admitted that the revisions which had to be introduced into classical statistics were only an indication of the need for a far-reaching modification of classical mechanics whose details were far from being understood. Planck expressed this feeling in the opening of his address to the Congress when he said: "The framework of classical dynamics, even if combined with the Lorentz-Einstein principle of relativity, is obviously too narrow to account also for all those physical phenomena which are not directly accessible to our coarse senses."[216] This statement by Planck is one of the earliest, if not the earliest, renunciations of the universal validity of classical mechanics. Eucken, the editor of the German edition of the *Proceedings* of the Congress, formulated the problem as follows:[217] "Is quantum theory a branch of conventional mechanics; or, conversely, is traditional mechanics a special case of a more general mechanics (which comprises, or is identical with, quantum theory); or, finally, is quantum theory a science altogether outside mechanics?"

The first step in this groping toward a new mechanics was made by Arnold Sommerfeld when he presented a paper on "The application of

[214] We have rendered the argument in a slightly modified version to fit its later formulation and considered only the case of one degree of freedom needed for the linear oscillator. The generalization to f degrees of freedom is obvious.

[215] REF. 211, p. 100; p. 274.

[216] *Ibid.*, p. 93.

[217] REF. 208 (Halle, 1914), p. 372.

Planck's theory of the quantum of action to molecular physics"[218] to the Solvay Congress. Referring to the term "quantum of action," which in his view was "a most fortunate" name, Sommerfeld suggested that Planck's h should be related, not merely nominally, to Hamilton's "action function" $\int L\, dt$, where L denotes the kinetic potential, viz., the difference between the kinetic energy T and the potential energy V of the dynamic system under consideration. Sommerfeld proposed the fundamental hypothesis that in every elementary process the atom gains or loses a definite amount W of action, where $W = \int_0^\tau L\, dt = h/2\pi$ and τ is the duration of the process. Like every basic principle the statement could not, of course, be proved in a rigorous manner. Sommerfeld could only vindicate it by a number of arguments such as its relativistic invariance or the fact that it easily explained the production of x-rays or γ rays, arguments which in his opinion were sufficiently convincing. Sommerfeld also showed that his hypothesis readily accounted for the photoelectric effect. Let us briefly describe Sommerfeld's derivation of Einstein's photoelectric equation. The electron, subject to a semielastic force $-fx$ with $x = 0$ as its equilibrium position at time $t = 0$, is supposed to be acted upon by the external electric force $F = E \cos nt$, where $\nu = n/2\pi$ denotes the frequency of the monochromatic incident radiation. The equation of motion is $m\ddot{x} + fx = eE \cos nt$, the kinetic energy $T = \frac{1}{2}m\dot{x}^2$, and the potential energy $V = \frac{1}{2}fx^2$. According to the principle proposed, the electron would be liberated as soon as W, which after partial integration and in view of the equation of motion can be written as $\frac{1}{2}mx\dot{x} - \frac{1}{2}e\int_0^\tau xF\, dt$, reaches the value $h/2\pi$. The first term $\frac{1}{2}mx\dot{x}$, which refers to the instant $t = \tau$ (it vanishes for $t = 0$), is the kinetic energy of the electron at that instant divided by the proper circular frequency $n_0 = 2\pi\nu_0 = (f/m)^{1/2}$. This can be seen as follows. W, like x and $\dot{x}$, is a fast-oscillating function with slowly varying amplitude and therefore attains the value $h/2\pi$ for the first time in the vicinity of a maximum (otherwise it would already have reached this value in the preceding oscillation). It follows for $t = \tau$ that $dW/dt = 0$ or $T = V$, which implies $\dot{x} = n_0 x$ and $\frac{1}{2}mx\dot{x} = T/n_0$. Sommerfeld then showed that for resonance, when $n = n_0$, the second term in the expression for W, viz., the time virial of the external force, can be neglected in comparison with T. Consequently, $W = h/2\pi = T/n_0 = T/n$ or $T = h\nu$.[219] But this, apart from the work function, is Einstein's photoelectric equation. The experimentally established fact that the energy of the ejected photoelectron

[218] A. Sommerfeld, "Rapport sur l'application de la théorie de l'élément d'action aux phénomènes moléculaires non périodiques," REF. 208 (Paris, 1912), pp. 313–372; "Das Plancksche Wirkungsquantum und seine allgemeine Bedeutung für die Molekularphysik," *Physikalische Zeitschrift 12*, 1057–1066 (1911).

[219] Cf. also A. Sommerfeld and P. Debye, "Theorie des lichtelektrischen Effektes vom Standpunkt des Wirkungsquantums," *Annalen der Physik 41*, 873–930 (1913).

depends on the frequency ν of the incident radiation, but not on its intensity, was thus for Sommerfeld a direct consequence of his newly proposed fundamental hypothesis or, conversely, could be regarded as evidence for its validity.

Sommerfeld's principle, although soon discarded in the subsequent development of the theory, has been described at some length because of its historical importance in having initiated an approach different from all methods proposed so far. For so far, quantum phenomena had been treated by what may be described most appropriately as quantum statistics of the harmonic oscillator. It originated from Planck's study of black-body radiation and accounted for this phenomenon. But almost no other field in physics was affected by it. The only exception, prior to the first Solvay Congress, will now be discussed in some detail.

1.5 Applications of Quantum Conceptions to the Molecular Kinetic Theory

Reviewing the status of quantum theory as discussed so far, we may say that Planck's theory, which, as we have seen, was from the mathematical point of view ultimately the outcome of an interpolation, was from the methodological point of view, until the first Solvay Congress, almost a theory *ad hoc.* In fact, until 1907 it was purely *ad hoc,* constructed to account for black-body radiation and without any relevance to other phenomena in physics. If it is agreed that a physical theory becomes significant only if it also accounts for phenomena other than those for whose explanation it was conceived, then it was Einstein's paper of 1907 on "Planck's theory of radiation and the theory of specific heats"[220] which endowed Planck's work with physical significance. For Einstein showed that quantum conceptions can be successfully applied also outside the area of radiation problems when he explained certain inconsistencies in the classical molecular kinetic theory on the basis of Planck's theory.

Referring to Planck's result that the mechanism of energy transfer admits only of energy values that are integral multiples of $h\nu$, Einstein declared: "We should not be satisfied, I believe, with this conclusion. For the following question forces itself upon our mind: If it is true that the elementary oscillators that are used in the theory of energy transfer between radiation and matter cannot be interpreted in terms of the present molecular kinetic theory, must we then not also modify our theory for the other oscillators which are used in the molecular theory of heat? In my opinion there can be no doubt about the answer. If Planck's theory of

[220] A. Einstein, "Die Plancksche Theorie der Strahlung und die Theorie der spezifischen Wärme," *Annalen der Physik 22,* 180–190 (1907).

radiation really strikes the core of the matter, then it should be expected that in other areas of the theory of heat contradictions also exist between the present molecular kinetic theory and experience which can be resolved by the method just proposed."[221]

Such a contradiction was known to exist in the theory of specific heats. During the fifty years following the statement by Dulong and Petit that "the atoms of all simple bodies have exactly the same heat capacity,"[222] nobody, it seems, doubted the correctness of this "law," and Regnault's[223] extensive measurements of specific heats only seemed to confirm it. It should be noted, however, that almost all these measurements were carried out at room temperature or above since analogous experiments at lower temperatures at that time would have posed insurmountable technical difficulties. The first conflict occurred in 1872 when Weber[224] measured the specific heat of diamond below room temperature and found that at $-50°C$ its molar heat was only 0.76 cal/mole deg. Prior to 1896 only two other experiments seem to have been significant in this respect, those performed by Pebal and Jahn[225] and those carried out by Schüz,[226] the former working with antimony and the latter with alloys and amalgams of low melting points. But in spite of the scarcity of empirical data of this kind, Boltzmann[227] and, particularly, Richarz[228] had already formulated a theory of specific heats which was to account also for the exceptions of the Dulong-Petit law. In 1895 in his ice-machine factory in Munich Linde invented[229] the process of liquefying air by utilizing the Joule-Thomson effect and thus made further progress possible. In fact, this invention advanced the technique of measuring specific heats at lower temperatures to such an extent that only three years later, in 1898, at the Institute of Physics of the Berlin University, Behn[230] was able to sum up his findings with these

[221] *Ibid.*, p. 184.

[222] "Les atomes de tous les corps simples ont exactement la même capacité pour la chaleur." P. L. Dulong and A. T. Petit, "Sur quelques points importants de la théorie de la chaleur," *Annales de Chimie et de Physique 10*, 395–413 (1819).

[223] V. Regnault, "Recherches sur la chaleur spécifique des corps simples et des corps composés," *Annales de Chimie et de Physique 73*, 5–72 (1840), *1*, 129–207 (1841).

[224] F. H. Weber, "Die specifische Wärme des Kohlenstoffs," *Poggendorff's Annalen der Physik 147*, 311–319 (1872). See also his paper "Die specifische Wärme der Elemente Kohlenstoff, Bor und Silicium," *ibid. 154*, 367–423, 553–582 (1875).

[225] L. Pebal and H. Jahn, "Ueber die specifische Wärme des Antimons und einiger Antimonverbindungen," *Poggendorff's Annalen der Physik 27*, 584–605 (1886).

[226] L. Schüz, "Ueber die specifische Wärme von leicht schmelzbaren Legierungen und Amalgamen," *Wiedemannsche Annalen der Physik 46*, 177–203 (1892).

[227] L. Boltzmann, "Analytischer Beweis des 2. Hauptsatzes der mechanischen Wärmetheorie aus den Sätzen über das Gleichgewicht der lebendigen Kraft," *Wiener Berichte 63* (part 2), 712–732 (1871), p. 731 et seq.

[228] F. Richarz, "Ueber das Gesetz von Dulong und Petit," *Wiedemannsche Annalen der Physik 48*, 708–716 (1893).

[229] C. Linde, "Erzielung niedrigster Temperaturen," *Wiedemannsche Annalen der Physik 57*, 328–332 (1896).

[230] U. Behn, "Ueber die specifische Wärme einiger Metalle bei tiefen Temperaturen," *Wiedemannsche Annalen der Physik 66*, 237–244 (1898).

words: "The graphical representation of the decrease of specific heats with temperature seems to suggest that all curves converge at absolute zero so that the specific heats at that temperature have the same extremely small value (0?). However, this result can here be expressed only as an assumption."[231] Further investigations by Tilden[232] and Dewar[233] seemed to conform with this result.

This, then, was the experimental situation prior to Einstein's paper of 1907. From the theoretical point of view, the Dulong-Petit rule according to which the product of the atomic or molecular weight and the specific heat is a constant for all solids (approximately 6 cal/mole) was, of course, perfectly accounted for by classical physics. For, according to the law of equipartition and on the assumption that for small vibrations the potential energy of the particles is a quadratic function of their displacements, each of the N particles with three degrees of freedom (N is Avogadro's number) possesses on the average the total energy $3kT$ at temperature T. The energy per mole is consequently $3RT$ and the specific heat per mole approximately 6 cal.

It was Einstein who realized at the end of 1906 that the systematic diminution of the specific heats of solids with decreasing temperature can be accounted for on the basis of Planck's quantum conceptions. Neglecting atomic interactions and assuming that all the atoms in the solid are oscillating with the same frequency, he applied Eq. (1.16) for the mean energy of the atom and obtained for the energy per mole the expression

$$E = \frac{3Nh\nu}{\exp(\beta\nu/T) - 1}$$

from which he determined the specific heat per mole

$$c_v = \frac{dE}{dT} = \frac{3R(\beta\nu/T)^2 \exp(\beta\nu/T)}{[\exp(\beta\nu/T) - 1]^2} \tag{1.20}$$

where $\beta = h/k$, a formula which shows that c_v decreases with T but is equal to $3R$ at sufficiently high temperatures.

Einstein's formula (1.20) was found to fit experimental data remarkably well. It was only at extremely low temperatures that it showed small deviations from the observed results. The reason for this discrepancy, as Einstein[234] later recognized himself, was the assumption that the vibrations of the atoms in the solid all have the same frequency ν. Debye's

[231] *Ibid.*, p. 243.

[232] W. A. Tilden, "The specific heats of metals and the relation of specific heat to atomic weight," *Philosophical Transactions of the Royal Society of London 201*, 37–43 (1903).

[233] J. Dewar, "Studies with the liquid hydrogen and air calorimeters," *Proceedings of the Royal Society of London* (*A*), *76*, 325–340 (1905).

[234] A. Einstein, "Elementare Betrachtungen über die thermische Molekularbewegung in festen Körpern," *Annalen der Physik 35*, 679–694 (1911).

"continuum theory"[235] of the specific heat, which treated the lattice of the solid as a continuum with a continuous spectrum of vibrational frequencies cut off at a characteristic maximum and which led to the famous T^3 law, and Born's modified cutoff procedure[236] in accordance with his theory of lattice vibrations, developed in collaboration with von Kármán, were essentially only refinements of Einstein's approach.

Einstein's work on specific heats was important not only, as we have pointed out, from the methodological point of view in so far as it extended the applicability of quantum conceptions to the molecular kinetic theory, but also for the following reason.[237] Walther Nernst, as we know from various sources,[238] initially disliked quantum theory, claiming that it was "really nothing else than an interpolation formula."[239] Later, however, when his attention was drawn to the exact determination of specific heats at low temperatures in connection with his discovery of the famous theorem which carries his name—and which, as we know today from quantum-statistical mechanics, is indeed but a macroscopic manifestation of quantum effects—and when his experiments and those of his collaborators, such as Eucken, seemed to agree with Einstein's formula (1.20), he changed his mind and called Planck's work "a most ingenious and fruitful theory."[240] In a paper presented to the Berlin Academy on January 26, 1911, Nernst declared: "At present, quantum theory is essentially only a rule for calculation, of apparently a very strange, one might even say grotesque, nature; but it has proven so fruitful by the work of Planck, as far as radiation is concerned, and by the work of Einstein, as far as molecular mechanics is concerned, . . . that it is the duty of science to take it seriously and to subject it to careful investigations."[241]

Thus, Nernst thought it advisable to organize an international meeting of prominent physicists with the exclusive purpose of discussing the basic problems and issues of quantum theory. When the Belgian industrialist Ernest Solvay offered his generous support, it was primarily due to Nernst's initiative that the meeting convened and became an important event in the history of physics. Ultimately, as we saw, Einstein's work on specific heat was, though only indirectly and in a roundabout way, one of the

[235] P. Debye, "Zur Theorie der spezifischen Wärme," *Annalen der Physik 39*, 789–839 (1912); *The Collected Papers of Peter J. W. Debye* (Interscience, New York, London, 1954), pp. 650–696.

[236] M. Born and T. von Kármán, "Über Schwingungen in Raumgittern," *Physikalische Zeitschrift 13*, 297–309 (1912); "Zur Theorie der spezifischen Wärme," *ibid. 14*, 15–19 (1913).

[237] Cf. M. J. Klein, "Einstein, Specific Heats, and the Early Quantum Theory" (paper read to the *Annual A A A S Meeting of the History of Science Society*, in Montreal, Dec. 29, 1964; M. Jammer, "Comments on Klein's paper 'Einstein, Specific Heats, and the Early Quantum Theory' "), *Science 148*, 173–180 (1965).

[238] For example, *Archive for the History of Quantum Physics*, Interview with P. Debye on Mar. 5, 1962.

[239] *Ibid.*

[240] W. Nernst, "Der Energieinhalt fester Stoffe," *Annalen der Physik 36*, 395–439 (1911).

[241] W. Nernst, "Über neuere Probleme der Wärmetheorie," *Berliner Berichte 1911*, p. 86.

factors responsible for the convocation of the first Solvay Congress.[242] In a similar way it also gave rise to the earliest attempt at a new approach to quantum theory, an approach which later proved to be important from the didactic point of view.

The trend of Nernst's work in physical chemistry did not direct his interest to the radiative aspects of quantum physics, aspects which, most probably, he never studied thoroughly. A careful analysis of Nernst's reactions and comments during the discussions of the various reports at the Congress would only corroborate this contention. On the other hand, Nernst was deeply interested in the question whether quantum physics could not be developed on a basis independent of the study of black-body radiation and in such a way that in its logical structure considerations pertaining to the molecular kinetic theory gain logical priority over those pertaining to radiation phenomena. Indeed, one of Nernst's papers, entitled "Specific heat and quantum theory,"[243] written in collaboration with Lindemann, begins with these words: "In a recently published investigation one of us [it was Nernst] has given a representation of quantum theory which, following Einstein, considers radiative phenomena as only secondary circumstances ('begleitende Umstände') and takes as its immediate point of departure the atomic vibrations."[244] It was the earliest attempt to approach quantum physics from the molecular kinetic theory. Compared with the traditional introduction to quantum physics via the theory of black-body radiation, this alternative approach is regarded as preferable from the didactic viewpoint by all those teachers of modern physics who consider it unfortunate for the logical development of our subject that its history began with so complicated a concept as Planck's quantization of the electromagnetic vibrations in an isothermal cavity.[245] Nernst's point of view did not gain prominence at the Congress, as the title of the official proceedings, "La Théorie du Rayonnement et les Quanta" or that of the German version, "Die Theorie der Strahlung und der Quanten," indicates.

The significance of this meeting for the conceptual development of our theory should have become evident from our preceding discussion of its deliberations. But apart from its general importance for its participants and its appeal to their concerted effort to find a new mechanics of a much

[242] For details on the Congress see M. de Broglie, *Les Premiers Congrès de Physique Solvay* (Editions Albin Michel, Paris, 1951), and REF. 237.

[243] W. Nernst and F. A. Lindemann, "Spezifische Wärme und Quantentheorie," *Zeitschrift für Elektrochemie 17*, 817–827 (1911).

[244] "In einer kürzlich erschienen Untersuchung hat der eine von uns eine Darstellung der Quantentheorie gegeben, die nach dem Vorgang Einsteins die Strahlungserscheinungen nur als begleitende Umstände auffasst und unmittelbar an die Atomschwingungen anknüpft." *Ibid.*, p. 817. The paper referred to is Nernst's "Zur Theorie der spezifischen Wärme und über die Anwendung der Lehre von den Energiequanten auf physikalisch-chemische Fragen überhaupt," *Zeitschrift für Elektrochemie 17*, 265–275 (1911), received Feb. 21, 1911. (See REF. 104.)

[245] See REF. 1.

wider scope than that of the harmonic oscillator, it seems also to have had a decisive effect in an indirect way on others who did not actively participate. Bohr[246] recalled how impressed he had been by Rutherford's "vivid account of the discussions" when he met Rutherford in Manchester in 1911, shortly after Rutherford had returned from Brussels. In his autobiographical notes,[247] Louis de Broglie described with what ardor and enthusiasm he studied the text of the discussions which his older brother Maurice, one of the scientific secretaries of the meeting, was editing for publication. As a result of this, he declared, "I decided to devote all my efforts to investigate the real nature of the mysterious quanta that Planck had introduced into theoretical physics ten years earlier."[248]

It is convenient to define as the first phase in the development of quantum theory the period in which all quantum conceptions and principles proposed referred exclusively to black-body radiation or harmonic vibrations. It is the period of the early quantum statistics of the harmonic oscillator. If we accept this definition, we may regard the Solvay Congress of 1911 as the closing act of this phase or as the prelude of a new period.

In concluding this chapter, we should note that the first stage in the conceptual development of quantum theory, although devoid of any far-reaching or powerful theories, serves as an outstanding example of human ingenuity. For it showed how far man's intellect can penetrate into the secrets of nature on the basis of comparatively inconspicuous evidence. Most of the concepts formed at this stage will be vindicated, as we shall see, by the later development of the theory, though on different grounds.

[246] See REF. 176.

[247] L. de Broglie, "Vue d'ensemble sur mes travaux scientifiques," in *Louis de Brolie—Physicien et Penseur* (Editions Albin Michel, Paris, 1953), 457–486.

[248] "Je m'étais promis de consacrer tous mes efforts à comprendre la véritable nature des mystérieux quanta que Max Planck avait introduits dans la Physique Théorique dix ans auparavant, mais dont on n'apercevait pas encore la signification profonde." *Ibid.*, p. 458.

2 *Early Applications of Quantum Conceptions to Line Spectra*

2.1 *Regularities in Line Spectra*

In the preceding chapter we saw how the study of the single physical phenomenon of black-body radiation led to the conceptions of quanta and to the quantum statistics of the harmonic oscillator, and thus to results which defied the principles of classical mechanics and, in particular, the equipartition theorem. It was generally agreed that classical physics was incapable of accounting for atomic or molecular processes. But how to modify classical physics by new conceptions was as yet an open question.

It was clear that further progress was contingent upon a broadening of the scope of pertinent physical phenomena. Indeed, it was the study of the discrete spectra of the chemical elements which constituted the next stage of development after the study of the continuous spectra[1] of thermal radiation and contributed decidedly to the development of quantum theory.

Ever since Newton laid the foundations of spectroscopy in 1666, the question concerning the origin of spectral lines had been regarded as an important problem in physics. Late-nineteenth-century physicists interpreted the spectrum of a given element in a way analogous to the acoustic vibrations of a sound-producing body. They assumed that the whole spectrum of a given element originated in the natural periods of oscillations of one and the same atom. Thus, to mention only one example, Herschel inferred from Balmer's famous formula that atomic processes resemble acoustic phenomena in organ pipes.[2]

[1] As shown in the beginning of the preceding chapter, Kirchhoff's law and, in its wake, the study of black-body radiation had their historical origin in the investigation of the Fraunhofer lines, that is, ultimately in a question pertaining also to discrete spectra.

[2] A. S. Herschel, "On a relation between the spectrum of hydrogen and acoustics," *Astrophysical Journal 7*, 150–155 (1898).

Mitscherlich[3] was the first to point out that spectroscopy should be regarded not only as a method of chemical analysis—spectroanalysis had just recorded its first spectacular achievements—but also as a clue to the secrets of the inner structure of the atom and the molecule. His own research, however, was confined to a merely qualitative comparison of different spectra of elements and compounds. Five years later, Mascart[4] was the first to draw attention to the existence of definite arithmetical relations between the wavelengths of certain lines in the same spectrum. He interpreted the double lines in the spectrum of sodium and the triplets in that of magnesium as harmonic vibrations. Inspired by Mascart's remarks, Lecoq de Boisbaudran[5] in 1869 thought he had found that the spectral lines of nitrogen, whose wavelengths he measured in units of 10^{-6} mm, exhibit similar harmonic relations. He tried to show that this spectrum is made up of two sets of bands, one extending from the red to the green and the other from the green to the violet, and that each band of the second set had a wavelength which was in the ratio of three to four with that of a corresponding band of the first set. Although soon disproved by R. Thalèn, a pupil of Ångström, Lecoq's paper inaugurated an intensive search for numerical regularities in spectra.[6] The first study of the spectrum of hydrogen, from this point of view, was made by G. Johnstone Stoney. In accordance with his basic assumption "that the lines in the spectra of gases are to be referred to periodic motions within the individual molecule, and not to the irregular journeys of the molecules amongst one another,"[7] Stoney declared that "one periodic motion in the molecule of the incandescent gas may be the source of a whole series of lines in the spectrum of the gas,"[8] an idea which, if mathematically formulated, would have led him to a Fourier analysis of the motion. To substantiate his claim, Stoney

[3] A. Mitscherlich, "Ueber die Spectren der Verbindungen und der einfachen Körper," *Poggendorff's Annalen der Physik 121*, 459–488 (1864).

[4] M. Mascart, "Sur les spectres ultraviolets," *Comptes Rendus 69*, 337–338 (1869). On p. 338 he wrote: "Un problème important, qui doit se proposer l'analyse spectrale, est de savoir s'il existe une relation entre les différentes raies d'une même substance ou bien entre les spectres de substances analogues. J'ai observé, en 1863, que les six raies principales du sodium, aperçues pour la première fois par MM. Wolf et Diacon, sont doubles, et que les deux raies qui constituent chacun des groupes sont à peu près à la même distance que celle de la double raie. Cela parait être la répétition d'un même phénomène en différents points de l'échelle spectrale." With reference to the magnesium triplets he declared: "Il me semble difficile que la reproduction d'un pareil phénomène soit un effet du hasard; n'est-il pas plus naturel d'admettre que ces groupes de raies semblables sont des harmoniques qui tiennent à la constitution moléculaire du gaz lumineux?" *Ibid.*

[5] Lecoq de Boisbaudran, "Sur la constitution des spectres lumineux," *Comptes Rendus 69*, 445–451, 606–615, 657–664, 694–700 (1869).

[6] Cf., for example, J. L. Soret, "On harmonic ratios in spectra," *Philosophical Magazine 42*, 464–465 (1871).

[7] G. J. Stoney, "The internal motions of gases compared with the motions of waves of light," *Philosophical Magazine 36*, 132–141 (1868).

[8] G. J. Stoney, "On the cause of the interrupted spectra of gases," *Philosophical Magazine 41*, 291–296 (1871).

referred to Angström's[9] famous measurements of the wavelengths of the four then known hydrogen spectral lines: 6562.10×10^{-7} mm, 4860.74×10^{-7} mm, 4340.10×10^{-7} mm, and 4101.2×10^{-7} mm. Stoney contended that three of these lines, namely, those with wavelengths, if reduced to vacuum, $\lambda_1 = 6563.93 \times 10^{-7}$ mm, $\lambda_2 = 4862.11 \times 10^{-7}$ mm, and $\lambda_3 = 4102.37 \times 10^{-7}$ mm, "are to be referred to the same motion in the molecule of the gas"[10] and are "the 20th, 27th and 32nd harmonics of a fundamental vibration whose wavelength in vacuo" is $\lambda_0 = 0.13127714$ mm.[11]

Although forced to admit that "the other harmonics (19th, 21st, 22nd) are not found in this spectrum of hydrogen," Stoney was confident that further examinations would eventually establish additional harmonics.[12] In carrying out this research, subsequently with the assistance of Dr. Reynolds of Dublin,[13] Stoney found it "convenient to refer the positions of all lines in the spectrum to a scale of reciprocals of the wave-lengths"[14] rather than to that of wavelengths used so far. His reasons for this innovation were as follows: "This scale has the great convenience, for the purposes of the investigation, that a system of lines with periodic times that are harmonics of one periodic time are equidistant upon it; and it has the further convenience, which recommends it for general use, that it resembles the spectrum as seen in the spectroscope much more closely than the scale of direct wave-lengths used by Ångström in his classic map." And Stoney continued: "If, then, k be the wave-number of a fundamental motion in the aether, its wave-length will be $1/k$ of a millimeter, and its harmonics will have the wave-lengths $1/2k$, $1/3k$, etc., in other words, they occupy the positions $2k$, $3k$, etc., upon the new map."

Stoney—the same Stoney who twenty-three years later introduced the term "electron" into physics—thus was the first to define the subsequently so important notion "wave-number." Even his original notation (k) is common usage in modern theoretical physics, where it generally denotes—for the sake of further computational convenience—the number of waves in 2π units of length (2π cm) rather than per millimeter as proposed by Stoney.

[9] A. Ångström, *Recherches sur le Spectre Solaire* (Upsala, 1868), p. 31. Ångström's table of wavelengths, published in this book, served for a long time as a standard for spectroscopists.

[10] REF. 8 (1871), p. 294.

[11] These are, of course, the lines H_α, H_β, and H_δ of the Balmer series which satisfy the relations $\lambda_1 = \lambda_0/20$, $\lambda_2 = \lambda_0/27$, and $\lambda_3 = \lambda_0/32$.

[12] The other lines of the Balmer series, down to 3699 Å, were discovered a few years later in the ultraviolet spectra of Sirius and other white stars by the English astronomer W. Huggins and published in his paper "Sur les spectres photographiques des étoiles," *Comptes Rendus* *90*, 70–73 (1880).

[13] G. J. Stoney and J. E. Reynolds, "An inquiry into the cause of the interrupted spectra of gases," *Philosophical Magazine* *42*, 41–52 (1871).

[14] G. J. Stoney, "On the advantage of referring the position of lines in the spectrum to a scale of wave-numbers," *British Association Reports, Edinburgh, 41*, pp. 42–43 (1871).

It was in the course of similar investigations that Liveing and Dewar[15] discovered in the spectra of sodium and potassium the existence of two series of lines, one of sharp and one of diffuse lines, and established[16] the existence of homologous series, that is, series of lines of the same type in chemically analogous elements. At the same time, Arthur Schuster stressed the importance of the search for "an empirical law connecting the different periods of vibration in which we know one and the same molecule to be capable of swinging."[17] He even applied the theory of probability to calculate the accidental occurrence of harmonic ratios on the assumption that no such law exists and that all lines are distributed at random. Having thus tested the statistical significance of the available data, Schuster[18] concluded that "most probably some law hitherto undiscovered exists which in special cases resolves itself into the law of harmonic ratios."

Three years later, such an empirical formula, the first to comprise correctly all lines of a spectral series, was published by Johann Jakob Balmer, a schoolteacher in Basel.[19] His first paper, published in the *Verhandlungen der Naturforschenden Gesellschaft in Basel* (*Proceedings of the Scientific Society of Basel*) for the year 1885, contained the prophetic words: "It appears to me that hydrogen . . . more than any other substance is destined to open new paths to the knowledge of the structure of matter and its properties. In this respect the numerical relations among the wavelengths of the first four hydrogen spectral lines should attract our attention particularly."[20] Balmer noticed that the wavelengths of the four hydrogen lines as measured by Ångström could be represented in terms of a "basic number" $h = 3645.6 \times 10^{-7}$ as $\frac{9}{5}h$, $\frac{4}{3}h$, $\frac{25}{21}h$, and $\frac{9}{8}h$ or, equivalently, as $\frac{9}{5}h$, $\frac{16}{12}h$, $\frac{25}{21}h$, and $\frac{36}{32}h$. He thus saw that the numerators form the sequence 3^2, 4^2, 5^2, 6^2 whereas the corresponding denominators are the differences of

[15] G. D. Liveing and J. Dewar, "On the spectra of sodium and potassium," *Proceedings of the Royal Society of London* (*A*), *29*, 398–402 (1879); reprinted in G. D. Liveing and J. Dewar, *Collected Papers on Spectroscopy* (Cambridge University Press, 1915), pp. 66–70.

[16] G. D. Liveing and J. Dewar, "The spectrum of magnesium," *Proceedings of the Royal Society of London* (*A*), *32*, 189–203 (1881); *Collected Papers on Spectroscopy* (REF. 15), pp. 118–132.

[17] A. Schuster, "On harmonic ratios in the spectra of gases," *Proceedings of the Royal Society of London* (*A*) *31*, 337–347 (1881).

[18] A. Schuster, "The genesis of spectra," *British Association Reports, Southhampton*, pp. 120–143 (1882).

[19] G. P. Thomson reports that when visiting Switzerland shortly after World War I, he was told by a young relative of J. J. Balmer that the latter (his great-uncle?) was a devotee of numerology, interested in such things as the number of the beast or the number of the steps of the pyramid. One day, chatting with a friend who happened to be a physicist or chemist, Balmer complained that he had "run out of things to do." Whereupon the friend replied: "Well, you are interested in numbers, why don't you see what you can make of this set of numbers that come from the spectrum of hydrogen?" and he gave him the wavelengths of the first few lines of the hydrogen spectrum. *Archive for the History of Quantum Physics*, Interview with G. P. Thomson on June 20, 1963.

[20] J. J. Balmer, "Notiz über die Spectrallinien des Wasserstoffs," *Verhandlungen der Naturforschenden Gesellschaft in Basel* *7*, 548–560 (1885).

squares $3^2 - 2^2$, $4^2 - 2^2$, $5^2 - 2^2$, $6^2 - 2^2$. He therefore expressed the wavelengths of these lines by the formula

$$\lambda = h \frac{m^2}{m^2 - 2^2} \quad \text{mm} \tag{2.1}$$

where $h = 3645.6 \times 10^{-7}$ and $m = 3, 4, 5, 6$. Balmer also predicted the existence of a fifth line with wavelength 3969.65×10^{-7} mm. Informed by von Hagenbach that this line and other additional lines had indeed been discovered by Huggins, Balmer showed in a second paper[21] that his formula applied to all twelve then known hydrogen lines. He also predicted correctly that in the series which subsequently carried his name, no lines of wavelength longer than 6562×10^{-7} mm would ever be discovered and that the series would converge at 3645.6×10^{-7} mm. The agreement between the calculated and the observed values of the wavelengths was extremely close in the visible region, but for shorter wavelengths slight systematic discrepancies were evident. In view of these discrepancies Balmer expressed some doubts whether the fault lay with the formula or with the data.

Balmer's discovery had immediate repercussions and gave new impetus to a most intensive search for other regularities in spectra. In an address delivered before the British Association for the Advancement of Science held in Bath in 1888, Runge[22] announced that in certain spectra he had discovered a number of harmonic series of lines whose wavelengths conform to formulas of the type

$$\frac{1}{\lambda} = a + \frac{b}{m} + \frac{c}{m^2} \quad \text{or} \quad \frac{1}{\lambda} = a' + \frac{b'}{m^2} + \frac{c'}{m^4}$$

where a, b, c, a', b', c' are constants and m is an integer. Balmer's formula (2.1), he pointed out, was a special case of the first type mentioned, namely, when $a = h^{-1}$, $b = 0$, and $c = -4h^{-1}$. In a detailed study on the structure of emission spectra, published in the *Transactions of the Royal Swedish Academy* in 1890, Rydberg[23] claimed he had been using Balmer's formula long before its publication. Stating as a fundamental principle that "in the spectra of all the elements analyzed so far there are series of rays whose wavelengths or wave numbers are functions of consecutive

[21] J. J. Balmer, "Zweite Notiz über die Spectrallinien des Wasserstoffs," *ibid.*, 750–752 (1885); *Wiedemannsche Annalen der Physik 25*, 80–87 (1885).

[22] C. Runge, "On the harmonic series of lines in the spectra of the elements," *British Association Reports, Bath*, pp. 576–577 (1888).

[23] J. R. Rydberg, "Recherches sur la constitution des spectres d'émission des éléments chimiques," *Kungliga Vetenskaps Akademiens Handlingar 23*, no. 11, 155 pp. (1890). Cf. also J. R. Rydberg, "On the structure of the line-spectra of the chemical elements," *Philosophical Magazine 29*, 331–337 (1890); "Sur la constitution des spectres linéaires des éléments chimiques," *Comptes Rendus 110*, 394–397 (1890); "Ueber den Bau der Linienspectren der chemischen Grundstoffe," *Zeitschrift für Physikalische Chemie 5*, 227–232 (1890).

integral numbers,"[24] Rydberg expressed these functions by a general formula $n = n_0 - N_0/(m + \mu)^2$, where n was the wave number ["nombre des longueurs d'onde sur l'unité de longueur (1 cm)"], n_0 and μ constants for each series, and N_0, later called "Rydberg's constant" and denoted by R, a constant common to all series and to all elements. Having shown that Balmer's formula was a special case with $n_0 = h^{-1}$, $N_0 = 4h^{-1}$, and $\mu = 0$, Rydberg calculated, on the basis of Ångström's measurements of the first four hydrogen lines, the value of N_0 as 109721.6 cm^{-1}.

Referring to the discoveries of Liveing and Dewar, who identified the principal, sharp ("second subordinate series") and diffuse ("first subordinate series") series in the spectra of the alkali metals, and to the work of Hartley,[25] who observed recurrent frequency differences for the components of certain doublet and triplet series ("Hartley's Law"), Rydberg showed that the wave numbers of the lines in these series can be expressed as functions of an integer m as follows: the component with the longer wavelength in each doublet of the principal series could be expressed by $n = n_0 - N_0/(m + p_1)^2$ and the other component by $n = n_0 - N_0/(m + p_2)^2$; correspondingly, the two components of each line in the first subordinate series could be expressed by $n = n_0' - N_0/(m + d)^2$ and $n = n_0'' - N_0/(m + d)^2$, and those of the second subordinate series by $n = n_0' - N_0/(m + s)^2$ and $n = n_0'' - N_0/(m + s)^2$. In 1896 Rydberg[26] and (independently) Schuster[27] discovered that the difference between the limit of the principal series and the common limit of the two other series coincided with the wave number of the first line of the principal series, that is, that the limits n_0' and n_0'' of the subordinate series coincided with $N_0/(2 + p_2)^2$ and $N_0/(2 + p_1)^2$, and, similarly, that the limit of the principal series coincided with $N_0/(1 + s)^2$. The series were therefore represented by the formulas

$$n = \frac{N_0}{(1 + s)^2} - \frac{N_0}{(m + p)^2}$$

with $p = p_1$ or p_2 and $m = 2, 3, \ldots$ for the two components of the principal series,

$$n = \frac{N_0}{(2 + p_2)^2} - \frac{N_0}{(m + d)^2} \quad \text{and} \quad n = \frac{N_0}{(2 + p_1)^2} - \frac{N_0}{(m + d)^2}$$

[24] "Dans les spectres de tous les éléments analysés il y a des séries de raies dont les longueurs d'onde sont des fonctions déterminées des nombres entiers consécutifs." *Ibid.* ("Recherches"), p. 33.

[25] W. N. Hartley, "On homologous spectra," *Journal of the Chemical Society 43*, 390–400 (1883).

[26] J. R. Rydberg, "Die neuen Grundstoffe des Cleveitgases," *Wiedemannsche Annalen der Physik 58*, 674–679 (1896).

[27] A. Schuster, "On a new law connecting the periods of molecular vibrations," *Nature 55*, 200–201 (1897); the article was written in 1896.

with $m = 3, 4, \ldots$ for those of the diffuse series, and finally

$$n = \frac{N_0}{(2 + p_2)^2} - \frac{N_0}{(m + s)^2} \quad \text{and} \quad n = \frac{N_0}{(2 + p_1)^2} - \frac{N_0}{(m + s)^2}$$

with $m = 2, 3, \ldots$ for those components of the sharp series. In 1907 Bergmann[28] found in the infrared spectra of potassium, rubidium, and cesium a fourth series whose convergence limit coincided with $N_0/(3 + d)^2$ and whose wave numbers were expressed by

$$n = \frac{N_0}{(3 + d)^2} - \frac{N_0}{(m + f)^2}$$

where $m = 4, 5, \ldots$. Some of the lines of this series, which was later called the "fundamental series" or "Bergmann series," had been discovered previously by Saunders.[29]

It occurred to Rydberg[30] and was later explicitly stated as a fundamental principle by Ritz[31] that the frequency of every spectral line of an element could be expressed as the difference between two terms or "spectral terms," each of which contained an integer. Ritz's principle, or, as it was subsequently called, "the combination principle," could not be accounted for by classical physics. In order to understand the reason for this incompatibility, it must be noted that the above-mentioned hypothesis according to which the spectrum as a whole was thought to be produced by the free vibrations of one single atom had meanwhile been refuted. For Conway[32] showed convincingly, in 1907, that each atom could give rise to only one spectral line at a time. Conway's contention was further elaborated by Bevan's[33] analysis of the anomalous dispersion by potassium vapor. Bevan showed that any theoretical explanation of this phenomenon, if worked out on the basis of the previous hypothesis, would necessarily imply an excessively great number of electrons per molecule. Each individual line in a spectrum had therefore to be associated with the periodic motion of an electron and the different lines of the spectrum with motions of excited electrons in different atoms. Classically, the spectrum of light

[28] A. Bergmann, *Beiträge zur Kenntnis der ultraroten Emissionsspektren der Alkalien*, Dissertation, Jena, 1907.

[29] F. A. Saunders, "Some additions to the arc spectra of the alkali metals," *Astrophysical Journal 20*, 188–201 (1904).

[30] J. R. Rydberg, "La distribution des raies spectrales," in *Rapports présentés au Congrès International de Physique* (Gauthier-Villars, Paris, 1900), vol. 2, pp. 200–224.

[31] W. Ritz, "Über ein neues Gesetz der Serienspektren," *Physikalische Zeitschrift 9*, 521–529 (1908); "On a new law of series spectra," *Astrophysical Journal 28*, 237–243 (1908). "Durch additive oder subtraktive Kombination, sei es der Serienformeln selbst, sei es der in dieselben eingehenden Konstanten, werden Formeln gebildet, die gewiße neu entdeckte Linien vollständig aus den früher bekannten zu berechnen gestatten." *Gesammelte Werke* (Gauthier-Villars, Paris, 1911), p. 162.

[32] A. W. Conway, "On series spectra," *Scientific Proceedings of the Royal Dublin Society 11*, 181–183 (1907).

[33] P. V. Bevan, "Dispersion of light by potassium vapour," *Proceedings of the Royal Society of London (A)*, *84*, 209–225 (1910).

had consequently to contain, together with the fundamental vibration, the higher harmonics whose frequencies have the form of a sum of integral multiples of the fundamental frequencies, a result inconsistent with Ritz's combination principle. We thus see that the study of the discrete spectra, like that of the continuous spectra, led to serious inconsistencies with classical physics.

2.2 Bohr's Theory of the Hydrogen Atom

Our preceding remarks on spectroscopy were confined to merely tracing the origin and early development of the combination principle. Other aspects of spectroscopic research and their influence on the conceptual development of quantum theory will be discussed in a different context. At present, it should be noted that the vast amount of work invested in the search for numerical relations among spectral lines was motivated by the hope that these relations, by analogy to certain problems in the theory of mechanical and acoustical vibrations, would throw new light upon the nature of the proper oscillations of the atom or its electrons and thus lead to an understanding of atomic structure and of microscopic processes. The combination principle, however, revealed the futility of such an approach. It gradually became clear that the establishment of merely mathematical relations devoid of any consistent theory would be labor in vain. The only way to avoid this impasse, it seemed, was to take refuge in a model of the atom whose structure could be based on independent evidence from other sources and to apply the mathematical relations to the model. The outcome of this conceptual development was, of course, Niels Bohr's synthesis of the combination principle with Rutherford's atomic model on the basis of Planck's quantum conception.

In fact, Bohr's work itself was intimately connected with the problem of finding a consistent model of the atom. Even external circumstances of his life were connected with it. For when Bohr left J. J. Thomson's laboratory in Cambridge, in March, 1912, to join Rutherford's team in Manchester, it was because of a disagreement with Thomson concerning the latter's "plum-cake" model of the atom—or, as Condon once phrased it, because "J. J. politely indicated that it might be nice if he [Bohr] left Cambridge and went to work with Rutherford."[34] It was a fortunate turn of fate, not only with respect to the outcome when eventually Bohr's brilliant and successful theory of the hydrogen atom contributed decidedly to the general acceptance of the Rutherford model, but also with respect to the whole time of Bohr's stay in Manchester; for there he found sympa-

[34] E. U. Condon, "60 years of quantum mechanics," *Physics Today 15*, 45 (1962).

thetic understanding and even encouragement from the very beginning of his work. More than fifteen years later Rutherford recollected: "On my side, the agreement with Planck's deduction of e early made me an adherent to the general idea of a quantum of action. I was in consequence able to view with equanimity and even to encourage Prof. Bohr's bold application of the quantum theory to explain the origin of spectra."[35]

For a full comprehension of Bohr's work we shall have to discuss briefly the development of atomic models at that time.

Nineteenth-century models of the atom, such as those proposed by Kelvin, Helmholtz, or Bjerknes, were primarily mechanical or hydrodynamical and were invalidated by the discovery of electrons and radioactive disintegration. One of the earliest models consistent with these discoveries was proposed by Perrin[36] in a popular lecture delivered to students and friends of the Sorbonne in 1901. It consisted of a positively charged particle surrounded by a number of electrons ("sorte de petites planètes") compensating the central charge. Perrin assumed that the internal electromagnetic forces could produce a dynamically stable system whose rotational periods would correspond to the frequencies or wavelengths of the lines in the emission spectrum of the atom.

Two years later, in December, 1903, J. J. Thomson[37] developed the atomic model to which reference was made in the preceding chapter. The hydrogen atom, for example, was represented by a positively charged sphere of a radius of about 10^{-8} cm with an electron oscillating at its center. Thomson's model had one great advantage: it implied without any additional assumptions a quasi-elastic binding of the electrons, a property which had been a basic hypothesis for Drude, Voigt, Planck, and Lorentz in their work on dispersion, absorption, and other phenomena. In particular, it was a convenient assumption for the explanation of monochromatic spectral lines whose frequencies were independent of the energy of vibration.

Thomson's model, however, failed to account for the large angle

[35] Note by Prof. E. Rutherford, *Die Naturwissenschaften 17*, 483 (1929). On Planck's calculation of the elementary charge e see p. 21.

[36] J. Perrin, "Les hypothèses moléculaires," *Revue Scientifique 15*, 449–461 (1901); reprinted in Jean Baptiste Perrin, *Œuvres Scientifiques* (Centre National de la Recherche Scientifique, Paris, 1950).

[37] J. J. Thomson, "The magnetic properties of systems of corpuscles describing circular orbits," *Philosophical Magazine 6*, 673–693 (1903); his main paper on this subject appeared in 1904: "On the structure of the atom—an investigation of the stability and periods of oscillation of a number of corpuscles arranged at equal intervals around the circumference of a circle; with application of the results to the theory of atomic structure," *ibid. 7*, 237–265 (1904). At the same time (1903) Philipp Lenard, experimenting on the penetration of cathode-ray particles through matter, suggested a model according to which the atom is built up of "dynamides," electric doublets, possessing mass, but of so small a magnitude ($\sim 10^{-12}$ cm) that only a vanishingly small part of the atomic volume is not empty. Cf. P. Lenard, "Über die Absorption von Kathodenstrahlen verschiedener Geschwindigkeit," *Annalen der Physik 12*, 714–744 (1903).

deflections, up to 150° with reference to the incident direction, in the scattering experiments with α particles which were carried out a few years later by Thomson's students, Geiger and Marsden.[38] When Thomson's explanations of these results on the basis of multiple scattering proved untenable,[39] Rutherford's theory of "single scattering"[40] and his well-known model of the atom became firmly established. Thomson's model also failed to explain the subsequently (1913) discovered Stark effect, which will be discussed later on. Finally—but this, of course, was not known until much later—if the Thomson model is quantized,[41] the resulting energy states, although almost identical with those of the Rutherford model, turn out to be generally nondegenerate. Thus, even had no Rutherford scattering experiments ever been performed, spectroscopic evidence alone would have decided in favor of the Rutherford model.

At the time when Thomson developed his atomic model, the Japanese physicist Nagaoka,[42] in an address before the Physico-Mathematical Society of Tokyo in December, 1903, proposed what he called a "Saturnian model." It consisted, like Perrin's, of a positively charged central particle surrounded by a number of equidistant electrons revolving with a common angular velocity. Nagaoka then tried to show that the lines in the emission spectrum have their origin in small transversal oscillations of the electron configuration. Referring to the treatment of similar oscillations in Maxwell's essay "On the stability of the motion of Saturn's rings,"[43] Nagaoka claimed he had proved the dynamical stability of his model and its compatibility with the spectroscopic observations of Runge and Rydberg. The correctness

[38] H. Geiger and E. Marsden, "On a diffuse reflection of the α-particles," *Proceedings of the Royal Society* (*A*),*82*,495–500 (1909); "The scattering of α-particles by matter," *ibid. 83*, 492–504 (1910).

[39] An analysis of the Thomson scattering leads to a probability formula with an exponential dependence on the angle of deflection φ, in disagreement with the observed $\sin^{-4}(\varphi/2)$ angle dependence. For details cf. G. P. Harnwell and W. E. Stephens, *Atomic Physics* (McGraw-Hill, New York, 1955), pp. 99–108.

[40] E. Rutherford, "The scattering of α and β particles by matter and the structure of the atom," *Philosophical Magazine 21*, 669–688 (1911); *The Collected Papers of Lord Rutherford of Nelson* (Interscience, New York, 1962), vol. 2, pp. 238–254; "The structure of the atom," *Philosophical Magazine 27*, 488–498 (1914); *Collected Papers*, vol. 2, pp. 445–455. The former paper, which already contained his famous scattering formula, has been reprinted in R. T. Beyer (ed.), *Foundations of Nuclear Physics* (Dover, New York, 1949), pp. 111–130, and in J. B. Birks (ed.), *Rutherford at Manchester* (Heywood, London, 1962; W. A. Benjamin, New York, 1963), pp. 182–204.

[41] Such a quantization has been performed by H. Zatzkis in his paper "Thomson atom," *American Journal of Physics 26*, 635–638 (1958).

[42] H. Nagaoka, "Motion of particles in an ideal atom illustrating the line and band spectra and the phenomena of radioactivity," *Bulletin of the Mathematical and Physical Society of Tokyo 2*, 140–141 (1904); "On a dynamical system illustrating the spectrum lines and the phenomena of radio-activity," *Nature* 69, 392–393 (1904); "Kinetics of a system of particles illustrating the line and band spectrum and the phenomena of radioactivity," *Philosophical Magazine 7*, 445–455 (1904).

[43] J. C. Maxwell, *Scientific Papers* (REF. 56 OF CHAP. 1), vol. 1, pp. 288–376.

of these computations and, in particular, the stability of the model, however, were soon challenged by Schott.[44]

Perrin's and Nagaoka's atomic models were further elaborated by the Cambridge astrophysicist Nicholson[45] in an attempt to explain the nature of a number of unidentified lines in the spectra of nebulae and in the spectrum of the solar corona. Associating these lines with the hypothetical elements "nebulium" and "protofluorine," Nicholson, closely following Nagaoka's suggestion, constructed atomic models to account for these lines. On the assumption of a central charge of four units surrounded by four electrons, Nicholson showed that the ratio between the frequencies corresponding to two different vibrational modes of the electronic ring coincided with the ratio between the frequencies of two lines in the nebular spectrum. This agreement, in his view, constituted sufficient evidence for the adequacy of his model. Nicholson then proceeded to give detailed calculations for other lines and to develop a general theory, in the course of which he made a number of important discoveries. His theory appeared to him to have been given even observational support. For the frequency of a third mode (λ4353), which he computed as a sequel to the two modes mentioned before and which, at the time of his first publication, was not known to correspond to any known spectral line, had actually been observed by W. H. Wright at the Lick Observatory and by M. Wolf at the Heidelberg Observatory.[46] Only in 1927 was the nature of the "nebulium" lines definitely clarified when Bowen showed that they are forbidden lines of highly ionized oxygen and nitrogen (NII, OII, and OIII).[47]

With respect to his model of the atom, Nicholson declared that "the main conception involved . . . is that of the nature of positive electricity. This is supposed to exist in small spherical volume distributions of uniform density, whose radius is small in comparison even with that of the electron, a reversal of the more generally accepted view. The mass of these positive units is very large in comparison with that of an electron, and gives rise to nearly the whole mass of an atom. . . ."[48] In his third essay, published in June, 1912, Nicholson suggested relating the occurrence of spectral series to Planck's constant of action. "The quantum theory," he stated, "has apparently not been put forward as an explanation of 'series' spectra. . . . Yet, in the belief of the writer, it furnishes the true explanation in certain cases, and we are led to suppose that lines of a series may not emanate from

[44] G. A. Schott, "A dynamical system illustrating the spectrum lines and the phenomena of radio-activity," *Nature 69*, 437 (1904). "On the kinetics of a system of particles illustrating the line and band spectra," *Philosophical Magazine 8*, 384–387 (1904).

[45] J. W. Nicholson, "The spectrum of nebulium," *Monthly Notices of the Royal Astronomical Society 72*, 49–64 (1912); "The constitution of the solar corona," *ibid.*, 139–150, 677–693, 729–739.

[46] J. W. Nicholson, "On the new nebular line at λ4353," *ibid.*, 693.

[47] I. S. Bowen, "The origin of the nebulium lines," *Nature 120*, 473 (1927).

[48] REF. 45, p. 49.

the same atom, but from atoms whose internal angular momenta have, by radiation or otherwise, run down by various discrete amounts from a standard value."[49]

Nicholson incorporated Planck's quantum of action into his theory by contending that the frequencies of different observed spectral lines can be accounted for on the following assumption: that the ratio between the energy of the system and the frequency of rotation of the ring of electrons is given by an integral multiple of Planck's constant. Nicholson's theory, based as it was on a correspondence between the frequencies of optical and of mechanical vibrations, was, of course, by no means compatible with the Ritz combination principle. Nor was his introduction of Planck's constant motivated by any considerations concerning the stability of his atomic model. In fact, with the exception of his casual remark to the effect that "the electrons in steady motion must lie in one plane, in order that their energy may not be dissipated by rapid radiation,"[50] Nicholson completely ignored the stability problem.

It is a curious coincidence in the history of our subject that Nicholson's most important innovations, namely, his idea of a heavy nucleus and his conception of spectra as quantal phenomena, were actually obtained at the same time by independent research. For it was in May, 1911, that Ernest Rutherford refuted Thomson's hypothesis of multiple scattering and established his well-known model of the atom in accordance with the large angle deflections of α particles by single encounters.[51] It was also in 1911 that Bjerrum, following a suggestion by Nernst,[52] made the first successful application of the quantum principle to molecular spectra when he showed[53] that the quantization of the rotational energy of molecules accounted for certain features in the absorption spectra of gaseous hydrochloric and hydrobromic acids. It seems beyond doubt that neither Rutherford nor Bjerrum knew of Nicholson's ideas and that the latter's conjectures were in no way influenced by the former. It should also be noted that Nicholson's anticipations of some of Bohr's conclusions were based, as Rosenfeld has pointed out,[54] on a most questionable and often even fallacious reasoning.

Bohr's first research in Manchester on the absorption of α particles by matter,[55] which he finished in the summer of 1912, proved highly instructive for his future work. Treating the problem, which had been taken

[49] *Ibid.*, p. 729.

[50] *Ibid.*, p. 50.

[51] See REF. 40.

[52] W. Nernst, "Zur Theorie der spezifischen Wärme und über die Anwendung der Lehre von den Energiequanten auf physikalisch-chemische Fragen überhaupt," *Zeitschrift für Elektrochemie 17*, 265–275 (1911).

[53] N. Bjerrum, "Über die ultraroten Absorptionsspektren der Gase," in *Nernst-Festschrift*, 1912, pp. 90–98.

[54] Cf. L. Rosenfeld's introduction to Niels Bohr's *On the Constitution of Atoms and Molecules* (reprint of Bohr's papers of 1913; Munksgaard, Copenhagen, and W. A. Benjamin, New York, 1963), pp. xii-xiii.

[55] N. Bohr, "On the theory of the decrease of velocity of moving electrified particles on passing through matter," *Philosophical Magazine 25*, 10–31 (1913).

up in terms of classical mechanics by J. J. Thomson as early as 1906, long before the discovery of the atomic nucleus, Bohr assumed an elastic binding of the atomic electrons. This binding, although entering his calculations only through the periods of motions which Bohr inferred from the characteristic resonances as known from optical dispersion, played an essential role in limiting the effective region of energy transfer around the path of the penetrating particle. Associated now with Rutherford's atomic model, the binding of electrons and the concomitant problem of the stability of the atom had to engage Bohr's attention from the outset of his work. For Bohr saw not only the merits of Rutherford's model but also the difficulties which it raised and, in particular, the fact that, in contrast to Thomson's model, dynamical principles alone could not provide any scale for its size. In June, 1912, in the first draft of his classic paper "On the constitution of atoms and molecules,[56] Bohr wrote:[57] "In the investigation of the configuration of the electrons in the atoms we immediately meet with the difficulty . . . that a ring, if only the strength of the central charge and the number of electrons in the ring are given, can rotate with an infinitely great number of different times of rotation, according to the assumed different radius of the ring; and there seems to be nothing . . . to allow, from mechanical considerations to discriminate between the different radii and times of vibration. In the further investigation we shall therefore introduce and make use of a hypothesis, from which we can determine the quantities in question. This hypothesis is: that for any stable ring (any ring occurring in the natural atoms) there will be a definite ratio between the kinetic energy of an electron in the ring and the time of rotation.[58] This hypothesis, for which no attempt at a mechanical foundation will be given (as it seems hopeless), is chosen as the only one which seems to offer a possibility of an explanation of the whole group of experimental results, which gather about and seems to confirm conceptions of the mechanismus of the radiation as the ones proposed by Planck and Einstein." This draft, composed of six manuscript sheets (one of which is unfortunately lost), was written by Bohr as a private communication to Rutherford and is naturally an important document for the understanding of the conceptual development of Bohr's atomic theory. Bohr realized that the stability of the Rutherford model could by no means be reconciled with the principles of Newton's mechanics and Maxwell's electrodynamics. For no system of point charges, according to these principles, ever admits of a stable equilibrium, and any dynamical equilibrium, based on the motions

[56] *Philosophical Magazine 26*, 1–25, 476–502, 857–875 (1913). For a recent reprint cf. REF. 54. German translation in N. Bohr, *Abhandlungen über Atombau aus den Jahren 1913–1916*, translated by H. Stintzing (Friedr. Vieweg, Braunschweig, 1921); reprinted in *Dokumente der Naturwissenschaft—Abteilung Physik* (Ernst Battenberg Verlag, Stuttgart, 1964), vol. 5, pp. 33–57, 58–83, 84–101.

[57] REF. 54, p. xxiii.

[58] Bohr obviously meant "angular velocity" instead of "time of rotation."

of electrons, must lead to a radiative dissipation of energy accompanied by a steady contraction of the system.

Since in Planck's discovery a fundamental limitation of classical physics had already been revealed, Bohr argued that the answer to the stability problem had to be sought in Planck's quantum of action. In contrast to his predecessors who related Planck's h to atomic models with the purpose of finding a mechanical or electromagnetic interpretation of h, Bohr recognized that Planck's constant should be applied to Rutherford's model not in order to elucidate the physical significance of the former, but rather to account for the stability of the latter. Bohr realized that to this end the theory should yield a constant of the dimension of length, characterizing the distance of the electron from the center of the stable orbit. But the only constant parameters that appeared in Rutherford's model of the atom were those of masses and charges, and from these no constant of the dimension of length could possibly be formed. In the fact, however, that the adjunction of h to m and e made it possible to construct an expression, such as h^2/me^2, which has the dimension of length—even with the required order of magnitude ($\approx 20 \times 10^{-8}$ cm)—Bohr saw additional evidence for the correctness of his assumption.

All previous applications of Planck's constant to atomic models referred to the Thomson atom and were generally based on the assumption of harmonic oscillations for which the rule of quantization could be derived from Planck's work. The most noteworthy of these attempts was Haas's conjecture, to which reference has been made before.[59] In an attempt to interpret the photoelectric effect as a resonance phenomenon between orbital frequencies and the frequencies of the incident radiation, Lindemann,[60] in 1911, applied Kepler's laws to the assumedly elliptical orbits of the electrons but otherwise adhered to Thomson's model. In March, 1912, Herzfeld[61] proposed a modification of Thomson's model by assuming circular electronic orbits and a nonuniform charge density of the positive sphere and derived from these assumptions the Balmer series by a quantization of energy in accordance with a rule formulated by Hasenöhrl[62] as a generalization of Planck's prescription for the quantization of the harmonic oscillator. But all these and similar[63] calculations, although leading to an agreement as to the order of magnitude between the calculated and observed values for the frequencies and dimensions of the atoms, lost their

[59] See REF. 168 OF CHAP. 1.

[60] F. A. Lindemann, "Über die Berechnung der Eigenfrequenzen der Elektronen im selektiven Photoeffekt," *Vehandlungen der Deutschen Physikalischen Gesellschaft 13*, 482–488 (1911).

[61] K. F. Herzfeld, "Über ein Atommodell, das die Balmer'sche Wasserstoffserie aussendet," *Wiener Berichte 121*, 593–601 (1912).

[62] F. Hasenöhrl, "Über die Grundlagen der mechanischen Theorie der Wärme," *Physikalische Zeitschrift 12*, 931–935 (1911).

[63] E. Wertheimer, "Zur Haberschen Theorie der Wärmetönung," *Verhandlungen der Deutschen Physikalischen Gesellschaft 14*, 431–437 (1912).

validity with the abandonment of the Thomson model on which they were based. Nicholson's hypotheses, with which Bohr became acquainted only toward the end of 1912, as we know from a Christmas card[64] sent by him to his brother Harald, were similarly invalidated, although on other grounds.

The basic problem which Bohr had to face was the question of how to apply Planck's quantum of action to the Rutherford model of the atom. On the assumption that the hydrogen atom consists of an electron revolving around a nucleus whose charge is equal and opposite to that of the electron and whose mass is very large in comparison with the mass m of the electron, Bohr first investigated how far classical mechanics could be employed. From Kepler's first law he knew that the orbit of the electron is an ellipse with the nucleus in one of its foci. Calling the major axis of the orbit $2a$, and the energy which has to be added to the system in order to remove the electron to an infinite distance from the nucleus W, the charge of the electron e, and that of the nucleus e', Bohr could easily show that on the basis of classical principles

$$\omega = \frac{\sqrt{2}}{\pi}\frac{W^{3/2}}{ee'\sqrt{m}} \qquad 2a = \frac{ee'}{W} \tag{2.2}$$

which indicates that the frequency of revolution ω and the major axis of the orbit depend only on the value of W and are independent of the eccentricity of the orbit. In the special case of a circular orbit with radius a, a direct proof of Eq. (2.2) can readily be established. In this case, $mv^2/a = ee'/a^2$ (centripetal force = force of attraction) shows that the total energy $U = \frac{1}{2}mv^2 - ee'/a = -ee'/2a = -W$ or $W = ee'/2a$; and from $v = 2\pi\omega a$ the first part of Eq. (2.2) follows immediately. Bohr now saw that by varying W all possible values for ω and $2a$ could be obtained. The existence of sharp spectral lines, however, shows that the latter quantities have to assume definite values characteristic for the system. Moreover, as a simple consideration of the dimensions verifies, no combination of e, e', and m can yield a quantity with the dimension of length, a condition necessary for its interpretation as the diameter of the atom. So far, Bohr ignored the radiation of energy which according to Maxwell's theory should be dissipated at a rate proportional to the square of the acceleration of the electron. If this radiation were taken into consideration, W would steadily increase and with it the frequency of revolution ω while the dimensions of the orbit would continually decrease, producing a continuous spectrum instead of discrete spectral lines as observed. Thus the atom would be not only indeterminate in its dimensions but also deprived of any stability whatever. Realizing the incompatibility of ordinary mechanics and electrodynamics with Rutherford's atomic model and in direct contradiction to Newton's

[64] REF. 54, p. xxxvi.

mechanics and Maxwell's electrodynamics, Bohr boldly postulated the following assumption in order to endow the atom with size and stability. There exists a discrete set of permissible or *stationary* orbits; and as long as the electron remains in any stationary orbit, no energy is radiated. It is interesting to note that the second part of this assumption, though not explicitly formulated, had been used a few years earlier by Langevin[65] in his theory of permanent magnetism.

Although convinced that the discreteness of the orbits (or of a) and, according to Eq. (2.2), also of W and ω was somehow connected with Planck's constant of action, Bohr was unable to identify the exact relationship until February, 1913. Two apparently unrelated phenomena gave him the clue to the solution of the problem. He had been acquainted from his Cambridge days with Whiddington's experiments.[66] Whiddington had shown that when an anticathode was bombarded by cathode rays of increasing velocity, sudden changes in the nature of the emitted radiation were produced at certain critical velocities. It seems that Whiddington's work suggested to Bohr the idea of energy levels.[67] In fact, Whiddington's results, as, of course, became clear only later on, constituted in the field of x-rays one of the most striking verifications of Bohr's theory, just as the subsequent Franck-Hertz experiments did in the range of visible light. The other clue came from spectroscopy. Even by the end of January, 1913, Bohr was not fully aware of the implications of spectroscopic research for his problem. For in a letter dated January 31, 1913, addressed to Rutherford from Copenhagen, where Bohr had served since September, 1912, as an assistant to Knudsen, Bohr declared: "I do not at all deal with the question of calculation of the frequencies corresponding to the lines in the visible spectrum. I have only tried, on the basis of the simple hypothesis, which I used from the beginning, to discuss the constitution of the atoms and molecules in their 'permanent' state."[68] A few days later, however, the spectroscopist H. M. Hansen, who had just arrived in Copenhagen from Göttingen, where he had been working under Voigt on the inverse Zeeman effect of lithium, drew his attention to Rydberg's work on the classification of spectral lines, a subject in which Bohr had not previously been interested. Asked by Hansen how the new model of the atom could account for the regularities discovered by Rydberg and Ritz, Bohr acquainted himself with the subject and soon recognized its importance for his problem. "As soon as I saw Balmer's formula," he stated repeatedly,[69] "the whole thing was immediately clear to me."

[65] P. Langevin, "Sur la théorie du magnétisme," *Comptes Rendus 139*, 1204–1207 (1904).

[66] R. Whiddington, "The production of characteristic Röntgen radiation," *Proceedings of the Royal Society of London* (*A*), *85*, 323–332 (1911).

[67] Cf. *Archive for the History of Quantum Physics*, Interview with Niels Bohr on Oct. 31, 1962.

[68] The full text of the letter is published in REF. 54, pp. xxxvi–xxxvii.

[69] REF. 54, p. xxxix.

How, indeed, Balmer's formula suggested to Bohr a way of incorporating quantum conceptions into the Rutherford model of the hydrogen atom will now be explained. For this purpose we shall closely analyze Bohr's exposition of this issue as presented in his epoch-making paper.[70] Balmer's formula and its immediate generalization can most conveniently be expressed, as we have seen in connection with the work of Ritz and Rydberg, by the equation[71]

$$\nu = Rc\left(\frac{1}{\tau_2^2} - \frac{1}{\tau_1^2}\right) \tag{2.3}$$

Bohr derived this formula in the first part[72] of the above-mentioned paper by three different methods. In his first derivation[73] he adopted the basic assumption of Planck's "second theory,"[74] according to which the amount of energy emitted by an atomic vibrator of frequency ν is $\tau h\nu$, τ being an integer, namely, the number of quanta emitted, each of frequency ν. Considering an electron at a great distance apart from the nucleus and "of no sensible velocity relative to the latter," Bohr studied the binding of this electron with the nucleus as a result of which the electron "has settled down in a stationary (circular) orbit around the nucleus" with orbital frequency ω. During this process, Bohr contended, a "homogeneous" (monochromatic) radiation is emitted of a frequency ν, equal to half the frequency of revolution of the electron in its final orbit—this relation suggesting itself in view of the fact that "the frequency of revolution of the electron at the beginning of the emission is 0." Thus, from

$$W = \frac{\tau}{2} h\omega$$

Bohr obtained by Eq. (2.2)

$$W = \frac{2\pi^2 m e^2 e'^2}{\tau^2 h^2} \qquad \omega = \frac{4\pi^2 m e^2 e'^2}{\tau^3 h^3} \qquad 2a = \frac{\tau^2 h^2}{2\pi^2 m e e'}$$

expressions which for $\tau = 1$, $|e| = |e'| = 4.7 \times 10^{-10}$, $e/m = 5.31 \times 10^{17}$, and $h = 6.5 \times 10^{-27}$ yield $W = 13$ eV, $\omega = 6.2 \times 10^{15}$ sec^{-1}, and $2a = 1.1 \times 10^{-8}$ cm in agreement with the empirical values of the ionization potential, the optical frequencies, and the linear dimensions of the hydrogen atom. If, as he found, the energy emitted by the formation of one of the stationary states is for the hydrogen atom

$$W = \frac{2\pi^2 m e^4}{h^2 \tau^2}$$

[70] REF. 56.
[71] Cf. REF. 23.
[72] REF. 56, pp. 1–25.
[73] *Ibid.*, pp. 5, 8, 9.
[74] REF. 195 OF CHAP. 1.

the "energy emitted by the passing of the system from a state corresponding to $\tau = \tau_1$ to one corresponding to $\tau = \tau_2$" turns out to be

$$W_{\tau_2} - W_{\tau_1} = \frac{2\pi^2 me^4}{h^2}\left(\frac{1}{\tau_2^2} - \frac{1}{\tau_1^2}\right)$$

Hence, since

$$W_{\tau_2} - W_{\tau_1} = h\nu$$

$$\nu = \frac{2\pi^2 me^4}{h^3}\left(\frac{1}{\tau_2^2} - \frac{1}{\tau_1^2}\right)$$

which is Balmer's formula.

In his second derivation[75] of the Balmer formula Bohr noted that it had not been necessary to assume that "a radiation is sent out corresponding to more than a single energy quantum $h\nu$." Since "as soon as one quantum is sent out the frequency is altered," it would be more consistent, he declared, to replace the previously assumed emission of τ quanta of frequency $\nu/2$ by that of one quantum of frequency $\tau\nu/2$. Bohr, in fact, generalized this relationship still further by putting the ratio between the emitted energy and the frequency of revolution of the electron for the different stationary states as

$$W = f(\tau)h\omega$$

where $f(\tau)$ is an as yet undetermined function of the integer τ. This he did because the particular choice $f(\tau) \equiv \tau/2$, as used in his first derivation, though plausible, did not seem to him to have sufficient logical cogency. Repeating now step by step the calculation of the first derivation, but with $\tau/2$ replaced by $f(\tau)$, Bohr obtained

$$\nu = \frac{\pi^2 me^2 e'^2}{2h^3}\left[\frac{1}{f^2(\tau_2)} - \frac{1}{f^2(\tau_1)}\right]$$

Whereas in his first derivation the particular specification $f(\tau) \equiv \tau/2$ led him directly to the Balmer formula, he had now to pay for the greater generality of his assumption with an additional step: he had to take recourse to the structure of the Balmer formula itself in which the variable factor has the form $1/\tau_2^2 - 1/\tau_1^2$. To obtain such a factor he concluded that $f(\tau) = c\tau$. Here c is a constant which has to be determined. For this purpose he considered the transition of the system between two successive stationary states corresponding to $\tau = N$ and $\tau = N - 1$. By a simple calculation he obtained for the frequency of the radiation emitted

$$\nu = \frac{\pi^2 me^2 e'^2}{2c^2 h^3}\frac{2N - 1}{N^2(N - 1)^2}$$

[75] REF. 56, pp. 12, 13.

and for the frequencies of revolution of the electron before and after the emission

$$\omega_N = \frac{\pi^2 m e^2 e'^2}{2c^3 h^3 N^3} \qquad \omega_{N-1} = \frac{\pi^2 m e^2 e'^2}{2c^3 h^3 (N-1)^3}$$

"If N is great," Bohr continued, "the ratio between the frequency before and after the emission will be very nearly equal to 1; and according to ordinary electrodynamics we should therefore expect that the ratio between the frequency of radiation and the frequency of revolution also is very nearly equal to 1. This condition will only be satisfied if $c = \frac{1}{2}$." In fact, since

$$\frac{\nu}{\omega_N} = \frac{cN^3(2N-1)}{N^2(N-1)^2}$$

tends to $2c$ for large N, it approaches 1 only if $c = \frac{1}{2}$. Thus the result of the first derivation is retrieved.

In his third derivation[76] he dispensed altogether with Planck's relation by replacing it by what he called "an interpretation of the emission by analogy with ordinary electrodynamics." For, he declared, "an electron rotating round a nucleus in an elliptical orbit will emit a radiation which according to Fourier's theorem can be resolved into homogeneous components, the frequencies of which are $n\omega$, if ω is the frequency of revolution of the electron." Taking consequently the frequency of the energy emitted during the transition from a state in which no energy is yet emitted to another stationary state as equal to a multiple of $\omega/2$, where ω is the frequency of revolution of the electron in the state considered, Bohr arrived at the same expression as before. "Consequently," he declared in the conclusion of his present discussion, "we may regard our preliminary considerations . . . only as a simple form of representing the results of the theory."[77]

Bohr's approach, as we saw, was essentially an application of what he later termed the "correspondence principle," according to which, in the limit where the action involved is sufficiently large to permit the neglect of the individual quanta, the fundamentally statistical account of quantum phenomena can be represented as a rational generalization of the classical physical description. Bohr must have been aware that, by postulating the quantization of the angular momentum, he could have presented his theory in a mathematically more compact way. But distrusting the legitimacy of employing the conceptions of the "old mechanics" and avoiding their use whenever possible, Bohr rejected this alternative and gave greater credence to the application of the correspondence principle. In fact, when he wrote

[76] *Ibid.*, p. 14.

[77] *Ibid.*

the first part of his classic paper, the quantization of the angular momentum was for him merely an "interpretation" in terms of "symbols taken from ordinary mechanics." For he declared: "While there obviously can be no question of a mechanical foundation of the calculations given in this paper, it is, however, possible to give a very simple interpretation of the results of the calculations by help of symbols taken from ordinary mechanics. Denoting the angular momentum of the electron round the nucleus by M, we have immediately for a circular orbit $\pi M = T/\omega$, where ω is the frequency of revolution and T the kinetic energy of the electron; for a circular orbit we further have $T = W$, and from $W = \tau h\omega/2$ we consequently get $M = \tau M_0$, where $M_0 = h/2\pi = 1.04 \times 10^{-27}$. . . . The angular momentum of the electron round the nucleus in a stationary state of the system is equal to an entire multiple of a universal value, independent of the charge on the nucleus."[78] Clearly, in this case the quantization of the angular momentum and the formula $W = \frac{1}{2}\tau h\omega$ are mathematically equivalent statements. In the concluding remarks of his paper[79] Bohr summarized all his assumptions employed so far in the following formulation: "(1) That energy radiation is not emitted (or absorbed) in the continuous way assumed in the ordinary electrodynamics, but only during the passing of the systems between different 'stationary' states. (2) That the dynamical equilibrium of the systems in the stationary states is governed by the ordinary laws of mechanics, while these laws do not hold for the passing of the systems between the different stationary states. (3) That the radiation emitted during the transition of a system between two stationary states is homogeneous, and that the relation between the frequency ν and the total amount of energy emitted E is given by $E = h\nu$, where h is Planck's constant. (4) That the different stationary states of a simple system consisting of an electron rotating round a positive nucleus are determined by the condition that the ratio between the total energy, emitted during the formation of the configuration, and the frequency of revolution of the electron is an entire multiple of $\frac{1}{2}h$. Assuming that the orbit of the electron is circular, this assumption is equivalent with the assumption that the angular momentum of the electron round the nucleus is equal to an entire multiple of $h/2\pi$. (5) That the 'permanent' state of any atomic system, i.e., the state in which the energy emitted is maximum, is determined by the condition that the angular momentum of every electron round the centre of its orbit is equal to $h/2\pi$."

It should be noted in this context that Ehrenfest,[80] a short time prior

[78] *Ibid.*, p. 15.

[79] *Ibid.*, p. 874.

[80] P. Ehrenfest, "Bemerkung betreffs der spezifischen Wärme zweiatomiger Gase," *Verhandlungen der Deutschen Physikalischen Gesellschaft 15*, 451–457 (1913) (this paper appeared on June 15, 1913, whereas the first part of Bohr's paper appeared in the July issue of the *Philosophical Magazine*, 1913); *Collected Scientific Papers* (REF. 77 OF CHAP. 1), pp. 333–339.

to Bohr, had already applied the quantization of angular momentum without, however, formulating it as a general principle. This he did in his refinement of the Einstein-Stern theory (see REF. 104 OF CHAP. 1) of the specific heat of diatomic gases when he showed that the assumption $\frac{1}{2}L(2\pi\nu)^2 = nh\nu/2$ (L is the moment of inertia, ν the frequency, and n an integer) accounted for the empirical temperature dependence of the rotational energy of hydrogen without the need of introducing a zero-point energy.

In addition to the close agreement between the calculated and the observed values of the Rydberg constant, Bohr found support for the correctness of his theory in the fact that Eq. (2.3) applied not only to the Balmer series with $\tau_2 = 2$ but also for $\tau_2 = 3$ (and consequently $\tau_1 = 4, 5, \ldots$). The case $\tau_2 = 3$ corresponded to a series in the infrared which, after having already been predicted by Ritz,[81] was actually observed in 1908 by Paschen[82] and which was later called the "Paschen series." Moreover, Bohr's prediction that "if we put $\tau_2 = 1$ and $\tau_2 = 4, 5, \ldots$, we get series respectively in the extreme ultraviolet and extreme ultrared, which are not observed, but the existence of which may be expected"[83] was soon verified. The series for $\tau_2 = 1$ was observed by Lyman[84] in 1914, that for $\tau_2 = 4$ by Brackett[85] in 1922, and that for $\tau_2 = 5$ by Pfund[86] in 1924.

In 1913 it could be rightfully argued that the fact that not all lines predicted by Eq. (2.3) had actually been observed was an indication of the deficiency of experimental procedures rather than a disproof of Bohr's theory. The existence, however, of a single hydrogen line at variance with Eq. (2.3) would have shown that Bohr's theory was at least incomplete, if not altogether faulty. It was therefore a challenge to Bohr when his attention was drawn to a paper, published in 1896, in which the American astronomer Pickering claimed he had found hydrogen lines not accounted for by the ordinary Balmer formula in the spectrum of the star ζ Puppis. Nor did these lines, as it then became clear, agree with Bohr's more general formula (2.3). Pickering claimed having observed "a series of lines whose approximate wave-lengths are 3814, 3857, 3923, 4028, 4203, and 4505, the last line being very faint. These six lines," he argued, "form a rhythmical series like that of hydrogen and apparently are due to some element not yet found in other stars or on the Earth. The formula of Balmer will not represent this series, but if we add a constant term and write $\lambda = 4650[m^2/(m^2 - 4)] - 1032$, we obtain for m equal to 10, 9, 8, 7, 6

[81] W. Ritz, "Über ein neues Gesetz der Serienspektren," *Physikalische Zeitschrift 9*, 521–529 (1908).

[82] F. Paschen," Zur Kenntnis ultra-roter Linienspektren," *Annalen der Physik 27*, 537–570 (1908).

[83] REF. 56, p. 9.

[84] T. Lyman, "An extension of the spectrum in the extreme-violet," *Physical Review 3*, 504–505 (1914).

[85] F. Brackett, "A new series of spectrum lines," *Nature 109*, 209 (1922).

[86] A. H. Pfund, "The emission of nitrogen and hydrogen in the infrared," *Journal of the Optical Society of America 9*, 193–196 (1924).

and 5 the wave-lengths 3812, 3858, 3928, 4031, 4199 and 4504."[87] One year later Pickering identified these lines as hydrogen lines for the following reason: "A remarkable relation exists between these two series, from which it appears that the second series, instead of being due to some unknown element as was at first supposed, is so closely allied to the hydrogen series, that it is probably due to that substance under conditions of temperature or pressure as yet unknown. The wave-length of the lines of hydrogen may be computed by the formula $3646.1[n^2/(n^2 - 16)]$ which is the formula of Balmer, slightly modifying the constant term so that the standard wave-length of Rowland shall be represented, and substituting $\frac{1}{2}n$ for m. The wave-length of the lines of hydrogen may be determined by this formula if we substitute for n the even integers 6, 8, 10, 12 etc. . . . If now we substitute for n the odd integers 5, 7, 9, 11 etc., we obtain the wave-lengths of the second series of lines in the spectrum of ζ Puppis."[88] Thus, one single formula, namely, $\lambda = 3646[n^2/(n^2 - 16)]$, applied both to the Balmer lines (for even n) and to the newly discovered lines (for odd n). Kayser[89] at first expressed some doubts about Pickering's identification of these lines but later[90] accepted the theory on the grounds that hydrogen would have two series, which "then corresponds admirably with the other elements." Kayser also pointed out: "That this series has never been observed before, can perhaps be explained by not sufficient temperatures in our Geissler tubes and most of the stars." When in July, 1897, the line λ5413.9, corresponding to $n = 7$, also was observed as "clearly visible" in the spectrum of ζ Puppis,[91] Pickering's contentions were generally accepted.

Rydberg,[92] in accordance with his general classification of series, interpreted Balmer's series as the diffuse (or first subordinate) and Pickering's as the sharp (or second subordinate) series of hydrogen, so that the former corresponded to $n = n_0 - N_0/(m + 1)^2$ for $m = 2, 3, \ldots$ and the latter to $n = n_0 - N_0/(m + 0.5)^2$. Now, in accordance with the Rydberg-Schuster law a "principal series" of hydrogen should exist whose wave numbers are given by

$$n = \frac{N_0}{(1 + 0.5)^2} - \frac{N_0}{(m + 1)^2}$$

with $m = 1, 2, \ldots$, and whose first[93] member consequently is λ4687.88. In fact, a line λ4686 was actually observed[94] in the spectrum of ζ Puppis and

[87] E. C. Pickering, "Stars having peculiar spectra," *Astrophysical Journal 4*, 369–370 (1896). Cf. also *ibid.*, 142–143.

[88] E. C. Pickering, "The spectrum of ζ Puppis," *Astrophysical Journal 5*, 92–94 (1897).

[89] H. Kayser, "On the spectrum of ζ Puppis," *ibid.*, 95–97.

[90] H. Kayser, "On the spectrum of hydrogen," *ibid.*, 243.

[91] E. C. Pickering, "The spectrum of ζ Puppis," *ibid. 6*, 259 (1897).

[92] J. R. Rydberg, "The new series in the spectrum of hydrogen," *ibid. 6*, 233–238 (1897), *7*, 233 (1899).

[93] Owing to atmospheric absorption, the other lines λ2734.55, λ2386.50, . . . would not be visible.

[94] Cf. the footnote to Rydberg's paper (REF. 92) by Hale and Keeler.

also, during the Indian solar eclipse of January 22, 1898, in the spectrum of the sun's chromosphere.[95] Furthermore, in 1912 Fowler observed in the spectrum of a Plücker discharge tube containing a mixture of hydrogen and helium four lines at positions corresponding to Rydberg's so-called "principal series of hydrogen" and a new series of lines in the ultraviolet with the same high-frequency limit.[96]

These were the facts which Bohr then had to explain on the basis of his theory. At first he showed that the Pickering series and the hypothetical "principal series of hydrogen," originally expressed by

$$\nu = cR\left(\frac{1}{2^2} - \frac{1}{(m + \frac{1}{2})^2}\right) \quad \text{and} \quad \nu = cR\left(\frac{1}{1.5^2} - \frac{1}{m^2}\right)$$

with $m = 2, 3, \ldots$ respectively, could be rewritten as

$$\nu = 4cR\left(\frac{1}{4^2} - \frac{1}{k^2}\right)$$

with $k = 5, 7, \ldots$, and

$$\nu = 4cR\left(\frac{1}{3^2} - \frac{1}{k^2}\right)$$

with $k = 4, 6, \ldots$ respectively, that is, as series whose Rydberg constant is four times that of hydrogen. Bohr could now easily show that his theory, if applied to the ionized helium atom with nuclear charge $2e$, leads exactly to these formulas.

In a letter to Rutherford, dated March 6, 1913, Bohr referred to these ideas and wrote that "the chemist Dr. Bjerrum suggested to me that if my point was right the lines might also appear in a tube filled with a mixture of helium and chlorine (oxygen, or other electro-negative substances); indeed it was suggested, that the lines might be still stronger in this case. Now, we have not in Copenhagen the opportunity to do such an experiment satisfactorily; I might therefore ask you, if you possibly would let it perform in your laboratory, or if you perhaps kindly would forward the suggestion to Mr. Fowler, which may have the arrangement still standing."[97] In fact, when Evans,[98] following this suggestion, published a short note in *Nature* in which he indicated that his experiments did support Bohr's claim that helium, and not hydrogen, is the origin of the lines in

[95] Sir Norman Lockyer, Chisholm-Batten, and A. Pedler, "Total eclipse of the sun, January 22, 1898," *Philosophical Transactions of the Royal Society of London 197*, 151–228 (1901).

[96] A. Fowler, "Observations of the principal and other series of lines in the spectrum of hydrogen," *Monthly Notices of the Royal Astronomical Society 73*, 62–71 (1912). Small discrepancies between observed and calculated frequencies were noted. Their explanation was later supplied by the correction of the Rydberg constant for finite mass of the nucleus.

[97] REF. 54, p. xxxix.

[98] E. J. Evans, "The spectra of helium and hydrogen," *Nature 92*, 5 (Sept. 4, 1913).

question, Fowler[99] argued against Evans that the observed wavelengths of these lines differ, though only by a small amount, from their theoretical values. Bohr,[100] however, soon pointed out that if the finite ratio between the mass m of the electron and the mass m_n of the nucleus is taken into consideration, and if therefore the value of m in his formula for the Rydberg constant R is replaced by that of the "reduced mass"

$$m' = \frac{m}{1 + m/m_n}$$

the observed discrepancies of about 0.04 percent are fully accounted for. When subsequent experiments carried out by Evans[101] at the University of Manchester, by Fowler,[102] and by Paschen[103] corroborated this contention, Fowler's original objection became another striking confirmation of Bohr's theory.

It will also be recalled that the dependence of R on m_n, as contended by Bohr, led Urey[104] and his collaborators in 1932 to the discovery of heavy hydrogen or deuterium H^2 with an isotope shift of 1.79 Å for the first Balmer line H_α and 1.32 Å for H_β.

The most direct confirmation of Bohr's interpretation of spectral terms as stationary energy levels and of his frequency condition (assumptions 1 and 3 of his summary) was afforded by a series of experiments performed by Franck and Hertz.[105] Electrons from a thermoionic source were accelerated to a known energy and directed against atoms of a gas or vapor at low pressure. At low electronic energies only elastic collisions occurred and no radiation was observed. As soon, however, as the electron's energy was equal to or exceeded a critical value (4.9 eV in the case of mercury vapor), inelastic collisions took place and radiation (λ2537, the mercury resonance line) was observed. The energy loss of the electron corresponded to the energy difference between the ground state and the excited state of the atom which, by transition to the ground state, re-emits this energy in the form of light in accordance with Bohr's frequency condition.

[99] A. Fowler, "The spectra of helium and hydrogen," *ibid.*, 95 (Sept. 25, 1913).

[100] N. Bohr, "The spectra of helium and hydrogen," *ibid.*, 231–232 (Oct. 23, 1913).

[101] E. J. Evans, "The spectra of helium and hydrogen," *Philosophical Magazine 29*, 284–297 (1915).

[102] A. Fowler, "Series lines in spark spectra," *Proceedings of the Royal Society of London* (*A*), *90*, 426–430 (1914).

[103] F. Paschen, "Bohr's Helium Linien," *Annalen der Physik 50*, 901–940 (1916).

[104] H. C. Urey, F. G. Brickwedde, and G. M. Murphy, "A hydrogen isotope of mass 2 and its concentration," *Physical Review 40*, 1–15 (1932).

[105] J. Franck and G. Hertz, "Über Zusammenstöße zwischen Elektronen und den Molekülen des Quecksilberdampfes und die Ionisierungsspannung desselben," *Verhandlungen der Deutschen Physikalischen Gesellschaft 16*, 457–467 (1914); "Über Kinetik von Elektronen und Ionen in Gasen," *Physikalische Zeitschrift 17*, 409–416 (1916); "Die Bestätigung der Bohrschen Atomtheorie im optischen Spektrum durch Untersuchungen der unelastischen Zusammenstöße langsamer Elektronen mit Gasmolekülen," *ibid. 20*, 132–143 (1919).

Our analysis of Bohr's contribution of 1913 would not do justice to his work if it confined itself to an exposition of the physical contents without due consideration of its philosophical foundations. First let us note that Bohr's theory was, in general, very favorably accepted. When Einstein, a severe censor as far as physical theoretizing was concerned, was informed in September, 1913, by Hevesy in Vienna that Evans's experiments had confirmed Bohr's ascription of the Pickering lines to helium, he described Bohr's theory as an "enormous achievement" and as "one of the greatest discoveries." In fact, Einstein recognized the importance of Bohr's theory as soon as he became acquainted with it. This happened when, in one of the weekly physics colloquia held in Zurich conjointly by the University and the Institute of Technology, a report on the first part of Bohr's 1913 paper was given shortly after its publication. At the end of the discussion von Laue protested: "This is all nonsense! Maxwell's equations are valid under all circumstances."[106] Whereupon Einstein rose and declared: "Very remarkable! There must be something behind it. I do not believe that the derivation of the absolute value of the Rydberg constant is purely fortuitous."[107]

Also in September, 1913, at the 83d meeting of the British Association for the Advancement of Science, Jeans, in "the most important discussion of Section A, if not of the whole meeting,"[108] gave a summarizing report on the subject of radiation.[109] Stating that "any discussion of the nature of radiation is of necessity inextricably involved in the larger question as to the ultimate form of the laws which govern the smallest processes of nature" and acknowledging the need for "a very extensive revision" of the laws which so far have been believed to be expressible in the form of differential equations, he spoke highly of Bohr's new theory. Although fully aware that this theory contains difficulties, still unsurmounted, "which appear to be enormous," such as the difficulty of explaining the Zeeman effect, Jeans regarded "the series of results obtained [by Bohr as] far too striking to be dismissed merely as accidental." He called Bohr's theory "a most ingenious and suggestive, and I think we must add convincing, explanation of the laws of spectral series."[110] The favorable acceptance of the new ideas was advanced by extremely good coverage of this meeting by the press and, in particular, by the report published in

[106] "Das ist Unsinn, die Maxwellschen Gleichungen gelten unter allen Umständen, ein Elektron auf Kreisbahn muß strahlen." For this quotation as well as the quotation in REF. 107 the author is indebted to Professor F. Tank (Zurich), who attended the colloquium (Letter of May 11, 1964, from Professor Tank to the author).

[107] "Sehr merkwürdig, da muß etwas dahinter sein; ich glaube nicht, daß die Rydbergkonstante durch Zufall in absoluten Werten ausgedrückt richtig herauskommt."

[108] "Physics at the British Association," *Nature 92*, 305 (1913).

[109] J. H. Jeans, "Discussion on Radiation," in *Report of the 83rd Meeting of the British Association for the Advancement of Science*, Birmingham, Sept. 10–17 (London, 1914), pp. 376–386.

[110] *Ibid.*, p. 376.

the Saturday issue of *The Times*[111] with its reference to "Dr. Bohr's ingenious explanation of the hydrogen spectrum," the article in *Nature*[112] which called Bohr's theory "convincing and brilliant," as well as Norman Campbell's résumé on the structure of the atom, also published in *Nature*,[113] which described Bohr's assumptions as "simple, plausible and easily amenable to mathematical treatment" and as leading to results "in exact quantitative agreement with observation."

Bohr himself, however, regarded his theory as merely a "preliminary and hypothetical" way of representing a number of experimental facts which defied all explanation on the basis of ordinary electrodynamics and classical mechanics. In an address delivered before the Physical Society in Copenhagen on December 20, 1913, Bohr[114] declared as his present objective not "to propose an explanation of the spectral laws" but rather "to indicate a way in which it appears possible to bring the spectral laws into close connection with other properties of the elements, which appear to be equally inexplicable on the basis of the present state of the science."[115] In fact, his intention was not to give a satisfactory answer to a definite question but to search for the right question to ask. For it was an established fact for him that ordinary electrodynamics and classical mechanics are inadequate to account for the stability of Rutherford's atomic model and that it is impossible "to obtain a satisfactory explanation of the experiments on temperature radiation with the aid of electrodynamics, no matter what atomic model be employed. The fact that the deficiencies of the atomic model we are considering stand out so plainly is therefore perhaps no serious drawback; even though the defects of other atomic models are much better concealed, they must nevertheless be present and will be just as serious."[116]

Not only did Bohr fully recognize the profound chasm in the conceptual scheme of his theory, but he was convinced that progress in quantum theory could not be obtained unless the antithesis between quantum-theoretic and classical conceptions was brought to the forefront of theoretical analysis. He therefore attempted to trace the roots of this antithesis as deeply as he could. It was in this search for fundamentals that he introduced the revolutionary conception of "stationary" states, "indicating thereby that they form some kind of waiting places between which occurs the emission of the energy corresponding to the various spectral lines."[117] For he clearly realized

[111] "British Association: Problems of Radiation (from our special correspondent)," *The Times* (London), No. 40,316, Sept. 13, 1913, p. 10.

[112] REF. 108.

[113] N. Campbell, "The structure of the atom," *Nature 92*, 586–587 (1913).

[114] N. Bohr, "Om Brintspektret," *Fysisk Tidsskrift 12*, 97–114 (1914); reprinted, *ibid. 60*, 185–202 (1963); English translation, "On the spectrum of hydrogen," in N. Bohr, *The Theory of Spectra and Atomic Constitution* (Cambridge University Press, 1922), pp. 1–19.

[115] *Ibid.*, p. 100; p. 4.

[116] *Ibid.*, p. 105; pp. 9–10.

[117] *Ibid.*, p. 107; p. 11.

that Planck's idea of a discontinuity in energy emission—the very idea which probably suggested to him the notion of "stationary states" as "waiting places" ("Holdepladser")—was foreign to classical physics, irreconcilable with it, but nevertheless legitimate and necessary. It was in this search for fundamentals that he generalized the derivation of Rydberg's constant from its first version, based on Planck's formula, to its last formulation, based on "a connection with the ordinary conceptions."[118] This analogy, however, he stressed time and again, should not be misleadingly interpreted as an analogy in the foundations. For "nothing has been said here about how and why the radiation is emitted."[119] Concluding his address, Bohr declared: "I hope I have expressed myself sufficiently clearly so that you appreciate the extent to which these considerations conflict with the admirably coherent group of conceptions which have been rightly termed the classical theory of electrodynamics. On the other hand, by emphasizing this conflict, I have tried to convey to you the impression that it may also be possible in the course of time to discover a certain coherence in the new ideas."[120]

We thus see that, contrary to Planck and Einstein, Bohr did not try to bridge the abyss between classical and quantum physics, but from the very beginning of his work, searched for a scheme of quantum conceptions which would form a system just as coherent, on the one side of the abyss, as that of the classical notions on the other side of the abyss.[121]

[118] *Ibid.*, p. 108; p. 13.

[119] *Ibid.*, p. 108; pp. 12–13.

[120] *Ibid.*, p. 114; p. 19.

[121] For a penetrating analysis of this aspect of Bohr's work, from the viewpoint of the Copenhagen interpretation, cf. K. M. Meyer-Abich, *Korrespondenz, Individualität und Komplementarität* (Dissertation, University of Hamburg, 1964), published as vol. 5 in the series "Boethius—Texte und Abhandlungen zur Geschichte der exakten Wissenschaften," edited by J. E. Hofmann, F. Klemm, and B. Sticker (Franz Steiner Verlag, Wiesbaden, 1965).

3 The Older Quantum Theory

3.1 Quantum Conditions and the Adiabatic Principle

Although the declared aim of Bohr's classic paper of 1913, as its title implied, was the development of a general theory on the constitution of atoms and molecules, Bohr's theory afforded a rigorous and adequate explanation only of hydrogen and hydrogen-like atoms, and all his immediate efforts at extending the theory to systems with more than one electron failed. Even at that time the spectrum of neutral helium with its two electrons could not be accounted for, nor was it possible to find a reasonable arrangement of the electron orbits from which the first ionization potential could be deduced in agreement with observation.

It will be recalled that Bohr referred at the beginning of his work to elliptic orbits of the electrons but later dealt almost exclusively with the special case of circular motion. He also confined his discussion to the nonrelativistic case assuming that the velocity of the electron is small compared with that of light. However, in 1914 when Curtis[1] found a small systematic disagreement between Bohr's theoretical values of the wavelengths of the hydrogen lines and their observed values, Bohr[2] introduced the relativistic variability of mass and derived the formula

$$\nu = \frac{2\pi^2 e^4 mM}{h^3(m + M)} \left(\frac{1}{n_1^2} - \frac{1}{n_2^2}\right)\left[1 + \frac{\pi^2 e^4}{c^2 h^2}\left(\frac{1}{n_1^2} + \frac{1}{n_2^2}\right)\right] \tag{3.1}$$

where e and m are the charge and the mass of the electron, M the mass of the nucleus, and n_1, n_2 positive integers, and where terms involving higher powers than the second of the ratio between the velocity of the

[1] W. E. Curtis, "Wavelengths of hydrogen lines and determination of the series constant," *Proceedings of the Royal Society of London* (*A*), *90*, 605–620 (1914).

[2] N. Bohr, "On the series spectrum of hydrogen and the structure of the atom," *Philosophical Magazine* *29*, 332–335 (1915).

electron and the velocity of light are neglected. The correction term depending on the sum of the inverse squares of the integers had the same sign as the deviation from the Balmer law found by Curtis but accounted for only one-third of the deviations observed.

The main problem which physics had to face at that time, however, was the question of how to extend the theory to more complex systems. In fact, as early as 1911 Poincaré had raised this question during a discussion at the Solvay Congress[3] when he asked how Planck's treatment of the harmonic oscillator and its quantum condition, according to which the elementary region of the a priori probability in the pq plane was determined by the equation $\iint p\, dq = h$, should be extended to systems with more than one degree of freedom. Planck, in reply, expressed his conviction that the formulation of a quantum theory for systems with more than one degree of freedom would soon be possible.[4] Four years later the answer was found primarily by Planck and by Sommerfeld.

Planck[5] solved the problem for dynamic systems of f degrees of freedom whose equations of motion admit f regular integrals. Generalizing his treatment of the harmonic oscillator, he divided the phase space by means of the surfaces $F(p_k, q_k) = \text{const}$, defined by the integrals, into regions of volume h^f and postulated that the stationary states correspond to the f-dimensional intersections of these surfaces. It was later shown by Epstein[6] and by Kneser[7] that Planck's conditions for the characterization of the stationary states are equivalent to the quantum conditions as formulated by Sommerfeld and obtained by a completely different approach.

For a full comprehension of Sommerfeld's approach the conceptual situation of the quantum theory in the years 1913 to 1915 has to be reviewed. The outstanding success of Bohr's theory as far as hydrogenic atoms are concerned suggested that it would be advisable to study not only the content of the theory but also its methodological approach. This study, now, made it increasingly clear that the quantum-theoretic treatment of a dynamical system consisted of three parts: first, the application of classical mechanics for the determination of the possible motions of the system; second, the imposition of certain quantum conditions for the selection of the actual or allowed motions; and, third, the treatment

[3] REF. 208 OF CHAP. 1, p. 120.

[4] Planck: "Une hypothèse des quanta pour plusieurs degrés de liberté n'a pas encore été formulée, mais je ne crois nullement impossible d'y parvenir." *Ibid.*, p. 120.

[5] M. Planck, "Die Quantenhypothese für Molekeln mit mehreren Freiheitsgraden," *Verhandlungen der Deutschen Physikalischen Gesellschaft 17*, 407–418, 438–451 (1915); *Physikalische Abhandlungen und Vorträge* (REF. 42 OF CHAP. 1), vol. 2, pp. 349–360, 362–375. "Die physikalische Struktur des Phasenraumes," *Annalen der Physik 50*, 385–418 (1916); *Physikalische Abhandlungen und Vorträge*, vol. 2, pp. 386–419.

[6] P. S. Epstein, "Über die Struktur des Phasenraumes bedingt periodischer Systeme," *Berliner Berichte 1918*, pp. 435–446.

[7] H. Kneser, "Untersuchungen zur Quantentheorie," *Mathematische Annalen 84*, 277–302 (1921).

of the radiative processes as transitions between allowed motions subject to the Bohr frequency formula. It was hoped that by applying this scheme of ideas to systems of arbitrarily many degrees of freedom, a systematic method of bringing abstract order to the increasing wealth of experimental results could be obtained. The detailed elaboration of this peculiar synthesis of classical and quantal conceptions, usually referred to as "the older quantum theory," led eventually to the establishment of two general principles which may be regarded as the foundations of the theory: the adiabatic principle and the correspondence principle.

The first major achievement of the older quantum theory was Arnold Sommerfeld's generalization of Bohr's theory of the hydrogen atom.[8] As early as 1891 Michelson[9] had discovered that the Balmer series was not composed of truly single lines. This discovery, incompatible, of course, with Bohr's theory, was either ignored or not regarded as a weighty argument against Bohr's theory in view of the small order of magnitude involved. Sommerfeld, however, suspected that Bohr's analysis of the hydrogen atom was only approximately correct, based as it was on only one quantum condition, the quantization of the angular momentum. Sommerfeld therefore hoped that a generalization to two degrees of freedom, corresponding to the two-dimensional motion of the electron in its orbital plane, would lead to a fuller agreement with experience and at the same time indicate how systems with more than one degree of freedom have to be treated.

Recognizing that the quantization of the angular momentum could be expressed, in terms of plane coordinates, by the requirement that $\oint p_\varphi \, d\varphi = n_\varphi h$, where p_φ is the (angular) momentum corresponding to the azimuthal angle φ and where the integration is extended over the period of φ, Sommerfeld postulated that the stationary states of a periodic system with f degrees of freedom are determined by the conditions that "the phase integral for every coordinate is an integral multiple of the quantum of action"[10] or that for $k = 1, 2, \ldots, f$

$$\oint p_k \, dq_k = n_k h \tag{3.2}$$

where p_k is the momentum corresponding to the coordinate q_k, n_k is a non-negative integer, and the integration is extended over a period of q_k.

These "Sommerfeld conditions," as the quantum conditions (3.2) were subsequently called, were also obtained—even a few weeks before their

[8] A. Sommerfeld, "Zur Theorie der Balmerschen Serie," *Münchener Berichte 1915*, pp. 425–458; "Die Feinstruktur der wasserstoff- und wasserstoffähnlichen Linien," *ibid.*, pp. 459–500; "Zur Quantentheorie der Spektrallinien," *Annalen der Physik 51*, 1–94, 125–167 (1916).

[9] A. A. Michelson, "On the application of interference-methods to spectroscopic measurements," *Philosophical Magazine 31*, 338–346 (1891), *34*, 280–299 (1892).

[10] REF. 8, p. 9.

establishment by Sommerfeld—by Wilson[11] and, independently, by Ishiwara[12] in their attempts to develop a general theory to cover in a unified way Planck's treatment of the harmonic oscillator and Bohr's discussion of the hydrogen atom. It was, of course, immediately obvious that the generalized quantum conditions (3.2) comprise in a single formula the correct quantization for these dynamic systems.[13]

Wilson considered a multiply periodic system whose kinetic energy is a linear function of the squares of the generalized velocities $\dot{q}_k$. Consequently, $2L = \sum \dot{q}_k(\partial L/\partial \dot{q}_k) = 2 \sum L_k$, where $2L_k = \dot{q}_k p_k$. Hence $2\int L_k\, dt = \int p_k\, dq_k$, where the integration is extended over the period $1/\nu_k$ corresponding to $\dot{q}_k$. Wilson now formulated his quantum conditions as follows: "The discontinuous energy exchanges always occur in such a way that the steady motions satisfy the equations: $\int p_k\, dq_k = n_k h$, where n_k are positive integers (including zero) and the integrations are extended over the values p_k and q_k corresponding to the period $1/\nu_k$."[14] In addition Wilson postulated the two following hypotheses: "(1) Interchanges of energy between dynamical systems and the aether, or between one dynamical system and another are 'catastrophic' or discontinuous. . . . (2) The motion of a system in the intervals between such discontinuous energy exchanges is determined by Hamiltonian dynamics as applied to conservative systems." Ishiwara formulated his quantum conditions as follows: "In nature, motions always occur in such a way that every phase plane $p_i q_i$ can be divided in those elementary regions of equal probability whose average value in a definite point of the phase space $h = (1/j) \sum_{i=1}^{j} \int q_i\, dp_i$ is equal to a universal constant." In contrast to Wilson and Ishiwara, who never applied their theory to the calculation of atomic spectra, Sommerfeld used his conditions for a relativistic treatment of the hydrogen atom. Resulting in the famous theory of the fine structure of hydrogen, Sommerfeld's discovery, in full agreement with observation, made a great impression at the time and overshadowed the work of the others.

According to classical mechanics the motion of the electron in the hydrogen atom is an elliptic orbit in a fixed plane. "Guided by the experience of the Balmer series"[15] as well as by Debye's extension[16] of Planck's treatment of the harmonic oscillator to arbitrary periodic motions of one degree of freedom, Sommerfeld regarded the Kepler motion as a two-dimensional

[11] W. Wilson, "The quantum theory of radiation and line spectra," *Philosophical Magazine 29*, 795–802 (1915).

[12] J. Ishiwara, "Die universelle Bedeutung des Wirkungsquantums," *Toyko Sugaku Buturigakkawi Kizi 8*, 106–116 (1915).

[13] For the oscillator

$$nh = 2\int_{q_1}^{q_2} [2m(E - \tfrac{1}{2}kq^2)]^{1/2}\, dq = 2\pi\left(\frac{m}{k}\right)^{1/2} = \frac{E}{\nu}$$

where q_1 and q_2 are the roots of the equation $E - \frac{1}{2}kq^2 = 0$.

[14] REF. 11, p. 796.

[15] REF. 8, p. 6.

[16] P. Debye, "Zustandsgleichung und Quantenhypothese," in *Vorträge über die kinetische Theorie der Materie* (Teubner, Leipzig, 1914), p. 27.

problem. Introducing plane polar coordinates, the radius vector r and the azimuthal angle ψ in the plane of the orbit, he postulated that not only ψ but also r has to be subjected to quantum conditions.[17] From the radial and azimuthal quantum conditions

$$\oint p_r\, dr = n'h \qquad \text{and} \qquad \oint p_\psi\, d\psi = kh \tag{3.3}$$

where n' is the radial quantum number and k the azimuthal quantum number, he deduced that $k/(k + n') = b/a$, where b is the minor and a the major semiaxis of the ellipse, and showed that the energy of the corresponding stationary orbit is given by the equation $E = -Rhc/(k + n')^2$ or, for hydrogen-like atoms with $Z > 1$, by $E = -RhcZ^2/(k + n')^2$. Since the case $k = 0$, corresponding to linear orbits through the nucleus at the origin, had to be excluded, the set of values for $k + n'$ coincided with the set of values for τ in Bohr's formula $E = -Rhc/\tau^2$ for circular orbits. Thus in spite of the greater multiplicity of orbital shapes, Sommerfeld's theory so far did not lead to any additional energy levels.

Disappointed by this result, Sommerfeld treated the same problem as a system of three degrees of freedom. For this purpose he introduced the polar coordinates r, θ, and φ with the nucleus as origin, where r as before denoted the radius vector, θ the latitude (or more precisely the colatitude) with reference to a given polar axis, and φ the azimuth. From the quantum conditions $\oint p_r\, dr = n'h$, $\oint p_\varphi\, d\varphi = n_1 h$, and $\oint p_\theta\, d\theta = n_2 h$, where n', n_1, and n_2 are the radial, equatorial, and latitudinal quantum numbers respectively, and from the comparison of the expression for the kinetic energy in terms of r, θ, φ with its expression in terms of r and ψ, he found that $k = n_1 + n_2$. Since the total angular momentum $p_\psi = kh/2\pi$ is normal to the orbital plane and its projection upon the polar axis is[18] p_φ, Sommerfeld concluded that $n_1 = k \cos \alpha$ or $\cos \alpha = n_1/(n_1 + n_2)$, where α is the angle between the direction of p_ψ and the polar axis. This equation indicated the existence of discrete inclinations of the orbital plane with respect to the polar axis, a relationship which was referred to as "space quantization" and which was physically significant whenever the polar axis was uniquely determined, as, for instance, by the direction of a magnetic field.

Yet even this treatment of the hydrogen atom as a system of three degrees of freedom did not increase the number of energy levels. The reason, as we shall see presently, was the two-fold degeneracy of the

[17] In his Munich lectures Sommerfeld used to say: "Was dem ψ recht ist, ist dem r billig!" Sommerfeld had already employed the term "quantum number" in 1916 whereas Bohr, as late as 1918, spoke only of "entire numbers" or "integers."

[18] This can be shown as follows. The Hamiltonian is given by

$$\frac{1}{2m}\left(p_r^2 + \frac{p_\theta^2}{r^2} + \frac{p_\varphi^2}{r^2 \sin^2 \theta}\right) - \frac{e^2}{r}$$

so that $\dot{\varphi} = p_\varphi/(mr^2 \sin^2 \theta)$ or $p_\varphi = m(r \sin \theta)(r \sin \theta \dot{\varphi}) =$ mass $\cdot$ (projection of the radius vector on the equatorial plane) $\cdot$ (rotational velocity of the end point of the projected radius vector).

system which Sommerfeld treated as a system of three degrees of freedom but which had only one degree of periodicity.

Finally, in the second part of his paper[19] Sommerfeld treated the problem relativistically. He showed that as in the case of every periodic motion under the influence of a central force, the electron with rest mass m describes a rosette or, more precisely, an ellipse with a slowly precessing perihelion and with one of its foci at the nucleus. Originally Sommerfeld solved the problem by a relativistic generalization of the nonrelativistic treatment of the Kepler motion. Later[20] he realized that the "royal road" for solving the problem is the method of separating the variables in the corresponding Hamilton-Jacobi differential equation. As shown by the theory of relativity, the Hamiltonian, which in this case is also the total energy E, is given by the expression

$$E = mc^2(\gamma^{-1} - 1) - \frac{Ze^2}{r}$$

where $\gamma = (1 - v^2/c^2)^{1/2}$, v is the velocity of the electron, and where plane polar coordinates r and ψ are used as in (3.3). Since $p_r{}^2 + p_\psi{}^2/r^2 = \gamma^{-2}m^2v^2$, the Hamilton-Jacobi equation reads

$$\left(\frac{\partial S}{\partial r}\right)^2 + \frac{1}{r^2}\left(\frac{\partial S}{\partial \psi}\right)^2 = 2mE + \frac{2mZe^2}{r} + \frac{1}{c^2}\left(E + \frac{Ze^2}{r}\right)^2$$

It can be separated in the coordinates r and ψ and only in these. The coordinate ψ being cyclic (ignorable), p_ψ is a constant and the azimuthal quantum condition yields $2\pi p_\psi = kh$. On the other hand, the radial quantum condition, in which, because of the advance of the perihelion, the range of integration extends from $\psi = 0$ to $\psi = 2\pi/\sqrt{1 - Z^2/p_\psi{}^2}$, leads to the equation $-2\pi i(\sqrt{C} - B/\sqrt{A}) = n'h$, where $C = -p_\psi{}^2 + Z^2e^4/c^2$, $B = mZe^2 + Ze^2E/c^2$, and $A = 2mE + E^2/c^2$. Introducing the fine-structure constant $\alpha = 2\pi e^2/hc$ and eliminating p_ψ, Sommerfeld obtained for the total energy $E = E_{nk}$ the expression

$$E_{nk} = mc^2\{[1 + \alpha^2Z^2(n - k + \sqrt{k^2 - \alpha^2Z^2})^{-2}]^{-1/2} - 1\}$$

where $n = n' + k$. Expanded in powers of α up to α^2

$$E_{nk} = -Z^2Rhc\left[\frac{1}{n^2} + \frac{\alpha^2Z^2}{n^4}\left(\frac{n}{k} - \frac{3}{4}\right)\right] \tag{3.4}$$

which for circular orbits ($n = k$) and $Z = 1$ agrees with Bohr's result (3.1).

The relativistic correction term, that is, the term containing the factor

[19] REF. 8, pp. 44–94.

[20] A. Sommerfeld, *Atombau und Spektrallinien* (Vieweg, Braunschweig, 1919), pp. 327–357, 520–522; *Atomic Structure and Spectral Lines*, translated from the third (1922) German edition by H. L. Brose (Methuen, London, 1923), pp. 467–496, 608–611. This book, a classic on the older quantum theory and by far the most-read text at its time, originated from Sommerfeld's lectures on atomic models during 1916/17 at the University of Munich.

α^2, is a function of n and k and the energy levels are therefore indeed multiplets. Sommerfeld obtained the result he expected: a theoretical account of the fine structure of hydrogen lines. As formula (3.4) shows, it should have been easier, owing to the factor Z^4, to observe the fine structure in the spectrum of ionized helium than in that of hydrogen. In fact, Paschen's precision measurements of the helium spectrum[21] appeared to be in perfect quantitative agreement with Sommerfeld's prediction, and Michelson's early discovery found its explanation. Incidentally, Paschen's confirmation of Sommerfeld's theory served also as an indirect confirmation of Einstein's relativistic formula for the velocity dependence of inertial mass. For Glitscher,[22] following a suggestion by W. Lenz, showed that the alternative mass-variation formula,[23] proposed by Abraham in accordance with his theory of the "rigid electron," would have implied a splitting of the helium lines radically different from that actually observed by Paschen.

Sommerfeld's theory of the fine structure soon became the subject of extensive studies and detailed elaborations. In his dissertation[24] entitled "Intensities of Spectral Lines" Kramers calculated the relative intensities of the fine-structure components on the basis of Bohr's correspondence principle (which will be discussed in Sec. 3.2) and obtained results which generally agreed with Paschen's measurements. A notable exception, as pointed out by Goudsmit and Uhlenbeck[25] and also by Landé,[26] was the component $4_1 - 3_1$, that is, the transition from the energy level $n = 4$, $k = 1$ to the level $n = 3$, $k = 1$ in the fine structure of helium and, as Hansen[27] observed, also in that of hydrogen. The selection rule $\Delta k = \pm 1$ forbids the appearance of this component and yet according to Paschen's measurements, the line had a relatively high intensity. Any explanation of its appearance on the basis of the Stark effect, which, as is known, invalidates the selection rule, was shown to be untenable.

In an attempt to render Sommerfeld's approach more rigorous from the relativistic point of view, Darwin[28] replaced the ordinary Coulomb potential in Sommerfeld's expression for the total energy by a retarded

[21] F. Paschen, "Bohr's Helium Linien," *Annalen der Physik 50*, 901–940 (1916); cf. also E. Laue, "Über die Frage der Feinstruktur ausgewählter Spektrallinien," *Physikalische Zeitschrift 25*, 60–68 (1924).

[22] K. Glitscher, "Spektroskopischer Vergleich zwischen den Theorien des starren und des deformierbaren Elektrons," *Annalen der Physik 52*, 608–630 (1917).

[23] Cf. M. Jammer, *Concepts of Mass in Classical and Modern Physics* (Harvard University Press, Cambridge, Mass., 1961; Harper and Brothers, New York, 1964), chap. 12, pp. 154–171.

[24] H. A. Kramers, "Intensities of spectral lines," REF. 112.

[25] S. Goudsmit and G. E. Uhlenbeck, "Opmerking over de spectra van waterstof en helium," *Physica 5*, 266–270 (1925).

[26] Cf. W. Pauli, "Quantentheorie," in *Handbuch der Physik*, edited by H. Geiger (Springer, Berlin, 1926), vol. 23, p. 128.

[27] G. Hansen, "Die Feinstruktur der Balmerlinien," *Annalen der Physik 78*, 558–600 (1925). Cf. also N. A. Kent, L. B. Taylor, and H. Pearson, "Doublet separation of fine structure of the Balmer lines of hydrogen," *Physical Review 30*, 266–283 (1927).

[28] C. G. Darwin, "The dynamical motions of charged particles," *Philosophical Magazine 39*, 537–551 (1920).

potential. Wilson's suggestion[29] to reformulate the quantum conditions in accordance with the general theory of relativity by postulating that $\int (P_s + eA_s)\, dq_s = n_s h$ for $s = 1, 2, 3$, and 4, where A_s is the vector potential, was shown by Richardson[30] to lead to inconsistencies after a similar attempt by Kar[31] had been shown by Smekal[32] to be untenable. Attempts at reconciling the quantum conditions with the general theory of relativity were made by Schrödinger[33] and, in particular, by Wereide,[34] who, encouraged by Bohr and Kramers, systematically studied the possibility of applying relativity to the Rutherford-Bohr model of the atom. All these relativistic considerations, however, were soon invalidated by the subsequent discovery of the spin, and thus had little impact on the later development of the theory.

The generalized quantum conditions (3.2) were for Sommerfeld the ultimate foundations of quantum theory or, as he expressed it, statements "unproved and perhaps incapable of being proved."[35] Strictly speaking, Sommerfeld was right as far as contemporary physics was concerned. For only in 1926, after the advent of wave mechanics, could Sommerfeld's conditions be rigorously deduced as approximations with the help of the Wentzel-Kramers-Brillouin (WKB) method.

Sommerfeld's conditions were not so arbitrary as they seemed to be at first. In fact, their deeper physical significance was soon revealed with the establishment of the "adiabatic principle," which, as stated previously, was one of the conceptual foundations of the older quantum theory. The origin of this principle and its role in the development of the older quantum theory will now be discussed.

The concept of "adiabatic changes" in purely mechanical systems originated in the early attempts[36] of Boltzmann[37] and Clausius[38] to reduce

[29] W. Wilson, "The quantum theory and electromagnetic phenomena," *Proceedings of the Royal Society of London* (*A*), *102*, 478–483 (1922).

[30] O. W. Richardson, "The generalized quantum conditions," *Philosophical Magazine 46*, 911–914 (1923).

[31] S. C. Kar, "On the theory of generalized quanta and the Balmer lines," *Philosophical Magazine 45*, 610–621 (1923).

[32] *Physikalische Berichte 4*, 1082–1083 (1923).

[33] E. Schrödinger, "Über eine bemerkenswerte Eigenschaft der Quantenbahnen eines einzelnen Elektrons," *Zeitschrift für Physik 12*, 13–23 (1922).

[34] T. Wereide, "The general principle of relativity applied to the Rutherford-Bohr atom-model," *Physical Review 21*, 391–396 (1923). Cf. also K. Ogura, "Sur le mouvement d'une particle dans le champ d'un noyeau chargé," *Japanese Journal of Physics 3*, 85–94 (1924); M. von Laue, "G. A. Schotts Form der relativistischen Dynamik und die Quantenbedingungen," *Annalen der Physik 73*, 190–194 (1924).

[35] REF. 8, p. 6.

[36] The earliest attempt of this kind was probably W. J. M. Rankine's "On the hypothesis of molecular vortices, or centrifugal theory of elasticity, and its connexion with the theory of heat," *Philosophical Magazine 10*, 354–363, 411–420 (1855).

[37] L. Boltzmann, "Über die mechanische Bedeutung des zweiten Hauptsatzes der Wärmetheorie," *Wiener Berichte 53*, 195–220 (1866); "Analytischer Beweis des zweiten Hauptsatzes der mechanischen Wärmetheorie aus den Sätzen über das Gleichgewicht der lebendigen Kraft," *ibid. 63*, 712–732 (1871).

[38] R. Clausius, "Über die Zurückführung des zweiten Hauptsatzes der mechanischen Wärmetheorie auf allgemeine mechanistische Principien," Memoir read on Oct. 9, 1870, before the Nieder-

the second law of thermodynamics to pure mechanics. Boltzmann[39] showed that, provided the laws governing the behavior of a mechanical system are subject to the principle of least action, the kinetic energy (in analogy to the addition of heat dQ) supplied to a simply periodic system of period τ is given by the expression $2d(\tau \bar{E}_{\text{kin}})/\tau$, where $\bar{E}_{\text{kin}} = (1/\tau) \int_0^\tau E_{\text{kin}}\, dt$ is the average kinetic energy of the system. Hence, if no external energy is supplied (corresponding to the adiabatic condition $dQ = 0$), the ratio $\bar{E}_{\text{kin}}/\nu$ is an invariant. At the same time, Clausius emphasized in this context the importance of the study of rheonomous systems acted upon by external forces whose law of action varies with time and whose potentials contain parameters which vary slowly compared with the natural periods of the system. In such processes the motions determining coordinates of the system were not directly affected by the external forces, just as in an adiabatic change of state in thermodynamics the coordinates which determine the heat motion are not affected by external agencies. The introduction of these ideas into the theory of so-called cyclic (in the topological sense of the word!) systems as studied primarily at the end of the century by Helmholtz[40] and Hertz[41] led to the definition of an "adiabatic motion" as the motion of a system which is acted upon by external forces in such a manner that these forces do not affect directly the coordinates of the system while certain parameters undergo slow variations.[42]

In 1902 Lord Rayleigh pointed out that in certain sinusoidally oscillating systems—such as a simple pendulum with its string being slowly shortened or a transversally vibrating string being slowly covered by a narrow ring or the standing waves in a slowly contracting cavity—such adiabatic motions take place and that during the process the ratio between the energy and the frequency remains unchanged.[43] At the Solvay Congress

---

rheinische Gesellschaft für Natur- und Heilkunde; *Poggendoff's Annalen der Physik 142*, 433–461 (1871). "On the reduction of the second axiom of the mechanical theory of heat to general mechanical principles," *Philosophical Magazine 42*, 161–181 (1871).

[39] L. Boltzmann, *Vorlesungen über die Principe der Mechanik* (J. A. Barth, Leipzig, 1904), vol. 2, sec. 48, p. 182.

[40] H. von Helmholtz, "Principien der Statik monozyklischer Systeme," *Journal für die reine und angewandte Mathematik 97*, 112–140, 317–336 (1884); "Studien zur Statik monozyklischer Systeme," *Berliner Berichte 1884*, pp. 159–177, 311–318, 755–789, 1197–1201.

[41] H. Hertz, *Principien der Mechanik in neuem Zusammenhang dargestellt* (Leipzig, 1894), sec. 560; *The Principles of Mechanics Presented in a New Form* (Dover, New York, 1956), p. 213.

[42] For a readable discussion of these aspects of classical mechanics see A. Brill, *Vorlesungen zur Einführung in die Mechanik raumfüllender Massen* (Teubner, Leipzig, Berlin, 1909).

[43] Lord Rayleigh, "On the pressure of radiation," *Philosophical Magazine 3*, 338–346 (1902); *Scientific Papers* (REF. 73 OF CHAP. 1), vol. 5, pp. 41–48. For more recent discussions on adiabatic invariants cf. W. B. Morton, "Simple examples of adiabatic invariance," *Philosophical Magazine 8*, 186–194 (1929), which treats, among other examples, the adiabatic invariant associated with Keplerian motion accompanied by an accretion of the orbiting mass and a gradual change in the strength of the center of force. Cf. also P. L. Bhatnagar and D. S. Kothari, "A note on the principle of adiabatic invariance," *Indian Journal of Physics 16*, 272–275 (1942), where adiabatic invariants for systems such as the compound pendulum, an oscillating magnet, and an electric oscillating circuit are discussed. For a simple mathematical analysis see

in 1911 Lorentz raised the question whether a quantized pendulum whose string is being shortened remained in its quantized state. Einstein answered unhesitatingly: "If the length of the pendulum is changed infinitely slowly, its energy remains equal to $h\nu$ if it was originally $h\nu$."[44]

At the same time Paul Ehrenfest already recognized the fundamental importance of the concept of adiabatic invariance for quantum theory. In his study of the statistical foundations of Planck's radiation law, a study which he began as early as 1906 if not earlier,[45] Ehrenfest was puzzled by the apparent paradox that Wien's displacement law "which is wholly derived from *classical* foundations" remained "unshaken in the midst of the world of radiant phenomena whose *anticlassical* character stood out ever more inexorably."[46] Wien's law, it will be recalled, asserted[47] the proportionality between the energy density $u_\nu(\nu,T)$ and $\nu^3F(\nu/T)$ or, since u_ν referred to $8\pi\nu^2/c^3$ proper vibrations within the frequency interval $d\nu$ at ν, the equality between the energy ϵ_ν of each proper vibration of frequency ν and $\nu\Phi(\nu/T)$, where $\Phi(\nu/T)$ differed from $F(\nu/T)$ merely by a constant factor. Writing this relation as $\epsilon_\nu/\nu = \Phi(\nu/T)$, Ehrenfest understood that Wien's displacement law established a relation between two adiabatic invariants,[48] ϵ_ν/ν and ν/T. But Ehrenfest now also recognized the reason why Wien's original derivation of his displacement law, based on an adiabatical and infinitely slow compression of the radiation cavity, remained valid in quantum theory. For, ϵ_ν/ν being an adiabatic invariant, Planck's quantum condition $\epsilon_\nu = nh\nu$ for each proper oscillation, if satisfied initially, will be satisfied at every moment during the process of compression. Hence the validity of Wien's law in quantum theory was assured by the fact that the ratio ϵ_ν/ν for the sinusoidal vibrations before the adiabatic change equals the ratio ϵ_ν'/ν' for the modified sinusoidal vibrations after the adiabatic change. Referring to the former state of vibrations as the "undeformed" state or motion and to the latter as the "deformed" state or motion, Ehrenfest pointed out that Wien's law implied that in the course of an adiabatic transformation an allowed (or stationary) undeformed

---

S. Tomonaga, *Quantum Mechanics* (North-Holland Publishing Company, Amsterdam, 1962), vol. 1, pp. 21–24, 290–298.

[44] REF. 208 OF CHAP. 1, p. 450.

[45] M. J. Klein, "The origins of Ehrenfest's adiabatic principle," *Proceedings of the Xth International Congress of History of Science, Cornell University, 1962* (Hermann, Paris, 1964), pp. 801–804.

[46] P. Ehrenfest, "Adiabatische Transformationen in der Quantentheorie und ihre Behandlung durch Niels Bohr," *Die Naturwissenschaften 11*, 543–550 (1923); *Collected Scientific Papers* (REF. 77 OF CHAP. 1), pp. 463–470.

[47] See p. 8.

[48] That ν/T is an adiabatic invariant can be seen as follows. For a cubic radiation cavity of volume V and energy density u it follows from the equation, radiation pressure $= -u/3 = \delta(uV)/\delta V$, that $u \propto V^{-4/3}$. On the other hand, by the Stefan-Boltzmann law, $u \propto T^4$. Therefore $T \propto V^{-1/3}$. Finally, since $\nu \propto V^{-1/3}$, as shown by the boundary conditions used to enumerate the number of proper vibrations, ν/T is adiabatically invariant. In fact, the preceding considerations afford a most convenient proof of the displacement law. Cf. L. Brillouin, *La Théorie des Quanta et l'Atom de Bohr* (Paris, 1922), pp. 177–178.

motion changes into an allowed deformed motion while the adiabatic invariant retains its initial value. The assumption that this conclusion applies quite generally, and not only for sinusoidal motion, forms the content of what Ehrenfest, following a suggestion of Einstein,[49] called the "adiabatic principle." The first full exposition of this principle, with reference to Boltzmann's "mechanical theorem" mentioned previously, appeared in a paper[50] which Ehrenfest published in 1913, shortly after his appointment as Lorentz's successor in Leyden. Three years later, in a comprehensive essay[51] on this subject, Ehrenfest formulated the principle as follows. Let the coordinates of the system, described by $q_1, \ldots, q_n$, contain certain "parameters" $a_1, a_2, \ldots, a_n$, the values of which can be altered slowly, and let the kinetic energy of the system be a homogeneous quadratic function of the velocities $\dot{q}_1, \dot{q}_2, \ldots, \dot{q}_n$, the coefficients of which are functions of the $q_1, \ldots, q_n$ and possibly also of the $a_1, \ldots, a_n$. If an infinitely slow change of the parameters transforms motion $\beta(a)$ into another motion $\beta(a')$, $\beta(a)$ and $\beta(a')$ are called "adiabatically related to each other" and the change affected "a reversible adiabatic affection." If it is assumed that $\beta(a_0)$, corresponding to the particular values $a_{10}, a_{20}, \ldots, a_{n0}$ of the parameters, is an allowed motion, then the adiabatic principle asserts that for general values of $a_1, a_2, \ldots$ of the parameters those and only those motions $\beta(a)$ are allowed which are adiabatically related to $\beta(a_0)$, that is, those motions which can be transformed into, or may be derived from, allowed motions by a reversible adiabatic affection.

For Ehrenfest, the main importance of the adiabatic principle—apart from the fact that it determined the formal applicability of classical mechanics for quantum theory in the case of time-dependent parameters—was its use for the determination of the stationary states of deformed systems which are adiabatically related to undeformed systems with known quantum conditions. Thus, as Ehrenfest pointed out, the adiabatic principle made it possible to determine the allowed motions of any periodic system of one degree of freedom if it is adiabatically related to the sinusoidal oscillator. For in order to find the quantum condition for the deformed system, it was only necessary to find some function J of the parameters

[49] A. Einstein, "Beiträge zur Quantentheorie," *Verhandlungen der Deutschen Physikalischen Gesellschaft 16*, 820–828 (1914). Einstein spoke of the "Adiabatenhypothese" (p. 826).

[50] P. Ehrenfest, "Een mechanische theorema van Boltzmann en zijne betrekking tot de quanta theorie," *Verslag van de Gewoge Vergaderingen der Wis-en Natuurkundige Afdeeling, Amsterdam, 22*, 586–593 (1913); "A mechanical theorem of Boltzmann and its relation to theory of energy quanta," *Proceedings of the Amsterdam Academy 16*, 591–597 (1914); *Collected Scientific Papers* (REF. 77 OF CHAP. 1), pp. 340–346.

[51] P. Ehrenfest, "Adiabatische Invarianten und Quantentheorie," *Annalen der Physik 51*, 327–352 (1916); "Adiabatic invariants and the theory of quanta," *Philosophical Magazine 33*, 500–513 (1917); *Collected Scientific Papers* (REF. 77 OF CHAP. 1), pp. 378–399.

and constants of motion which is invariant under the transformation and which coincided with ϵ_ν/ν, the ratio between the energy and the frequency of the oscillator, at the beginning of the transformation. Once such an adiabatic invariant is found, the quantum condition for the deformed system reads $J = nh$, where n is a nonnegative integer.

The existence of adiabatic invariants for systems of f degrees of freedom was proved in statistical mechanics as early as 1910 by P. Hertz,[52] who showed that the $2f$-dimensional volume $V = \int \cdots \int dq_1 \cdots dp_n$ which in phase space is enclosed by the $(2f - 1)$-dimensional hypersurface $\epsilon(q,p,a) = \text{const}$ is invariant under reversible adiabatic transformations. For $f = 1$, Ehrenfest pointed out, this result coincided with Boltzmann's "mechanical theorem." For in this case

$$\frac{2\bar{E}_{\text{kin}}}{\nu} = 2\int_0^\tau E_{\text{kin}}\, dt = \oint \dot{q}\left(\frac{\partial E_{\text{kin}}}{\partial \dot{q}}\right) dt = \oint \dot{q}p\, dt = \iint dq\, dp = V$$

Since for the harmonic oscillator $2\bar{E}_{\text{kin}} = \epsilon_\nu = nh\nu$ and the adiabatic invariant $J = \epsilon_\nu/\nu = nh$, the quantum condition for any periodic system of one degree of freedom which is adiabatically related to the oscillator is, according to Ehrenfest, $\oint p\, dq = nh$. Although the adiabatic principle thus seemed at first to afford a systematic approach to the quantum-theoretic treatment of a great variety of dynamical systems, its practical application quite frequently met with insurmountable difficulties. For the method was of no avail whenever the adiabatic transformation had to involve a state in which a period of motion tends to infinity (as, for example, in the transformation from a pendulum to a rotator). For in such a case the condition of an infinitely slow variation of the parameters in comparison to the natural periods of the system could no longer be satisfied.

In 1916 Ehrenfest[53] established the conformance of Sommerfeld's quantum conditions (3.3) for the hydrogen atom with the adiabatic principle by showing that the phase integrals in (3.3) are indeed adiabatically invariant. Integrating Lagrange's azimuthal equation of motion, he obtained that $p_\psi = mr^2\dot{\psi}$ is a constant which corresponds to the coordinate $q_2 = \psi$ with the period $\nu_\psi = \dot{\psi}/2\pi$. The adiabatic invariant $2\bar{E}_{\text{kin}}^{(\psi)}/\nu_\psi$ is therefore equal to

$$\frac{p_\psi \dot{\psi}}{\dot{\psi}/2\pi} = 2\pi p_\psi = \oint p_\psi\, d\psi$$

which is Sommerfeld's first phase integral. Further, if $\chi(r,a_1,a_2, \ldots)$ is the potential of the central force, depending on the parameters $a_1, a_2, \ldots,$

[52] P. Hertz, "Über die mechanischen Grundlagen der Thermodynamik," *Annalen der Physik 33*, 537–552 (1910). Hertz was led to the discovery of the adiabatic invariance of volumes enclosed by isoenergic manifolds by a hint contained in chap. 13 ("Effect of various processes on an ensemble of systems") of J. W. Gibbs's *Elementary Principles in Statistical Mechanics* (Scribner, New York; Arnold, London, 1902), pp. 157–158.

[53] REF. 51, p. 338.

Lagrange's radial equation $m\ddot{r} - mr\dot{\psi}^2 + d\chi/dr = 0$ or $m\ddot{r} = p_\psi{}^2/mr^3 - d\chi/dr$ may be interpreted as the equation of a linear motion, between the limits $r_1 < r < r_2$, subject to a force with the potential $\Phi = -p_\psi{}^2/2mr^2 + \chi$, and the corresponding adiabatic invariant is $2\bar{E}_{\text{kin}}^{(r)}/\nu_r = \oint p_r\, dr$, which is Sommerfeld's second phase integral.

The adiabatic principle thus revealed the mystery of the quantum conditions. In fact, Bohr's early quantum condition according to which "the angular momentum of the electron round the nucleus is equal to an entire multiple of $h/2\pi$" (see page 81), or $2\pi mvr = nh$, now lost its air of magic. For $2\pi mvr = 2\cdot\frac{1}{2}mv^2\cdot(2\pi r/v) = 2\bar{E}_{\text{kin}}/\nu$ is—already according to Boltzmann's theorem—an adiabatic invariant. Sommerfeld's result, at first unexpected, according to which the spectrum of the hydrogen atom remains unaltered whether the atom is treated as possessing only circular orbits or, more generally, as having elliptic orbits with a great variety of eccentricities, lost its moment of surprise. For it could be shown that the system with a circular orbit is adiabatically related (in the sense of Ehrenfest) to a system with an elliptic orbit of the same time period and hence $\bar{E}_{\text{kin}}$ must be the same for both systems. Now, according to the virial theorem of classical mechanics, for inverse-square-law forces, $2\bar{E}_{\text{kin}} = -\bar{E}_{\text{pot}}$ so that both systems with the same period possess also the same total energy. Consequently, in consideration of Kepler's third law, an allowed circular orbit in the hydrogen atom is energetically equivalent to an allowed elliptic orbit whose major axis equals the diameter of the former. But this was exactly Sommerfeld's result.

The adiabatic principle, as we see, made an important contribution to a profounder understanding of what had happened so far in the development of quantum theory. But certain questions on basic aspects still remained obscure. One of the problems so far unsolved was the choice of the proper coordinates. The Kepler problem, for example, could be treated in parabolic coordinates as well as in polar coordinates, but the quantum conditions and the quantized orbits that resulted from these two treatments were different for different systems of coordinates.

The question of this ambiguity was successfully solved and a criterion for the correct selection of coordinates was established by Paul Sophus Epstein, a former student of Sommerfeld, and by Karl Schwarzschild, the director of the Astrophysical Observatory at Potsdam. Although intended to solve a specific question in quantum theory, their work was essentially the conclusion of a problem in theoretical mechanics which had a long history of its own. When Jacobi solved the "one-body problem," that is, the determination of the motion of a mass particle in the gravitational field of a stationary central body, he showed[54] that this Kepler problem can be solved by separating the Hamilton, later called the "Hamilton-

[54] C. G. J. Jacobi, *Vorlesungen über Dynamik* (Reimer, Berlin, 1866), 24th and 25th lectures, pp. 183–198.

Jacobi differential," equation in two different sets of coordinates. Since then questions concerning the existence and uniqueness of separable variables and the characteristics of motions associated with them had engaged the attention of theoretical physicists and mathematicians.

In 1887 Staude,[55] for systems with two degrees of freedom, and later Stäckel,[56] for systems with an arbitrary finite number of degrees of freedom, showed that the motion of a system whose Hamilton-Jacobi differential equation can be integrated by separation of variables is "multiply periodic" or, in the terminology of astronomy, "conditionally periodic." In other words, the generalized coordinates q_k can be represented for a system with f degrees of freedom as f-fold infinite Fourier series of the form

$$q_k = \sum_{\tau_1,\ldots,\tau_f=-\infty}^{\infty} C^{(k)}_{\tau_1\cdots\tau_f} \exp\left[2\pi i(\tau_1 w_1 + \cdots + \tau_f w_f)\right] \tag{3.5}$$

where $w_k = \nu_k t$ and where the summation extends over all integral values of the τ_k. The motion, in general, is not periodic, its orbit having the characteristics of an "open" Lissajous figure. Only if the ν_k are commensurable does the current point of the orbit ever return strictly to its starting point. Stäckel also showed that for systems with a separable Hamilton-Jacobi differential equation the integral of action $2\int E_{\text{kin}}\,dt$ can be separated into a sum $\sum_{k=1}^{f} \int [F_k(q_k)]^{1/2}\,dq_k$. Each of the f terms in this sum depends only on one of the generalized coordinates, and these oscillate between two fixed limits (*libration*) or are such that their increase by a certain constant value leaves the configuration of the system unchanged (*rotation*).

Meanwhile action-angle variables were introduced and their use was recognized as a powerful technique for the determination of the frequencies of periodic motions. If q_k, p_k are a pair of separation variables, the action variable was defined as $J_k = \oint p_k\,dq_k$, where the integration extends over a complete libration or rotation of the coordinate q_k, whatever the case may be. The generalized coordinate conjugate to J_k was known as the "angle variable" w_k and satisfied $w_k = \partial W_k/\partial J_k$, while W is the generating function of the contact transformation. From the canonical equation

$$\dot{w}_k = \frac{\partial H(J_1, \ldots, J_f)}{\partial J_k} = \nu_k$$

[55] O. Staude, "Über eine Gattung doppelt reell periodischer Funktionen zweier Veränderlicher," *Mathematische Annalen 29*, 468–485 (1887).

[56] P. Stäckel, "Über die Integration der Hamilton-Jacobischen Differentialgleichung mittels Separation der Variablen," *Habilitationsschrift* (Halle, 1891); "Über die Bewegung eines Punktes in einer n-fachen Mannigfaltigkeit," *Mathematische Annalen 42*, 537–544 (1893); "Über die Integration der Hamilton-Jacobischen Differentialgleichung mittels der Separation der Variablen," *ibid.*, 545–563 (1893); "Sur une classe de problèmes de dynamique," *Comptes Rendus 116*, 485–487 (1893); "Sur l'intégration de l'équation différentielle de Hamilton," *ibid. 121*, 489–492 (1895); "Über die Gestalt der Bahncurven bei einer Klasse dynamischer Probleme," *Mathematische Annalen 54*, 86–90 (1901).

where the ν_k are constant functions of the action variables, and from the fact that the w_k change by unity when the q_k go through a complete period [since $\Delta w_k = \oint (\partial w_k/\partial q_j)\, dq_j = \oint (\partial^2 W/\partial q_j\, \partial J_k)\, dq_j = (\partial/\partial J_k) \oint p_j\, dq_j = \partial J_j/\partial J_k = \delta_{jk}$] it follows, if T_k is the period associated with q_k, that $\Delta w_k = 1 = \nu_k T_k$, so that ν_k is the frequency associated with the periodic motion of q_k. Thus, as we see, the method of action-angle variables provided the frequencies of periodic motions without affording their complete solution. It was therefore quite natural that this technique developed originally in astronomy and was of little importance in general theoretical mechanics. In fact, Delauney, who introduced this method in his *Théorie du Mouvement de la Lune*,[57] in which he accounted for most of the lunar accelerations which Laplace's theory, as corrected by J. C. Adams, left unexplained, made no mention of it in his *Traité de Mécanique Rationelle.*[58] Nor did other texts on theoretical mechanics, prior to the work of Schwarzschild and Epstein, give any detailed discussion on this subject. In astronomy, however, the importance of action-angle variables was meanwhile fully recognized. Poincaré, in his *Leçons de Mécanique Céleste*,[59] and Charlier, in his classic textbook on celestial mechanics *Die Mechanik des Himmels*,[60] made systematic use of the method.

It was Schwarzschild's great merit[61] to have introduced into the quantum theory the analytical method of employing action-angle variables or "uniformizing variables," as Bohr later used to call them. Although originally intended as theoretical accounts of the Stark effect on the basis of Sommerfeld's generalization of Bohr's theory, Schwarzschild's and Epstein's[62] papers not only clarified the problem concerning the choice of the coordinates to be used in expressing the quantum conditions but showed quite generally how quantum-theoretic problems concerning periodic or conditionally periodic systems have to be linked up with the Hamilton-Jacobi theory in classical mechanics. In fact, it was due to their work that in its subsequent elaborations by Bohr[63] and Born[64] it almost seemed as if Hamilton's method had expressly been created for treating quantum-mechanical problems. According to Epstein the general quantum conditions (3.2) should be formulated with reference to those coordinate systems in which the Hamilton-Jacobi differential equation allows itself to be separated. Epstein justified this rule by geometrical reasons based on certain properties of the envelopes of the orbit in configuration space.

[57] C. E. Delauney, *Théorie du Mouvement de la Lune* (Paris, 1860–1867).

[58] Paris, 4th ed. 1873, 6th ed. 1878.

[59] H. Poincaré, *Leçons de Mécanique Céleste* (Gauthier-Villars, Paris, 1905).

[60] C. V. L. Charlier, *Die Mechanik des Himmels*, 2 vols. (Veit, Leipzig, 1907).

[61] K. Schwarzschild, "Zur Quantenhypothese," *Berliner Berichte 1916*, pp. 548–568. The paper was published on the day of his death, May 11, 1916.

[62] P. S. Epstein, "Zur Theorie des Starkeffektes," *Annalen der Physik 50*, 489–520 (1916); "Zur Quantentheorie," *ibid. 51*, 168–188 (1916).

[63] N. Bohr, "On the application of the Quantum Theory to atomic structure," *Proceedings of the Cambridge Philosophical Society* (Supplement), 1924.

[64] M. Born, *Vorlesungen über Atommechanik* (Springer, Berlin, 1925).

According to Schwarzschild, who like Epstein confined his treatment to conditionally periodic systems, the general quantum conditions should be formulated in terms of action-angle variables. Showing that this is always possible for separable systems, he paid special attention to those cases in which the Hamilton-Jacobi differential equation can be separated in various sets of coordinates. From the necessary and sufficient conditions of separability, found in 1904 by Levi-Civita,[65] it followed immediately that with a given set of conjugate variables q_k, p_k satisfying the conditions also all those canonical variables are separable which can be obtained from the former by point transformations. A dynamical system which in addition to this class of variables is separable in still another set was called by Schwarzschild a "degenerate" system. Schwarzschild showed that nondegeneration occurs when all the frequencies ν_k in (3.5) are incommensurable with one another. If, however, these frequencies satisfy n linear equations with integral coefficients, the system has an n-fold degeneration. In this case the orbit does not, as in the nondegenerate case, fulfill in the configuration space an f-fold, but only an $(f - n)$-fold continuum and the system can be transformed by an appropriate change of variables into a system of $f - n$ degrees of freedom. For degenerate systems, according to Schwarzschild, the number of quantum conditions $J_k = \oint p_k\, dq_k = n_k h$ should be equal to the degree of periodicity, the number of incommensurable frequencies in (3.5). Planck's quantum conditions,[66] as Epstein[67] showed, embrace those of Schwarzschild. A mathematically rigorous investigation of the equivalence between Planck's, Schwarzschild's, and Epstein's quantum conditions was carried out by Kneser.[68]

Burgers,[69] a collaborator of Ehrenfest in Leyden, and Born[70] studied the conditions which have to be imposed upon (not necessarily separable) dynamical systems with time-independent Hamiltonians in order to assure the possibility of introducing action-angle variables appropriate for the formulation of quantum conditions. They had to assume (1) that—just as in separable systems—a canonical transformation exists which transforms the q_k, p_k into new variables w_k, J_k by means of a generating function $S(q_1, J_1, \ldots, q_f, J_f)$ such that the configuration of the system is

[65] T. Levi-Civita, "Sulla integrazione della equazione di Hamilton-Jacobi per separazione di variabili," *Mathematische Annalen 59*, 383–397 (1904). Levi-Civita's conditions are the following set of $f(f - 1)/2$ partial differential equations:

$$\frac{\partial H}{\partial p_k}\frac{\partial H}{\partial p_s}\frac{\partial^2 H}{\partial q_k\, \partial q_s} - \frac{\partial H}{\partial p_k}\frac{\partial H}{\partial q_s}\frac{\partial^2 H}{\partial q_k\, \partial p_s} - \frac{\partial H}{\partial q_k}\frac{\partial H}{\partial p_s}\frac{\partial^2 H}{\partial p_k\, \partial q_s} + \frac{\partial H}{\partial q_k}\frac{\partial H}{\partial q_s}\frac{\partial^2 H}{\partial p_k\, \partial p_s} = 0$$

for $k = 1, 2, \ldots, f$ and $s = 1, 2, \ldots, k - 1, k + 1, \ldots, f$.

[66] REF. 5.

[67] P. S. Epstein, "Über die Struktur des Phasenraumes bedingt periodischer Systeme," *Berliner Berichte 1918*, pp. 435–446.

[68] REF. 7.

[69] J. M. Burgers, "Het Atoommodel van Rutherford-Bohr," Dissertation (Leyden, 1918).

[70] M. Born, REF. 64, paragraph 15, pp. 98–108.

a periodic function in the angle variables w_k with the primitive period 1; (2) that the Hamiltonian transforms thereby into a function W which depends only on the action variables J_k; and (3) that the function

$$S^* = S - \sum_k w_k J_k$$

for which

$$p_k = \frac{\partial S^*(q_1, w_1, \ldots)}{\partial q_k} \qquad J_k = -\frac{\partial S^*}{\partial w_k}(q_1, w_1, \ldots)$$

is periodic in the w_k with period 1. It could then be shown that the J_k (in the degenerate case their number is the degree of periodicity) are uniquely determined up to a homogeneous linear transformation with integral coefficients and a determinant ± 1.

In addition Burgers[71] could show that the action variables are adiabatic invariants. Since the generating function has now to be considered as time-dependent, the transformed Hamiltonian $\bar{H}$ satisfies $\bar{H} = H + \partial S^*/\partial t$, where H regarded as a function of the action-angle variables is independent of the w_k. The transformed canonical equation for the action variable reads

$$\dot{J}_k = -\frac{\partial \bar{H}}{\partial w_k} = -\frac{\partial}{\partial w_k}\left(\frac{\partial S^*}{\partial t}\right)$$

or since S^*, like S, depends on t only through the parameter a,

$$\dot{J}_k = -\frac{\partial}{\partial w_k}\left(\frac{\partial S^*}{\partial a}\right)\dot{a}$$

In view of the slow variation of a Burgers deduced that

$$\frac{\Delta J_k}{\dot{a}} = -\int \frac{\partial}{\partial w_k}\left(\frac{\partial S^*}{\partial a}\right) dt$$

and, utilizing a method introduced by Delauney and elaborated by Poincaré,[72] showed that the left-hand side of the last equation is of the order of magnitude $\dot{a}(t_2 - t_1)$, so that for infinitely slow variations of the parameters $a(\dot{a} - 0)$, provided $\dot{a}(t_2 - t_1)$ remains finite, $\Delta J_k = 0$. Certain difficulties connected with the violation of the incommensurability conditions were later removed by von Laue,[73] and finally Levi-Civita[74] studied

[71] J. M. Burgers, "Adiabatische invarianten bij mechanische systemen," *Verslag van de Gewone Vergaderingen der Wis- en Natuurkundige Afdeeling, Koninklijke Akedemie van Wetenschappen te Amsterdam 25*, 849–857, 918–922, 1055–1061 (1916); "Adiabatic invariants of mechanical systems," *Proceedings of the Amsterdam Academy 20*, 149–157, 158–162, 163–169 (1917); "Die adiabatischen Invariaten bedingt periodischer Systeme," *Annalen der Physik 52*, 195–202 (1917). In his dissertation (REF. 69) he gave a simpler proof.

[72] REF. 59, vol. 1, pp. 353–357.

[73] M. von Laue, "Zum Prinzip der mechanischen Transformierbarkeit (Adiabatenhypothese)," *Annalen der Physik 76*, 619–628 (1925).

[74] T. Levi-Civita, "Drei Vorlesungen über adiabatische Invarianten," *Hamburger Abhandlungen aus dem Mathematischen Seminar der Universität 6*, 323–366 (1928); "A general survey of the theory of adiabatic invariants," *Journal of Mathematics and Physics 13*, 18–40 (1934).

rigorously the mathematical foundations of the theory of adiabatic invariants.

The adiabatic principle was not only a constructive method for the establishment of quantum conditions and the determination of allowed motions. It also served as a criterion for the correctness of a given quantization. Planck's treatment[75] of the asymmetric top, whose moments of inertia were subjected to an adiabatic change during which the boundaries of the elementary regions in phase space remained unaltered, is a classic example of this use of the principle. Finally, the adiabatic principle could also be applied directly to the treatment of atomic systems which were exposed to the influence of external forces, provided the degree of periodicity was not changed by the presence of these forces. One of the most brilliant applications of the principle, in this sense, and, at the same time, one of the most striking achievements of the older quantum theory was the explanation of the Stark effect by Schwarzschild and Epstein. Let us then briefly recall the history of this effect.

According to the principles of classical physics a homogeneous external electric field should not affect the type of motion of an electron in the atom. A constant field superimposed upon a quasi-elastically bound electron will merely shift the center of the orbit by an amount proportional to the intensity of the field, as the equation of motion immediately shows. Yet in spite of these arguments and in spite of the absence of any observational evidence at that time, Voigt[76] predicted in 1901, a few years after Zeeman's discovery of the splitting of spectral lines in magnetic fields, an analogous effect for electrostatic fields. Voigt also construed a theory to account for such an effect but showed that even for rather strong field intensities the splitting would be too small to be experimentally observable.

When Stark,[77] however, about ten years later, studied atomic ions and the spectra of their canal rays and came to the conclusion that the change of the electric state of an atom due to ionization entails a change of its "optical frequencies," he decided to investigate the problem experimentally, in defiance of Voigt's discouraging remarks. Since Geissler tubes, the conventional sources for hydrogen lines, proved incapable of sustaining powerful electrostatic fields, Stark applied such fields to canal rays and investigated the luminescence of a canal-ray tube in a layer directly behind the perforated cathode. Thus, in his laboratory at the Technische Hochschule in Aachen, Stark[78] discovered in 1913 that an electric field splits

[75] M. Planck, "Zur Quantelung des asymmetrischen Kreisels," *Berliner Berichte 1918*, pp. 1166–1174 (1918); *Physikalische Abhandlungen und Vorträge* REF. 42 OF CHAP. 1), vol. 2, pp. 489–497.

[76] W. Voigt, "Über das elektrische Analogon des Zeeman-effectes," *Annalen der Physik 4*, 197–208 (1901).

[77] J. Stark, *Die Atomionen chemischer Elemente und ihre Kanalstrahlenspektra* (Springer, Berlin, 1913).

[78] J. Stark, "Beobachtungen über den Effekt des elektrischen Feldes auf Spektrallinien," *Berliner Berichte 1913*, pp. 932–946 (Nov. 20); "Beobachtungen über den Effekt des elektrischen Feldes auf

the lines of the Balmer series into a number of components. This phenomenon, which, as Stark already recognized, was by no means peculiar to hydrogen, was subsequently called the "Stark effect" or sometimes also the "Stark-LoSurdo effect."

For at the same time—or possibly even earlier[79]—LoSurdo,[80] studying the Doppler effect and examining the radiation emitted by the gas molecules in the Crookes dark space between the cathode and the negative glow in a discharge tube, found, like Stark, that the hydrogen lines split into several components. Although LoSurdo's experimental setup was much simpler than Stark's, he could obtain only qualitative results, having no direct method at his disposal for a precise measurement of the steep potential gradient as a function of position inside the tube.

Stark's discovery that each Balmer line is split up into components whose number increases with the series number of the line, that the resolution is proportional to the strength of the electric field, and that the components are linearly polarized when viewed transversely to the direction of the field soon became the subject of intensive theoretical research. The first attempt to account for these results was made by Warburg,[81] who succeeded in interpreting the general features of the effect on the basis of Bohr's theory. Bohr's own early work[82] on the Stark effect resulted in a good qualitative agreement with experience, but the finer details could not be explained.

In his first paper on the subject, published in 1914, Schwarzschild[83] discussed the Stark effect along classical lines and showed that the mathematical treatment of the effect can be regarded as a limiting case of Jacobi's[84] solution of the motion under the influence of two fixed centers of Newtonian (or Coulombian) attraction, a case which according to

---

Spektrallinien," *Annalen der Physik 43* 965–982 (1914) (I. Quereffekt), 983–990 (II. Längseffekt), in collaboration with G. Wendt, 991–1016 (III. Abhängigkeit von der Feldstärke), in collaboration with H. Kirschbaum, 1017–1047 (IV. Linienarten, Verbreiterung), also in collaboration with H. Kirschbaum; "Beobachtungen über den Effekt des elektrischen Feldes auf Spektrallinien," *Göttinger Berichte 1914*, pp. 427–444 (V. Feinzerlegung der Wasserstoffserie).

[79] "E facilmente potei persuadermi che il fenomeno da me prima osservato era identico a quello di Stark." A. LoSurdo, "Sul fenomeno analogoa quello di Zeeman nel campo elettrico," *Atti della Reale Accademia dei Lincei 22*, 664–666 (Dec. 21, 1913).

[80] A. LoSurdo, "Su l'analogo elettrico del fenomeno di Zeeman: efetto longitudinale," *Atti della Reale Accademia dei Lincei 23*, 82–84 (1914); "Su l'analogo elettrico del fenomeno di Zeeman: le varie righe della serie di Balmer presentano diverse forme di scomposizione," *ibid.*, 143–144.

[81] E. Warburg, "Bemerkungen zu der Aufspaltung der Spektrallinien im elektrischen Feld," *Verhandlungen der Deutschen Physikalischen Gesellschaft 15*, 1259–1266 (1913).

[82] N. Bohr, "On the effect of electric and magnetic fields on spectral lines," *Philosophical Magazine 27*, 506–524 (1914); "On the quantum theory of radiation and the structure of the atom," *ibid. 30*, 394–415 (1915).

[83] K. Schwarzschild, "Bemerkungen zur Aufspaltung der Spektrallinien im elektrischen Feld," *Verhandlungen der Deutschen Physikalischen Gesellschaft 16*, 20–24 (1914).

[84] REF. 54, p. 221.

Jacobi is separable in elliptical coordinates (i.e., in the parameters of families of confocal ellipses and hyperbolas whose foci are at the fixed centers of force). Schwarzschild now pointed out that the situation arrived at if one center of force is removed to infinity while its attractive power is correspondingly increased is that of the Stark effect. The families of ellipses and hyperbolas become in the limit confocal parabolas whose focus lies at the remaining fixed center which coincides with the nucleus. The Hamilton-Jacobi differential equation for the Stark effect, Schwarzschild concluded, is therefore separable in parabolic coordinates.

Thus, using parabolic coordinates ξ, η and the azimuthal angle ψ with respect to the direction of the field, Schwarzschild[85] and Epstein,[86] in their classic papers of 1916, knew from the beginning that the problem under discussion is that of a conditionally periodic motion and that each of the corresponding momenta p_ξ, p_η, p_ψ is the square root of a simple rational function of the corresponding position coordinate. Owing to the external field, the motion is nondegenerate and subject to the three quantum conditions

$$\oint p_\xi\, d\xi = n_1 h \qquad \oint p_\eta\, d\eta = n_2 h \qquad \oint p_\psi\, d\psi = n_3 h$$

where the parabolic quantum numbers n_1 and n_2 are nonnegative integers and the equatorial quantum number n_3 is a positive integer (zero has to be excluded since it would correspond to a collision between the electron and the nucleus). $p_\xi = \sqrt{f_1(\xi)}$ and $p_\eta = \sqrt{f_2(\eta)}$ each contain the energy constant W, an integration constant β, and the constant p_ψ which by the third quantum condition is proportional to n_3. Solving for W, which consequently depends on all three quantum numbers, Schwarzschild and Epstein obtained for the energy, approximated to the first power of the field intensity F, the expression

$$-W = \frac{2\pi^2 m Z^2 e^4}{h^2} \frac{1}{(n_1 + n_2 + n_3)^2} + \frac{3h^2 F}{8\pi^2 m Z e} (n_2 - n_1)(n_1 + n_2 + n_3)$$

where Ze is the charge of the nucleus. The agreement between the wavelengths of the components of the Balmer lines as calculated on the basis of this formula and their observed positions was excellent. In fact, Epstein could declare:[87] "We believe that the reported results prove the correctness of Bohr's atomic model with such striking evidence that even our conservative colleagues cannot deny its cogency. It seems that the potentialities of quantum theory as applied to this model are almost miraculous and far from being exhausted."

The older quantum theory, as we understand today, could apparently

[85] REF. 61.

[86] P. S. Epstein, "Zur Theorie des Starkeffekts," *Physikalische Zeitschrift 17*, 148–150 (1916), and REF. 62.

[87] *Ibid.*, p. 150.

achieve this brilliant success because the interaction between the spin of the electron and the applied field is negligibly small.[88] It should also be recalled that the first-order Stark effect is peculiar to hydrogen. For it is only in hydrogen that the accidental (Coulomb field) degeneracy is removed by the external field; in all other atoms it is already removed by the internal fields owing to the other electrons in the atom.

3.2 *The Correspondence Principle*

The formal applicability of classical mechanics to quantum theory as far as a given stationary state is concerned was, as we have seen, the subject of the adiabatic principle. The formal applicability of classical mechanics to quantum theory as far as transitions between such states are concerned is the subject of the correspondence principle, the second fundamental principle of the older quantum theory. Relating the kinematics of the electron to the properties of the emitted radiation, the correspondence principle became a conceptual instrument indispensable for the quantum-theoretic treatment of problems concerning the finer details of the emitted radiation, such as the intensity of spectral lines or their polarization. Even for problems which could not be treated by the theory of periodic or conditionally periodic systems this principle produced results in fair agreement with experience.

The conceptual foundation of the correspondence principle was based on the assumption that quantum theory or at least its formalism contains classical mechanics as a limiting case. This idea was expressed by Planck[89] as early as 1906, when he showed that in the limit $h \rightarrow 0$ quantum-theoretic conclusions converge toward classical results, or, as he said: "The classical theory can simply be characterized by the fact that the quantum of action becomes infinitesimally small."[90] In fact, it was easy to see that for h tending to zero Planck's radiation formula goes over into the classic Rayleigh-Jeans formula.

The same result, however, is also obtained if for constant h the frequency ν approaches zero. This was the idea chosen by Bohr in his establishment of the correspondence principle. For, according to his frequency law, the convergence of ν toward zero implies that the energy differences become arbitrarily small or, in other words, that in the limit the energies form almost a continuum. Since for periodic or conditionally periodic

[88] Cf. R. Schlapp, "The Stark effect of the fine-structure of hydrogen," *Proceedings of the Royal Society of London* (*A*), *119*, 313–334 (1928), where it is shown that the inclusion of the spin-field interaction leads only to a very small additional splitting of certain energy levels.

[89] M. Planck, *Vorlesungen über die Theorie der Wärmestrahlung* (REF. 11 OF CHAP. 1).

[90] *Ibid.*, p. 143.

systems the energy can be expressed as a function of the phase integrals and these, in turn, as functions of the quantum numbers n,[91] in virtue of the quantum conditions, quantum theory approaches classical physics whenever the differences Δn in the quantum numbers are small compared with their absolute values n.

As pointed out on page 79, Bohr applied this reasoning, though only as a procedure *ad hoc* in his treatment of the hydrogen spectrum, as early as 1913 when he identified the classical frequency of revolution of the electron with the radiation frequency of the emitted spectral line for transitions between the $(n + 1)$st and the nth stationary state for very large n. In his address before the Physical Society in Copenhagen, Bohr[92] declared explicitly in this context that his recourse to classical physics should not be interpreted as a quest for causal explanation but merely as a heuristic device. "You understand, of course," he said, "that I am by no means trying to give what might ordinarily be described as an explanation; nothing has been said here about how and why the radiation is emitted. On one point, however, we may expect a connection with the ordinary conceptions, namely, that it will be possible to calculate the emission of slow electromagnetic oscillations on the basis of classical electrodynamics."[93]

In his important memoir of 1918 "On the quantum theory of line-spectra" in which Bohr discussed the application of the quantum theory to the determination of the line spectrum of a given atomic system, he remarked that even without introducing detailed assumptions as to the mechanism of transition between stationary states, one can show that the frequencies calculated by the frequency law "in the limit where the motions in successive stationary states comparatively differ very little from each other, will tend to coincide with the frequencies to be expected on the ordinary theory of radiation from the motion of the system in the stationary states."[94] Bohr added that these considerations would be shown to throw light on the question of the polarization and intensity of the different lines of the spectrum of a given system. This was Bohr's first explicit formulation of the correspondence principle. In the memoir under discussion he did not yet call it by this name but referred to it as "a formal analogy between the quantum theory and the classical theory."

In Section Three of Part One of his memoir Bohr justifies his contention for conditionally periodic systems as follows.[95] According to the classical theory of such systems $\nu_k = \partial E/\partial J_k$, where $E = H$ is the total energy of the system—the notation of page 102 is used—whereas according

[91] For the sake of convenience we call the integers in (3.2) "quantum numbers" although this designation is of a later date.

[92] REF. 114 OF CHAP. 2.

[93] REF. 119 OF CHAP. 2.

[94] N. Bohr, "On the quantum theory of line-spectra," *Kongelige Danske Videnskabernes Selskabs Skrifter Naturvidenskabelig og mathematisk afdeling, series 8, IV, 1*, 1–118 (1918–1922); quotation on p. 8; *Über die Quantentheorie der Linienspektren*, translated by P. Hertz (Vieweg, Braunschweig, 1923).

[95] *Ibid.*, pp. 27–36.

to the frequency formula $h\nu_{\text{qu}} = \Delta E$. Let $\Delta J_k = (n_k' - n_k)h = \tau_k h$, where n_k' and n_k are the quantum numbers characterizing the stationary states before and after the transition. On the assumption of large quantum numbers $\Delta E = \sum (\partial E/\partial J_k)\ \Delta J_k = \sum \tau_k \nu_k h$. Hence $\nu_{\text{qu}} = \sum \tau_k \nu_k = \nu_{\text{cl}}$, where ν_{cl} is the classical frequency (combination of higher harmonics) appearing in one of the terms of the multiple Fourier expansion (3.5) on page 102. In the limit, therefore, the quantum-theoretic frequency ν_{qu} coincides with the classical mechanical frequency ν_{cl}. By demanding that this correspondence remain approximately valid also for moderate and small quantum numbers, Bohr generalized and modified into a principle what in the limit may be regarded formally as a theorem.

The expression "correspondence" as a *terminus technicus* in this sense appeared for the first time in Bohr's paper "On the series spectra of the elements"[96] in the following statement: "Moreover, although the process of radiation cannot be described on the basis of the ordinary theory of electrodynamics, according to which the nature of the radiation emitted by an atom is directly related to the harmonic components occurring in the motion of the system, there is found, nevertheless, to exist a far-reaching *correspondence* between the various types of possible transitions between the stationary states on the one hand and the various harmonic components of the motion on the other hand. This correspondence is of such a nature that the present theory of spectra is in a certain sense to be regarded as a rational generalization of the ordinary theory of radiation."[97] Having meanwhile recognized the far-reaching methodological importance of this idea for the further development of quantum theory, Bohr named it "the correspondence principle" in the "Appendix to Part Three" of his Academy memoir.[98] Without the principle, Bohr realized, the quantum theory in contrast to the atomic theory of classical physics was an incomplete theory. Newton's mechanics and Maxwell's electrodynamics allowed calculating not only the frequencies but also the intensities and states of polarization of spectral lines. For according to classical physics the absolute value of the square of the (vectorial) Fourier coefficient $\mathbf{D}_{\tau_1 \cdots \tau_f}$ in the multiple expansion representing the (vectorial) dipole moment served as a measure of the intensity of the corresponding partial vibration, characterized by the integers $\tau_1, \ldots, \tau_f$ in (3.5).[99] The ratio of the x, y, and z components of this (vectorial) Fourier coefficient specified the polarization. Classical physics was therefore a complete theory of atomic spectra affording full information on the nature of the emitted

[96] N. Bohr, "Über die Linienspektren der Elemente," *Zeitschrift für Physik 2*, 423–469 (1920); English translation in *The Theory of Spectra and Atomic Constitution* (REF. 114 OF CHAP. 2), pp. 20–60.

[97] REF. 114 OF CHAP. 2, pp. 23–24.

[98] REF. 94, p. 112.

[99] The intensity is

$$[(2\pi\nu_{\text{cl}})^4/3c^2]\ |\mathbf{D}_{\tau_1} \cdots {}_{\tau_f}|^2$$

where $\nu_{\text{cl}} = |\tau_1\nu_1 + \cdots + \tau_f\nu_f|$.

radiation—though at variance with experience. Bohr's theory, however, which expressly disclaimed any knowledge concerning the mechanism of transitions between stationary states, renounced thereby from the beginning any possibility of a rational foundation for deriving the intensities and polarizations of spectral lines. The only way out of this impasse, as Bohr recognized, was via the correspondence principle.

But instead of establishing a direct correspondence between the intensity of a spectral line and the respective Fourier coefficient in the classical representation of the dipole moment, Bohr incorporated into his theory an additional idea recently conceived by Einstein. For in 1916 Einstein derived Planck's radiation law on the assumption that the transitions are governed by certain probability laws similar to those postulated in the theory of radioactivity.[100] Denoting the probability for the occurrence of an externally not excited transition from the higher energy level E_m to the lower E_n, during the time interval dt, by $A_m{}^n\,dt$ and the probability for the occurrence of an induced or stimulated transition between the same energy levels in the radiation field by $B_m{}^n\rho\,dt$, where ρ is the energy density (which on page 5 was denoted by u_ν), and finally the probability of the occurrence of the inverse transition (absorption) by $B_n{}^m\rho\,dt$, Einstein deduced from the preservation of the canonical distribution of states under equilibrium conditions that

$$B_n{}^m\rho \exp\left(-\frac{E_n}{kT}\right) = (B_m{}^n\rho + A_m{}^n) \exp\left(-\frac{E_m}{kT}\right) \tag{3.6}$$

Assuming that ρ increases to infinity with T, Einstein obtained $B_n{}^m = B_m{}^n$, the first, though implicit, application of what later was called "the principle of detailed balancing." Comparing the solution for ρ,

$$\rho = \frac{A_m{}^n/B_m{}^n}{e^{(E_m-E_n)/kT} - 1}$$

with the Wien radiation law in the limit of high frequencies and with the Rayleigh-Jeans formula in the limit of small frequencies, Einstein concluded that

$$E_m - E_n = h\nu$$

which is Bohr's frequency condition, and that

$$A_m{}^n = \frac{8\pi h\nu^3}{c^3} B_m{}^n \tag{3.7}$$

and thus obtained Planck's radiation law, Eq. (1.16) on page 21. It

[100] A. Einstein, "Zur Quantentheorie der Strahlung," *Mitteilungen der Physikalischen Gesellschaft, Zürich, 18*, 47–62 (1916); "Strahlungsemission und -absorption nach der Quantentheorie," *Verhandlungen der Deutschen Physikalischen Gesellschaft 18*, 318–323 (1916); "Quantentheorie der Strahlung," *Physikalische Zeitschrift 18*, 121–128 (1917).

should be noted in the present context that the very existence of the induced emission was suggested to Einstein—one is almost tempted to say, in accordance with the correspondence principle—by the classical theory concerning the interaction between the radiation field and a system of oscillating charges.[101] Had Einstein omitted the induced radiation, he would have obtained Wien's, and not Planck's, radiation law.

In presenting these probabilistic considerations which, as Einstein phrased it, lead "in an amazingly simple and general way"[102] to Planck's law, Einstein—contrary to Planck[103]—did not explicitly define his conception of probability. In view of this fact it has often been claimed[104] that, ironically as it sounds, Einstein, the declared protagonist of causality in physics and the philosopher who rejected the "dice-playing god,"[105] was the first to introduce the modern quantum-theoretic notion of probability, not as an expression of insufficient knowledge of fine-scale parameters, as conceived in the kinetic theory, but as an affirmation in a fundamental manner of the contingency of a single event. A closer analysis of Einstein's paper, however, will show that this imputation is not justified. First of all, Einstein's comparison of his statistical law with that of radioactive disintegration[106] cannot be used as an argument since at that time radioactive disintegration was generally considered, as, for example, by Planck, as a process involving as yet unknown parameters. Second, the absence of a declaration in favor of causality cannot be interpreted as a declaration in favor of the absence of causality. Third, referring to the related problem of the reaction of an atom or molecule during emission, Einstein stated expressly that "in the present state of the theory the direction of recoil is only 'statistically' determined,"[107] alluding thereby to the merely preliminary character of such an approach.

Bohr, however, in agreement with his own basic ideas, saw in Einstein's approach a renunciation of any inquiry into the causal structure of transitions. Einstein's statement that a molecule may pass from one state to an energetically lower state "without excitation by an external cause"[108]—a statement which, in principle at least, leaves room for the activity of

[101] On this point cf. J. H. van Vleck, "The absorption of radiation by multiply periodic orbits, and its relation to the correspondence principle and the Rayleigh-Jeans law," *Physical Review 24*, 330–346, 347–365 (1924).

[102] "... in verblüffend einfacher und allgemeiner Weise." REF. 100 (1917), p. 121.

[103] See p. 49.

[104] See, for example, REF. 121 OF CHAP. 2 (1964), p. 74.

[105] "Der große anfängliche Erfolg der Quantentheorie kann mich doch nicht zum Glauben an das fundamentale Würfelspiel bringen" ("The great initial success of quantum theory cannot convert me to believe in that fundamental game of dice"). Einstein in a letter to Born of Nov. 7, 1944. Cf. M. Born, *Natural Philosophy of Cause and Chance* (Oxford University Press, 1949; Dover, New York, 1964), p. 122.

[106] REF. 100 (1917), p. 123.

[107] "Das Molekül erleidet in einer beim jetzigen Stande der Theorie nur durch den 'Zufall' bestimmten Richtung bei dem Elementarprozeß der Ausstrahlung einen Rückstoß von der Größe $h\nu/c$." *Ibid.*, p. 127.

[108] "... ohne Anregung durch äußere Ursache." *Ibid.*, p. 123.

"inner" parameters—was interpreted by Bohr as meaning that the system "will start *spontaneously* to pass to the stationary state of smaller energy,"[109] where "spontaneously" is taken as essentially synonymous with "acausal." In fact, following Bohr, Einstein's $A_m{}^n$ is often referred to as the "probability factor of spontaneous emission" although Einstein never used this term in his paper.

Extending the domain of the applicability of the correspondence principle, Bohr associated Einstein's a priori probabilities with the abovementioned Fourier coefficients in the multiple Fourier expansion of the dipole moment, the correspondence being

$$A_m{}^n \leftrightarrow \frac{(2\pi)^4\nu^3}{3c^3h}\,|D_{m-n}|^2$$

One of the important results which Bohr could derive directly from the correspondence principle referred to conditionally periodic systems of f degrees of freedom possessing an axis of symmetry and subject to a central force. Calling the separable variables $q_1, q_2, \ldots, q_f$ and taking $q_1 \equiv \varphi$ as the cyclic variable, Bohr showed that a straightforward application of the Hamilton-Jacobi theory[110] led for $k = 1, \ldots, f$ to the equations $x_k + iy_k = f_k(q_2, \ldots, q_f)e^{i\varphi}$, where the function f_k does not contain the cyclic variable and where x_k, y_k are the cartesian coordinates of the kth point mass in the invariable plane drawn through the center of gravity. For the components of the dipole moment, therefore,

$$P_x + iP_y = \sum \pm e(x_k + iy_k) = F_1(q_2, \ldots, q_f)e^{i\varphi}$$

and $P_z = \sum ez_k = F_2(q_2, \ldots, q_f)$. Since for $k = 2, \ldots, f$,

$$w_k = f_k(q_2, \ldots, q_f, J_1, \ldots, J_f)$$

or inversely

$$q_k = g(w_2, \ldots, w_f, J_1, \ldots, J_f)$$

where $w_k = \nu_k t + \delta_k$ and $w_\varphi = \nu t + \delta$, the expansion of F_1 and F_2 into a Fourier series yields

$$P_x + iP_y = \exp\,(2\pi i\nu t) \sum_{\tau_2\ldots\tau_f} D^{(r)}_{\tau_2\ldots\tau_f} \exp\,[2\pi it(\tau_2\nu_2 + \cdots + \tau_f\nu_f)]$$

and

$$P_z = \sum_{\tau_2\ldots\tau_f} D^{(z)}_{\tau_2\ldots\tau_f} \exp\,[2\pi it(\tau_2\nu_2 + \cdots + \tau_f\nu_f)]$$

Since τ_1 assumes in the Fourier expansion of $P_x + iP_y$ only the value 1 and does not appear at all in the Fourier expansion of P_z and recalling that $\tau_1 = n_1' - n_1$, Bohr derived from the correspondence principle that

[109] REF. 94, p. 7. Italics by Bohr.

[110] REF. 94, pp. 32–34. For a more detailed elaboration see REF. 20 (1923), pp. 584–585.

the azimuthal quantum number n_1 or k, as it was usually denoted, could decrease or increase only by one unit for circularly polarized radiation whereas it had to remain unchanged for radiation polarized parallel to the axis of the system. The correspondence principle, as this calculation showed, thus led to a theoretical explanation of certain restrictions in the use of Ritz's combination principle, restrictions which subsequently were called "selection rules."

One of the examples to which Bohr immediately applied these considerations was Planck's linear harmonic oscillator,[111] whose mechanical frequency of motion may be called ν_{cl}. As shown in REF. 13 on page 92, its quantum-theoretic energy levels are $nh\nu_{cl}$ and, according to Bohr's frequency condition, its quantum-theoretic optical frequencies ν_{qu} are therefore $(\Delta n)\nu_{cl}$. Since the Fourier expansion of the linear harmonic motion reduces to only one term, namely, that with frequency $1 \cdot \nu_{cl}$, the only surviving τ is unity and Bohr's correspondence principle requires that $\Delta n = \pm 1$ or $\nu_{qu} = \nu_{cl}$. Now, as a rule, the quantum-theoretic frequency ν_{qu} coincides with the classical frequency ν_{cl} only in the case of high energy levels. For Planck's oscillator, however, as we see, the equation $\nu_{qu} = \nu_{cl}$ holds strictly for all energy levels, even those characterized by the smallest possible quantum numbers. It now became clear that this exceptional proper[illegible] of the harmonic oscillator was a fortunate feature for the conceptual development of quantum theory. Otherwise, the theory would have had to revoke its own foundations, namely, the radiation law for the derivation of which Planck identified the mechanical frequency of the oscillator with its optical frequency.

On the basis of the correspondence principle Kramers,[112] a pupil and later associate of Bohr, calculated in his dissertation on the "Intensities of spectral lines" the relative intensities of the components of the fine structure and of the Stark effect with special consideration of the first four Balmer lines in the hydrogen spectrum and showed that the theoretical results agree surprisingly well with the observational data as known at the time.[113] Kramers summarized his conclusions as follows: "... the results obtained as regards the applications to the Stark effect and to the fine structure of the hydrogen lines must be considered as affording a general support of Bohr's fundamental hypothesis of the connection between the intensity of spectral lines and the amplitudes of the harmonic vibrations into which the motion of the electron in the atom may be resolved, the more so because it seemed possible to obtain a natural understanding of certain marked deviations of the observed intensities from the

[111] REF. 94, pp. 32–33.

[112] H. A. Kramers, "Intensities of spectral lines," *Kongelige Danske Videnskabernes Selskabs Skrifter Naturvidenskabelig og mathematisk afdeling*, series 8, III.3 (1919).

[113] Cf. also L. Bloch, "La structure des atomes" (V. Utilisation du principe de correspondance), *Journal de Physique 3*, 110–124 (1922).

preliminary theoretical estimate of the intensity distribution obtained on the basis of this hypothesis. It seems therefore justifiable to conclude that Bohr's considerations offer a sound basis for a further development of the theory of intensities of spectral lines."[114]

The correspondence principle turned out to be a most versatile and productive conceptual device for the further development of the older quantum theory—and, as we shall see in due course, even for the establishment of modern quantum mechanics. For periodic and conditionally periodic systems it lent additional support to the formulation of the general quantum conditions, and this on the following grounds: if for high quantum numbers the quantum-theoretic frequency

$$\frac{\Delta E}{h} = \frac{1}{h}\frac{dE}{dJ}\Delta J = \frac{\nu\,\Delta J}{h}$$

is identified with the classical mechanical frequency $\tau\nu$, then $\tau h = \Delta J$ and, since τ is an integer, $J = nh$ up to an additional constant. Even for systems which are neither periodic nor conditionally periodic a method of quantization was worked out by Smekal[115] on the basis of the principle, in contrast to the approach of Born and Pauli,[116] who based their calculations on the classical theory of perturbations as developed in astronomy. Indeed, in the hands of Bohr and his school the correspondence principle was like "a magic wand that allowed the results of the classical wave theory to be of use for the quantum theory."[117] It was due to this principle that the older quantum theory became as complete as classical physics. But a costly price had to be paid. For taking resort to classical physics in order to establish quantum-theoretic predictions, or in other words, constructing a theory whose corroboration depends on premises which conflict with the substance of the theory, is of course a serious inconsistency from the logical point of view. Being fully aware of this difficulty, Bohr attempted repeatedly to show that "the correspondence principle must be regarded purely as a law of the quantum theory, which can in no way diminish the contrast between the postulates and electrodynamic theory."[118] The earliest allusion to such a conception may perhaps be found as early as 1921 in a paper[119] in which Bohr briefly discussed the function of the principle. In his statement that the principle originated "in an effort to

[114] REF. 112, p. 384.

[115] A. Smekal, "Bemerkungen zur Quantelung nicht bedingt periodischer Systeme," *Zeitschrift für Physik 11*, 294–303 (1922), *15*, 58–60 (1923).

[116] M. Born and W. Pauli, "Über die Quantelung gestörter mechanischer Systeme," *Zeitschrift für Physik 10*, 137–158 (1922); reprinted in M. Born, *Ausgewählte Abhandlungen* (Vandenhoek & Ruprecht, Göttingen, 1963), vol. 2, pp. 1–22.

[117] REF. 20 (1923), p. 583.

[118] N. Bohr, "On the application of the quantum theory to atomic structure," *Proceedings of the Cambridge Philosophical Society* (Supplement), part I (Cambridge University Press, 1924), p. 22.

[119] N. Bohr, "Zur Frage der Polarisation der Strahlung in der Quantentheorie," *Zeitschrift für Physik 6*, 1–9 (1921).

attain a simple asymptotic agreement between the spectrum and the motion of an atomic system in the limiting region where the stationary states differ relatively only little from each other,"[120] he apparently tried to describe the principle without reference to conceptions extraneous to quantum theory. In accordance with his insistence on the irreconcilability of quantum theory with classical mechanics and electrodynamics, he regarded the correspondence principle as merely affirming a formal analogy of heuristic value. Although his numerous and often somewhat conflicting statements, made from 1920 to 1961, on the essence of the correspondence principle[121] make it difficult, if not impossible, to ascribe to Bohr a clear-cut unvarying conception of the principle,[122] it seems that at the time under discussion, that is, in the early twenties, he did not view the principle as implying the inclusion of classical physics within quantum theory. Not only would such a conception have contradicted, of course, his fundamental dialectics of the irreconcilability just mentioned, but it would also have implied that a quantum-theoretic theorem makes an assertion on classical physics. Bohr's attitude, at least at that time, was well described by Kramers[123] when he wrote in 1923: "Bohr expressed himself in his talks somewhat as follows: classical as well as quantum theory are each as a description of nature merely a caricature; it allows, so to speak, in two extreme regions of phenomena an asymptotic presentation of physical reality."[124]

Later, Bohr's conception of complementarity, which gave a deeper significance to his previous ideas on the irreconcilable disparity between classical and quantum theory, precluded, now on epistemological grounds, the possibility to interpret the correspondence principle as asserting the inclusion of classical mechanics within quantum mechanics. But at that time, as we shall see, the correspondence principle had accomplished its

[120] "Nach diesem Prinzip, deren Erkennung aus den Bestrebungen entstanden ist, eine einfache asymptotische Übereinstimmung zwischen dem Spektrum und der Bewegung eines Atomsystems in dem Grenzgebiet, wo die stationären Zustände nur vehältnismäßig wenig voneinander abweichen, zu erreichen. . . ." *Ibid.*, p. 2.

[121] For a detailed analysis of these statements cf. REF. 121 OF CHAP. 2.

[122] Thus, in *Atomic Theory and the Description of Nature* (Cambridge University Press, 1961), p. 8, he declared: ". . . the necessity of making an extensive use . . . of the classical concepts, upon which depends ultimately the interpretation of all experience, gave rise to the formulation of the so-called correspondence principle . . . ." But only in 1922 he expressed for the first time the need of using classical concepts when he said: "Beim jetzigen Standpunkt der Physik muß jedoch jede Naturbeschreibung auf eine Anwendung der in der klassischen Theorie eingeführten und definierten Begriffe gegründet werden." In "Über die Anwendung der Quantentheorie auf den Atombau," *Zeitschrift für Physik 13,* 117–164 (1922). Quotation on p. 117.

[123] H. A. Kramers, "Das Korrespondenzprinzip und der Schalenbau des Atoms," *Die Naturwissenschaften 11,* 550–559 (1923).

[124] "Bohr hat sich in Gesprächen wohl folgendermaßen ausgedrückt: Sowohl die klassische Theorie wie die Quantentheorie sind beide als Naturbeschreibung nur eine Karikatur; sie gestattet sozusagen in zwei extremen Erscheinungsgebieten eine asymptotische Darstellung des wirklichen Geschehens." *Ibid.*, p. 559.

major function: it had contributed decisively to the establishment of modern quantum mechanics.

In 1924 Bohr wrote: "As frequently emphasized, these principles, although they are formulated by the help of classical conceptions, are to be regarded purely as laws of the quantum theory, which give us, notwithstanding the formal nature of the quantum theory, a hope in the future of a consistent theory, which at the same time reproduces the characteristic features of the quantum theory, important for its applicability, and, nevertheless, can be regarded as a rational generalization of classical electrodynamics."[125] About three years later this consistent theory was accomplished. In fact, there was rarely in the history of physics a comprehensive theory which owed so much to one principle as quantum mechanics owed to Bohr's correspondence principle.

3.3 *The Zeeman Effect and Multiplet Structure*

It was, of course, by no means a pure coincidence in the development of the older quantum theory that in the same year in which Schwarzschild and Epstein explained the Stark effect Debye and Sommerfeld interpreted the normal Zeeman effect. The Zeeman effect not only "opened a new world of facts which interest the physicist, the chemist, and even the astronomer," as Becquerel and Deslandres[126] realized in less than two years after its discovery, but its study also contributed—to an extent much greater than the study of the Stark effect—to the conceptual development of quantum theory; we shall therefore discuss the Zeeman effect and the problems it raised in greater detail.

The first to anticipate a possible influence of magnetic forces on spectral lines was Michael Faraday. His philosophy of the essential unity of all natural forces and particularly his discovery of the magnetic rotation of the plane of polarization[127] in 1845 convinced him of the existence of an intimate relationship between optical and magnetic phenomena. It was only in 1862, however, five years before his death, that Faraday performed an experiment in order to confirm his anticipation: he investigated the spectra of sodium, barium, strontium, and lithium with Steinheil's newly invented prismatic spectroscope and an electromagnet. In a letter dated March 12, 1862, Faraday described the outcome as follows: "The colourless gas flame ascended between the poles of the magnet and the salts of sodium, lithium etc., were used to give colour. A Nicol's polariser was placed just

[125] REF. 118, p. 42.

[126] H. Becquerel and H. Deslandres, "Contribution á l'étude du phénomène de Zeeman," *Comptes Rendus 126*, 997–1001 (1898).

[127] Cf. M. Faraday, *Experimental Researches in Electricity* (Quaritch, London, 1839–1855), vol. 3, p. 1.

before the intense magnetic field and an analyser at the other extreme of the apparatus. Then the electro-magnet was made, but not the lightest trace of effect on or change in the lines in the spectrum was observed in any position of polariser or analyser."[128]

In his sketch of Faraday's life Maxwell mentioned that Faraday "made the relation between magnetism and light the subject of his very last experimental work. He endeavoured, but in vain, to detect any change in the lines of the spectrum of a flame when the flame was acted on by a powerful magnet."[129]

Tait was probably the first to resume, after Faraday, the study of the effect of magnetism on spectra when he discussed,[130] in 1875, the possibility of an influence of magnetic forces on the absorption of light. Ten years later Fievez[131] reported a definite broadening (and partial reversal) of the D lines of sodium and of some other principal series lines in the spectra of potassium, lithium, and thallium if exposed to magnetic fields. However, as a critical examination of Fievez's description of his experimental method shows, it is almost certain that in his observations the true effect must have been masked by disturbing factors such as pressure broadening. Referring to Fievez's experiments, Lodge[132] pointed out: " . . . it appears likely that a variety of unimportant causes of disturbance must have been present, and that if the true effect was seen at all, it was so mixed up with spurious effects as to be unrecognizable in its simplicity, and so remained at that time essentially undiscovered." It was Pieter Zeeman who resorted to a method of experimentation which eliminated such objections.

In the course of his study of the Kerr effect it occurred to Zeeman in 1893 to find out "whether the light of a flame if submitted to the action of magnetism would perhaps undergo any change."[133] He therefore studied the spectrum of a sodium flame placed between the poles of a Ruhmkorff electromagnet—but again without positive results. Two years later, when reading Maxwell's above-mentioned reference to Faraday's experiment on this problem, Zeeman said to himself: "If a Faraday thought of the possibility of the above-mentioned relation, perhaps it might be yet worthwhile to try the experiment again with the excellent auxiliaries of spectroscopy of the present time."[134] Thus, in August, 1896, in his laboratory at Leiden University, he repeated the experiment with a powerful Ruhmkorff

[128] Dr. Bence Jones, *The Life and Letters of Faraday* (Longmans, Green & Co., London, 1870), vol. 2, p. 444.

[129] J. C. Maxwell, *Scientific Papers* (REF. 56 OF CHAP. 1), vol. 2, p. 790.

[130] G. P. Tait, "On a possible influence of magnetism on the absorption of light," *Proceedings of the Royal Society of Edinburgh 1875–1876*, p. 118.

[131] C. Fievez, "De l'influence du magnétisme sur les caractères des raies spectrales," *Bulletin de l'Académie des Sciences de Belgique 9*, 381–385 (1885).

[132] O. Lodge, "The latest discoveries in physics," *The Electrician 38*, 568–570 (Feb. 26, 1897).

[133] P. Zeeman, "On the influence of magnetism on the nature of the light emitted by a substance," *Philosophical Magazine 43*, 226–239 (1897).

[134] *Ibid.*

electromagnet, energized by a battery of accumulators (27 amp), with a Rowland grating of 10-foot radius and 14,983 lines per inch, and with a bunsen flame fed with small quantities of sodium from a piece of asbestos impregnated with common salt, and placed between the poles. He now found that whenever "the current was put on, the two D lines were distinctly widened."[135] In a second series of experiments Zeeman enclosed heated sodium in a porcelain tube, closed at both ends by parallel glass plates and continuously rotated round its axis to avoid temperature variations. "Excitation of the magnet caused immediate widening of the lines," which regained their original appearance as soon as the current was cut off. "It thus appears very probable that the period of sodium light is altered in the magnetic field," concluded Zeeman's report.[136] Thus ten years after Hertz demonstrated the fundamental role of electric and magnetic forces for the *propagation* of light, Zeeman established their role for the *production* of light as well.

Lorentz, to whom Zeeman reported the outcome of his experiments before their publication, saw at once how to interpret these results on the basis of his electron theory. Resolving the motion of the quasi-elastically bound charged particle into a linear vibration in the direction of the field (and hence unaffected by the field) and into two superposed circular motions, executed in opposite directions, in the plane perpendicular to the direction of the field, Lorentz predicted that the light from the edges of the broadened line, if viewed longitudinally (i.e., in the direction of the lines of force), should be circularly polarized, and if viewed transversally (i.e., at right angles to the field), should be plane-polarized, a conclusion which Zeeman immediately confirmed with the help of a quarter-wavelength plate and an analyzer. Lorentz's classic explanation of the Zeeman effect, based on the equation $mv^2/r = kr + (e/c)vH$, where H denotes the intensity of the magnetic field, $k = 4\pi^2 m\nu_0^2$, and $v = 2\pi r\nu$, and leading to $\nu = \nu_0 \pm eH/4\pi mc$ for the frequencies of the two circular motions, was published[137] in 1897, at a time when Zeeman had already successfully resolved the "broadened line" λ4800 of cadmium into doublets for longi-

[135] *Ibid.*, p. 227. "If the current was put on, the two D lines were distinctly widened. If the current was cut off, they returned to their original position. The appearing and disappearing of the widening was simultaneous with the putting on and off of the current. The experiment could be repeated an indefinite number of times."

[136] P. Zeeman, "Over den invloed eener magnetisatie op den aard van het door een stof uitgezonden licht," *Verslag van de Gewone Vergaderingen der Wis- en Natuurkundige Afdeeling, Koninklijke Akademie van Wetenschappen te Amsterdam 5*, 181–185, 242–248 (1896). The paper quoted in REF. 133 is an English translation of this article. Cf. also P. Zeeman, "Über einen Einfluß der Magnetisirung auf die Natur des von einer Substanz emittirten Lichtes," *Verhandlungen der Physikalischen Gesellschaft zu Berlin 7*, 128–130 (1896), and *Nature 57*, 192, 311 (1897).

[137] H. A. Lorentz, "Über den Einfluß magnetischer Kräfte auf die Emission des Lichtes," *Wiedemannsche Annalen der Physik 63*, 278–284 (1897).

tudinal observation and triplets for transversal observation, employing a field of about 32,000 gauss and a grating spectroscope.[138]

It soon became clear that Lorentz's result was but an application of a general electrodynamic theorem which Larmor[139] had proved in 1897 and two years later employed for the explanation of the Zeeman effect.[140] According to Larmor's theorem the effect of a magnetic field on a group of rotating ions or electrons for all of which the ratio e/m of electric charge to mass has the same value is to superimpose on the orbital motion a precessional motion about the direction of the field with angular velocity ω given by $eH/2mc$, or, as he expressed it in his *Philosophical Magazine* article, "the oscillation thus modified will be brought back to its original aspect if the observer is attached to a frame which revolves with angular velocity $eH/2mc$ round the axis of the magnetic field."[141] Larmor's proof of his famous theorem was based on a straightforward derivation from the equation of motion (the term proportional to c^{-2} being neglected), its physical meaning being essentially the balancing of the Coriolis force by the Lorentz force. Thus, applying the theorem to the above-mentioned circular motions and keeping in mind that the third linear vibration was unaffected, he obtained immediately the splitting of the line into the Lorentz triplet. It is interesting to note that Larmor knew before Zeeman published his experiments and before the electron was discovered that a magnetic effect would depend on the charge-to-mass ratio of the charged particles. Assuming that nothing smaller than an atom could radiate, Larmor obtained a very small separation (about 2000 times too small) and concluded—just as Voigt did with respect to the Stark effect—that the effect would be too small to be detectable.

As soon, however, as the electron was discovered and the charge-to-mass ratio of cathode rays and β rays established (J. J. Thomson, 1897), the agreement of this ratio with e/m as obtained from the analysis of the Zeeman effect with known magnetic intensity lent decisive support to the hypothesis that the electron was indeed a constituent part of the atom and

[138] P. Zeeman, "Over doubletten en tripletten in het spectrum, teweeggebracht door uitwendige magnetische krachten," *Verslag van de Gewone Vergaderingen der Wis- en Natuurkundige Afdeeling, Koninklijke Akademie van Wetenschappen te Amsterdam 6*, 13–18, 99–102, 260–262 (1897); "Doublets and triplets in the spectrum produced by external magnetic forces," *Philosophical Magazine 44*, 55–60, 255–259 (1897). At about the same time also Michelson separated the components of sodium and cadmium lines with his interferometer; cf. A. Michelson, "Radiation in a magnetic field," *ibid.*, 109–115.

[139] J. Larmor, "On the theory of the magnetic influence on spectra; and on the radiation from moving ions," *Philosophical Magazine 44*, 503–512 (1897).

[140] J. Larmor, "On the dynamics of a system of electrons or ions; and on the influence of a magnetic field on optical phenomena," *Transactions of the Cambridge Philosophical Society 18*, 380–407 (1900); reprinted in Sir Joseph Larmor, *Mathematical and Physical Papers* (Cambridge University Press, 1929), vol. 2, pp. 158–191.

[141] REF. 139, p. 504. Reprinted in *Mathematical and Physical Papers* (REF. 140), vol. 2, p. 141.

responsible for the emission of light. Moreover, from the fact that the shorter wavelength component was circularly polarized in the same sense as the current in the electromagnet, König[142] and Cornu[143] inferred the sign of the charge of the electron ("resinous"). It was also rather fortunate for the development of the theory that Zeeman had been unable to resolve the real pattern of the broadened sodium lines but did so with some lines in the spectra of cadmium and zinc, lines which belong to the singlet systems. Thus, in the fall of 1897 the agreement between experience and theory was perfect.

A few weeks later, however, on December 22, 1897, Preston reported to the Royal Dublin Society on his experiments on the Zeeman effect, performed with Barrett's powerful electromagnet of the Royal College of Science in Dublin and a concave grating with a radius of 21.5 feet and 14.438 lines per inch, and declared: "It is interesting to notice that the two lines of sodium and the blue line 4800 of cadmium do not belong to the class which show as triplets. In fact, the blue cadmium line belongs to the weak-middled quartet class, while one of the D lines (D_2) shows as a sextet of fine bright lines, i.e. form sharp and equally intense lines enclosed by two somewhat less sharp on the outside. On the other hand, the other D line (D_1) shows as a quartet, not of the weak-middled class, such as 4800 of cadmium, but of the doublet type, that is, of the type which would result from the complete absorption of the central line combined with the sharp reversal of each of the side lines."[144] Preston's observation of what was subsequently called the "anomalous Zeeman effect"—in contrast to the "normal effect" which "conforms" to theory—was soon confirmed by Cornu,[145] who showed that the central component of the "triplet" of the D_1 line in the sodium spectrum was really a doublet and that each component of the "triplet" of the D_2 line was also a doublet or, in other words, that the D_1 line was a quadruplet and the D_2 line a sextuplet. From 1898 on, when Lorentz attempted unsuccessfully to interpret[146] Preston's and Cornu's observations by generalizing his theory of the normal effect, until the end of the older quantum theory, the anomalous Zeeman effect remained an unsolved problem.

[142] C. G. W. König, "Beobachtung des Zeeman'schen Phänomens," *Wiedemannsche Annalen der Physik 62*, 240–248 (1897).

[143] A. Cornu, "Sur l'observation et l'interprétation cinématique des phénomènes découverts par Mr. le Dr. Zeeman," *Comptes Rendus 125*, 555–561 (1897).

[144] T. Preston, "Radiative phenomena in a strong magnetic field," *Scientific Transactions of the Royal Dublin Society 6*, 385–391 (1897).

[145] A. Cornu, "Sur quelques résultats nouveaux relatifs au phénomène découvert par M. le Dr. Zeeman," *Comptes Rendus 126*, 181–186 (1889).

[146] H. A. Lorentz, "Beschouwingen over den invloed van een magnetisch veld op de uitstraling van licht," *Verslag van de Gewone Vergaderingen der Wis- en Natuurkundige Afdeeling, Koninklijke Akademie van Wetenschappen te Amsterdam 7*, 113–116 (1898); "Zur Theorie des Zeemaneffektes," *Physikalische Zeitschrift 1*, 39–41 (1899).

With the introduction of improved optical apparatus with higher resolving power it was soon realized, particularly through the work of Runge and Paschen in the early years of this century, that the normal triplet was the exception rather than the rule. It would lead us too far to describe in detail the progress in the study of the anomalous effect, such as Preston's discovery that analogous spectral lines of the same series, even if belonging to different elements, exhibit the same pattern ("Preston's rule"[147]) or Runge's empirical law[148] according to which all Zeeman patterns known at the time could be expressed in terms of integral multiples of the ratio a/r, where a is the normal triplet interval and r an integral number called the Runge denominator.[149]

With the advent of Bohr's theory of the hydrogen atom and the development of the older quantum theory it was, of course, an interesting and important question whether the new arsenal of concepts could also be applied to the Zeeman effect and if so, whether complete clarification could then be achieved. As mentioned in the beginning of the present section, Debye[150] and Sommerfeld[151] succeeded in interpreting the normal Zeeman effect on the basis of the new conceptions. The equations of motion for the electron of the hydrogen atom in a homogeneous magnetic field H along the z axis are

$$m\ddot{x} = \frac{e}{c} H\dot{y} - \frac{\partial V}{\partial x} \qquad m\ddot{y} = -\frac{e}{c} H\dot{x} - \frac{\partial V}{\partial y} \qquad m\ddot{z} = -\frac{\partial V}{\partial z}$$

where V is the electrostatic potential $-e^2/r$. The Hamiltonian is given

[147] T. Preston, "On the modification of the spectra of iron and other substances radiating in a strong magnetic field," *Proceedings of the Royal Society of London* (*A*), *43*, 26–31 (1898); "General law of the phenomena of magnetic perturbation of spectral lines," *Nature 59*, 248 (1899). Preston's rule was extremely useful for the determination of the series to which a given spectral line belongs.

[148] C. Runge, "Über die Zerlegung von Spektrallinien im magnetischen Felde," *Physikalische Zeitschrift 8*, 232–237 (1907).

[149] For further details on the Zeeman effect the reader is referred to W. Voigt, *Magneto- und Elektrooptik* (Teubner, Leipzig, 1908); P. Zeeman, *Researches in Magneto-optics* (Macmillan, London, 1903); P. Zeeman, *Magnetooptische Untersuchungen* (Leipzig, 1914); P. Zeeman, *Verhandelingen over magneto-optische Verschijnselen* (Leiden, 1921). The Zeeman issue of *Physica 1* (1921) with papers by H. A. Lorentz, "De theoretische Beteekenis van het Zeeman-effect"; H. Kamerlingh Onnes, "Zeeman's Ontdekking van het naar hem genoemde effect"; T. van Lohnizen, "Het anomale Zeeman-effect"; H. R. Woltjer, "Herinnering aan het Laboratorium van Prof. Zeeman"; and others. The Zeeman issue of *Die Naturwissenschaften 1921* with papers by A. Sommerfeld and E. Back, "Fünfundzwanzig Jahre Zeemaneffekt"; A. Landé, "Über den anomalen Zeemaneffekt"; and others. E. Back and A. Landé, *Zeemaneffekt und Multiplettstruktur der Spektrallinien* (Springer, Berlin, 1925). J. B. Spencer, "An historical investigation of the Zeeman effect" (Dissertation, University of Wisconsin, 1964).

[150] P. Debye, "Quantenhypothese und Zeeman-Effekt," *Göttinger Nachrichten 1916* (June 3), pp. 142–153; *Physikalische Zeitschrift 17*, 507–512 (Sept. 7, 1916).

[151] A. Sommerfeld, "Zur Theorie des Zeeman-Effekts der Wasserstofflinien, mit einem Anhang über den Stark-Effekt," *Physikalische Zeitschrift 17*, 491–507 (Sept. 7, 1916).

by the expression

$$\frac{1}{2m}\left[\left(p_1 + \frac{e}{2c}Hy\right)^2 + \left(p_2 - \frac{e}{2c}Hx\right)^2 + p_3{}^2\right] + V$$

where the coordinates x, y, z and the momenta $p_1 = m\dot{x} - (e/2c)Hy$, $p_2 = m\dot{y} + (e/2c)Hx$, $p_3 = m\dot{z}$ are canonical variables. Transforming to spherical polar coordinates (point transformation), so that the new canonical variables are r, θ, φ, p_r, p_θ, p_φ, Debye obtained the new Hamiltonian

$$\frac{1}{2m}\left(p_r{}^2 + \frac{p_\theta{}^2}{r^2} + \frac{p_\varphi{}^2}{r^2 \sin^2\theta} - \frac{e}{c}Hp_\varphi\right) - \frac{e^2}{r}$$

where the term proportional to c^{-2} has been neglected. Since φ is cyclic, $p_\varphi = \alpha_3$ (a constant) and the Hamilton-Jacobi equation reads

$$\left(\frac{\partial W}{\partial r}\right)^2 + \frac{1}{r^2}\left(\frac{\partial W}{\partial \theta}\right)^2 + \frac{\alpha_3{}^2}{r^2 \sin^2\theta} - 2\omega m\alpha_3 - \frac{2me^2}{r} + 2m\alpha_1 = 0$$

where $\omega = eH/2mc$ and α_1 is the negative constant energy. The variables being separable,

$$p_r = \frac{\partial W}{\partial r} = \left(2\omega m\alpha_3 + \frac{2me^2}{r} - 2m\alpha_1 - \frac{\alpha_2{}^2}{r^2}\right)^{1/2} \qquad p_\theta = \frac{\partial W}{\partial \theta} = \left(\alpha_2{}^2 - \frac{\alpha_3{}^2}{\sin^2\theta}\right)^{1/2}$$

where $\alpha_2{}^2$ is the third constant of integration. Integration of $2\int_{r_1}^{r_2} p_r\,dr = m_1h$, $2\int_{\theta_1}^{\theta_2} p_\theta\,d\theta = m_2h$, and $\oint p_\varphi\,d\varphi = 2\pi\alpha_3 = m_3h$ yields

$$(\pi e^2 \sqrt{2m}/\sqrt{\alpha_1 - \omega\alpha_3}) - 2\pi\alpha_2 = m_1h \qquad \text{and} \qquad 2\pi(\alpha_2 - \alpha_3) = m_2h$$

Calling $m_1 + m_2 + m_3 = n$, Debye obtained

$$\alpha_1 = \frac{2\pi^2 me^4}{n^2h^2} + \frac{m_3h\omega}{2\pi}$$

and, by Bohr's frequency condition,

$$\nu = \frac{2\pi^2 me^4}{h^3}\left(\frac{1}{n'^2} - \frac{1}{n''^2}\right) + \frac{\omega}{2\pi}\,(m_3' - m_3'')$$

where the primed quantum numbers refer to the final orbit and the double-primed quantum numbers to the initial orbit. The magnetic field, as Debye thus could show, modifies the frequency of the emitted spectral line by a shift $\Delta\nu = \Delta m_3 \cdot (eH/4\pi mc)$ which accounts for the normal Lorentz triplet provided $\Delta m_3 = \pm 1$ or 0. That the classic Lorentz theory was able to arrive at exactly the same result was readily understood in view of the fact that the quantum of action h has canceled out. Sommerfeld, whose method of interpreting the Zeeman effect did not differ essentially from Debye's, summarized the situation as follows: "In the present state the quantum treatment of the Zeeman effect achieves just as much as

Lorentz's theory, but not more. It can account for the normal triplet . . . but hitherto it has not been able to explain the complicated Zeeman types."[152] Nor did Sommerfeld's relativistic treatment of the Zeeman effect improve the situation.[153]

The magnetic quantum number m_3, as it was called by Sommerfeld, coincided, of course, with the equatorial quantum number n_1 (see page 93) which had been previously introduced by exactly the same definition. As $n_1 = k \cos \alpha$, the H-dependent term in the energy expression, namely, $-m_3 h\omega/2\pi$, could be written as $-(kh/2\pi)(e/2mc)H \cos \alpha$ or, in vectorial notation, $-(\boldsymbol{\mu}\cdot\mathbf{H})$, which was known to be the interaction energy of a magnetic moment $\boldsymbol{\mu}$ in a magnetic field $\mathbf{H}$. That $kh/2\pi \cdot e/2mc$ is indeed the magnetic moment generated by the circling of an electron of mechanical angular momentum $p_\psi = kh/2\pi$ followed easily, as is well known, from classical electrodynamics. The magnetic moment due to the electron's orbital motion in the smallest of the Bohr orbits in the hydrogen atom ($k = 1$), i.e., $eh/4\pi mc$, was called by Pauli[154] in 1920 the "Bohr magneton μ_B"—in contradistinction to the "Weiss magneton," which Weiss,[155] in the course of his investigations on ferromagnetics, had introduced in 1911 as the greatest common divisor of the magnetic moments of molecules. Although the "Weiss magneton" was approximately five times smaller than the "Bohr magneton," it was the latter, and not the former, which was soon recognized as the natural unit of magnetic moment.

The problem of the anomalous Zeeman effect was early recognized as intimately related to the question concerning the origin of multiplet structures of spectral lines. For, as spectroscopic evidence clearly showed, all components of a multiplet exhibit the anomalous Zeeman effect whereas singlets show only the normal effect. More generally, the type of the Zeeman pattern of a component evidently depended upon the multiplicity with which it was associated. Thus, the solution of one of these problems seemed to be the solution of both. In the course of these investigations new concepts had to be introduced which proved of fundamental importance for the further development of the theory. It is primarily for this reason that the study of multiplets and the anomalous Zeeman effect will be discussed in the sequel.

The application of Ritz's conception of terms to the alkali spectra and their line series led, as we have seen, to the conclusion that the terms

[152] REF. 20 (1923), p. 304.

[153] REF. 151, pp. 495–498.

[154] W. Pauli, "Quantentheorie und Magneton," *Physikalische Zeitschrift 21*, 615–617 (1920); reprinted in W. Pauli, *Collected Scientific Papers*, edited by R. Kronig and V. F. Weisskopf (Interscience, New York, 1964), vol. 2, pp. 36–38.

[155] P. Weiss, "Sur la rationalité des rapports des moments magnétiques moléculaires et le magnéton," *Journal de Physique 1*, 900–912, 965–988 (1911). "J'appelle cette partie aliquote commune aux moments magnétiques des atomes-grammes le magnéton-gramme," p. 967. "Über die rationalen Verhältnisse der magnetischen Momente der Moleküle und das Magneton," *Physikalische Zeitschrift 12*, 935–952 (1911).

of the sharp series were given by $Rc/(n + s)^2$, those of the principal series by $Rc/(n + p_1)^2$ or $Rc/(n + p_2)^2$, those of the diffuse series by $Rc/(n + d)^2$, and so on. The so-called "Rydberg corrections" $s, p_1, p_2, d, \ldots$ measured the deviation of each term from its analogue in the hydrogen spectrum. According to Bohr's theory the alkali spectra originate by a mechanism which differs from that of the hydrogen spectrum essentially only in one aspect: the optical electron (valence electron) moves in a field which, owing to the influence of the inner or core electrons, is not strictly Coulombian. Nonpenetrating orbits, that is, those of small eccentricity, had to be characterized by small Rydberg corrections since for them the conditions of the hydrogen atom were approximately satisfied.

As Sommerfeld had shown, the orbits could be classified by two quantum numbers: the principal quantum number n $(= n' + k)$, defining the main energy level, and the azimuthal quantum number k, defining a sublevel of the main energy level. Whereas n was any positive integer, the associated k was restricted to the condition $n \geq k > 0$ and subject to the selection rule $|\Delta k| = 1$. In accordance with the geometrical interpretation of k/n for the hydrogen atom as the ratio of minor to major semiaxis of the orbital ellipse, the orbit of greatest eccentricity—and hence with the greatest Rydberg correction, which on spectroscopic evidence was s—had to be assigned the smallest possible value of k, that is, $k = 1$. Vice versa, the greatest possible value of k, that is, $k = n$, had to be assigned to the term with the smallest Rydberg correction for a given n. Terms for which $k = 1, 2, 3, 4, \ldots$ were called $s, p, d, f, \ldots$ terms[156] in accordance with the first letter in the name of the corresponding (sharp, principal, diffuse, fundamental) series.

The question was now raised whether the two quantum numbers n and k were sufficient to characterize all known terms or energy levels. The introduction of the magnetic quantum number already showed that the answer had to be negative. Moreover, it was even at that time well established that in the alkali spectra the principal series consisted of doublets with decreasing separation as n increased, whereas the lines of the sharp series were doublets of constant separation. Resolution with high-dispersion spectrometers revealed that also the diffuse and fundamental series consisted of doublets with so-called satellites in addition to the chief lines. It was therefore obvious that the principal, diffuse, and fundamental terms themselves were doublets and the sharp terms simple or singlets. Similar spectroscopic evidence showed that the terms in the alkaline-earth spectra were singlets and triplets, with the exception of the s terms, which were always only singlets.

Until 1922 only singlet, doublet, and triplet terms were known. The

[156] Sommerfeld and others originally called f terms "b terms" in accordance with the name "Bergman series."

conception of multiplicity was considerably extended in 1923, particularly through Catalan's[157] investigations of the spectra of manganese and chromium. It became clear that complex terms up to octets do exist and that in such complex-term systems the multiplicity of levels for the sequence of s, p, d terms, etc., begins with 1 for the s terms and increases always by two units for the subsequent terms until the "permanent" level number, or "permanence number" ("Permanenznummer") as Sommerfeld called it, was obtained which gave the system its name (e.g., "sextet system"). Thus, for a sextet, for example, the multiplicities of the s, p, d, f, g terms were found to be, respectively, 1, 3, 5, 6, 6. Groups of spectral lines which arise from transitions between such complex-term systems were called "multiplets," as suggested by Catalan.[158] It was now also an established fact that atoms of the elements in each column of the periodic table have characteristic multiplicities: doublets for the alkali elements, singlets and triplets for the alkaline-earth elements, doublets and quartets for the elements of the third column, and so on.

Spectroscopic evidence now showed that in such multiplets—as, for example, in the group of lines corresponding to transitions between p and d terms—not all the lines are present which are allowed by Ritz's combination principle. This nonappearance of certain lines was for Sommerfeld[159] an important clue. For it suggested to him the operation of an as yet undetermined selection principle which prohibits the appearance of lines otherwise permissible. In order to discover this selection principle, he assigned tentatively to each spectral term a qualifying number n_i which he called "inner quantum number" and which he later, following a suggestion by Bohr, denoted by the letter j. Each term was thus characterized by three quantum numbers and, as Sommerfeld proposed, was denoted as, for example, in the case $n = 5$, $k = 1$, $j = 3$, by the notation ${}^5s_3$. Analyzing the doublets and triplets of diffuse series, Sommerfeld found it possible to assign numerical values to the inner quantum number so as to establish a selection rule $\Delta j = \pm 1$ or 0. Landé[160] soon adduced additional evidence to the effect that the transition $j = 0$ to $j = 0$ has to be excluded. In the case of odd multiplicities Sommerfeld assigned to s terms $j = 0$ for singlets, $j = 1$ for triplets, $j = 2$ for quintets, etc., to p terms $j = 1$ for singlets, $j = 2, 1, 0$ for triplets, $j = 3, 2, 1$ for quintets, etc., to d terms $j = 2$ for singlets, $j = 3, 2, 1$, for triplets, $j = 4, 3, 2, 1, 0$ for quintets, and so on. Once the full number of the multiplicity has been reached—which

[157] M. A. Catalan, "Series and other regularities in the spectrum of manganese," *Philosophical Transactions of the Royal Society of London 223*, 127–173 (1923).

[158] The term "multiplet" was introduced as a *terminus technicus* in this sense for the first time on p. 146 of Catalan's paper, *ibid.*

[159] A. Sommerfeld, "Allgemeine spektroskopische Gesetze, insbesondere ein magnetooptischer Zerlegungssatz," *Annalen der Physik 63*, 221–263 (1920).

[160] A. Landé, "Anomaler Zeemaneffekt und Seriensysteme bei Ne und Hg," *Physikalische Zeitschrift 22*, 417–422 (1921).

occurs for each series with the appearance of $j = 0$—no further increase in the number of j values takes place. Sommerfeld was fully aware that the assignment of j values as proposed was not uniquely determined since the addition of an arbitrary constant did not violate the prescribed selection rule involving only differences in j. For even multiplicities Sommerfeld proposed a similar scheme in which, however, as we shall see later on, the values of j had to be half integral. In any case, the selection rule did not so far throw any light on the physical significance of the inner quantum number, which, for the time being, was only a numbering device.

Now, as a rule, all multiplet spectral lines exhibited an anomalous Zeeman effect. On the basis of van Lohuizen's[161] investigations and the combination principle it was clear that this line splitting had its origin in a corresponding splitting of the multiplet energy levels. Thus it was obvious that further progress was contingent upon obtaining a successful explanation of the Zeeman splitting of multiplet levels.

To this end Sommerfeld[162] and Landé[163] formulated, in agreement with the empirical data available at that time, what has been called the "magnetic-core hypothesis." According to this hypothesis the atomic core, i.e., the nucleus and the inner (nonoptical) electrons, possesses an angular momentum of s units of $h/2\pi$ and a corresponding magnetic moment. The latter produces an axially symmetric magnetic field whose symmetry axis coincides with the direction of the core's angular momentum. In other words, the optical electron is subject to what might be called an internal Zeeman effect, its angular-momentum vector being allowed to assume only discrete inclinations with respect to the axis of the core.

For the sake of historical accuracy it should be pointed out that an interpretation of doublets and triplets as due to an internal Zeeman effect had already been proposed in 1919 by Roschdestwensky (Rojdestvensky).[164] On the basis of a theory, specially construed for this purpose, Roschdestwensky calculated the doublet separation of the first line in the principal series of lithium and found a value about five times too large. His theory, as he later realized himself, was incompatible with Bohr's correspondence principle, of which he knew nothing because of the political situation at the time.

[161] T. van Lohuizen, "The anomalous Zeeman-effect," *Proceedings of the Amsterdam Academy 22*, 190–199 (1919).

[162] A. Sommerfeld, "Über die Deutung verwickelter Spektren (Mangan, Chrom, usw.) nach der Methode der inneren Quantenzahlen," *Annalen der Physik 70*, 32–62 (1923); "Zur Theorie der Multipletts und ihrer Zeemaneffekte," *ibid. 73*, 209–227 (1924).

[163] A. Landé, "Über den anomalen Zeemaneffekt," *Zeitschrift für Physik 5*, 231–241 (1921), *7*, 398–405 (1921); "Termstruktur und Zeemaneffekt der Multipletts," *ibid. 15*, 189–205 (1923), *19*, 112–123 (1923).

[164] D. Roschdestwensky, *Das innere Magnetfeld des Atoms erzeugt die Dublette und Triplette der Spektralserien* (Kniga Verlag, Berlin, 1922); "Address held before the Annual Convention of the Petrograd Optical Institute," Dec. 15, 1919, originally published in *Transactions of the Optical Institute in Petrograd 1*, 1–20 (1920) (in Russian).

In analogy with the vectorial representation of angular momenta in classical mechanics which Landé[165] had adopted for quantum theory as early as 1919, the angular-momentum vector **R** of the core ("Rumpf" = trunk) was assumed to combine with the angular-momentum vector **K** of the optical electron to form a resultant vector **J**, the total angular momentum of the atom. In accordance with classical mechanics **R** and **K** were assumed to precess around **J** and the latter, in the presence of an external magnetic field, was described as oscillating, like a gyroscope, about the direction of the field. The magnitude of **J** depended, of course, not only on the size of the components **K** and **R** but also on the angle between them. In fact, the different orientations of **K** with respect to the axis of **R** were correctly interpreted by Landé as the cause of the energy differences among the various sublevels of the same multiplet level. It was easy, of course, to see that the magnetic interaction between two magnetic moments is proportional to the scalar product of the respective angular momenta. Finally, Sommerfeld's inner quantum number j was identified with the magnitude of **J** in units of $h/2\pi$ and the physical significance of j as representing the total angular momentum of the atom was thereby established.

The vector model, however, soon gave rise to serious difficulties. If j, l, and s express,[166] in units of $h/2\pi$, the magnitudes, respectively, of the vectors **J**, **K**, and **R**, the geometry of the vector diagram implied that $|l - s| \leq j \leq |l + s|$. Thus, if j, l, and s are integers, the inner quantum number j, for given values of l and s, can obviously assume $r = l + s - |l - s| + 1$ different values. For $l \leq s$, therefore, a multiplet level should be composed of $2l + 1$ sublevels and for $l \geq s$ of $2s + 1$ sublevels. Since for $l \leq s$ the number of levels increases with increasing l, the multiplicities $r = 2l + 1$ of $s, p, d, \ldots$ levels reveal immediately the magnitude l of the corresponding angular-momentum vector **K**. In accordance with the accepted model of the atom one had to expect, of course, that for s levels $l = 1$, for p levels $l = 2$, etc. Spectroscopic experience, however, showed irrefutably that, contrary to these predictions, for s terms $(k = 1)$, being singlets (i.e., $r = 1$), $l = 0$, and quite generally $l = k - 1$. For a given principal quantum number n the number l, it became clear, could assume the values $0, 1, \ldots, n - 1$. The older quantum theory could never resolve this inconsistency.

It should be clear that Sommerfeld's assignment of quantum numbers was purely empirical and open to modifications. In fact, a different assignment which proved useful for many purposes and was widely adopted

[165] A. Landé, "Eine Quantenregel für die räumliche Orientierung von Elektronenringen," *Verhandlungen der Deutschen Physikalischen Gesellschaft 21*, 585–588 (1919).

[166] Here the notation has been somewhat modernized to prevent confusion and to show the correspondence with the later quantum-mechanical significance of these quantities.

was proposed by Landé[167] as follows: $R = \frac{1}{2}, \frac{2}{2}, \frac{3}{2}, \ldots$ for singlet, doublet, triplet, ... systems, $K = \frac{1}{2}, \frac{3}{2}, \frac{5}{2}, \ldots$ for $s, p, d, \ldots$ states, and J, half integral for odd multiplicities and integral for even multiplicities, as the quantum number belonging to the resultant of the vectors associated with R and K. Obviously, $R = s + \frac{1}{2}$ and $K = l + \frac{1}{2}$.[168]

Landé[169] applied his scheme to his penetrating analysis of the anomalous Zeeman effect and arrived at an important, though not complete, kinematical interpretation of this phenomenon. In the normal Zeeman effect, it will be recalled, the displacement of a resolved level was given by the expression $m_3 h\omega/2\pi$, where $\omega = eH/2mc$ and where, in accordance with the correspondence principle, a transition with $\Delta m_3 = 0$ corresponds to a line component polarized parallel to the field (π component) and $\Delta m_3 = \pm 1$ to a polarization normal to the field (σ component).

It will also be recalled that the displacement or energy shift of a multiplet line component in the anomalous Zeeman effect could be expressed according to Runge as an integral multiple of a/r, where r is an integer and $a = h\omega/2\pi$ (see page 123). Since the resolution of a multiplet component led always (in the cases under consideration) to a symmetric pattern, a specification of one-half of the (transversal) pattern contained all the information needed. A Zeeman pattern could thus conveniently be described by writing the Runge denominator r under a horizontal line and above it, as numerators, separated by commas, the displacements of the various spectral components in units of a, π components usually being enclosed in parentheses.[170] Now, just as in the discussion of the normal Zeeman effect of the hydrogen atom (page 124) m_3 was the projection of k on the direction of the external field, a quantum number m was introduced to characterize the projection of J on the direction of the field. In accordance with Landé's notation the numerical value of m, which for even multiplicities had to be half integral and for odd multiplicities integral, was subject to the condition $-|J - \frac{1}{2}| \leq m \leq |J - \frac{1}{2}|$, so that $m_{\max} = |J - \frac{1}{2}|$. Landé now assigned to each multiplet level appropriate m values in agreement with the selection and polarization rules. Thus, in the case of the (four-line) Zeeman pattern of the D_1 line of sodium, a

[167] A. Landé, "Über den anomalen Zeemaneffekt," *Zeitschrift für Physik 5*, 231–241 (1921); "Zur Theorie der anomalen Zeeman- und magnetomechanischen Effekte," *ibid. 11*, 353–363 (1922). Cf. also A. Landé, *Die neuere Entwicklung der Quantentheorie* (Steinkopff, Dresden, Leipzig, 1926, 2d ed.), pp. 64–83.

[168] In contrast to Sommerfeld's scheme, according to Landé the vectors R and K are not allowed to be collinear. For, since K is always half integral, an integral R for *even* multiplicity would lead, if collinear with K, to a half integral J associated with an *odd* multiplicity.

[169] REF. 163.

[170] The Zeeman pattern of the familiar D_1 line of sodium ($3\,^2S_{1/2} - 3\,^2P_{1/2}$) (5896 Å) could thus be written $\pm \frac{(2),4}{3}$ and of the D_2 line ($3\,^2S_{1/2} - 3\,^2P_{3/2}$) (5890 Å) as $\pm \frac{(1),3,5}{3}$ (six lines).

line which according to Landé's notation[171] originates by a transition from the level $^2S_1$ to $^2P_1$, these terms were each given the two m values $-\frac{1}{2}$ and $\frac{1}{2}$:

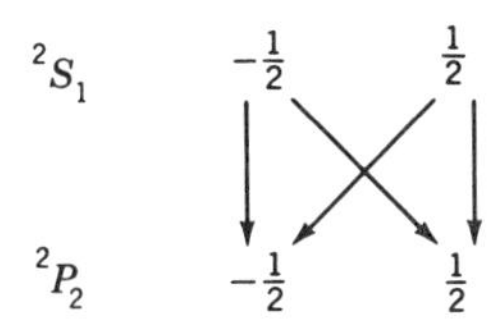

Vertical arrows indicate π components, oblique arrows indicate σ components

Similarly, the m values assigned to the $^2P_2$ term (according to Landé's notation) from which the transition associated with the D_2 line initiates were $-\frac{3}{2}$, $-\frac{1}{2}$, $\frac{1}{2}$, $\frac{3}{2}$. By assigning to each unresolved multiplet level a set of integral space m values, Landé could account for the number of lines and their polarizations in the Zeeman pattern in accordance with the selection rule $\Delta m = \pm 1$ or 0. But in order to characterize also the displacement of the various Zeeman terms, an additional set of numbers was required. These numbers in units of a give the energy difference between the Zeeman level under discussion and the unresolved multiplet level. For the D lines of the sodium spectrum Landé found the following values:

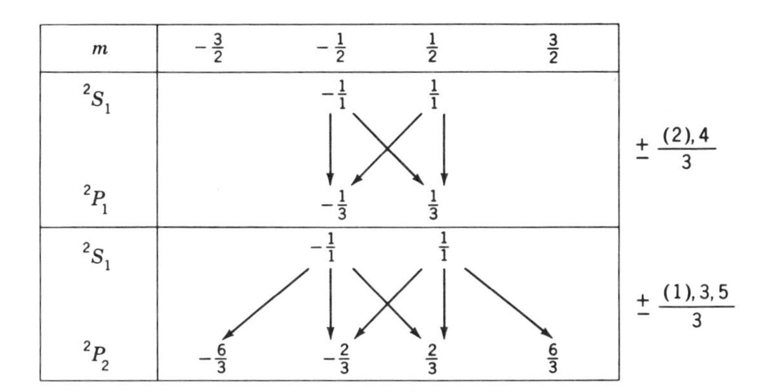

That is, with $^2S_1$ and $m = -\frac{1}{2}$ he associated $-\frac{1}{1}$, with $^2S_1$ and $m = \frac{1}{2}$ he associated $\frac{1}{1}$, and so on. Analyzing this empirical array, Landé recognized that the new numbers assigned to each term were just the products of the m value by a constant number g which was independent of m but depended on the numbers characterizing the unresolved level ($^2S_1$ or $^2P_1$ or $^2P_2$ in

[171] In Landé's term notation the upper index denotes (as usually) the maximum order of multiplicity of the group to which the term belongs. The lower index, to avoid fractions, is the greatest integer contained in J.

the case of the D lines). More specifically, Landé showed that the empirical formula

$$g = \tfrac{3}{2} + \frac{R^2 - K^2}{2(J^2 - \frac{1}{4})} \qquad \text{or} \qquad g = 1 + \frac{\bar{J}^2 + \bar{R}^2 - \bar{K}^2}{2\bar{J}^2}$$

where $\bar{J}$ denotes $\sqrt{(J + \frac{1}{2})(J - \frac{1}{2})}$ and similarly for $\bar{R}$ and $\bar{K}$, describes exactly the dependence of g, which he called the "splitting factor," on R, K, and J. For $^2S_1$ ($R = \frac{2}{2}$, $K = \frac{1}{2}$, $J = 1$), for example, $g = 2$, in agreement with the newly assigned numbers for this term.

In his search for a theoretical explanation of the splitting factor Landé reverted to the vector diagram, replaced R, K, J by $\bar{R}$, $\bar{K}$, $\bar{J}$, and deduced from the preceding expression for g that $g = 1 + [\bar{R} \cos(J, R)/\bar{J}]$. Since $m = \bar{J} \cos(J, H)$, where H indicates the direction of the external field, he obtained $mg = \bar{J} \cos(J, H) + \bar{R} \cos(J, R) \cos(J, H)$, and as the product of the last two cosine functions is precisely the average of $\cos(R, H)$, $mg = \bar{K}\, \overline{\cos(K, H)} + 2\bar{R}\, \overline{\cos(R, H)}$, where bars over the cosine functions denote average values over complete cycles. The result differed from $m = \bar{K}\, \overline{\cos(K, H)} + \bar{R}\, \overline{\cos(R, H)}$,which should have been expected were Larmor's theorem applicable to R and K.

Comparing the last two equations, Landé suggested that the appearance of the splitting factor g might be due to an anomalous ratio of the mechanical angular momentum of the core to its magnetic moment, to a ratio which is half the value to be expected on classical grounds. He proposed therefore to "modify" Larmor's theorem but admitted the lack of any convincing theoretical explanation for such a "modification."[172] Landé's perspicacious analysis of the anomalous Zeeman effect thus led to a spectroscopic confirmation of the anomalous gyromagnetic ratio whose existence after Beck's[173] careful experiments on the "Einstein–de

[172] A. Landé, "Über den anomalen Zeemaneffekt" (Part 2), *Zeitschrift für Physik* 7, 398–405 (1921). "Einen theoretischen Grund für diese Modifikation können wir nicht angeben," p. 398.

[173] E. Beck, "Zum experimentellen Nachweis der Ampèreschen Molekularströme," *Annalen der Physik 60*, 109–148 (1919). O. W. Richardson, in "A mechanical effect accompanying magnetization," *Physical Review 26*, 194–195, 248–253 (1908), was the first to suggest, on the basis of Lorentz's electron theory and Langevin's theory of magnetism, the existence of a definite relation between the angular momentum and the magnetic moment of ferromagnetic materials. A. Einstein and J. W. de Haas, in "Experimenteller Nachweis der Ampèreschen Molekularströme," *Verhandlungen der Deutschen Physikalischen Gesellschaft 17*, 152–170 (1915), investigated experimentally for the first time the gyromagnetic ratio, i.e., the ratio between the angular momentum and the magnetic moment, by imparting a rotational impulse to a rod by a given change of magnetization and found its value as predicted by theory. Shortly thereafter, however, J. S. Barnett, in "Magnetization by rotation," *Physical Review 6*, 239–270 (1915), "Magnetization of iron, nickel and cobalt by rotation and the nature of the magnetic molecule," *ibid. 10*, 7–21 (1917), and J. Q. Stewart, in "The moment of momentum accompanying magnetic moment of iron and nickel," *ibid. 11*, 100–120 (1918), obtained serious discrepancies, Stewart's value for iron being only 49 percent, of nickel 50 percent of the theoretical value. Beck, finally, showed in 1919 by a series of most careful experiments, carried out at the Swiss Federal Institute of Technology (E.T.H.), that the gyromagnetic ratio is half the value to be expected if the effect resulted from the electron orbits in the atom.

Haas effect" could no longer be doubted. Like Beck, Landé envisaged a rotation of the core or nucleus, in addition to the Larmor precession, as a possibility worthy of further investigation.

Another obscure feature which Landé's analysis involved was his introduction of the geometric means of angular momenta, the introduction of $\sqrt{(J+\frac{1}{2})(J-\frac{1}{2})}$ for J, or in Sommerfeld's notation[174] of $\sqrt{(j+1)j}$ for j, of whose legitimacy Landé entertained serious doubts in view of their "irrationality in the scale of quanta." However, he felt justified in publishing these results "because they may serve as a significant clue toward a final solution of the problem of the anomalous Zeeman splitting."[175]

3.4 *Exclusion Principle and Spin*

The Sommerfeld-Landé magnetic-core theory of the fine structure of spectral lines seemed to accord completely with the outcome of an important experiment which Otto Stern[176] proposed in 1921 as a test of Sommerfeld's space quantization in a magnetic field. Since Dunoyer's[177] pioneer work molecular beams had been investigated to some extent, but mainly in connection with problems of gas kinetics. The development of a general method for the study of molecular beams and its application to atomic physics was due mainly to the work of Stern, which he carried out first in his laboratory at the University of Frankfurt, in collaboration with Gerlach, and subsequently at the Institute for Physical Chemistry in Hamburg, to whose chairmanship he was appointed in 1923. In the first experiment, performed at the end of 1921, Stern and Gerlach[178] produced a narrow beam of silver atoms by evaporation in a heated oven and directed the beam in a high vacuum (10^{-4} to 10^{-5} mm Hg) through collimating slits along the strong field gradient at the sharp edge of the pole piece of a DuBois electromagnet. Impinging on a plate of glass, the atoms formed a thin deposit. A broadening of the deposit on the target was clearly detectable. In a second series of experiments, taking exposures of eight hours and more, Stern and Gerlach[179] established unequivocally that the atomic beam split in the magnetic field into precisely two beamlets.

[174] In Sommerfeld's notation: $g = 1 + \dfrac{j(j+1) + s(s+1) - j_a(j_a+1)}{2j(j+1)}$

[175] A. Landé, "Termstruktur und Zeemaneffekt der Multipletts," *Zeitschrift für Physik 15*, 189–205 (1923), p. 200.

[176] O. Stern, "Ein Weg zur experimentellen Prüfung der Richtungsquantelung im Magnetfeld," *Zeitschrift für Physik 7*, 249–253 (1921).

[177] L. Dunoyer, "Sur la théorie cinétique des gaz et la réalisation d'un rayonnement matériel d'origine thermique," *Comptes Rendus 152*, 529–595 (1911).

[178] O. Stern and W. Gerlach, "Der experimentelle Nachweis des magnetischen Moments des Silberatoms," *Zeitschrift für Physik 8*, 110–111 (1922).

[179] O. Stern and W. Gerlach, "Der experimentelle Nachweis der Richtungsquantelung im Magnetfeld," *Zeitschrift für Physik 9*, 349–355 (1922). "Die Aufspaltung des Atomstrahles im Magnetfeld erfolgt in zwei diskrete Strahlen. Es sind keine unabgelenkten Atome nachweisbar," p. 351.

From the geometry of the apparatus and from a careful measurement of the gradient of the magnetic-field strength they could even confirm that each silver atom possesses a magnetic moment in the direction of the field of the amount of one Bohr magneton with an error of not more than 10 percent. Similar experiments were subsequently carried out also with other atoms, such as hydrogen, sodium, potassium, cadmium, thallium, zinc, copper, and gold.[180] The conspicuous absence of any silver atoms at the center of the deflection pattern in the Stern-Gerlach experiment was a striking confirmation of space quantization. For according to classical physics, which viewed atoms as elementary magnets susceptible of assuming a continuum of orientations, the pattern should have exhibited a Gaussian distribution with its maximum at the center, the position of the undeflected beam. But in the Stern-Gerlach experiment no undeflected atoms were detectable at all! Moreover, if measurements of specific heats are ignored, it was the first experiment of a nonoptical nature with no reference to radiative energy to confirm a fundamental conclusion in quantum theory. Indeed, it was a triumph for quantum theory and Sommerfeld could rightfully declare: "With their bold experimental method Stern and Gerlach demonstrated not only the existence of space quantization, they also proved the atomistic nature of the magnetic moment, its quantum-theoretic origin and its relation to the atomic structure of electricity."

Now, it was known from spectroscopic evidence that the ground state of silver is an s state belonging to a group whose multiplicity is 2. Hence, according to the Sommerfeld-Landé theory it followed from the equation $2 = 2s + 1$ that the angular momentum of the core and, since $j_a = 0$, also the total angular momentum of the atom is $\frac{1}{2}$ in units of $h/2\pi$. The components of the total angular momentum along the direction of the field were therefore $+\frac{1}{2}$ and $-\frac{1}{2}$ and the beam should split into exactly two beamlets—in full agreement with the experiment.

Yet in spite of this apparently[181] perfect agreement with theory, the Stern-Gerlach experiment, when scrutinized closely, led to grave difficulties. In fact, Einstein and Ehrenfest,[182] in a penetrating discussion of the possible mechanisms of alignment of the atoms in the field, pointed out that no satisfactory answer could be given to the question: "How do the atoms

[180] W. Gerlach and O. Stern, "Über die Richtungsquantelung im Magnetfeld," *Annalen der Physik 74,* 673–699 (1924). J. B. Taylor, "Magnetic moments of the alkali metal atoms," *Physical Review 28,* 576–583 (1926). A. Leu, "Versuche über die Ablenkung von Molekularstrahlen im Magnetfeld," *Zeitschrift für Physik 41,* 551–562 (1927). E. Wrede, "Untersuchungen zur Molekularstrahlenmethode," *ibid.,* 569–575.

[181] The reader is aware, of course, that the subsequent discovery of the spin showed that what Stern and Gerlach actually measured, without realizing it at the time, was the intrinsic magnetic moment of the electron and that therefore the present interpretation of the experiment was erroneous.

[182] A. Einstein and P. Ehrenfest, "Quantentheoretische Bemerkungen zum Experiment von Stern und Gerlach," *Zeitschrift für Physik 11,* 31–34 (1922); *Collected Scientific Papers* (REF. 77 OF CHAP. 1), pp. 452–455.

take up their orientations?" The high vacuum of the field eliminated the possibility of explaining the process in terms of collisions, and radiative energy exchanges could not be considered for this purpose since they require 10^9 to 10^{11} sec whereas the alignment occurred in less than 10^{-4} sec. Only two alternatives were left: either "the real mechanism is such that atoms never assume states in which they are not completely quantized" or "at rapid changes states occur which, with respect to the orientation, violate the quantum rules." Either assumption, they showed, leads to insurmountable conceptual difficulties, such as to a violation of the energy principle or to the conclusion that only systems capable of emitting radiation can be quantized. Although the deficiencies of proposed models of atoms and of suggested pictorial representations of atomic processes had been repeatedly stressed before, the present paper by Einstein and Ehrenfest was an early presentation of the conceptual impasse encountered in any attempt at forming a picture of the precise behavior of atoms. It is particularly interesting to note that this "writing on the wall," years before the advent of modern quantum mechanics, had as its coauthor Albert Einstein, who persistently refused to accept the basic tenets of this greatest departure from classical idealization and who, in particular, renounced its aspect of complementarity which disclaims the possibility of universally consistent models because of the need of contradictory descriptions.

The theory which predicted space quantization thus, as we see, brought upon itself—through the very experimental verification of its prediction—insuperable difficulties. But more trouble lay ahead.

The difficulties to be faced in solving the Zeeman effect, on the one hand, and the importance of a satisfactory explanation of precisely this phenomenon for the progress of theoretical physics, on the other, were already recognized by Wolfgang Pauli in his early days. After receiving the Ph.D. degree at the University of Munich, where he had studied under Sommerfeld, he spent the winter semester 1921/22 in Göttingen as an assistant to Born. On the importance of this period of his future work Pauli wrote as follows: "A new phase of my scientific life began when I met Niels Bohr for the first time. This was in 1922, when he gave a series of guest lectures at Göttingen,[183] in which he reported on his theoretical investigations on the periodic system of elements. I shall recall only briefly that the essential progress made by Bohr's considerations at that time was the explaining by means of the spherically symmetric atomic model the formation of the intermediate shells of the atom and the general properties of the rare earths. The question as to why all electrons for an atom in its ground state were not bound in the innermost shell had already been emphasized by Bohr as a fundamental problem in his earlier works. In his Göttingen lectures he treated particularly the closing of this inner-

[183] See in this context REF. 107 OF CHAP. 1.

most K shell in the helium atom and its essential connection with the two noncombining spectra of helium, the ortho- and parahelium spectra. However, no convincing explanation for this phenomenon could be given on the basis of classical mechanics. It made a strong impression on me that Bohr at that time and in later discussions was looking for a *general* explanation which should hold for the closing of *every* electron shell and in which the number 2 was considered as essential as 8, in contrast to Sommerfeld's approach."[184]

In the fall of 1922 Pauli accepted Bohr's invitation to visit Copenhagen and to assist him in a German edition of his works. It was on this occasion that Pauli made his first serious effort to explain the anomalous Zeeman effect. In an address[185] delivered in 1945 at the Princeton Institute for Advanced Study, at a dinner given in his honor on the occasion of his Nobel Prize award (1945), Pauli referred to this visit and recalled how frustrated he had felt at that time. "The anomalous type of splitting was on the one hand especially fruitful because it exhibited beautiful and simple laws, but on the other hand it was hardly understandable, since very general assumptions concerning the electron, using classical theory as well as quantum theory, always led to the same triplet. A closer investigation of this problem left me with the feeling that it was even more unapproachable. A colleague who met me strolling rather aimlessly in the beautiful streets of Copenhagen said to me in a friendly manner, 'You look very unhappy'; whereupon I answered fiercely, "How can one look happy when he is thinking about the anomalous Zeeman effect?' "

In a paper[186] written in Copenhagen and in a sequel[187] written in Hamburg, where in 1923 he accepted the position of Privatdozent, Pauli generalized certain results of Sommerfeld's treatment of the Zeeman effect of the alkali and alkaline-earth spectra and derived the values of certain

[184] W. Pauli, *Exclusion Principle and Quantum Mechanics*, Nobel Prize Lecture delivered on Dec. 13, 1946, in Stockholm, and published by Editions du Griffon, Neuchatel (1947), pp. 9–10. The German version is reprinted in W. Pauli, *Aufsätze und Vorträge über Physik und Erkenntnistheorie* (Vieweg, Braunschweig, 1961), under the title "Das Ausschließungsprinzip und die Quantenmechanik," pp. 129–146; also reprinted in W. Pauli, *Collected Papers* (REF. 154), vol. 2, pp. 1080–1096. Cf. also *Nobel Lectures—Physics (1942–1962)*, published for the Nobel Foundation (Elsevier, Amsterdam, London, New York, 1964), pp. 27–43. The last words in this quotation refer to Sommerfeld's Keplerian attempt to connect the number 8 with the number of corners of a cube.

[185] W. Pauli, "Remarks on the History of the Exclusion Principle," *Science 103*, 213–215 (1946); *Collected Papers* (REF. 154), vol. 2, pp. 1073–1075. In a letter to Sommerfeld, written in Copenhagen on June 6, 1923, Pauli described his numerous unsuccessful efforts to explain the anomalous effect; it ended with the words: "Eine Zeit Lang war ich ganz verzweifelt."

[186] W. Pauli, "Über die Gesetzmäßigkeiten des anomalen Zeemaneffekts," *Zeitschrift für Physik 16*, 155–164 (1923); *Collected Papers* (REF. 154), vol. 2, pp. 151–160.

[187] W. Pauli, "Zur Frage der Zuordnung der Komplexstrukturterme in starken und in schwachen äußeren Feldern," *Zeitschrift für Physik 20*, 371–387 (1924); *Collected Papers* (REF. 154), vol. 2, pp. 176–192.

multiplet terms previously investigated by Landé. For this purpose he associated without theoretical justification[188] appropriate magnetic moments with the angular momenta of the atom and succeeded in tracing the complete transition of the splitting from the case of weak fields to that of strong fields. However, convinced of an intimate relation between the theory of multiplet structure and the problem of the building up of the periodic system of elements, Pauli did not regard his work as final or conclusive. Referring[189] to his inaugural lecture at the University of Hamburg, he recalled: "The contents of this lecture appeared very unsatisfactory to me, since the problem of the closing of the electronic shells had been clarified no further. The only thing that was clear was that a closer relation of this problem to the theory of multiplet structure must exist. I therefore tried to examine again critically the simplest case, the doublet structure of the alkali spectra. I arrived at the result that the point of view then orthodox—according to which a finite angular momentum of the atomic core was the cause of this doublet structure—must be given up as incorrect."

In fact, in the fall of 1924 Pauli advanced some striking arguments against this "orthodox" theory or, as we called it, the magnetic-core theory, and showed its inconsistency with experience. Calculating[190] the relativistic mass variation of K-shell electrons in homologous elements, Pauli found that the ratio of the magnetic moment to the angular momentum, as obtained on the basis of the core theory, had to be multiplied by a correction factor γ, namely, the time-average value of $(1 - v^2/c^2)^{1/2}$ taken over a complete orbit of the revolving electron. He found that for Ba ($Z = 56$) $\gamma = 0.924$, for Hg ($Z = 80$) $\gamma = 0.817$, for Tl ($Z = 81$) $\gamma = 0.812$. As these results exemplify, the above-mentioned ratio (i.e., the inverse gyromagnetic ratio), according to the core theory, had to be a slowly decreasing function of the atomic number; hence Landé's g factor and the Zeeman splitting had to depend on the atomic number, a result contrary to experience.

In view of the incontestable absence of such a dependence and in view of the completely unexplained exceptional position of the K shell which the core theory regarded as possessing nonzero momentum in contrast to all other closed shells, Pauli rejected the assumption that an inner closed shell participates at all, through its contribution to the core momentum, in the formation of the multiplet structure of optical spectra and their Zeeman splittings. "We incline therefore," he declared,[191] "to doubt the

[188] Pauli emphasized this unsatisfactory feature repeatedly. "Eine befriedigende modellmäßige Deutung der dargelegten Gesetzmäßigkeiten ist uns nicht gelungen." REF. 186, p. 164; p. 160.

[189] REF. 185, p. 214; p. 1074.

[190] W. Pauli, "Über den Einfluß der Geschwindigkeitsabhängigkeit der Elektronenmasse auf den Zeemaneffekt," *Zeitschrift für Physik 31*, 373–385 (1925); *Collected Papers* (REF. 154), vol. 2, pp. 201–213.

[191] *Ibid.*, p. 383, p. 211.

correctness of the contention that rare gas configurations partake in the complex structure and the anomalous Zeeman effect in form of core momenta." Assuming that for a closed shell both the angular momentum and the magnetic moment vanish, Pauli concluded that in the case of the alkali atoms the angular momenta of the atoms as well as their energy changes in an external magnetic field are due to the valence electron alone. "In particular, the angular momenta of alkali atoms and their energy changes in an external magnetic field have to be considered as due essentially to the exclusive action of the optical electron which also has to be regarded as the source of the magnetomechanical anomaly. The doublet structure of the alkali spectra as well as the deviation from Larmor's theorem are due, according to this view, to a peculiar, classically not describable two-valuedness ("eine eigentümliche, klassisch nicht beschreibbare Art von Zweideutigkeit") in the quantum-theoretic properties of the optical electron."[192]

Pauli thus showed that the relativistically refined core theory was incompatible with experience. In this context it should perhaps be recalled that ever since Pauli wrote, at the age of twenty, at Sommerfeld's request, his comprehensive article on the theory of relativity for the *Encyklopädie der mathematischen Wissenschaften*,[193] he was not only an expert on the theory and a master of its technique but also an ardent advocate of its fundamental tenets, which he unhesitatingly espoused. In fact, it was Pauli's strong conviction in the absolute validity of the theory of relativity which made him reject the "orthodox" core theory and thus prepared the way for the conception of the spin.

In the fall of 1924, just when Pauli advanced these arguments against the magnetic-core theory, Edmund C. Stoner's classic paper on "The distribution of electrons among atomic levels"[194] appeared and gave Pauli an important clue for the progress of his work. For a full understanding of the conceptual situation at that time some details concerning the theory of the shell structure of the atom have to be recalled.

Since from optical spectra conclusions could be drawn only regarding the processes taking place in the exterior portions of the atom and since the study of the constitution of atoms had to start from the inner regions for its explanation of the building-up process, it was only natural that the theory of the shell structure had its beginning in the study of the x-ray spectra, the most important source of information concerning the internal

[192] *Ibid.*, p. 385; p. 213.

[193] *Relativitätstheorie* in *Encyklopädie der mathematischen Wissenschaften*, vol. 5, part 2 (Teubner, Leipzig, 1921), pp. 539–775; reprinted, with "Supplementary Notes by the Author," under the title *The Theory of Relativity* (Pergamon Press, New York, 1958); *Collected Papers* (REF. 154), vol. 1, pp. 1–263.

[194] E. C. Stoner, "The distribution of electrons among atomic levels," *Philosophical Magazine 48*, 719–726 (Oct. 1, 1924).

structure of the atom. In 1911 Barkla,[195] in his work on x-rays, introduced the notion of K, L, . . . "series of radiation." Two years later Moseley[196] published the results of his famous experiments according to which "every element from aluminium to gold is characterized by an integer N which determines its X-ray spectrum . . . this integer N, the atomic number of the element, is identified with the number of positive units of electricity contained in the atomic nucleus . . . ," and according to which "the frequency of any line in the X-ray spectrum is approximately proportional to $A(N - b)^2$, where A and b are constants."

When Kossel[197] soon afterward explained these experimental results on the basis of Bohr's model of the atom, he connected Barkla's energy levels with different groups of electrons in the atom, and since then the same letters K, L, M, . . . have been used to denote the various groups or shells of electrons in the atom. According to Kossel's theory the emission of characteristic x-ray lines is caused by the transition of electrons from higher quantum orbits to replace ejected inner electrons; his theory thus accounted for the fact that—contrary to optical spectra—characteristic x-ray lines manifest themselves only in emission and never in absorption spectra. It is obvious that Kossel's theory was already based on the assumption, although not explicitly stated, that to every inner orbit or shell there corresponds a maximum number of electrons and that each shell becomes particularly stable when occupied by this maximum number of electrons.

It was primarily Bohr who on the basis of his previous work succeeded in the early twenties in formulating an acceptable theory of the building up of the whole periodic system of elements. In agreement with Kossel's ideas Bohr[198] proposed in 1921 the following distribution of electrons. Referring to the closed shells of the inert gases, he declared: "For the atoms of these elements we must expect the constitutions indicated by the following symbols: Helium (2_1), Neon $(2_1\ 8_2)$, Argon $(2_1\ 8_2\ 8_2)$ . . . , where the large figures denote the number of electrons in the groups starting from the innermost one, and the small figures the total number of quanta characterising the orbits of electrons within each group."

[195] C. G. Barkla, "The spectra of the fluorescent Röntgen radiations," *Philosophical Magazine 22*, 396–412 (1911). In an earlier paper, entitled "Phenomena of Röntgen-ray transmission," *Proceedings of the Cambridge Philosophical Society 15*, 257–268 (1909), Barkla called the first two series A and B, a notation which he revised in 1911 with the remark: "The letters K and L are, however, preferable, as it is highly probable that series of radiations both more absorbable and more penetrating exist." Cf. footnote on p. 406 of his 1911 paper.

[196] H. G. J. Moseley, "The high-frequency spectra of the elements," *Philosophical Magazine 26*, 1024–1034 (1913), *27*, 703–713 (1914).

[197] W. Kossel, "Bemerkung zur Absorption homogener Röntgenstrahlen," *Verhandlungen der Deutschen Physikalischen Gesellschaft 16*, 898–909, 953–963 (1914); "Bemerkungen zum Seriencharakter der Röntgenspektren," *ibid. 18*, 339–359 (1916).

[198] N. Bohr, "Atomic structure," *Nature 107*, 104–107 (1921).

Using all the evidence furnished by physics and chemistry at that time and employing half theoretical and half empirical methods, Bohr and others suggested more detailed schemes of distribution. The first attempt, of course, was to consider also the azimuthal quantum number k which, being subject to the condition $0 < k \leq n$, would have led to the conclusion that K terms, the energy terms corresponding to $n = 1$ and $k = 1$, are single, L terms, corresponding to $n = 2$ and $k = 1$ or 2, are doublets (which, as we have seen, should remain unresolved in the case of a perfectly Coulombian field), M terms triplets, etc. The analysis of x-ray spectra, however, showed that L terms are triplets, M terms quintets, etc. It was therefore obvious that the two quantum numbers n and k were insufficient for a classification of the electrons in their orbits. Introducing the inner quantum number[199] j $((j + \frac{1}{2}))$ which, for doublets, is equal either to k or to $k - 1$ (for s terms j is always equal only to k), Landé proposed in 1922 a scheme[200] which assigned to each sublevel a unique set of the three quantum numbers n, k, j $((n, l + 1, j + \frac{1}{2}))$:

	K	L			M					
		L_{I}	L_{II}	L_{III}	M_{I}	M_{II}	M_{III}	M_{IV}	M_{V}	
n	1	2	2	2	3	3	3	3	3	
k	1	1	2	2	1	2	2	3	3	etc.
j	1	1	1	2	1	1	2	2	3	

Thus, for example, for the electrons in sublevel L_{II}: $n = 2, k = 2$, and $j = 1$.

For the distribution of the electrons with respect to the different shells the scheme proposed by Bohr and Coster[201] was still generally accepted. The table on page 141—it suffices for our purpose to give only its beginning—indicates this scheme.

Meanwhile, however, careful studies of the x-ray absorption bands and precision measurements of the relative intensities of x-rays (e.g., of the K_{α_1} line, corresponding to the transition $L_{III} \rightarrow K$, and of the K_{α_2} line, corresponding to $L_{II} \rightarrow K$) were carried out, and from these additional information could be inferred. The results of these experiments, performed

[199] For the convenience of the reader who may wish to refer to the original papers, the original notation is given and the modern notation is added between double parentheses.

[200] A. Landé, "Zur Theorie der Röntgenspektren," *Zeitschrift für Physik 16*, 391–395 (1922).

[201] N. Bohr and D. Coster, "Röntgenspektren und periodisches System der Elemente," *Zeitschrift für Physik 12*, 342–374 (1923).

N \ n_k	1_1	2_1	2_2	3_1	3_2	3_3
1 H	1					
2 He	2					
3 Li	2	1				
4 Be	2	2				
...						
10 Ne	2	4	4			
11 Na	2	4	4	1		

primarily by de Broglie and Dauvillier,[202] proved to be incompatible with the assumed distribution. Full agreement between theory and experience was achieved only when Stoner proposed his well-known scheme of electron distribution in the article which was so influential in Pauli's work. Even prior to its theoretical vindication by quantum mechanics Stoner's scheme was considered as an important improvement over Bohr's. Sommerfeld, who immediately after its publication advocated its use both in his scientific papers and in the subsequent editions of his *Atombau und Spektrallinien*, characterized its advantages in 1925 as follows: "Being based on the incontestable experience as to number and order of x-ray levels and on the association of quantum numbers with these, Stoner's scheme is much more trustworthy ("vertrauenserweckend") than Bohr's. It has an arithmetic rather than a geometric-mechanical character; without assuming any symmetry of orbits it exploits not some, but all available data of x-ray spectroscopy."[203]

What, then, was Stoner's new idea and "arithmetic" approach? Referring to Landé's classification, Stoner declared: "The number of electrons in each completed level is equal to double the sum of the inner quantum numbers as assigned, there being in the K, L, M, N levels, when completed, 2, 8 $(= 2 + 2 + 4)$, 18 $(= 2 + 2 + 4 + 4 + 6)$, . . . electrons. It is suggested that the number of electrons associated with each sub-level separately is also equal to double the inner quantum number."[204] Thus, according to Stoner, who called the azimuthal quantum number k_1 and the

[202] L. de Broglie and A. Dauvillier, "Le système spectral des rayons Röntgen et structure de l'atome," *Journal de Physique* *5*, 1–9 (1924). A. Dauvillier, "Sur la distribution des électrons entre les niveaux L des éléments," *Comptes Rendus 178*, 476–479 (1924).

[203] A. Sommerfeld, "Zur Theorie des periodischen Systems," *Physikalische Zeitschrift 26*, 70–74 (1925).

[204] REF. 194, p. 722.

inner quantum number k_2, the L_{I} sublevel, characterized by $n = 2$, $k_1 = 1$, and $k_2 = 1$ ($(n = 2, l = 0, j = \frac{1}{2})$), contains $2k_2 = 2$ electrons, the L_{II} sublevel, characterized by $n = 2$, $k_1 = 2$, and $k_2 = 1$, contains $2k_2 = 2$ electrons, and the L_{III} sublevel, characterized by $n = 2$, $k_1 = 2$, and $k_2 = 2$, contains $2k_2 = 4$ electrons, in striking contrast to Bohr's classification as proposed in his summarizing article "Linienspektren und Atombau"[205] which appeared in 1923. A greater concentration of electrons in the outer subgroups and, accordingly, the closing of the inner subgroups at an earlier stage distinguished Stoner's from Bohr's scheme. The total number of $s, p, d, \ldots$ electrons in each shell—in other words, the maximum capacity for such electrons in each shell—is given, as in Bohr's scheme, by $4k_1 - 2$ ($(2(2l + 1))$), in accordance with the equation $2 \sum k_2 = 4k_1 - 2$.

From the scientific as well as historical point of view it is interesting to note that Stoner's suggestions, at least in part, had been proposed earlier on the basis of purely chemical considerations. In fact, six months before the appearance of Stoner's paper, Main Smith published an article[206] in the *Journal of the Society of Chemical Industry* in which he not only declared, in opposition to Bohr's original scheme, that "it is certain that the 2 quanta group or L level consists of at least 3 levels, the 3 quanta group or M level of 5 levels," but also inferred from chemical evidence a detailed distribution of electrons similar to Stoner's. Still, Stoner's article, as compared with Smith's, was not only more appealing to the mind of physicists who rarely read, if at all, the official organ of the Federal Council for Pure and Applied Chemistry (London), but also more important because it connected the distribution of electrons with the problem of multiplet structure. It was this aspect of Stoner's paper which attracted Pauli's attention—apart from the fact that it challenged Bohr's review article, just referred to,[207] which happened to be one of the papers edited by Pauli for Bohr in Copenhagen. Most stimulating for Pauli was Stoner's remark that "twice the inner quantum number does give the observed term multiplicity as revealed by the spectra in a weak magnetic field. . . . In other words, the number of possible states of the (core + electron) system is equal to twice the inner quantum number, these $2j$ states being always possible and equally probable, but only manifesting themselves separately in the presence of the external field,"[208] together with his previously quoted suggestion "that the number of electrons associated with

[205] N. Bohr, "Linienspektren und Atombau," *Annalen der Physik 71*, 228–288 (1923); see, in particular, the table on p. 260.

[206] J. D. Main Smith, "Atomic structure," *Journal of the Society of Chemical Industry* (*Review*) *43*, 323–325 (Mar. 28, 1924), 437, 490, 548–549 (1924). Cf. Main Smith's claim of priority in his letter to the editor "Distribution of electrons in atoms," *Philosophical Magazine 50*, 878–879 (1925), where he complains: "On the continent and in America my priority is acknowledged, but in my own country my work has failed even to be cited in papers in your magazine."

[207] REF. 205.

[208] REF. 194, p. 725.

each sub-level separately is equal to double the inner quantum number k_2," or as Pauli later[209] formulated this statement: "For a given value of the principal quantum number, the number of energy levels of a single electron in the alkali-metal spectra in an external magnetic field is the same as the number of electrons in the closed shell of the rare gases which correspond to this principal quantum number."

Indeed, Stoner's remark led Pauli to a far-reaching conclusion. Accepting Stoner's suggestion that the number of stationary states in a magnetic field for given n, k_1, and k_2 is just $2k_2$ $((2j+1))$ and employing—in addition to the quantum numbers n, k_1, and k_2—also a quantum number m_1 $((m_j))$ to represent the component of the angular momentum in the direction of the field, Pauli recognized that the shell structure of atoms finds its natural explanation if each possible orbit or state is labeled by the four quantum numbers n, k_1, k_2, and m_1 $((n, l, j, m_j))$, where $-(k_2 - \frac{1}{2}) \leq m_1 \leq (k_2 - \frac{1}{2})$ $((-j \leq m_j \leq j))$, and if it is assumed that only *one* electron is allowed to occupy each of these states.

In fact, on the basis of these assumptions, as Pauli pointed out, a sublevel with given n, k_1, and k_2 contains

$$(k_2 - \tfrac{1}{2}) + (k_2 - \tfrac{1}{2}) + 1 = 2k_2 \qquad ((2j+1))$$

electrons, as suggested by Stoner in contrast to Bohr; a subgroup with given n and k_1 contains

$$\sum_{k_2=k_1-1}^{k_1} 2k_2 = 4k_1 - 2 \qquad \left(\left(\sum_{j=l-\frac{1}{2}}^{l+\frac{1}{2}} (2j+1) = 2(2l+1) \right)\right)$$

electrons; and a shell with given n contains

$$\sum_{k_1=1}^{n} (4k_1 - 2) = 2n^2 \qquad \left(\left(\sum_{l=0}^{n-1} 2(2l+1) = 2n^2 \right)\right)$$

electrons, as suggested by Stoner in agreement with Bohr.

Being constantly aware of Bohr's fundamental problem as to why not all electrons, for the ground state of an atom, occupy the innermost shell, Pauli realized that his idea gains general importance if interpreted as operating as a principle of prohibition. It explained the shell structure of the atom and the periods of the periodic system of the elements because it *excluded* the possibility that more than one electron occupies an orbit or state as defined above.

One way to test the principle, Pauli immediately understood, was to consider the case of triplet s terms for two "equivalent" electrons (i.e., electrons with the same n and k_1) as in the alkaline-earth atoms. In this case the set of all four quantum numbers n, k_1, k_2, and m_1 $((n, l, j, m_j))$ would be the same for both electrons and hence the principle would de-

[209] REF. 184 (1947), pp. 13–14.

mand the exclusion of such terms. Their actual nonappearance in nature—a fact previously unaccounted for—was for Pauli a striking evidence for the validity of his exclusion principle, which he formulated as follows: "There never exist two or more equivalent electrons in an atom which, in strong fields, agree in all quantum numbers n, k_1, k_2, m_1. If there exists in the atom an electron for which these quantum numbers (in the external field) have definite values, this state is 'occupied.' "[210] With regard to this formulation it should be recalled that the physical significance of k_2 was still, to say the least, problematic and that Pauli, at that time, was able to define the four quantum numbers only in the case of strong magnetic fields when the coupling between the electrons was completely broken down. However, showing on the basis of thermodynamic reasoning[211] that the number of states is invariant under a transition from strong to weak fields, Pauli was able to establish the general validity of the exclusion principle. In concluding his paper, Pauli expressed the hope that at some future date a more profound understanding of the fundamental principles of quantum theory would enable us to put the exclusion principle on firmer foundations. For the time being, however, he had to admit that "no deeper motivation of the rule can be provided."[212]

With the subsequent development of the formalism of quantum mechanics and especially through the work of Heisenberg,[213] Dirac[214]—Dirac was the first to introduce a determinantal formulation for the acceptable wave functions, a formulation which after the appearance of Slater's paper[215] was usually referred to as the "Slater determinant"—and Pauli,[216] it was recognized that the essence of the principle is the requirement that all state functions, with the inclusion of the spin functions, of a system of similar particles, provided they are fermions, must be antisymmetric with respect to particle exchange. Thus it became clear that the

[210] "Es kann niemals zwei oder mehrere äquivalente Elektronen im Atom geben, für welche in starken Feldern die Werte aller Quantenzahlen n, k_1, k_2, m_1 übereinstimmen. Ist ein Elektron im Atom vorhanden, für das diese Quantenzahlen (im äußeren Felde) bestimmte Werte haben, so ist dieser Zustand 'besetzt.' " W. Pauli, "Über den Zusammenhang des Abschlußes der Elektronengruppen im Atom mit der Komplexstruktur der Spektren," *Zeitschrift für Physik 31*, 765–785 (1925), quotation on p. 776; *Collected Papers* (REF. 154), vol. 2, p. 225.

[211] It was, of course, essentially the principle of the invariance of the statistical weights of quantum states.

[212] "Eine nähere Begründung für diese Regel können wir nicht geben," p. 776, REF. 210. Fourteen years later, before leaving Zurich for Princeton, Pauli derived the principle field-theoretically as a consequence of relativistic invariance. Cf. his paper "The connection between spin and statistics," *Physical Review 58*, 716–722 (1940); *Collected Papers* (REF. 154), vol. 2, pp. 911–918.

[213] W. Heisenberg, "Mehrkörperproblem und Resonanz in der Quantenmechanik," *Zeitschrift für Physik 38*, 411–426 (1926); "Über die Spektra von Atomsystemen mit zwei Elektronen," *ibid.* 39, 499–518 (1926).

[214] P. A. M. Dirac, "On the theory of quantum mechanics," *Proceedings of the Royal Society of London* (*A*), *112*, 661–677 (1926); especially p. 669.

[215] J. C. Slater, "The theory of complex spectra," *Physical Review 34*, 1293–1322 (1929).

[216] W. Pauli, "Über Gasentartung und Paramagnetismus," *Zeitschrift für Physik 41*, 81–102 (1927); *Collected Papers* (REF. 154), vol. 2, pp. 284–305.

principle of exclusion, like that of relativity, is not merely another theorem in physics but rather a general precept regulating the very formulation of physical laws. It was therefore no surprise that the range of its applications proved extremely wide: it was important for an understanding of the shell structure of atoms, of the energies of electrons in metals, of chemical valency, of ferromagnetism in crystals, and many other phenomena.

With the exception of a study by Henry Margenau[217] philosophical analysis has so far shown little interest in the exclusion principle in blatant contrast to its inquisitiveness concerning the relativity principle. This indifference, as Margenau mentioned, was certainly due to the fact that the exclusion principle was "born amid a frenzy of factual discoveries," that its results appeared less appealing and less paradoxical than those of relativity, and that the problems solved by it were either too new or too old to stir excitement. An additional reason, however, which Margenau seems to have overlooked, was undoubtedly the fact that the invariance requirement imposed by the relativity principle led to radical revisions of time-honored conceptions such as space and time, whereas the corresponding requirement of the exclusion principle, that is, the postulated antisymmetry of state functions (for fermions) seemed to have little bearing on the physical or philosophical significance of these functions per se. Margenau's discussion on the relation between the Pauli principle, on the one hand, and the principle of causality, on the other, deserves still today serious consideration on the part of every physicist interested in the foundations of his science.

In addition to the above-mentioned classification of the electrons in the atom by the quantum numbers n, k_1, k_2, and m_1 Pauli suggested, as an alternative in the case of strong magnetic fields, replacing k_2 by a magnetic quantum number m_2 which represented, in appropriate units, the magnetic interaction energy of the atom with the external field. Pauli thus showed not only that it was possible to determine the maximum occupation number of atomic shells but that one also could predict—with the help of the total angular momentum in the direction of the field $\sum m_1$ and the total magnetic interaction energy with the external field $\sum m_2$—the type of multiplets that can arise if the number of electrons for given n and k_1 is known. For alkali doublets the total angular momentum j of the atom (in units of $h/2\pi$) is $k_2 - \frac{1}{2}$ or, since k_2 is k_1 or $k_1 - 1$, either $k_1 - \frac{1}{2}$ or $k_1 - \frac{3}{2}$, i.e., in modern notation $j = l + s$, where $s = \pm\frac{1}{2}$. In addition, Pauli showed that in strong fields the component of the total angular momentum in the direction of the field is given by $m_l + m_s$ (in modern notation) whereas

[217] H. Margenau, "The exclusion principle and its philosophical importance," *Philosophy of Science 11*, 187–208 (1944). Cf. also H. Margenau, *The Nature of Physical Reality* (McGraw-Hill, New York, 1950), pp. 427–447.

the interaction energy (in units of μ_B H) is given by $m_l + 2m_s$. Here m_l represents the component of the orbital angular momentum l in the direction of the field and $m_s = \pm\frac{1}{2}$. Since Pauli, as we have seen, did not associate s, as had been done hitherto, with the core but rather with the valence electron itself and its "classically not describable two-valuedness," it should have been an easy matter for him to ascribe to the electron an intrinsic angular momentum $m_s = \pm\frac{1}{2}$ (in units of $h/2\pi$) and a corresponding intrinsic magnetic moment $2m_s$ (in units of μ_B). The historical question as to why Pauli was not the first to make these assumptions is the subject of a careful study by van der Waerden,[218] to which the interested reader is referred for further details.

Nevertheless, just as Stoner's paper on the distribution of electrons was instrumental for Pauli's discovery of the exclusion principle, so, in turn, was Pauli's paper on the exclusion principle influential for the discovery of the spin. In fact, the concept of the spinning electron was conceived—only to be rejected—by Kronig and was later rediscovered, so to speak, by Goudsmit and Uhlenbeck under the influence of Pauli's work.

In January, 1925, R. Kronig, at that time a traveling fellow of Columbia University, visited Tübingen, the Mecca of spectroscopists, to work with Landé, Gerlach, and Back.[219] Landé had just received "a long and very interesting letter" from Pauli with details on the exclusion principle and the assignment of the four quantum numbers. When Kronig was shown this letter and read Pauli's remarks concerning the angular momentum j and its possible values $l + \frac{1}{2}$ and $l - \frac{1}{2}$, it occurred to him immediately that the difference between the total angular momentum j and the orbital angular momentum l "might be considered as an intrinsic angular momentum" of the electron. In addition he assumed that this intrinsic angular momentum, which he interpreted dynamically as originating from a spinning motion of the electron about its own axis, is accompanied by a magnetic moment of one Bohr magneton. The same afternoon he put his new ideas to test by computing the doublet separations in the fine structure of alkali spectra. The size of these separations, as, for example, the 6 Å between the two D lines in the sodium spectrum, was a puzzle at that time. Recalling that according to the Lorentz transformations for electromagnetic field components the electrostatic field in the rest frame of the nucleus gives rise to a magnetic field in the instantaneous rest frame of the electron, he calculated the energy of interaction between the intrinsic magnetic moment of the electron and the magnetic field which, of course, apart from a factor of proportionality turned out to be the scalar product of the orbital momentum and the spin vector. Kronig thus obtained a doublet splitting

[218] B. L. van der Waerden, "Exclusion principle and spin," in *Theoretical Physics in the Twentieth Century*, edited by M. Fierz and V. F. Weisskopf (Interscience, New York, 1960), pp. 199–244 (particularly pp. 209–216).

[219] Cf. R. de L. Kronig, "The turning point," *ibid.*, p. 5 et seq.

proportional to the fourth power of the effective nuclear charge or, more precisely, proportional to the product of the squares of the screened nuclear charges inside and outside the atomic core. His result was in full agreement with experience and also in accord with Landé's semiempirical "relativistic splitting rule"[220] but in striking contrast to the magnetic-core theory, according to which the splitting should increase with only the third power of the effective nuclear charge.

Encouraged by this qualitative result, Kronig decided to put the new hypothesis also to a quantitative test by applying it to the interpretation of the simplest of all spectra, the spectrum of hydrogen or of hydrogen-like atoms. This, however, raised a serious conceptual difficulty because Sommerfeld's (spin-free) relativistic treatment, as we have seen, accounted extremely well for the observed fine structure of such spectra. The assumption of an additional spin-orbit interaction could therefore only destroy this agreement unless, for some reason or other, Sommerfeld's relativistic formula (3.4), page 94, could be differently interpreted as the result of a compensation between the spin-orbit coupling and the relativistic precession of the electronic orbit in its plane. Now, the observed violation of the selection rule $\Delta k = \pm 1$, as mentioned on page 95, and the meanwhile well-established selection rule $\Delta j = \pm 1$ or 0 for the inner quantum number as well as other arguments[221] suggested that the azimuthal quantum number k in Sommerfeld's formula (3.4) should be replaced by $j + \frac{1}{2}$. Consequently, Kronig tried to find out whether the spin-orbit interaction does, in fact, compensate the relativistic precession in such a way that for every given j the two energy levels for $l = j + \frac{1}{2}$ and $l = j - \frac{1}{2}$ coincide. For only such a degeneracy, Kronig presumed, could lead "to the same number and pattern of energy levels as predicted by Sommerfeld's original treatment." His attempts to obtain this agreement with experience, however, failed. The result of his calculation always differed from the observed value by a factor of 2. In addition, a simple calculation showed that a point on the surface of the spinning electron of classical radius and intrinsic angular momentum $h/4\pi$ would necessarily have a velocity many times greater than that of light. Finally, when also Pauli, Kramers, and Heisenberg, with whom he discussed this matter, rejected his hypothesis, Kronig decided to leave his ideas unpublished. Thus, although the atom was still unhesitatingly regarded as a tiny solar system, Kronig's suggestion

[220] A. Landé, "Die absoluten Intervalle der optischen Dubletts und Tripletts," *Zeitschrift für Physik 25*, 46–57 (1924). For details on these relativistic or "regular" doublets as well as on the "irregular" doublets and their relation to x-ray spectroscopy, the reader is referred to A. Rubinowicz, "Ursprung und Entwicklung der älteren Quantentheorie," in *Handbuch der Physik*, vol. 24 (Quantentheorie), edited by A. Smekal (Springer, Berlin, 2d ed. 1933; Edwards Brothers, Ann Arbor, Mich., 1943).

[221] J. C. Slater, "Interpretation of the hydrogen and helium spectra," *Proceedings of the Washington Academy of Sciences 11*, 732–738 (1925).

of the spin was considered a rather strange and peculiar idea; Pauli reportedly called it "a flash of wit" ("ein ganz witziger Einfall"). It seems that the main reason for the rejection of the spin was the relativistic prohibition of velocities greater than that of light.

Although Kronig, as far as we know, was the first to apply the notions of spin and intrinsic magnetic moment to the interpretation of spectra, the very idea of a magnetic electron was not new. In fact, even ten years earlier, Parsons'[222] speculative theory of magnetons anticipated the existence of intrinsically magnetic elementary particles spinning about their own axis with an angular momentum such that "the velocity at the circumference" of the particles is that of light. In 1921 A. H. Compton, who in collaboration with O. Rognley studied the intensity of x-rays reflected from crystal surfaces, found that this intensity is unaffected by a variation of the electronic distribution on the surfaces by the application of magnetic fields. On the basis of this evidence he thought it difficult to avoid the conclusion "that the elementary magnet is not the atom as a whole . . . . Since neither the molecule nor the atom gives a satisfactory explanation of the experiments, the view suggests itself that it is something within the atom, presumably the electron, which is the ultimate magnetic particle."[223] At the end of his paper Compton declared: "May I then conclude that the electron itself, spinning like a tiny gyroscope, is probably the ultimate magnetic particle." Compton apparently never applied his hypothesis to the anomalous Zeeman effect, most probably because, as it seems, he was never particularly interested in spectroscopy as such. But even had he done so, it is unlikely that prior to Landé's penetrating analysis a satisfactory understanding of this subject would have been obtained.

But it was precisely in spectroscopy where the need for an additional degree of freedom for the electron was most urgently felt and particularly so in the interpretation of the alkali doublets. For it was found that although their separations could formally be accounted for fairly well by Sommerfeld's relativistic treatment, as mentioned previously, the relativistic mechanism underlying this approach could not possibly apply. For such a mechanism presupposed orbits of radically different eccentricities whereas, as was shown by Millikan and Bowen, there was abundant evidence that the same azimuthal quantum number has to be assigned to both components of these doublets. Millikan and Bowen thus declared early in 1924: "The only way in which it appears to be possible to avoid

[222] A. L. Parsons, "A magneton theory of the structure of the atom," *Smithsonian Miscellaneous Collections 65*, No. 11 (1915).

[223] A. H. Compton, "The magnetic electron," *Journal of the Franklin Institute 192*, 145–155 (1921), based on a paper read before the American Association for the Advancement of Science, Dec. 27, 1920. Cf. also Compton's paper "Possible magnetic polarity of free electrons," *Philosophical Magazine 41*, 279–281 (1921). That "the electron is subject to rotations" he had already inferred in 1919 from his studies of x-ray scattering; cf. REF. 20 OF CHAP. 4.

the foregoing serious difficulties is to throw overboard altogether the relativity explanation of the 'relativity doublet' and to assume that the amazing success of this relativity formula . . . is not due at all to differences in the shapes of elliptical and circular orbits, as postulated by the relativity theory of doublet separations, but that there is some other cause which by mere chance leads exactly to this relativity formula without actually necessitating relativity conceptions."[224] The need for an additional degree of freedom is clearly expressed in their concluding remarks: " . . . some way must be found to permit two orbits which have the same shape (azimuthal quantum numbers) but different orientations (inner quantum numbers) to possess widely different screening constants, i.e. widely different energies. This would seem to require the introduction of a dissymmetry not heretofore contemplated into atomic models."[225]

How this fourth degree of freedom of the electron was finally, in the wake of Pauli's work, introduced by Goudsmit and Uhlenbeck, independently of Kronig's investigations, and became part of the conceptual apparatus of modern physics has been told in detail by Uhlenbeck himself:

"Goudsmit and myself hit upon this idea by studying a paper of Pauli, in which the famous exclusion principle was formulated and in which, for the first time, *four* quantum numbers were ascribed to the electron. This was done rather formally; no concrete picture was connected with it. To us this was a mystery. We were so conversant with the proposition that every quantum number corresponds to a degree of freedom, and on the other hand with the idea of a point electron, which obviously had three degrees of freedom only, that we could not place the fourth quantum number. We could understand it only if the electron was assumed to be a small sphere that could rotate . . . .

"Somewhat later we found in a paper of Abraham (to which Ehrenfest drew our attention) that for a rotating sphere with surface charge the necessary factor two (in the magnetic moment) could be understood classically. This encouraged us, but our enthusiasm was considerably reduced when we saw that the rotational velocity at the surface of the electron had to be many times the velocity of light! I remember that most of these thoughts came to us on an afternoon at the end of September 1925. We were excited, but we had not the slightest intention of publishing anything. It seemed so speculative and bold, that something ought to be wrong with it, especially since Bohr, Heisenberg and Pauli, our great authorities, had never proposed anything of the kind. But of course we told Ehrenfest. He was impressed at once, mainly, I feel, because of the visual character of our hypothesis, which was very much in his line. He

[224] R. A. Millikan and I. S. Bowen, "Some conspicuous successes of the Bohr atom and a serious difficulty," *Physical Review 24*, 223–228 (1924).

[225] *Ibid.*, p. 228.

called our attention to several points (e.g. to the fact that in 1921 A. H. Compton already had suggested the idea of a spinning electron as a possible explanation of the natural unit of magnetism) and finally said that it was either highly important or nonsense, and that we should write a short note for *Naturwissenschaften* and give it to him. He ended with the words 'und dann werden wir Herrn Lorentz fragen.' This was done. Lorentz received us with his well known great kindness, and he was very much interested, although, I feel, somewhat sceptical too. He promised to think it over. And in fact, already next week he gave us a manuscript, written in his beautiful handwriting, containing long calculations on the electromagnetic properties of rotating electrons. We could not fully understand it, but it was quite clear that the picture of the rotating electron, if taken seriously, would give rise to serious difficulties. For one thing, the magnetic energy would be so large that by the equivalence of mass and energy the electron would have a larger mass than the proton, or, if one sticks to the known mass, the electron would be bigger than the whole atom! In any case, it seemed to be nonsense. Goudsmit and myself both felt that it might be better for the present not to publish anything; but when we said this to Ehrenfest, he answered: 'Ich habe Ihren Brief schon längst abgesandt; Sie sind beide jung genug um sich eine Dummheit leisten zu können!' "[226]

This, then, is the story of how eventually, on November 20, 1925, the idea of the spinning electron was published in *Die Naturwissenschaften* in a short paper entitled "Replacement of the hypothesis of the nonmechanical stress by a postulate concerning the intrinsic behavior of every single electron."[227] As the title indicates, it was the authors' intention to eliminate from atomic physics the hypothesis of the "nonmechanical stress" in terms of which Bohr[228] had attempted to explain the doublet structure. According to this assumption the coupling between the valence electron and the atomic core should not be regarded as a coupling between two periodicity systems but was supposed, owing to the "mechanically not describable stability conditions," to impose upon the core a stress ("Zwang") which, though not reducible to the effect of an external field of force, enables the core to assume two orientations rather than one. Referring explicitly to Pauli's "classically not describable two-valuedness" and his insistence on the use of four quantum numbers, Uhlenbeck and Goudsmit proposed to reinterpret the vector R, which hitherto was thought to repre-

[226] G. E. Uhlenbeck, *Oude en nieuwe vragen der natuurkunde*, Address delivered at Leiden on Apr. 1, 1955 (North-Holland Publishing Co., Amsterdam, 1955). The quotation has been translated into English by van der Waerden. See also S. A. Goudsmit, "Die Entdeckung des Elektronenspins," *Physikalische Blätter 21*, 445–453 (1965).

[227] G. E. Uhlenbeck and S. Goudsmit, "Ersetzung der Hypothese vom unmechanischen Zwang durch eine Forderung bezüglich des inneren Verhaltens jedes einzelnen Elektrons," *Die Naturwissenschaften 13*, 953–954 (1925).

[228] REF. 205.

sent the angular momentum of the core, as due to the "intrinsic rotation" ("eigene Rotation") of the electron. In addition, to obtain full agreement between theory and experience, they postulated that "the ratio of the magnetic moment of the electron to its mechanical angular momentum for the intrinsic rotation is twice as great as for the orbital motion," referring the reader to Abraham's classic calculations.[229] They also admitted, though only in a footnote, that their proposal would lead to a "peripheral velocity" of the electron which is many times that of light.

In a letter to the editor of *Nature*,[230] in which Uhlenbeck and Goudsmit reasserted their arguments for the introduction of the spin, Bohr added a few lines expressing his acceptance of the new ideas. With reference to his own hypothesis, now challenged by these authors, he declared: "In my article expression was given to the view that these difficulties were inherently connected with the limited possibility of representing the stationary states of the atom by a mechanical model. The situation seems, however, to be somewhat altered by the introduction of the hypothesis of the spinning electron which, in spite of the incompleteness of the conclusions that can be derived from models, promises to be a very welcome supplement to our ideas of atomic structure . . . . Indeed, it opens up a very hopeful prospect of our being able to account more extensively for the properties of elements by means of mechanical models, at least in the qualitative way characteristic of applications of the correspondence principle."[231]

Bohr's optimistic reaction did not yet prepare the way for a general acceptance of the new hypothesis. A few weeks after the publication in *Nature* Kronig, in a reply,[232] reemphasized the main arguments against the acceptance of the spin hypothesis and pointed to a new additional difficulty. For Pauli[233] had already suggested in August, 1924, explaining the hyperfine structure in the Zeeman pattern of heavy atoms by the assumption of a nonvanishing nuclear angular momentum, an assumption which established the first connecting link between (atomic) spectroscopy and nuclear dynamics. Since the study of the Zeeman effect clearly shows, Kronig argued, that electrons possess invariably a magnetic moment of one Bohr magneton, no matter in what orbit or in what atom they are found, one should expect that the same holds also for the electron when it

[229] M. Abraham, "Principien der Dynamik des Elektrons," *Annalen der Physik 10*, 105–179 (1903).

[230] G. E. Uhlenbeck and S. Goudsmit, "Spinning electrons and the structure of spectra," *Nature 117*, 264–265 (1926).

[231] *Ibid.*, p. 265.

[232] R. de L. Kronig, "Spinning electrons and the structure of spectra," *Nature 117*, 555 (1926).

[233] W. Pauli, "Zur Frage der theoretischen Deutung der Satelliten einiger Spektrallinien und ihrer Beeinflußung durch magnetische Felder," *Die Naturwissenschaften 12*, 741–743 (1924); *Collected Papers* (REF. 154), vol. 2, pp. 198–200. On S. Goudsmit's and E. Back's independent suggestion of nuclear spin cf. S. A. Goudsmit, "Pauli and nuclear spin," *Physics Today 14*, 18–21 (1961).

is part of the nucleus.[234] If according to the spin hypothesis every electron had really an intrinsic moment of one magneton, Kronig contended, then the conclusion cannot be avoided that the magnetic moment of the nucleus is of the same order of magnitude, "unless the magnetic moments of all nuclear electrons just happened to cancel." But in his view there is no place in the theory of the Zeeman effect for such an additional moment of the nucleus and the probability that in all atomic nuclei the magnetic moments just cancel each other is a priori almost nil. Kronig thus concluded his reply with the words: "The new hypothesis, therefore, appears rather to effect the removal of the family ghost from the basement to the sub-basement, instead of expelling it definitely from the house."

The situation changed, however, in the spring of 1926, when Thomas[235] and subsequently Frenkel[236] succeeded in correcting the error which led to the disturbing factor of 2 in the calculation of the doublet separations. It was Thomas's paper and the correct computation of the doublet separation which in March, 1926, converted even Pauli to accept the spin hypothesis—Pauli, who still in January, in a conversation with Bohr, had called it an "Irrlehre" ("heresy"). From our modern point of view Pauli's obstinate opposition to the notion of spin was fully justified as was his original conception of it as a "classically not describable two-valuedness." For as he said himself: "Although at first I strongly doubted the correctness of this idea because of its classical mechanical character, I was finally converted to it by Thomas's calculations on the magnitude of doublet splitting. On the other hand, my earlier doubts as well as the cautious expression 'classically not describable two-valuedness' experienced a certain verification during later developments, as Bohr was able to show on the basis of wave mechanics that the electron spin cannot be measured by classically describable experiments (as, for instance, deflection of molecular beams in external electromagnetic fields) and must therefore be considered as an essentially quantum-mechanical property of the electron."[237]

How Pauli[237a] succeeded a year later in formulating within the framework of nonrelativistic quantum mechanics a consistent theory of the spinning electron with the use of his celebrated spin matrices and how his two-component equations—in the presence of some external field—soon turned out to be the nonrelativistic limit of Dirac's famous relativistic

[234] Until 1932, when J. Chadwick published his work on the discovery of the neutron [*Nature* 129, 312; *Proceedings of the Royal Society of London* (*A*), *136*, 692–708], electrons were thought to be constituent parts of the nucleus.

[235] L. H. Thomas, "The motion of the spinning electron," *Nature 117*, 514 (1926) (written February, 1926).

[236] J. Frenkel, "Die Elektrodynamik des rotierenden Elektrons," *Zeitschrift für Physik 37*, 243–262 (1926). Frenkel treated the question as a four-dimensional problem in Minkowski space, after having received from Pauli a draft of Thomas's paper.

[237] REF. 184 (1947), pp. 15–16.

[237a] W. Pauli, "Zur Quantenmechanik des magnetischen Elektrons," *Zeitschrift für Physik 43*, 601–623 (1927).

equation of the electron[237b] is well known to every student of modern physics. What is not so well known, however, is the following. The so often repeated statement—suggested, probably, by this peculiar historical development—that "the spin is a purely relativistic effect" is a misconception. For it can be shown[237c] that a consistent theory of spin $\frac{1}{2}$ with the correct value of the intrinsic magnetic moment can be established without *ad hoc* assumptions within the framework of Galileo-invariant, and not necessarily Lorentz-invariant, wave equations, though with the exclusion of spin-orbit interactions and the Darwin term (which thus are truly relativistic effects). Pauli's description of the spin "as an essentially quantum-mechanical property" remains true, however, as can easily be shown by taking the limit $h \rightarrow 0$.

As we have seen, the notion of the spin of the electron raised, within the framework of the older quantum theory and its revised vector model, serious conceptual difficulties—but it worked. It was typical, in this respect, for many other notions and relations in terms of which the older quantum theory described atomic processes surprisingly well without, however, being able to provide strictly logical proofs. In fact, the major part of the descriptive contents of the older theory could be carried over without difficulties into the new quantum mechanics as a way of describing by models, though only to a limited extent, the rigorous results of the theory.

Reviewing the development of the older quantum theory, we recall that it was essentially a restriction of classical dynamics allowing only for orbits or states which satisfy the quantum conditions. These quantum conditions, although to some extent deprived of their mysterious nature by Ehrenfest's adiabatic principle, were never given a deeper logical justification. In addition, the range of their applicability, it turned out, was severely limited. The following considerations will clarify this point. In analogy to the methods of classical dynamics in its treatment of the many-body problem also the older quantum theory developed, mainly through the work of Bohr, Born, Brody, and Epstein, a theory of perturbations. At first such an approach seemed rather futile since, contrary to celestial mechanics, where the attraction of the central body has a much higher order of magnitude than the perturbations by the planets among themselves, the repulsive forces among the electrons are of the same order of magnitude as the force of attraction exerted on them by the nucleus. These presumably insurmountable difficulties, however, were overcome precisely by that feature of quantum theory by which it differs from

[237b] P. A. M. Dirac, "The quantum theory of the electron," *Proceedings of the Royal Society of London* (*A*), *117*, 610–624 (1928), *118*, 351–361 (1928).

[237c] See, for example, A. Galindo and C. Sanchez del Rio, "Intrinsic magnetic moment as a nonrelativistic phenomenon," *American Journal of Physics 29*, 582–584 (1961), and J.-M. Lévy-Leblond, "Nonrelativistic particles and wave equations" (to be published) and *Thesis* (Paris, 1965; unpublished).

classical dynamics, namely, the quantum conditions. Yet in spite of the successful development of an intrinsically consistent theory of perturbations, the results of its applications were frustrating as the study of the simplest possible case, the three-body problem of the neutral helium atom, clearly showed. Bohr, in fact, abandoned early all hope for a strictly analytic study of polyelectronic systems. His very conception of "penetrating orbits" which he used to obtain qualitative results was, strictly speaking, the first renunciation of his own conception of atomic dynamics as a theory of periodic systems. The penetrating orbit could no longer be treated as a periodic process because of the strong interaction with the atomic core. Moreover, Bohr's introduction of the conception of "penetrating orbits" initiated the treatment of the atom by models, a method which in its further elaborations by Sommerfeld and particularly by Landé gradually lost contact with the theory of periodic or conditionally periodic motions. The necessity of introducing half odd-integral quantum numbers, the appearance of $l = k - 1$ instead of the theoretically required k, the still obscure relationship between the spin-orbit interaction and the relativity correction in the case of the hydrogen atom, and above all the absence of any definite methodology—all these deficiencies were a clear indication that the older quantum theory could be regarded as only a first step toward an as yet unknown satisfactory theory.

Even conservative physicists such as Mie agreed that fundamental revisions of classical dynamics were imperative. In an address delivered in 1925 in Freiburg, Mie[238] explained that quantum theory owes its origin—and its difficulties—to the attempt "to connect the physics of matter with the physics of the ether." He continued: "The problem of matter cannot be expected to find its solution until the hitherto accepted Hamiltonian principle is replaced by a principle which can serve as the foundation of a theoretical edifice of both a new mechanics and a new physics of the ether."

If Mie meant to say that the foundations of the new mechanics would have to be a coalescence of optics and particle dynamics, his prediction was certainly correct. In fact, the very term "quantum mechanics," which made its first appearance as a *terminus technicus* in the title of Born's paper "On quantum mechanics"[239] in 1924, had already been coined to mean a "quan-

[238] Gustav Mie, *Die Grundlagen der Quantentheorie* (Speyer and Kaerner, Freiburg in Baden, 1926).

[239] M. Born, "Über Quantenmechanik," *Zeitschrift für Physik 26*, 379–395 (1924), received June 13, 1924; reprinted in M. Born, *Ausgewählte Abhandlungen* (Vandenhoek & Ruprecht, Göttingen, 1963), vol. 2, pp. 61–77; also reprinted in *Dokumente der Naturwissenschaft—Abteilung Physik* (Ernst Battenberg Verlag, Stuttgart, 1962), vol. 2, pp. 13–29. In Born's *Vorlesungen über Atommechanik* (Springer Verlag, Berlin, 1925), completed in 1924 and based on his lectures at the University of Göttingen in the winter semester 1923/24, only the terms "quantum theory" ("Quantentheorie"), "atomic mechanics" ("Atommechanik"), in analogy to "celestial mechanics" ("Himmelsmechanik"), but not the term "quantum mechanics" ("Quantenmechanik"), appear in the text. Born's *Vorlesungen über Atommechanik* has been translated into English by J. W. Fisher and D. R. Hartree under the title *The Mechanics of the Atom* (G. Bell and Sons, London, 1927; republished by F. Ungar, New York, 1960).

tum mechanics of coupling" or of interaction and was employed by Bohr to denote his attempt to apply Kramers's treatment[240] of the interaction between light waves and electrons to the mutual interaction between the electrons themselves. It was to denote a theory of mechanics consistent with Kramers's dispersion formula. The formation of the new quantum mechanics, with respect to both its conceptual contents and its methodological procedures, did indeed originate in the study of optical problems. Our analysis of the conceptual development of modern quantum mechanics has therefore to begin with a discussion of these problems.

Before we start this discussion in the following chapter, let us make a few remarks concerning the term "quantum mechanics." In his Nobel Prize Lecture, delivered in Stockholm on December 11, 1954, on "The Statistical Interpretation of Quantum Mechanics"[241] Born stated that his 1924 paper[242] introduced "probably for the first time" ("wohl zum ersten Male") the term "quantum mechanics" ("Quantenmechanik"), as we have just mentioned. This statement was repeated by Arnim Hermann in his biography of Born when he said that "the new word appeared for the first time in Born's paper of 1924 . . . ."[243] It should be noted, however, that the almost identical expression "mechanics of quanta" had already been used earlier. Thus, for example, H. A. Lorentz, in his address "The Old and the New Mechanics,"[244] delivered at the Sorbonne on December 10, 1923, at the occasion of the Jubilee Celebration of the Société Française de Physique, declared: "All this has great beauty and extreme importance, but unfortunately we do not really understand it. We do not understand Planck's hypothesis concerning oscillators nor the exclusion of nonstationary orbits, and we do not understand, how, after all, according to Bohr's theory, light is produced. There can be no doubt, a mechanics of quanta, a mechanics of discontinuities, has still to be made."[245] These words have been quoted not only because of the terminological issue but also because Lorentz's statement reflected a general feeling that was prevailing at the time among many theoreticians and expressed their assessment of the situation of theoretical physics prior to the advent of quantum mechanics.

[240] H. A. Kramers, "The law of dispersion and Bohr's theory of spectra," *Nature 113*, 673–674 (1924).

[241] M. Born, "Die statistische Deutung der Quantenmechanik," *Les Prix Nobel en 1954* (Stockholm, 1955), pp. 79–90; *Ausgewählte Abhandlungen* (REF. 239), vol. 2, pp. 430–441; *Dokumente der Naturwissenschaft—Abteilung Physik* (REF. 239), vol. 2, pp. 1–12; *Nobel Lectures—Physics (1942–1962)* (REF. 184), pp. 256–267.

[242] REF. 239.

[243] *Dokumente der Naturwissenschaft—Abteilung Physik* (REF. 239), vol. 1, p. 14.

[244] H. A. Lorentz, "L'ancienne et la nouvelle mécanique," *Le Livre du Cinquantenaire de la Société Française de Physique* (Editions de la Revue d'Optique Théorique et Instrumentale, Paris, 1925), pp. 99–114; reprinted in H. A. Lorentz, *Collected Papers* (Marinus Nijhoff, The Hague, 1934), vol. 7, pp. 285–302.

[245] "Tout cela est d'une grande beauté et d'une extrême importance, mais malheureusement nous ne le comprenons pas. Nous ne comprenons ni l'hypothèse de Planck sur les vibrateurs, ni l'exclusion des orbits non stationaires, et nous ne voyons pas, dans la théorie de Bohr, comment, en fin de compte, la lumière est produite. Car, il faut bien l'avouer, la mécanique des quanta, la mécanique des discontinuites, doit encore être faite." *Ibid.*, p. 110; pp. 285–286.

In the context of our present discussion of Born's introduction of the term "quantum mechanics" we wish to make a few remarks on our use of this term in the present text. We shall call "quantum physics" that branch of experimental and theoretical physics which studies quantum phenomena, that is, physical processes which involve Planck's constant h, and "quantum theory" its theoretical part. Quantum theory thus comprises "quantum field theory" and "quantum mechanics." The latter, as used in our work, will denote the general abstract theory irrespective of its particular representation such as matrix mechanics or wave mechanics. These, consequently, will be regarded as special formulations of quantum mechanics.

Whether the suggested terminology, ultimately a matter of definition, is appropriate depends on the particular viewpoint adopted. Those who, like de Broglie[246] and his school, consider undulatory processes as of primary importance would probably ascribe to wave mechanics a more preferential status. Those who, like Landé or, on different grounds, Mackey, view quantum mechanics as essentially a development of statistical mechanics may even object to the term itself, as, indeed, Mackey did when he wrote: "We should like to emphasize that in quantum mechanics the quantum rules are a deduction and not a fundamental feature of the formulation of the theory. In this sense the name quantum mechanics is unfortunate. It is basically a revision of statistical mechanics in that one studies the change in time of probability measures but no longer supposes that the motion of these measures is that induced by a motion of points in phase space. The laws of quantum mechanics place certain restrictions on the possible *simultaneous* probability distributions of various observables and give differential equations which may be integrated to show how they change in time. Everything else is deducible from these laws."[247]

[246] Cf. L. de Broglie, *Non-linear Wave Mechanics—A Causal Interpretation*, translated by A. J. Knodel and J. C. Miller (Elsevier, Amsterdam, London, New York, Princeton, 1960), p. 61.

[247] G. W. Mackey, *Lecture Notes on the Mathematical Foundations of Quantum Mechanics* (mimeographed, Harvard University, 1960), p. 95; *Mathematical Foundations of Quantum Mechanics* (Benjamin, New York, Amsterdam, 1963), p. 61.

4 The Transition to Quantum Mechanics

4.1 Applications of Quantum Conceptions to Physical Optics

Since the inception of quantum theory—in fact, since Planck's "compromise"[1] prior to his discovery of the elementary quantum of action—the conflict between the undulatory and the corpuscular conceptions of radiation became more and more acute, particularly so, as we have seen, under the impact of Einstein's ideas on light quanta and energy fluctuations. Initially regarded as merely a working hypothesis or, as Einstein called it apologetically, "a heuristic viewpoint,"[2] the quantum-corpuscular view on the nature of radiation became eventually a serious challenge to the undulatory theory of light. It proved particularly important for the interpretation of processes in physical optics which involved interactions of radiation and matter.

It was primarily Arthur Holly Compton's experiment which put the quantum-corpuscular view on firm empirical foundations. In fact, Compton contributed to the acceptance of the quantum view of radiation just as much as Fresnel had done a century earlier for the acceptance of the classical wave theory of light. It should be noted, however, that whereas Fresnel's famous memoir[3] of 1818 was a theoretical vindication of Young's experimental work, Compton's contribution was an experimental confirmation of Einstein's theoretical conclusions. In view of its decisive importance for the basic conceptions of modern quantum mechanics the Compton experiment will be discussed in greater detail.

[1] See pp. 18 and 46.
[2] See REF. 117 OF CHAP. 1.
[3] A. Fresnel, "Mémoire sur la diffraction de la lumière (1818)," published in 1826 in the *Mémoires de l'Académie des Sciences* (Paris), vol. 5.

As Bartlett[4] recently pointed out, the qualitative aspects of the basic facts underlying the Compton effect had been well known from the investigations of scattered γ rays long before Compton began his study of the scattering of x-rays. In fact, as early as 1904 Eve[5] investigated the scattering of γ radiation emitted from radium and found that the secondary rays differ in their power of penetrability from the primary radiation. That secondary radiation has much greater coefficients of absorption than primary radiation was shown in 1908 by Kleeman.[6] Similar results were obtained by Madsen[7] and Florance.[8] The latter studied the scattered radiation by means of an electroscope which was moved along an arc at constant distance from the scatterer. The experiments revealed that the wavelength of the secondary radiation depends on the angle of scattering. That also for x-rays scattered radiation is less penetrating than primary radiation was shown by Sadler and Mesham[9] in 1912. Since the measurement of absorption coefficients was at that time the only technique used for investigating the quality of these radiations, no accurate quantitative results could be obtained.

The theory of high-frequency radiation scattering, prevailing at that time, was developed primarily by J. J. Thomson[10] on the basis of Maxwell's electrodynamics. According to this theory, whenever an electromagnetic pulse traverses an atom, it sets the electrons into forced oscillations and the radiation emitted by these electrons in virtue of their accelerations is identified with the scattered radiation. As a simple calculation shows,[11] the intensity I_θ observed at a distance r from the scatterer and at a scattering angle θ relative to the direction of the primary beam of intensity I is given, per unit volume of matter, by the equation

$$I_\theta = \frac{Ine^4(1 + \cos^2 \theta)}{2m^2r^2c^4} \tag{4.1}$$

where n is the number of electrons per unit volume, m and e the mass and charge of the electron, and c the velocity of light. Thus, the energy scattered

[4] A. A. Bartlett, "Compton effect: historical background," *American Journal of Physics 32*, 120–127 (1964).

[5] A. S. Eve, "On the secondary radiation caused by the beta and gamma rays of radium," *Philosophical Magazine 8*, 669–685 (1904).

[6] R. D. Kleeman, "On the different kinds of gamma rays of radium, and the secondary gamma rays which they produce," *Philosophical Magazine 15*, 638–663 (1908).

[7] J. P. V. Madsen, "Secondary gamma radiation," *Philosophical Magazine 17*, 423–448 (1909).

[8] D. C. H. Florance, "Primary and secondary gamma rays," *Philosophical Magazine 20*, 921–938 (1910).

[9] D. A. Sadler and P. Mesham, "The Röntgen radiation from substances of low atomic weight," *Philosophical Magazine 24*, 138–149 (1912).

[10] J. J. Thomson, *Conduction of Electricity through Gases* (2d ed., Cambridge University Press, 1906), pp. 321–330 (theory of the secondary radiation).

[11] For a simple derivation of this formula see A. H. Compton, *X-rays and Electrons* (Van Nostrand, New York, 1926), pp. 58–61.

is independent of the wavelength of the incident rays. The formula agreed fairly well with experiments on soft x-rays and not too small scattering angles, as shown, for example, by Barkla and Ayres.[12]

The first who questioned the validity of Thomson's theory was Gray, who, incidentally, was also one of the first to establish that γ rays and x-rays are of the same nature. Reexamining the experimental results obtained by Madsen and Florance, Gray suggested as early as 1913 "that the usual explanation of the scattering of gamma rays cannot be quite correct . . . as it appeared that when the intensity of the primary rays was diminished by lead, the softer scattered rays were not cut down so quickly as one would expect."[13] Gray also emphasized that "the quality and quantity of the scattered radiation is approximately independent of the nature of the radiator."[14] In 1917 Ishino[15] showed that the total secondary γ-radiation is much less than Thomson's formula predicted.

When Gray[16] resumed his experiments on the scattering of γ rays of radium in 1920, he recognized the inadequacy of merely comparing the qualities of the primary and scattered radiations by absorption measurements. Referring to his own measurements, he stated that "the results we have obtained would be explained if we could always look on a beam of x- or gamma-rays as a mixture of waves of definite frequencies, and if rays of a definite frequency were altered in wave-length during the process of scattering, the wave-length increasing with the angle of scattering."[17] It is interesting to note that the Braggs,[18] in their classic paper on x-ray diffraction, had already given in 1913 a detailed description of the well-known x-ray spectrometer which carries their name—they referred to it as "an apparatus resembling a spectrometer in form, an ionisation chamber taking the place of the telescope"—and that so far no use of it had been made for the study of scattering.

That the intensity of scattered x-rays of very short wavelength is considerably less than predicted by the Thomson formula (4.1) has meanwhile been confirmed irrefutably by numerous experiments. All attempts at explanations on the basis of the classical electron theory—for example, Schott's[19] introduction of constraining and damping forces on the electrons

[12] G. Barkla and T. Ayres, "The distribution of secondary x-rays and the electromagnetic pulse theory," *Philosophical Magazine 21*, 275–278 (1911).

[13] J. A. Gray, "The scattering and absorption of the gamma rays of radium," *Philosophical Magazine 26*, 611–623 (1913).

[14] *Ibid.*, p. 622.

[15] M. Ishino, "The scattering and the absorption of the gamma rays," *Philosophical Magazine 33*, 129–146 (1917).

[16] J. A. Gray, "The scattering of x- and gamma-rays," *Journal of the Franklin Institute 190*, 633–655 (1920).

[17] *Ibid.*, p. 644.

[18] W. H. Bragg and W. L. Bragg, "The reflection of x-rays from crystals," *Proceedings of the Royal Society of London (A), 88*, 428–238 (1913), *89*, 246–248 (1913).

[19] G. A. Schott, "The scattering of x- and gamma rays by rings of electrons—a crucial test of the electron theory of atoms," *Proceedings of the Royal Society of London (A), 96*, 395–423 (1920).

or Compton's[20] theory of the "large electron" whose dimensions are comparable with the wavelength of hard γ rays and may therefore give rise to interference effects between the rays scattered from the different parts of the electron—failed to account for the observed facts. A radically new approach was necessary and this approach was worked out by Compton.

Compton, like Gray, began his work with the study of γ-ray scattering. As his early paper on "The degradation of gamma-ray energy"[21] showed, he was interested not only in the experimental aspects of "the nature and the general characteristics of secondary gamma rays" but also in an explanation of "the mechanism whereby comparatively soft secondary radiation is excited by relatively hard primary radiation." From the beginning of his work, as we see, the search for a mechanism of the scattering was at the forefront of his investigations. In 1921 Compton[22] employed for the first time the Bragg spectrometer "in combination with a recording device to register the intensity of the secondary radiation for different angles of the ionization chamber." This made it possible for him to realize that "in addition to scattered radiation there appeared in the secondary rays a type of fluorescent radiation, whose wavelength was nearly independent of the substance used as a radiator, depending only upon the wavelength of the incident rays and the angle at which the secondary rays are examined."[23] Compton found that the wavelength of the rays was increased in the scattering process by about 0.03 Å over the wavelength of the primary ray. After repeated measurements with the x-ray spectrometer Compton corrected this result to 0.02 Å. At first he attempted to account for this observed softening of the secondary radiation within the framework of classical conceptions and thus took recourse to the Doppler effect. He found, however, that in order to account for the observed shift, all of the electrons in the scatterer would have to be moving in the direction of the incident radiation with a velocity of about half that of light. Since such an assumption was obviously unacceptable, Compton concluded that the classical theory was irreconcilable with experience.

At that time Compton was a member of the Committee on X-ray Spectra of the Division of Physical Sciences in the National Research Council. The Committee, which under the chairmanship of William Duane (Harvard University) included Bergen Davis (Columbia University), A. W. Hull (General Electric Research Laboratories), and D. L. Webster (Leland Stanford University), was just about to publish a report on this

[20] A. H. Compton, "The size and shape of the electron," *Journal of the Washington Academy of Sciences 8*, 1–11 (1918); *Physical Review 14*, 20–43 (1919). As he stated in this paper, Compton deemed it necessary to assume that "the electron is subject to rotations as well as translations" (p. 32); cf. REF. 223 OF CHAP. 3.

[21] *Philosophical Magazine 41*, 749–769 (1921).

[22] A. H. Compton, "Secondary high frequency radiation," *Physical Review 18*, 96–97 (1921).

[23] A. H. Compton, "The spectrum of secondary rays," *Physical Review 19*, 267–268 (1922).

field of research. Duane objected to including Compton's results in this report, claiming that the evidence was inconclusive. Only owing to the insistence of Hull, Compton's results were published in the report.[24]

Since Thomson's theory and the Doppler effect did not explain the experimental results, Compton examined "what would happen if each quantum of x-ray energy were concentrated in a single particle and would act as a unit on a single electron."[25] On the basis of this quantum-corpuscular assumption and on the basis of the conservation principles of energy and momentum, Compton derived in a way well known to every student of modern atomic physics his famous formula for the change of wavelength

$$\Delta\lambda = \left(\frac{h}{mc}\right)(1 - \cos\theta) \tag{4.2}$$

where θ is the scattering angle, his formula for the energy of the recoil electron

$$E_{\text{kin}} = h\nu\frac{2\alpha\cos^2\varphi}{(1+\alpha)^2 - \alpha^2\cos^2\varphi} \tag{4.3}$$

where φ is the angle of recoil and $\alpha = h\nu/mc^2$, as well as the relation between the scattering angle and the angle of recoil

$$\cot\varphi = -(1+\alpha)\tan\tfrac{1}{2}\theta \tag{4.4}$$

Using the K_α line of molybdenum as primary radiation and graphite as scatterer, Compton measured repeatedly the angle dependence of the scattered radiation in his laboratory at Washington University in St. Louis and found full agreement with formula (4.2). He thus felt justified in making the first report on his new interpretation of x-ray scattering at the Washington meeting of the American Physical Society, which took place on April 20 and 21, 1923.[26] One month later the *Physical Review* published his celebrated paper "A quantum theory of the scattering of x-rays by light elements,"[27] in which he declared: "The present theory depends essentially upon the assumption that each electron which is effective in the scattering scatters a complete quantum. It involves also the hypothesis that the quanta of radiation are received from definite directions and are scattered in definite directions. The experimental support of the theory indicates

[24] A. H. Compton, "Secondary radiations produced by x-rays, and some of their applications to physical problems," *Bulletin of the National Research Council 4* (part 2), No. 20 (October, 1922).

[25] A. H. Compton, "The scattering of x-rays as particles" (Paper delivered as part of a program on "Topics in the history of modern physics" on Feb. 3, 1961, at a joint session of the American Physical Society and the American Association of Physics Teachers, New York City), *American Journal of Physics 29*, 817–820 (1961).

[26] A. H. Compton, "Wave-length measurements of scattered x-rays," *Physical Review 21*, 715 (1923).

[27] *Ibid.*, 483–502 (1923).

very convincingly that a radiation quantum carries with it directed momentum as well as energy."[28]

For the sake of historical accuracy it should be noted that about the same time also Peter Debye, then at Zurich, suggested, like Compton, that x-ray scattering can be explained by the assumption that an x-ray quantum loses energy by collision with an electron. In fact, Debye's paper[29] containing this suggestion had already been received by the editor of the *Physikalische Zeitschrift* in March, 1923, one month before Compton made his first report.

Compton's results became the subject of passionate discussions at the December, 1923, meeting of the American Physical Society in Washington. Compton recalls that "having frequently repeated the experiments I entered the debate with confidence, but was nevertheless pleased to find that I had support from P. A. Ross of Stanford and M. de Broglie of Paris, who had obtained photographic spectra showing results similar to my own. Duane at Harvard with his graduate students had been able to find not the same spectrum of the scattered rays, but one which they attributed to tertiary x-rays excited by photoelectrons in the scattering material. I might have criticized his interpretation of his results on rather obvious grounds, but thought it would be wiser to let Duane himself find the answer. Duane followed up this debate by visiting my laboratory (at that time in Chicago) and invited me to his laboratory at Harvard, a courtesy that I should like to think is characteristic of the true spirit of science. The result was that neither of us could find the reason for the difference in the results at the two laboratories, but it turned out that the equipment that I was using was more sensitive and better adapted than was Duane's to a study of the phenomenon in question."[30]

During the second week of August, 1924, the British Association for the Advancement of Science had its 92d annual meeting in Toronto. A full afternoon (Friday, August 8) was dedicated to a discussion on x-ray scattering.[31] Compton defended his theory while Gray and Duane referred to experiments performed by Allison, Clark, Stifler, and others as evidence against Compton's conclusions. As Compton reports: "The result was inconclusive. It was summarized by Sir C. V. Raman by this statement to

[28] *Ibid.*, p. 501.

[29] P. Debye, "Zerstreuung von Röntgenstrahlen und Quantentheorie," *Physikalische Zeitschrift 24*, 161–166 (1923). Debye remembers that he suggested to Scherrer the performance of an experiment to confirm these ideas. However, when he was publishing his paper, he found "that Compton had done it" (*Archive for the History of Quantum Physics*, Interview on May 3, 1962). Debye also objected to calling the effect "Compton-Debye effect," *ibid.*

[30] REF. 25, p. 818.

[31] A. H. Compton, "The quantum theory of the scattering of x-rays"; J. A. Gray, "Scattering of x- and gamma rays and the production of tertiary x-rays"; W. Duane, "On secondary and tertiary radiation," *British Association Reports, Toronto, 1925*, p. 363.

me privately after the meeting. 'Compton,' he said, 'you are a good debater, but the truth is not in you.' Nevertheless, it seems to have been this discussion that stimulated Raman to the discovery of the effect which now bears his name. Duane followed up this meeting by a new interpretation of the change in wavelength which he attributed to what he called a 'box' effect, explaining that surrounding the scattering apparatus with a lead box had in some way altered the character of the radiation. This interpretation I answered by repeating the experiment out of doors with essentially the same results, and at the same time Duane and his collaborators in a repetition of their own experiments began to find the spectrum line of the changed wavelength in accord with my collision theory. At the next meeting of the American Physical Society they reported a very good measurement of this change in wavelength."[32]

In fact, during the Washington meeting (December 29 to 31, 1924) referred to by Compton, his experimental results and his hypothesis finally found unanimous consent. Ross and Webster[33] declared: "Duane's tertiary peak has not been found in any of the experiments"; Compton, Bearden, and Woo[34] refuted the "box effect," stating that "we thus find no measurable effect due to an enclosing box"; and, most important of all, Allison and Duane[35] admitted in their paper on "The Compton effect" that "Well marked broad peaks appeared on the curves, the positions of which corresponded approximately with Compton's equation."

During the years 1921 to 1924, a period in which, as we have seen, the study of x-ray scattering led to the inescapable result of a quantum-corpuscular hypothesis, it became clear that this hypothesis could be applied also to other optical phenomena which till then had been regarded as irrefutable evidence in favor of the undulatory conception of light.

In 1921 Emden[36] showed that the assumption of quanta of light accounted not only for the Stefan-Boltzmann law, Wien's displacement law, and Planck's radiation law, but also for the Doppler effect, one of the most typically undulatory phenomena in optics. A more rigorous derivation of the Doppler effect on the basis of the quantum-corpuscular view was given in June, 1922, by Schrödinger,[37] who, incidentally, attributed physical reality exclusively to waves after the advent of wave mechanics. The elementary Doppler formula could indeed be easily obtained on these assumptions, as the following rough calculation indicates. If it is assumed

[32] REF. 25, p. 819.

[33] P. A. Ross and D. L. Webster, "The Compton and Duane effects," *Physical Review 25*, 235 (1925).

[34] A. H. Compton, J. A. Bearden, and Y. H. Woo, "Tests of the effect of an enclosing box on the spectrum of scattered x-rays," *ibid.*, 236.

[35] S. K. Allison and W. Duane, "The Compton effect," *ibid.*, 235–236.

[36] R. Emden, "Über Lichtquanten," *Physikalische Zeitschrift 22*, 513–517 (1921).

[37] E. Schrödinger, "Dopplerprinzip und Bohrsche Frequenzbedingung," *Physikalische Zeitschrift 23*, 301–303 (1922).

that the light quantum $h\nu$ is emitted in the direction of the motion of the atom of mass m and velocity v_1 before the emission and velocity v_2 after the emission, the principle of energy conservation entails that $h\,d\nu = (m/2)(v_1^2 - v_2^2)$ and that of momentum conservation that $mv_1 = mv_2 + (h\nu/c)$. A combination of these two equations leads to the result that the relative frequency shift $d\nu/\nu$ is given by $(v_1 + v_2)/2c$, which is Doppler's formula, a positive velocity corresponding to a motion toward the observer and resulting in an apparent frequency increase.

Soon Maxwell's equations themselves, the very foundation of the undulatory theory, became the target of similar attacks. Only a month after the publication of Schrödinger's paper Oseen[38] demonstrated that Maxwell's equations admit solutions which agree to an arbitrary degree of approximation with the corpuscular view in so far as the emitted energy can be regarded as being concentrated within an extremely narrow solid angle. Oseen showed that this angle can be taken so small that, up to a fractional part of the order of 10^{-10}, the total energy emitted from a point source on the earth converges at a distance like that of Sirius on an area not larger than that of a small coin ("Fünfpfennigstück").

The next undulatory property to yield to corpuscular interpretations was the diffraction by an infinite (reflecting) line grating as shown, curiously enough, by Duane when he was still a staunch opponent of Compton. Taking as x axis a direction in the plane of the grating perpendicular to its lines, Duane[39] treated the translational motion of the grating along this axis under the impact of the incident quanta of light as a periodic motion with a period defined by the translation over the grating constant a. Applying the quantum condition $\int_0^a p\,dx = nh$, Duane concluded that a transfer of momentum p to the grating occurs only in integral multiples of h/a. Since the energy loss of the light quantum $h\nu$, impinging at an angle of incidence θ, can be neglected in view of the large mass of the grating, the quantum will undergo no change in its frequency when reflected at an angle θ'. From the principle of conservation of momentum it then follows that $(h\nu/c)(\sin\theta - \sin\theta') = nh/a$, which is precisely Bragg's famous formula $a(\sin\theta - \sin\theta') = n\lambda$. Duane could easily generalize his result for the reflection of x-rays from infinite three-dimensional crystals. "In the above discussion of the reflection of x-rays by crystals," he declared, "nothing has been specifically said about wave-lengths or frequencies of vibration, although in order to conform with usage the energy of the incident radiation quantum was put equal to $h\nu$, and in order to compare the formulas obtained with the usual expressions of them, the wavelength

[38] C. W. Oseen, "Die Einsteinsche Nadelstichstrahlung und die Maxwellschen Gleichungen," *Annalen der Physik 69*, 202–204 (1922).

[39] W. Duane, "The transfer in quanta of radiation momentum to matter," *Proceedings of the National Academy of Sciences 9*, 158–164 (1923).

λ was introduced. We might just as well have called the energy ε and introduced it into the equation."[40]

As Duane's work and Compton's[41] elaborations of it showed, Fraunhofer diffraction phenomena could be accounted for in terms of the quantum-corpuscular theory of light. All attempts, however, to obtain similar results for finite diffracting systems or Fresnel diffraction phenomena proved unsuccessful. Epstein and Ehrenfest,[42] who made an extensive study of this problem, failed to solve it, mainly because they could not endow light quanta with characteristics corresponding to the undulatory properties of phase relations and coherence.

The conflict between undulatory and corpuscular conceptions of light was further accentuated by the difficulty of applying quantum conceptions to optical dispersion. For according to the perturbation theory, if applied to Bohr's model of the atom, resonance with incident radiation should occur at the frequencies[43] $\sum \tau_k \nu_k$ of the Fourier components of motion. But this conclusion was incompatible with experience, such as, for example, Wood's[44] famous experiments on sodium vapor and Bevan's[45] work on potassium vapor.

Thus, a decision between the two competing theories on the nature of radiation could not be enforced. For the interpretation of optical processes involving the interaction between light and matter, the quantum-corpuscular view seemed indispensable whereas phenomena like interference and diffraction seemed to require the conceptual apparatus of the classical wave theory of light. This state of affairs was well characterized by Sir William Bragg when he said that physicists are using on Mondays, Wednesdays, and Fridays the classical theory, and on Tuesdays, Thursdays, and Saturdays the quantum theory of radiation. The use of two such mutually exclusive, and therefore contradictory, notions was rightfully regarded as a serious dilemma at the very foundations of physical science. In fact, the situation was even worse! For it seemed unavoidable to define the corpuscular quantum of radiation in terms of frequency, frequency in terms of wavelength, and wavelength could be measured only by applying undulatory notions such as interference or diffraction. In other words, a hypothesis, supported by incontestable experimental evidence, became physically significant only by the use of its own negation.

[40] *Ibid.*, p. 161.

[41] A. H. Compton, "The quantum integral and diffraction by a crystal," *Proceedings of the National Academy of Sciences 9*, 359–362 (1923).

[42] P. S. Epstein and P. Ehrenfest, "The quantum theory of the Fraunhofer diffraction," *Proceedings of the National Academy of Sciences 10*, 133–139 (1924); "Remarks on the quantum theory of diffraction," *ibid. 13*, 400–408 (1927) (written 1924).

[43] See p. 111.

[44] R. W. Wood, "A quantitative determination of the anomalous dispersion of sodium vapour in the visible and ultraviolet regions," *Philosophical Magazine 8*, 293–324 (1904).

[45] See REF. 33 OF CHAP. 2.

4.2 *The Philosophical Background of Nonclassical Interpretations*

The older quantum theory was regarded either as "a rational generalization" of classical mechanics, as Bohr repeatedly called it, or as "a restriction" of classical mechanics, as Sommerfeld and Ehrenfest in view of the restrictive quantum conditions preferred. In any case, its metaphysical and epistemological foundations were generally held to be those of classical physics. Thus, with the possible exception of Bohr's insistence on methodological considerations, no specific philosophical study of its conceptual foundations seemed to be required. In fact, before the mid-twenties, philosophers of nature or of science regarded quantum theory as of no particular interest from their point of view; it was for them merely a particular branch of physics, just like, say, acoustics or electrochemistry. Nor did physicists recognize any need for philosophical analysis. As long as optical phenomena were relegated exclusively to the physics of waves or fields, and mechanical phenomena to the physics of particles or point masses, the classical dichotomy of particles and waves with its intrinsic problems was simply carried over from the Newtonian-Maxwellian picture of reality into that of quantum theory.

As soon, however, as it became clear that the interpretation of optical phenomena required, as we have seen, the use of mutually exclusive notions, it was recognized that a critical examination and epistemological analysis of the foundations of the theory was a necessity. For if, as Cassirer once remarked, "each answer which physics imparts concerning the character and the peculiar nature of its fundamental concepts assumes for epistemology the form of a question," how much more so if physics itself faces inconsistencies in its foundations.

Physicists traditionally refrain from declaring themselves as subscribing to a particular school of philosophical thought, even if they are conscious of belonging to it. The influence of a particular philosophical climate on their scientific work, although often of decisive importance for the formation of new conceptions, is generally ignored. It is certainly true that philosophical considerations in their effect upon the physicist's mind act more like an undercurrent beneath the surface than like a patent well-defined guiding line. It is the nature of science to obliterate its philosophical preconceptions, but it is the duty of the historian and philosopher of science to recover them under the superstructure of the scientific edifice. For this task biographies, correspondence, and autobiographical remarks supply more information in general than the scientific publications themselves.

Such documentary material, if studied with scrutiny, will show that certain philosophical ideas of the late nineteenth century not only prepared the intellectual climate for, but contributed decisively to, the formation of the new conceptions of the modern quantum theory. In particular, the writings of Renouvier, Boutroux, Kierkegaard, and Høffding seem to have been influential in this respect. Following Cournot,[46] for whom natural laws as construed by science are only approximations since physical reality itself is subject to contingency[47]—an event being contingent if its opposite involves no contradiction—Renouvier was one of the earliest modern thinkers who questioned the strict validity of the causality principle as a regulative determinant of physical processes. Renouvier's point of departure was his rejection of actual infinity, which was for him "a logical self-contradiction and an empirical falsehood."[48] With it he also rejected continuity since continuity presupposes an actual infinity of gradations. Furthermore, since the establishment of a causal relation between two distinct events, especially if separated in space and time, rests on the possibility of conceiving these two events as connected with each other by a continuous chain of intermediate events, Renouvier directed his polemics against the acceptance of causality both as an idealistic category of the understanding in the Kantian sense, that is, as a principle without which experience of an intelligible world would be impossible, and as a realistic principle of order in the cosmos. To account for the incontestably cognitive function of science, in spite of this rejection of causality, Renouvier proposed a phenomenalism according to which all that we immediately know is but a particular phenomenon or "representation." Every representation has a two-fold character: it is a "representing" and a "represented" ("représentatif" and "représenté"); it is an experience *of* something and it is something *experienced.* Realism, Renouvier contends, is erroneous in so far as it assumes that the object can be divorced from its representation, and idealism is erroneous in so far as it assumes that there exist representations with nothing to represent. To avoid these errors, experience, thus doubly qualified, should be taken as self-sufficient and as constituting ultimate reality itself.[49]

[46] Antoine Augustin Cournot, *Essai sur les Fondements de nos Connaissances et sur les Caractères de la Critique Philosophique* (Hachette, Paris, 1851); *Traité de l'Enchaînement des Idées Fondamentales dans les Sciences et dans l'Histoire* (Hachette, Paris, 1861). Cf. Jean de La Harpe, *De l'Ordre et du Hasard—le Réalisme Critique d'A. A. Cournot* (Neuchâtel, Secrétariat de l'Université, 1936).

[47] "Contingens est, cuius oppositum nullam contradictionem involvit, seu quod necessarium non est," C. Wolff, *Ontologia*, paragraph 294. On the historical development of the philosophy of contingency see F. Pelikan, *Entstehung und Entwicklung des Kontingentismus* (Simion, Berlin, 1915), and A. Levi, *L'Indeterminismo nella Filosofia Francese Contemporana* (Firenze, 1905).

[48] "Je dis que si toute la suite des nombres entiers était actuellement donnée il y aurait deux nombres égaux dont l'un serait plus grand que l'autre, ce qui est une contradiction formelle, *in terminis,*..." Charles Renouvier, *Les Principes de la Nature* (Colin, Paris, 1864), p. 37.

[49] Cf. G. Séailles, *La Philosophie de Charles Renouvier* (Alcan, Paris, 1905); G. S. Milhaud, *La Philosophie de Charles Renouvier* (Vrin, Paris, 1927).

Similar ideas and, in particular, a philosophy of nature based on contingency were advanced by Boutroux, who declared even in his thesis: "Analysing the notion of natural law, as seen in the sciences themselves, I found that this law is not a first principle but rather a result; that life, feeling, and liberty are true and profound realities, whereas the relatively invariable and general forms apprehended by science are but the inadequate manifestations of these realities."[50] "All experimental finding is reduced, in the end, to confining within as close limits as possible the value of the measurable elements of phenomena. We never reach the exact points at which the phenomenon really begins and ends. Moreover, we cannot affirm that such points exist, except, perhaps, in indivisible instants; a hypothesis which, in all probability, is contrary to the nature of time itself. Thus we see, as it were, only the containers of things, not the things themselves. We do not know if things occupy, in their containers, an assignable place. Supposing that phenomena were indeterminate, though only in a certain measure insuperably transcending the range of our rough methods of reckoning, appearances would none the less be exactly as we see them. Thus, we attribute to things a purely hypothetical if not unintelligible determination when we interpret literally the principle by which any particular phenomenon is connected with any other particular phenomenon."[51]

The rejection of classical determinism at the atomic level played an important role also in the philosophical system of Charles Sanders Peirce, who contended, as early as 1868, that "nature is not regular" and added immediately: "No disorder would be less orderly than the existing arrangement."[52] His theory of tychism (τυχή = chance) according to which "chance is a factor in the universe,"[53] formulated in the late eighties and publicized in the early nineties,[54] was for Peirce a logical conclusion from the fact that deterministic mechanics with its necessarily reversible laws cannot account for the undeniable existence of growth and evolution in nature. "The law of the conservation of energy," he wrote,[55] "is equivalent to the proposition that all operations governed by mechanical laws are reversible; so that an immediate corollary from it is that growth is not explicable by these laws, even if they be not violated in the process of

[50] Emile Boutroux, *De la Contingence des Lois de la Nature* (Baillière, Paris, 1874), preface.

[51] *Ibid.*, translated into English by F. Rothwell under the title *The Contingency of the Laws of Nature* (Open Court Publishing Co., Chicago, London, 1920), p. 28. *Die Kontingenz der Naturgesetze* (Diederichs, Jena, 1911). Cf. also O. Boelitz, *Die Lehre vom Zufall bei Emile Boutroux* (Quelle und Meyer, Leipzig, 1907).

[52] C. S. Peirce, "Grounds of validity of the laws of logic," *Journal of Speculative Philosophy 2*, 193–208 (1868); reprinted in *Collected Papers of Charles Sanders Peirce*, edited by C. Hartshorne and P. Weiss (Harvard University Press, Cambridge, Mass., 1935), vol. 5, pp. 190–222; quotation on p. 213.

[53] *Collected Papers, ibid.*, vol. 6, p. 137.

[54] C. S. Peirce, "The architecture of theories," *The Monist 1*, 161–176 (1891); "The doctrine of necessity examined," *ibid. 2*, 321–337 (1892); "Reply to the necessitarians," *ibid. 3*, 526–570 (1893); *Collected Papers* (REF. 52), vol. 6, pp. 11–27, 28–45, 46–92.

[55] *Ibid.*, p. 165.

growth." The deterministic or, as he often called it, necessitarian philosophy is therefore alone unable to explain reality.

Another incontestable argument for the doctrine of tychism was in his opinion the incapability of the necessitarians to prove their position empirically by the observation of nature. The essence of their position, he maintained, was "that certain continuous quantities have certain exact values." But how can observation determine the value of such a quantity "with a probable error absolutely *nil?*" And now—almost anticipating Heisenberg's uncertainty principle—he analyzed the process of experimental observation and concluded that absolute chance, and not an indeterminacy arising merely from our ignorance, is an irreducible factor in physical processes and hence an ultimate category for the existential universe. "Try to verify any law of nature, and you will find that the more precise your observations, the more certain they will be to show irregular departures from the law. We are accustomed to ascribe these, and I do not say wrongly, to errors of observation; yet we cannot usually account for such errors in any antecedently probable way. Trace their causes back far enough and you will be forced to admit they are always due to arbitrary determination, or chance."[56]

For the time being ideas of this kind were generally regarded as merely philosophical speculations and scientists did not take them too seriously. A notable exception before the advent of modern quantum mechanics was the Viennese physicist Exner, who proposed in 1919 a statistical interpretation of the apparent deterministic behavior of macroscopic bodies as resulting from a great number of probabilistic processes at the submicroscopic level. In the ninety-third of his *Lectures on the Physical Foundations of Science* Exner[57] declared: "From a multitude of events . . . laws can be inferred which are valid for the average state ('Durchschnittszustand') of this multitude (macrocosm) whereas the individual event (microcosm) may remain undetermined. In this sense the principle of causality holds for all macroscopic occurrences without being necessarily valid for the microcosm. It also follows that the laws of the macrocosm are not absolute laws but rather laws of probability; whether they hold always and everywhere remains to be questioned. . . ."[58] He continued: ". . . every single, however specialized, physical measurement produces only an average value resulting from billions of individual motions . . . but to predict in physics the outcome of an individual process is impossible."[59]

It should be clear, of course, that such probabilistic conceptions as

[56] *Ibid.*, p. 526; p. 46. It is interesting to note that, in accordance with these conclusions, he even claimed that the application of the theory of probability in the study of "the behavior of particles of gas is evidence for the existence of real chance."

[57] F. Exner, *Vorlesungen über die Physikalischen Grundlagen der Naturwissenschaften* (Deuticke, Wien and Leipzig, 1919; 2d ed. 1922).

[58] *Ibid.*, p. 705.

[59] *Ibid.*, p. 706.

advanced by Boutroux and Exner differ fundamentally from the traditional notions of probability as used, for example, in classical statistical mechanics. In classical physics probability statements were but an expression of human ignorance of the exact details of the individual event, either because of the insufficient resolving power of our measuring instruments or because of the huge number of events involved; the individual physical process, however, was always regarded as strictly obeying the law of cause and effect, and its result was always considered as uniquely determined. The new conception of probability, on the other hand, assumed not only that macroscopic determinism is a statistical effect but also that the individual microscopic or submicroscopic event is purely contingent.

Although only a small number of scientists subscribed to these new conceptions, their effect on physical thought should not be underestimated. Even such a staunch proponent of classical thought as Henri Poincaré was not completely immune against their impact. For when Poincaré, perhaps the last of the great theoreticians in classical physics, returned from the 1911 Solvay Congress in Brussels, he wrote: "It is well known to what hypothesis Planck was led by his researches on the laws of radiation. According to these the energy of radiators of light varies in a discontinuous manner. It is this hypothesis which is called the theory of quanta. It is hardly necessary to remark how these ideas differ from traditional conceptions; physical phenomena would cease to obey laws expressible by differential equations and this, undoubtedly, would be the greatest and most radical revolution in natural philosophy since the time of Newton."[60] If this statement was obviously provoked in response to the discussions at the Congress, a passage in Poincaré's *Valeur de la Science*, which had already attracted the attention of de Broglie,[61] shows that probabilistic considerations engaged the French mathematician at least as early as 1904. For he wrote: "Facts which appear simple to us will be only the result of a very large number of elementary facts which the law of chance alone will lead to a single goal. In that case a physical law will take on a wholly new aspect; it would no longer be a differential equation; it would take on the character of a statistical law."[62] Similarly, in his *Dernières Pensées*, six months before his death, Poincaré declared: "We now wonder not only whether the differential equations of dynamics must be modified, but whether the laws of motion can still be expressed by means of differential equations. . . . It is being asked whether it is not necessary to introduce discontinuities into the natural laws, not apparent ones but essential ones."[63]

[60] H. Poincaré, "Sur la théorie des quanta," *Journal de Physique 2*, 1–34 (1912).

[61] Louis de Broglie, *Savants et Découvertes* (Albin Michel, Paris, 1951), p. 251.

[62] H. Poincaré, *La Valeur de la Science* (Flammarion, Paris, 1904), p. 210.

[63] H. Poincaré, *Dernières Pensées* (Flammarion, Paris, 1913); in part translated into English by J. W. Boldue under the title *Last Essays* (Dover, New York, 1963), pp. 75–76.

It is clear that Poincaré's question whether differential equations are still the appropriate instrument for the mathematical formulation of physical laws was but a mathematician's way of expressing his doubts in the validity of the causality principle. For, obviously, the positing of a differential equation presupposes a continuous change or a continuous chain of events as implied in the conception of causality.

There seems to be no doubt that Renouvier's and Boutroux's works had been read by the young Poincaré just as there is no doubt that Poincaré's writings on the philosophy of science had been read by the originators of modern quantum physics.[64] The influence of Poincaré's ideas on the younger generation of mathematical physicists is well exemplified by Charles Galton Darwin, whose father, Sir George Darwin, already referred to the French mathematician as "the presiding genius—or, shall I say, my patron saint?" Charles Galton Darwin was so much impressed by Poincaré's article "Sur la théorie des quanta"[65] that in August, 1914, he translated it into English.[66] Whenever facing a problem of general importance, Darwin asked himself how Poincaré would have thought about it. In a letter to Bohr, for example, in which he stressed the need for sound analytical conclusions and inductive reasoning, Darwin wrote in 1919: "I wish Poincaré were alive, as he could do that kind of argument like no one else."[67]

C. G. Darwin was one of the first to search for a logically consistent formulation of the theoretical foundations of quantum theory. In an unpublished paper entitled "Critique of the foundations of physics"[68] he wrote in 1919: "I have long felt that the fundamental basis of physics is in a desperate state. The great positive successes of the quantum theory have accentuated all along, not merely its value, but also the essential contradictions over which it rests. . . . It may be that it will prove necessary to make fundamental changes in our ideas of time and space, or to abandon the conservation of matter and electricity, or even in the last resort to endow electrons with free will." Because of the war only now was the physical significance of Millikan's[69] brilliant experimental confirmation of the quantum-corpuscular interpretation of the photoelectric effect fully recognized. Since an explanation in terms of the classical wave theory conflicted with the principle of energy conservation,[70] it was only natural that Darwin studied at first the possibilities "to be deduced from denying

[64] De Broglie (REF. 61, p. 45) declared: "Tous les jeunes gens de ma génération qui s'intéressaient à la physique mathématique se sont nourris des livres d'Henri Poincaré."

[65] REF. 60.

[66] The original typescript of the English text "On the theory of quanta" is in the *Archive for the History of Quantum Physics*, Library of the American Philosophical Society, Philadelphia.

[67] Letter to Niels Bohr, dated July 20, 1919.

[68] Manuscript in the Library of the American Philosophical Society.

[69] See p. 36.

[70] See, for example, F. K. Richtmyer, E. H. Kennard, and T. Lauritsen, *Introduction to Modern Physics* (McGraw-Hill, New York, 5th ed., 1955), p. 98.

this principle," arguing "that the most promising outlook for the reconciliation of our difficulties is to suppose that energy is not exactly conserved." And Darwin continued: "Now, no one can doubt that energy is approximately conserved, but ordinary dynamical and electrical experiments only establish it statistically, and so really only put it on the same footing as the second law of thermodynamics." In paragraph 5 of the unpublished paper Darwin discussed the consequences of denying the conservation of energy but reached no satisfactory results.

In the following years Darwin became particularly interested in the problem of how to reconcile the phenomena of optical dispersion with the theory of quanta.[71] It was in this connection that, probably for the first time in the history of physics, a solution was proposed which was based on extending the existing conceptual apparatus of theoretical physics. "It must be taken as absolutely certain that both the electromagnetic theory and the quantum theory are valid in their respective fields, and equally certain that the two descriptions are incompatible. We can only conclude that they are parts of an overriding system, which would give rise to mathematical formulae identical with those of the present theory."[72] It is hard to say whether this statement suggested a synthesis of the two disparate conceptions, particle and wave, at a higher level, as the use of the term "overriding" seems to imply,[73] or whether it referred merely to the establishment of an abstract mathematical formalism which comprises quantum theory as well as wave theory, as his emphasis on "mathematical formulae" seems to indicate. But, whatever its meaning, Darwin's suggestion of extending the existing conceptual scheme was not heeded at the time and had to wait for its realization until further experimental and theoretical research prepared the way for the conception of an entity which manifests itself as a wave or as a particle depending upon the experimental conditions under which it is observed. The development of this conception of physical reality, which is intimately connected with the name of Niels Bohr and the so-called "Copenhagen interpretation" of quantum physics, is the culmination of certain epistemological ideas whose historical origin and philosophical background will now be discussed.

There can be no doubt that the Danish precursor of modern existentialism and neoorthodox theology, Søren Kierkegaard, through his influence on Bohr, affected also the course of modern physics to some extent. This can be gathered not only from certain references and allusions in Bohr's philosophically oriented writings but also from the very fact that Harald

[71] C. G. Darwin, "A quantum theory of optical dispersion," *Nature 110*, 841–842 (1922); *Proceedings of the National Academy of Sciences 9*, 25–30 (1923).

[72] C. G. Darwin, "The wave theory and the quantum theory," *Nature 111*, 771–773 (1923).

[73] In his unpublished paper of 1919 (REF. 68) the possibility of modifying theoretical physics by "a superstructure of Maxwellian or of Planckian dynamics" is already mentioned.

Høffding,[74] an ardent student and brilliant expounder of Kierkegaard's teachings, was the principal authority on philosophical matters for the young Bohr. We know that Bohr attended Høffding's lectures at the University of Copenhagen and read Høffding's writings[75] in which Kierkegaard's philosophy occupied a prominent place. Furthermore, Høffding and Niels's father, Christian Bohr, a professor of physiology at the University of Copenhagen, were not only colleagues but also intimate friends who had much in common; Christian Bohr's anti-Haeckelism and Harald Høffding's anti-Hegelianism placed both of them in opposition to contemporary thought.

Kierkegaard's philosophy of life and religion, his so-called "qualitative dialectic," his antithesis between thought and reality, his contradictory conceptions of life, and his insistence on the necessity of choice had apparently left a deep impression on Bohr's youthful mind. In particular Kierkegaard's emphasis on the practical value of thought, his opposition to the construction of systems, and his insistence that thought could never attain reality, for as soon as it thought to have done so it falsified reality by having changed it into imagined reality—all these ideas contributed to the creation of a philosophical climate which facilitated the surrender of classical conceptions.

Of particular importance for Niels Bohr was Kierkegaard's idea, repeatedly elaborated by Høffding, that the traditional speculative philosophy, in its claim to being capable of explaining everything, forgot that the originator of the system, however unimportant he may be, forms part of the being which is to be explained. "A system can be conceived only if one could look back on completed existence—but this would presuppose that one no longer exists.[76] Man cannot without falsification conceive of himself as an impartial spectator or impersonal observer; he always necessarily remains a participant. Thus man's delimitation between the objective and the subjective is always an arbitrary act and man's life a series of decisions. Science is a determinate activity and truth a human product, not only because it is man who created knowledge, but because the very object of knowledge is far from being a thing ready-made from all eternity.

One cannot fail to note that Høffding's discussion of the problem of knowledge seems to foreshadow certain conceptual traits in later quantum mechanics, as, for example, when he declares: "In an earlier connection I made use of Schiller's words: 'Wide is the brain and narrow is the world'; and now the sentence can be reversed: 'Wide is the world and narrow is the

[74] H. Høffding, *Søren Kierkegaard als Philosoph* (Frommann, Stuttgart, 1896).

[75] H. Høffding, *A History of Modern Philosophy* (Macmillan, London, New York, 1900); *The Problems of Philosophy* (Macmillan, London, 1905); *Modern Philosophers—Lectures Delivered at the University of Copenhagen* (Macmillan, London, 1915).

[76] "An das System könnte man erst denken, wenn man auf die abgeschlossene Existenz zurückblicken könnte—das würde voraussetzen, daß man nicht mehr existierte!" REF. 74, p. 66.

brain.' Knowledge, however rich and powerful it may be, is after all only a *part* of Being; and the problem of knowledge would be soluble, only if Being as a totality (in so far as we can conceive it as such a totality) could be expressed by means of a single one of its parts. In any event, our expression must always remain symbolic, even if our knowledge reaches its climax; it gives us only an extract from a more inclusive whole. Among all the possibilities of thought, only a single one appears in the reality as recognized by us. The reality which we recognize is, however, only a part of a greater whole;—and here we are not in a position to determine the relation between the parts and the whole. An exhaustive concept of reality is not given us to create."[77]

In life only sudden decisions, leaps, or jerks can lead to progress. "Something decisive occurs always only by a jerk, by a sudden turn which neither can be predicted from its antecedents nor is determined by these."[78] Referring to Kierkegaard's indeterministic theory of "leaps," Høffding called the Danish philosopher "the only indeterministic thinker who attempted to describe the leap"[79] but added later: "It seems to be clear that if the leap occurs between two states or two moments, no eye can possibly observe it, and since it therefore can never be a phenomenon, its description ceases to be a description."[80] Consequently, also "causality cannot be described."[81]

The preceding remarks and quotations do not, of course, intend to present a coherent and comprehensive exposition of the ideas of Kierkegaard or Høffding; their purpose is only to sketch the nature of some of these philosophical conceptions which later have been transferred into the realm of natural philosophy and have stimulated a new approach for the epistemological foundations of quantum theory.

That conceptions of this kind did indeed affect the new epistemological view of scientific inquiry was admitted by Bohr himself when he described[82] how he was influenced by Poul Martin Møller's *Adventures of a Danish Student*,[83] a novel "still read with delight by the older as well as the younger

[77] H. Høffding, *The Problems of Philosophy*, translated by G. M. Fisher (Macmillan, New York, 1905), pp. 114–115.

[78] "Das Leben und die Wirklichkeit . . . geht durch stets wiederholte Sprünge vorwärts. Etwas Entscheidendes tritt immer nur durch einen Ruck ein, durch eine plötzliche Schwenkung, die sich weder aus dem, was vorausgeht, vorhersehen läßt, noch dadurch bestimmt wird." REF. 74, p. 76.

[79] *Ibid.*, p. 83.

[80] "Es scheint klar zu sein, daß, wenn der Sprung zwischen zwei Zuständen oder zwischen zwei Augenblicken liegt, ihn kein Auge beobachten kann. Die Beschreibung desselben wird dann eigentlich keine Beschreibung, da er nie Phänomen werden kann." *Ibid.*, p. 83.

[81] *Ibid.*, p. 84.

[82] N. Bohr, "The Unity of Human Knowledge," Address delivered at the Congress in Copenhagen, October, 1960, arranged by La Fondation Européenne de la Culture; reprinted in N. Bohr, *Essays 1958–1962 on Atomic Physics and Human Knowledge* (Interscience, New York, London, 1963), pp. 8–16.

[83] P. M. Møller, *En Dansk Students Eventyr* (Gyldendal, Copenhagen, 1st ed. 1893, 4th ed. 1946).

generation in this country." In this novel Bohr found a "vivid and suggestive account of the interplay between the various aspects of our position, illuminated by discussions within a circle of students with different characters and divergent attitudes of life."[84] Møller described how a licentiate of a critical, cautious, and contemplative character and attitude of mind is reproached by his practical-minded cousin for not having taken advantage of the opportunities of finding a practical job; in reply to these admonitions the author made the licentiate declare: "My endless inquiries made it impossible for me to achieve anything. Moreover, I get to think about my own thoughts of the situation in which I find myself. I even think that I think of it, and divide myself into an infinite retrogressive sequence of 'I's who consider each other. I do not know at which 'I' to stop as the actual, and as soon as I stop at one, there is indeed again an 'I' which stops at it. I become confused and feel giddy as if I were looking down into a bottomless abyss, and my ponderings result finally in a terrible headache." Whereupon the cousin replied: "I cannot help you a bit in sorting your numerous 'I's. This is beyond my sphere of action, and I would either be or become as mad as you if I let myself take part in your superhuman reveries. My line is to stick to tangible things and walk along the broad highway of common sense; thus my egos never get confused."

It can hardly be doubted that such an introspective analysis, derisive of some current trends in European philosophy and thus akin to Kierkegaard's critique, had an impact on Bohr's "rediscovery of the dialectical process of cognition which had so long been obscured by the unilateral development of epistemology on the basis of Aristotelian logic and Platonic idealism," as L. Rosenfeld[85] once expressed it. Whether Rosenfeld is right with his remark that "Poul Møller had little imagined that his light-handed banter would one day start a train of thought leading to the elucidation of the most fundamental aspects of atomic theory and the renovation of the philosophy of science"[86] or whether it was originally Kierkegaard's and Høffding's insistence on the arbitrary but necessary act of drawing a boundary line between subject and object in the process of scientific inquiry, these ideas, as we shall see later on, were of importance for the future development of modern quantum theory. In spite of the lucid and logical presentation of scientific argumentations in his writings, Bohr was never fond of system constructing or axiomatization. Like Høffding, who "gave more consideration to the proposition of problems than to their solution," Bohr declared repeatedly: "Every sentence I say must be understood not as an affirmation, but as a question."[87] Kierkegaard's and Høffding's

[84] REF. 82, p. 13.

[85] L. Rosenfeld, "Niels Bohr's contribution to epistemology," *Physics Today 16*, 47–54 (1963).

[86] *Ibid.*, p. 48.

[87] L. Rosenfeld, *Niels Bohr—An Essay* (North-Holland Publishing Co., Amsterdam, 1945), p. 3.

stress on the practical, pragmatic significance of truth[88] reverberated in Bohr's frequent remark: "It is not the question at present whether this view is true or not, but what arguments we can honestly draw with respect to it from the available information."[89] Kierkegaard's and Høffding's insistence on the inevitable participation of the observer was unrestrictedly accepted by Bohr, for whom "man as an analyzing subject always occupies the central position."[90] Just as in psychology, Bohr pointed out, "it is clearly impossible to distinguish sharply between the phenomena themselves and their conscious perception,"[91] so also in physical research the traditional distinction between the observer and the object observed has to be abandoned in atomic physics. With it also the validity of the classical description of atomic phenomena in terms of space and time becomes questionable.

In an interesting exchange of letters between Høffding and Niels Bohr,[92] following Høffding's presentation of the galleys of a new edition of his *Formel Logik*[93] to Bohr, in which a discussion of the principle of the excluded third played an important role, Bohr wrote[94] to Høffding in 1922: "On the other hand we encounter difficulties which lie so deep that we do not have any idea of the way of their solution; it is my personal opinion that these difficulties are of such a nature that they hardly allow us to hope that we shall be able, inside the world of the atoms, to carry through a description in space and time of the kind that corresponds to our ordinary images."[95]

As we shall see, primarily in our discussion on the Copenhagen interpretation of quantum mechanics, Bohr was strongly influenced also by William James. Bohr repeatedly admitted how impressed he was particularly by the psychological writings of this American philosopher. He became acquainted with the teachings of James probably through Høffding, who had visited James in Cambridge (Massachusetts) when attending the St. Louis International Congress of Arts and Science in the fall of 1904.[96] James, it will be remembered, brought C. S. Peirce's pragmatic philosophy

[88] When Bohr's attention was drawn in the early thirties to William James's writings, he found them surprisingly congenial to his own ways of thinking.

[89] REF. 87, p. 8.

[90] *Ibid.*

[91] Niels Bohr, *Atomfysik og Menneskelig Erkendelse* (Schultz Forlag, Copenhagen, 1957), p. 39; translated into English under the title *Atomic Physics and Human Knowledge* (Wiley, New York, London, 1958), p. 27.

[92] Bohr's scientific correspondence in *Archive for the History of Quantum Physics* (original in Copenhagen).

[93] Nordiske Forlag, Copenhagen, 1st ed. 1903.

[94] Letter of Sept. 22, 1922.

[95] "På den anden Side møder vi Vanskeligheder af en saa dybtliggende Natur, at vi ikke aner Vejen til deres Løsning; efter min personlige Opfattelse er disse Vanskeligheder af en saadan Art, at de neppe lader os haabe indenfor Atomernes Verden at gennemføre en Beskrivelse i Rum og Tid af den Art, der svarer ti vore Saedvanlige Sansebilleder." *Ibid.*

[96] Cf. James's letter to F. C. S. Schiller of Oct. 26, 1904, in *The Letters of William James*, edited by his son Henry James (Atlantic Monthly Press, Boston, 1920), vol. 2, p. 216.

in metaphysics, ethics, religion, and the theory of knowledge to its logical conclusion. James's well-known maxim "We must find a theory that will work" still has its place in modern physics.

It is interesting to note, from our point of view, that Renouvier's contingentism and phenomenalism with its implications on the problem of free will was a turning point in James's intellectual and personal history—in fact, to such an extent that, in 1870, Renouvier's philosophy relieved him from his phobic panic and from his "monistic superstition."[97] James dedicated his posthumously published book *Some Problems of Philosophy*[98] to Renouvier, calling him "one of the greatest of philosophic characters."[99]

An interesting discussion on discontinuity and indeterminism and on the relation of these notions to the principle of energy conservation is found in the correspondence between Renouvier and James.[100] "Pluralism and indeterminism," wrote James,[101] "seem to be but two ways of stating the same thing." Renouvier fully agreed.[102] In a note enclosed in a letter to James[103] Renouvier explained how in his view the concept of discontinuous motion contradicts the idea of energy conservation; for the latter, he contended, is a consequence of the laws of motion; but these do not admit of discontinuous solutions.[104] These remarks were written in opposition to Joseph Delbœuf's provocative essay on "Determinism and Freedom—Freedom demonstrated by Mechanics,"[105] in which the Belgian mathematician attempted to show, on the basis of Newtonian mechanics, not only that discontinuous motions do exist, but also that their existence can

[97] *Ibid.*

[98] Longmans, Green & Co., New York, 1911.

[99] *Ibid.*, Dedication. Cf. also REF. 96, vol. 1, p. 186.

[100] R. B. Perry (ed.), "Correspondance de Charles Renouvier et de William James," *Revue de Métaphysique et de Morale 36*, 1–35, 193–222 (1929).

[101] Letter of Dec. 6, 1882, *ibid.*, p. 32. Cf. also R. B. Perry, *The Thought and Character of William James* (Little, Brown, Boston, 1935), vol. 1, p. 686.

[102] "Continuité et nécessité ou solidarité, c'est encore pour moi la même chose, de même que liberté et discontinuité s'accordent. . . ." Letter of Dec. 28, 1882, *ibid.*, p. 33.

[103] Letter of Jan. 26, 1883, *ibid.*, p. 195.

[104] "J'ai toujours pensé que les thèses de la nécessité et de la continuité absolue—ou de la liberté et de la discontinuité—s'impliquent réciproquement. Mais la discontinuité, et par suite, la possibilité de la liberté, peut-elle exister *dans* un système de mouvements soumis au régime de la conservation de l'énergie? Il y a des raisons mathématiques de la nier. On ne peut pas supposer *le temps* indéterminé, pour un lieu à occuper par une molécule, sans supposer *le lieu* de cette molécule indéterminé pour un temps donné. La discontinuité existe donc pour *le mouvement* (qui implique temps et espace). Or l'équation générale du mouvement d'un système de molécules, quelle qu'elle soit, est toujours une relation entre les lieux et les temps, elle détermine à volonté les lieux par les temps ou les temps par les lieux;—de plus elle n'admet que des quantités continues, ne s'applique qu'à des mouvements continus. Donc le mouvement d'une molécule qui serait libre à un moment donné de se mouvoir aussi bien que de ne pas se mouvoir, ne peut pas entrer dans une telle équation. L'équation n'existant pas pour cette molécule, la constance de l'énergie ne peut pas non plus se poser en tenant compte de cette molécule, puisque le théorème des forces vives (constance de l'énergie) est une propriété déduite de cette équation générale du mouvement." *Ibid.*, pp. 195–196.

[105] J. Delbœuf, "Déterminisme et Liberté—La Liberté démontrée par la Mécanique," *Revue Philosophique 13*, 454–480, 608–638 (1882), *14*, 156–189 (1882).

be reconciled with the principles of energy conservation and increase of entropy.[106]

As we shall see later on, James's work in psychology, which gave rise to a great variety of new developments in this field, and especially his monumental *Principles of Psychology*,[107] for many years one of the most popular textbooks on this subject, was destined, interestingly enough, to play an important role also in the development of quantum mechanics. Written in a rich and vivid literary style and containing a wealth of new ideas and suggestions, the *Principles*, and in particular its ninth chapter, "The Stream of Thought," was one of the major factors which influenced, wittingly or unwittingly, Bohr's formation of new conceptions in physics. It is not impossible that Bohr's idea of stationary states, as conceived in 1913, had already been influenced by the following passage from "The Stream of Thought": "Like a bird's life, [the stream of our consciousness] seems to be made of an alternation of flights and perchings. The rhythm of language expresses this, where every thought is expressed in a sentence, and every sentence closed by a period. The resting-places are usually occupied by sensorial imaginations of some sort, whose peculiarity is that they can be held before the mind for an indefinite time, and contemplated without changing; the places of flight are filled with thoughts of relations, static or dynamic, that for the most part obtain between the matters contemplated in the periods of comparative rest. *Let us call the resting-places the 'substantive parts,' and the places of flight the 'transitive parts,' of the stream of thought.* It then appears that the main end of our thinking is at all times the attainment of some other substantive part than the one from which we have just been dislodged. And we may say that the main use of the transitive parts is to lead us from one substantive conclusion to another. Now it is very difficult, introspectively, to see the transitive parts for what they really are. If they are but flights to a conclusion, stopping them to look at them before the conclusion is reached is really annihilating them. Whilst if we wait till the conclusion *be* reached, it so exceeds them in vigor and stability that it quite eclipses and swallows them up in its glare. Let anyone try to cut a thought across in the middle and get a look at its section, and he will see how difficult the introspective observation of the transitive tracts is. The rush of the thought is so headlong that it almost always brings us up at the conclusion before we can arrest it. Or if our purpose is nimble enough and we do arrest it, it ceases forthwith to be itself. As a snow-flake crystal caught in the warm hand is no longer a crystal but a drop, so, instead of catching the feeling of relation moving to its term,

[106] "Je suis parti des données des sciences naturelles; je n'ai contesté ni la loi de la conservation de la force, ni celle de la diminution progressive de l'énergie transformable; et j'ai démontré que ni l'une ni l'autre de ces lois n'impliquait le déterminisme." *Ibid.*, p. 188.

[107] W. James, *The Principles of Psychology* (H. Holt, New York, 1890; reprinted by Dover, New York, 1950).

we find we have caught some substantive thing, usually the last word we were pronouncing, statically taken, and with its function, tendency, and particular meaning in the sentence quite evaporated. The attempt at introspective analysis in these cases is in fact like seizing a spinning top to catch its motion, or trying to turn up the gas quickly enough to see how the darkness looks."[108]

This vivid description of the impossibility of observing introspectively the "transitive parts" in the stream of thought and feelings left a strong impression on Bohr's mind, as he admitted himself. Numerous allusions to the difficulties of introspective observation and frequent comparisons between physics and psychology in Bohr's later writings seem to indicate that some of his ideas—for instance his conceptions concerning the arbitrary boundary line between the observer and the object of observation, the incontrollable interaction between these two and its epistemological implications—have been inspired by James's philosophy.

To complete our survey on the philosophical background of modern quantum physics, the rise of the logical positivism in the twentieth century also has to be considered, being one of the factors without which the development of the new conceptual situation in quantum physics can hardly be understood. Although the Vienna Circle ("Wiener Kreis") entered publicity as an independent school of philosophy only at the end of the twenties,[109] the impact of logical empiricism on theoretical physics can be recognized much earlier.

Ever since a chair for the philosophy of the inductive sciences was established in 1895 at the University of Vienna, theoretical physics has been subjected to critical examinations by empiricistic philosophy. After Ernst Mach it was primarily Moritz Schlick—Schlick's doctor's thesis, under the guidance of Max Planck, was on optical reflection in nonhomogeneous media—who exerted great influence in this respect, especially after the publication of his *Space and Time in Contemporary Physics*.[110] This monograph stimulated great interest in the epistemological foundations of physics and, although it dealt primarily with the foundations of relativity, its influence on theoretical physics in general should not be ignored. Schlick entertained personal contact with such leading physicists and mathematicians as Einstein, Planck, and Hilbert, and his conceptions concerning epistemological questions in physics, in particular, his insistence that a statement is meaningful only if it is empirically verifiable, became

[108] *Ibid.* (Dover edition), pp. 243–244.

[109] The Prague meeting of the German Physical Society and of the German Mathematical Association in September, 1929, was the external occasion on which the members of the Vienna Circle proclaimed themselves officially as a separate school and organization. Their international standing dates from 1930, when Carnap and Reichenbach took over the *Annalen der Philosophie* and continued their publication under the name *Erkenntnis*.

[110] M. Schlick, *Raum und Zeit in der gegenwärtigen Physik* (Berlin, 1917); an English translation appeared in 1920.

influential among physicists in the early twenties. Since 1921 Wittgenstein's *Tractatus Logico-Philosophicus*[111] had evoked much discussion not only among professional philosophers alone. Its famous concluding statement "Whereof one cannot speak, thereof one must be silent" was soon to find, as we shall see in due course, an unexpected application in Heisenberg's new approach to the study of atomic phenomena.

Although only the later development of quantum mechanics gave rise to the present unprecedented interest in the philosophy of science, philosophy's original contributions to the development of the physical theory should not be ignored. It made physics disposed toward a logical clarification and methodological purification and forced physicists "to come to grips with philosophical problems to a greater degree than was ever the case in earlier generations."[112]

The philosophical schools we have mentioned, contingentism, existentialism, pragmatism, and logical empiricism, rose in reaction to traditional rationalism and conventional metaphysics whose ultimate roots were in the philosophy of Descartes. Their affirmation of a concrete conception of life and their rejection of an abstract intellectualism culminated in their doctrine of free will, their denial of mechanical determinism or of metaphysical causality. United in rejecting causality, though on different grounds, these currents of thought prepared, so to speak, the philosophical background for modern quantum mechanics. They contributed with suggestions to the formative stage of the new conceptual scheme and subsequently promoted its acceptance.

In concluding this section we wish to point out that much has been published on the influence of physics and, in particular, of quantum mechanics on current philosophical thought, but the converse problem, the study of how far philosophy contributed to current physical thought, has rarely been dealt with. But physical thought, even at its great turning points when treading unwonted paths in its formation of new conceptions, never works in a vacuum. Even "discontinuities" in its intellectual development are somehow related to general contemporary thought. We are fully aware that the present treatment of this problem is far from complete and, to some extent, even conjectural in its assertions. It is hoped that further research will throw additional light on these still obscure relations between recent philosophy and modern physics.

[111] L. Wittgenstein, "Logisch-Philosophische Abhandlung," *Annalen der Naturphilosophie 14*, 185–262 (1921). Originally published, as we see, in the final number of Ostwald's *Annalen*, Wittgenstein's essay was republished one year later with the German text and the English translation on opposite pages under the title *Tractatus Logico-Philosophicus* (Kegan Paul, Trench, Trubner and Co., London; Harcourt, Brace and Co., New York, 1922). More recently, a new translation by D. F. Pears and B. F. McGuinness has been published (Humanities Press, New York, 1961).

[112] A. Einstein, "Remarks on Bertrand Russell's Theory of Knowledge" in P. A. Schilpp (ed.), *The Philosophy of Bertrand Russell* (Tudor, New York, 1951), p. 279.

4.3 Nonclassical Interpretations of Optical Dispersion

The transition from the older quantum theory to modern quantum mechanics occurred within a surprisingly short period of time. But it was not one single conceptual process. It was not even "a straight staircase upwards," but rather "a tangle of interconnected alleys," as Born once characterized[113] it so fittingly. Correspondingly, also the surrender of classical notions and their replacement by fundamentally new conceptions, whose philosophical background was the subject of the preceding section, did not precipitate in one single movement, but took place rather piecemeal, though at a rapid pace.

The first step in this development, though only soon to be retraced, was in connection with the phenomenon of dispersion, which, as we have seen, was one of the most difficult problems to reconcile with quantum conceptions. But—perhaps just because of this difficulty—it proved most instrumental for the formation of new conceptions.

The glaring contrast between the discrete features of the quantum theory of atomic structure, on the one hand, and the continuous aspects of the electromagnetic theory of light, on the other, was one of the most challenging problems of the time. It was early realized that at the root of the difficulties to reconcile these two branches of physics was the principle of the exact conservation of energy and momentum. For it was hard to understand how, for example, in a system composed of an electromagnetic radiation field, susceptible of only continuous changes of energy, and an aggregate of atoms, emitting or absorbing only discrete quanta of energy, the sum total of a continuous and of a discrete amount of energy could be a constant. Viewed from this angle, Einstein's conception of discrete light quanta, by which he explained the energy balance for the interaction between matter and radiation, was essentially a redefinition of energy.

But there was an alternative: one could reject the energy principle as an exact law and regard it instead as merely a statistical law. This alternative, as we have seen,[114] had already been envisaged in 1919 by Darwin as a possibility to harmonize Bohr's quantum theory of the atom with Maxwell's theory of the field. In 1922 he applied this idea for the formulation of his theory of dispersion.[115] "It is, of course, a fact of observation," he declared, "that, in the gross, energy is conserved, but this means an averaged energy; and as pure dynamics has failed to explain many atomic phenomena, there seems no reason to maintain the exact conservation of energy, which is only one of the consequences of the dynamical equations"—an argumentation strikingly similar to Renouvier's reasoning in his letter

[113] M. Born, *Natural Philosophy of Chance and Cause* (Oxford University Press, 1949), p. 86.

[114] REF. 68.

[115] REF. 71.

to William James.[116] According to Darwin's theory of dispersion an atom, if struck by an incident wave of light, acquires a probability of emitting a spherical wave train which interferes with the exciting light, the probability being a function of the intensity of the incident light. When Bohr, however, in a private letter, directed Darwin's attention to certain inconsistencies and when it became clear that the dispersion in the case of very weak radiation cannot be accounted for, Darwin tried to modify his theory.[117] It was in the course of these attempts that he made the statement concerning the "overriding system" which we have mentioned above.[118]

The statistical element in Darwin's approach seems to have eventually inspired J. C. Slater to new far-reaching ideas. In January, 1924, Slater published a note on "Radiation and atoms."[119] Its declared purpose was "to build up a more adequate picture of optical phenomena than has previously existed." Convinced of the correctness of Compton's results concerning the existence of light quanta, Slater attempted "to tie together" light quanta and waves by constructing, on the basis of Poynting's vector and the energy density of the electromagnetic field, a quantity which transforms like a vector and thus could represent the velocity of the light quantum. After a careful study of this problem and, in particular, after having read Cunningham's treatise on *Relativity, the Electron Theory and Gravitation*,[120] the second edition of which had just been published, Slater saw the futility of this program. He therefore concluded that the only possibility to establish such a relation would be to assume a statistical connection between light quanta and waves.

The note itself, however, hardly referred to light quanta but suggested that each atom should be supposed to "communicate" constantly with all the other atoms as long as it is in one of its stationary states, this communication being established by a virtual field of radiation which acts as if it originates from oscillators whose frequencies are those of the possible quantum transitions of the given atom. "The part of the field originating from the given atom itself is supposed to induce a probability that that atom lose energy spontaneously, while radiation from external sources is regarded as inducing additional probabilities that it gain or lose energy, much as Einstein has suggested. The discontinuous transition finally resulting from these probabilities has no other external significance than simply to mark the transfer to a new stationary state, and the change from the continuous radiation appropriate to the old state to that of the new."

Thus, by assuming a virtual radiation field through which distant

[116] REF. 104.
[117] REF. 72.
[118] *Ibid.*
[119] J. C. Slater, "Radiation and atoms," *Nature 113*, 307–308 (Jan. 28, 1924). Cf. *Archive for the History of Quantum Physics*, Interview with J. C. Slater on Oct. 3, 1963.
[120] E. Cunningham, *Relativity, the Electron Theory and Gravitation* (Longmans, Green and Co., London, 2d ed., 1921).

atoms are capable of communicating with each other even before transitions between stationary states take place, Slater attempted to reconcile the discrete theory of light quanta with the continuous wave theory of the electromagnetic field. However, when Slater, visiting Bohr's Institute at that time, expounded the details of his theory in Copenhagen, Kramers pointed out that instead of leading to a strong coupling between emission and absorption processes among distant atoms, Slater obtained, contrary to his intentions, a greater independence of these processes than previously conceived. Furthermore, Kramers, like Bohr, still regarded Einstein's theory of light quanta, in spite of its acknowledged heuristic value as shown by its confirmation through the photoelectric effect, as an unsatisfactory solution of the problem of light propagation, if only because "the radiation 'frequency' ν appearing in the theory is defined by experiments on interference phenomena which apparently demand for their interpretation a wave constitution of light." Although certain details of Slater's hypothesis had therefore to be rejected, a theory of virtual oscillators and of an associated virtual radiation field was worked out so as to make it possible to harmonize the physical picture of the continuous electromagnetic field with the physical picture, not as Slater proposed of light quanta, but of the discontinuous quantum transitions in the atom. For, having renounced the possibility of any time-spatial description of transitions between stationary states, these transitions, "jumps" or "leaps," were for Bohr and Kramers discontinuous processes whereas at the same time the propagation of light, as just explained, was to be regarded as a continuous phenomenon. A statistical connection, as Slater proposed, was the only conceivable possibility of correlating continuity with discreteness. Moreover, such a correlation had also to renounce any causal connection for, as explained in the preceding section, causality presupposes continuity. Finally, the replacement of causal connectivity by statistical considerations made it logically unavoidable to degrade the principles of energy and momentum conservation to the status of statistical theorems.

Now the problem was how to formulate on the basis of such assumptions a consistent theory which could account for the observed phenomena of dispersion, refraction, and reflection. The program of such a theory was published by Bohr, Kramers, and Slater[121] in a paper entitled "The quantum theory of radiation."[122] Although soon disproved, as we shall see, on the basis of experimental evidence, this paper proved extremely important for the conceptual development of quantum theory for the following reasons: (1) it was the first major paper in physics which deliberately and programmatically renounced methods of explanation as well as fundamental

[121] Cf. *Archive for the History of Quantum Physics*, Interview with J. C. Slater on Oct. 8, 1963.

[122] *Philosophical Magazine 47*, 785–802; "Über die Quantentheorie der Strahlung," *Zeitschrift für Physik 24*, 69–87 (1924).

principles of classical theory; (2) it evoked, consequently, a great amount of discussion among physicists and directed thereby the attention of quantum theorists to questions concerning the epistemological foundations of atomic physics;[123] and (3) it was the point of departure of Kramers's detailed theory of dispersion whose further elaboration by Kramers and Heisenberg led the latter to the discovery of matrix mechanics, the first formulation of quantum mechanics.

Bohr and Kramers assumed with Slater that every atom, as soon and as long as it occupies a given stationary state, communicates "continually with other atoms through a time-spatial mechanism which is virtually equivalent with the field of radiation which on the classical theory would originate from the virtual harmonic oscillators corresponding with the various possible transitions to other stationary states." In other words, with each stationary state of a given atom a virtual radiation field is associated composed of as many monochromatic spherical waves as there are possible transitions to lower energy levels in the given atom in accordance with Bohr's frequency condition. The radiation field is not causally connected with transitions between stationary states. It is rather assumed that "the occurrence of transition processes for the given atom itself, as well as for the other atoms with which it is in mutual communication, is connected with this mechanism by probability laws," so that there exists for each frequency component of the field a temporally invariant probability, proportional to the time element dt, that the transition, corresponding to the particular frequency, occurs within dt. "The occurrence of a certain transition in a given atom will depend on the initial stationary state of this atom itself and on the states of the atoms with which it is in communication through the virtual radiation field, but not on the occurrence of transition processes in the latter atoms."

[123] How the Bohr-Kramers-Slater paper stimulated philosophical interest among physicists is illustrated by E. Schrödinger's paper "Bohr's neue Strahlungshypothese und der Energiesatz," *Die Naturwissenschaften 12*, 720–724 (1924). In this exposition and analysis of the Bohr-Kramers-Slater hypothesis—which, in Schrödinger's words, "is of equally great interest from the viewpoint of the physicist as from that of the philosopher"—Schrödinger mentioned nowhere that the proposed interatomic field is to be conceived as "virtual," that is, "effective but not factual." Moreover, Schrödinger ascribed to this field the capability of transferring energy, a conception hardly compatible with the Bohr-Kramers-Slater point of view, since energy or momentum transfer was always held as a criterion of physical reality in the full sense of the word. Schrödinger, it seems, was primarily interested in the Bohr-Kramers-Slater paper in view of its implications concerning the statistical nature of energy. In fact, the idea of energy being merely a statistical concept haunted Schrödinger throughout his life, ever since he was impressed by Exner's comparison of the alleged statistical validity of the energy conservation principle (REF. 57) with that of the second law in thermodynamics. Three years before his death, at a meeting of the Austrian Physical Society in Vienna on Mar. 26, 1958, he discussed this idea in his address "Might perhaps energy be merely a statistical concept?" *Il Nuovo Cimento 9*, 162–170 (1958). It thus seems likely that his preoccupation with the energy idea made him ignore the virtual nature of the Bohr-Kramers-Slater radiation field—an interesting item for the student of the psychology of scientific research.

Bohr, Kramers, and Slater were able to show that in the limit, where the correspondence principle holds, the resulting relation between the virtual radiation field, on the one hand, and the motions of the particle, on the other, coincided with the state of affairs as described by the classical radiation theory. For characteristic quantum processes, however, they abandoned "any attempt at a causal connexion between the transitions in distant atoms, and especially a direct application of the principles of conservation of energy and momentum." A transition in an atom A, according to their view, is not directly caused by a transition in a distant atom B nor is the energy difference between the initial and the final stationary states of A equal to that of B. "On the contrary, an atom which has contributed to the induction of a certain transition in a distant atom through the virtual radiation field conjugated with the virtual harmonic oscillator corresponding with one of the possible transitions to the other stationary state, may nevertheless itself ultimately perform another of these transitions." For individual atomic processes neither energy[124] nor momentum is thus preserved. However, by relating the probability factor to the intensity of radiation precisely in such a way that the radiative energy of a given frequency, emitted during the average lifetime of the excited state, is equal to the product of the actual energy loss of the atom and the probability (relative number of cases of occurrence) of the corresponding transition, energy is conserved on the average.

Applying their theory to the Compton scattering of radiation by free electrons, the authors of the paper under discussion had to conclude that the "illuminated electron possesses a certain probability of taking up in unit time a finite amount of momentum in any given direction."[125] Thus, the statistical interpretation of this process implied that the direction of the recoil electron is not uniquely determined as it would have to be according to the momentum-conservation law but displays a wide statistical distribution. That this conclusion opened a possibility for an experimental test of the theory was recognized by Bothe and Geiger[126] in June, 1924. In their experiment, which they performed a few months later[127] and which was a pioneer work in those days before the development of modern electronic coincidence techniques, they used two point counters, one for the reception of the scattered x-rays and the other for the recording of the recoil electrons; they found "that approximately every eleventh light quantum coincides with a recoil electron," a result which, owing to the

[124] For philosophical implications of this assumption cf. M. Jammer, "Energy," in *The New Catholic Encyclopedia* (The Catholic University of America, 1966).

[125] REF. 122, pp. 799–800.

[126] W. Bothe and H. Geiger, "Ein Weg zur experimentellen Nachprüfung der Theorie von Bohr, Kramers und Slater," *Zeitschrift für Physik 26*, 44 (1924).

[127] W. Bothe and H. Geiger, "Experimentelles zur Theorie von Bohr, Kramers und Slater," *Die Naturwissenschaften 13*, 440–441 (1925); "Über das Wesen des Comptoneffekts; ein experimenteller Beitrag zur Theorie der Strahlung," *Zeitschrift für Physik 32*, 639–663 (1925).

conditions of the experiment, showed that the chances of these coincidences being accidental are of the order 10^{-5}. Bothe and Geiger thus concluded that their result "is incompatible with Bohr's interpretation of the Compton effect."[128] Meanwhile, also Compton and Simon,[129] employing the technique of the cloud-chamber expansion for the determination of the direction and time of ejection of the recoil electrons, verified that "on the average there is produced one recoil electron for each quantum of scattered x-rays. Each recoil electron produces a visible track, and occasionally a secondary track is produced by the scattered x-ray. When but one recoil electron appears on the same plate with the track due to the scattered rays, it is possible to tell at once whether the angle satisfies" Eq. (4.4), page 161, as required by Compton's theory of light quanta. It was therefore possible to obtain an "unequivocal answer to the question whether the energy of a scattered x-ray quantum is distributed over a wide solid angle or proceeds in a definite direction." Compton and Simon found, just as Bothe and Geiger did, that "the results do not appear to be reconcilable with the view of the statistical production of recoil and photo-electrons by Bohr, Kramers and Slater. They are, on the other hand, in direct support of the view that energy and momentum are conserved during the interaction between radiation and individual electrons."[130]

It is noteworthy to recall in this context that about ten years later an apparently negative result of the Compton-Simon experiment[131] led to speculations over a possible revival of the Bohr-Kramers-Slater statistical theory and stimulated renewed interest in carrying out the Compton-Simon or Bothe-Geiger kind of experiments with improved instrumentation for time resolution. Thus, Hofstadter and McIntyre,[132] using stilbene scintillation counters, demonstrated irrefutably "that the recoil electrons and the scattered photons are emitted together within a time interval of less than 1.5×10^{-8} sec." In the same year, Cross and Ramsey[133] showed that the Compton electrons are ejected within ± 1 deg of the direction required by the conservation laws when the scattering direction of the associated quanta is specified.

In this context it should be noted that Bohr, for the first time, probably under the impact of Compton's discovery, had no objection to admitting and employing in the Bohr-Kramers-Slater paper the notion of momentum transfer of radiation and to associating it with quantum transitions,[134] and

[128] *Ibid.*, p. 641.

[129] A. H. Compton and A. W. Simon, "Directed quanta of scattered x-rays," *Physical Review 26*, 289–299 (1925). Cf. also A. H. Compton, "On the mechanism of x-ray scattering," *Proceedings of the National Academy of Sciences 11*, 303–306 (1925).

[130] *Ibid.*, p. 299.

[131] R. S. Shankland, "An apparent failure of the photon theory of scattering," *Physical Review 49*, 8–13 (1936).

[132] R. Hofstadter and J. A. McIntyre, "Simultaneity in the Compton effect," *Physical Review 78*, 24–28 (1950).

[133] W. G. Cross and N. F. Ramsey, "The conservation of energy and momentum in Compton scattering," *Physical Review 80*, 929–936 (1950).

[134] REF. 122, pp. 799–801.

this in spite of his persistent refusal to accept the idea of light quanta. Bohr's insistence on the exclusive validity of wave conceptions is well illustrated by his reaction to a letter in which Einstein, shortly after the publication of the paper under discussion, slightly criticized Bohr's opposition to the conception of light quanta. Bohr is said to have replied that even if Einstein had sent him a telegram informing him that by now he (Einstein) had found an irrevocable proof of the physical existence of light quanta, even then, Bohr continued, "the telegram could only reach me by radio on account of the waves which are there."[135]

It is hard to find in the history of physics a theory which was so soon disproved after its proposal and yet was so important for the future development of physical thought as the theory of Bohr, Kramers, and Slater. Now it will be understood that not because of its specific physical contents was it important, but because of its radically new approach. By interpreting Einstein's spontaneous emission as a process "induced by the virtual field of radiation" and Einstein's induced transitions as occurring "in consequence of the virtual radiation in the surrounding space due to other atoms,"[136] it paved the way for the subsequent quantum-mechanical conception of probability as something endowed with physical reality and not as merely a mathematical category of reasoning.[137] Also of epistemological importance was the very notion of "virtual" harmonic oscillators. For it was an early example of "some intermediate kind of reality," a conception which for some time seemed to have played a not negligible role in some quarters. For, as Heisenberg[138] later phrased it, when it became gradually clear that "no cheap solutions" could lead to a comprehensive understanding of quantum theory, but that one had to pay a high price, the "idea of having such intermediate kinds of reality was just the price one had to pay." Connected with this is also Heisenberg's contention that the idea of an atom as a set of oscillators or, as Landé once called it, "a virtual orchestra"[139] "prepared the way for the later idea that the assembly of oscillators is nothing but a matrix."[140]

However this may be, the Bohr-Kramers-Slater paper was undoubtedly instrumental for the development of quantum mechanics in view of the fact that it made it possible to establish a dispersion theory which offered a new approach to the fundamental problem concerning the interaction of

[135] *Archive for the History of Quantum Physics*, Interview with W. Heisenberg on Feb. 15, 1963.

[136] REF. 122, p. 791.

[137] REF. 135, Interview on Feb. 19, 1963. According to Heisenberg Born's statistical interpretation of the Schrödinger wave function had its ultimate root in the Bohr-Kramers-Slater paper.

[138] *Ibid.* By the way, it would be interesting to investigate how far this idea of a "time-spatial mechanism" (REF. 122, pp. 790, 791) of "virtual" oscillators, an idea which can be traced back to Ladenburg's "substitution electrons" ("Ersatzelektronen"), may be regarded as a forerunner of the recently so controversial conception of "hidden parameters."

[139] A. Landé, "Neue Wege der Quantentheorie," *Die Naturwissenschaften 14*, 455–458 (1926).

[140] REF. 135, Interview on Feb. 13, 1963.

matter and radiation. For, if the emission or absorption frequencies of the atom, in accordance with Bohr's frequency condition, have to be incorporated into the dispersion formula, the theory has to come to grips with the dispersion process itself. A confrontation with the basic problem of the quantum theory, the problem of the interaction of matter and radiation, could no longer be evaded—and it was this kind of confrontation which eventually led to the birth of the earliest formulation of modern quantum mechanics. A discussion of the development of modern dispersion theory, leading to the Kramers-Heisenberg formulation, is therefore indispensable for a profound comprehension of our subject.

According to the classical[141] electron theory of dispersion the equation of motion of a quasi-elastically bound particle of mass m and charge e under the influence of an external sinusoidally varying field $E = E_0 \exp(2\pi i\nu t)$ is

$$m\ddot{p}_k + m\omega_{0k}{}^2 p_k = e^2 E \tag{4.5}$$

where $p_k = er_k$ is the electric moment[142] of the particle and $\nu_{0k} = \omega_{0k}/2\pi$ is the proper frequency with which the particle would oscillate in the absence of external forces. The solution of the differential equation (4.5) is

$$p_k = \frac{e^2 E/m}{4\pi^2(\nu_{0k}{}^2 - \nu^2)} \tag{4.6}$$

Hence, from the well-known equation $D = E + 4\pi P$, where $P = \sum N_k p_k$ is the polarization, N_k the number, per unit of volume, of charged particles with proper frequency ν_{0k} or "dispersion electrons" of the kind k, and from the equation $D = \varepsilon E$, where $\varepsilon = n^2$ is the dielectric constant, it follows that the index of refraction n is given by the following dispersion formula:

$$n^2 = 1 + \frac{1}{\pi} \sum \frac{N_k e^2/m}{\nu_{0k}{}^2 - \nu^2} \tag{4.7}$$

[141] For the development of the early mechanical, electromagnetic, and phenomenological theories of dispersion the reader is referred to the following publications: B. de St. Venant, "Sur les diverses manières de présenter les théories des ondes lumineuses," *Annales de Chimie et de Physique 25*, 335–381 (1872); F. Rosenberger, *Die Geschichte der Physik*, vol. 3 (F. Vieweg und Sohn, Braunschweig, 1887–1890), pp. 309–312, 712–717; E. Carvallo, "Sur les théories et formules de dispersion," *Rapports du Congrès International de Physique, Paris, 1900* (Gauthier-Villars, Paris, 1900), vol. 2, pp. 175–199; P. Drude, "Zur Geschichte der electromagnetischen Dispersionsgleichungen," *Annalen der Physik 1*, 437–440 (1900); A. Breuer, *Uebersichtliche Darstellung der mathematischen Theorieen über die Dispersion des Lichtes*, vol. 1 (Hannover, 1890), vol. 2 (Erfurt, 1891); A. Pflüger, "Dispersion," in H. Kayser, *Handbuch der Spectroscopie* (Hirzel, Leipzig, 1908), vol. 4, pp. 267 et seq.; E. Mach, *Die Prinzipien der physikalischen Optik* (J. A. Barth, Leipzig, 1921), translated into English by J. S. Anderson, *The Principles of Physical Optics* (Methuen, London, 1926), chap. 6 and appendix; E. Whittaker, *A History of the Theories of Aether and Electricity* (T. Nelson and Sons, Edinburgh, London, 1953; Philosophical Library, New York, 1954).

[142] For simplicity we assume that the incident wave is linearly polarized with its electric vector parallel to the direction of the atomic dipole moment so that only scalar quantities need to be considered.

an equation which agrees essentially with the dispersion formulas derived on the basis of the elastic-solid theory by Sellmeier and Helmholtz and on the basis of the electromagnetic theory by Maxwell.[143] If each atom contains f_k dispersion electrons of the kind k, the polarizability α, defined by the equation $P = \alpha E$, is given by

$$\alpha = \frac{e^2}{4\pi^2 m} \sum \frac{f_k}{\nu_{0k}{}^2 - \nu^2} \tag{4.8}$$

These classical formulas agreed fairly well with experience and offered a satisfactory interpretation of dispersion and also of absorption if velocity-dependent terms, representing damping forces and leading to complex-valued indices of refraction, were introduced in Eq. (4.1). When Bohr's theory of stationary states, however, superseded the classical conception of elastically bound electrons, these formulas, in spite of their *de facto* validity, suddenly lost completely their theoretical justification. It became therefore a major project for theoretical research to reconcile these equations with the basic assumptions of quantum theory. The early attempts at formulating a theory of dispersion in terms of quantum-theoretic conceptions, as made by Debye,[144] Sommerfeld,[145] and Davisson,[146] proved unsatisfactory, primarily because of their application of the classical perturbation theory to Bohr's atomic model by which they were led to the inevitable conclusion that the Fourier frequencies $\tau_1\nu_1 + \cdots + \tau_f\nu_f$ are also the frequencies of the optical resonance lines contrary to experience.

The first step toward the formulation of a quantum-theoretic interpretation of dispersion was made by Ladenburg.[147] According to the classical electron theory the energy emitted per second by N oscillators of mass m, charge e, and frequency ν_0 is $J_{\text{cl}} = UN/\tau$, where $\tau = 3mc^3/8\pi^2e^2\nu_0{}^2$ is the time in which the oscillator's energy is reduced to ε^{-1} of its original value (ε is the base of the natural logarithms)[148] and where U is the average

[143] W. Sellmeier, "Ueber die durch die Aetherschwingungen erregten Mitschwingungen der Körpertheilchen und deren Rückwirkung auf die ersteren, besonders zur Erklärung der Dispersion und ihrer Anomalie," *Poggendorff's Annalen der Physik 145*, 399–421, 520–548, *147*, 386–404, 525–554 (1872); H. Helmholtz, "Zur Theorie der anomalen Dispersion," *ibid. 154*, 582–596 (1875); J. C. Maxwell, *Mathematical Tripos Examination*, *Cambridge Calendar*, 1869; cf. Lord Rayleigh, "The theory of anomalous dispersion," *Philosophical Magazine 48*, 151–152 (1899).

[144] P. Debye, "Die Konstitution des Wasserstoff-Moleküls," *Münchener Berichte 1915*, pp. 1–26.

[145] A. Sommerfeld, "Die Drudesche Dispersionstheorie vom Standpunkte des Bohrschen Modelles und die Konstitution von H_2, O_2 und N_2," *Annalen der Physik 53*, 497–550 (1917).

[146] C. Davisson, "The dispersion of hydrogen and helium on Bohr's theory," *Physical Review 8*, 20–27 (1916).

[147] R. Ladenburg, "Die quantentheoretische Deutung der Zahl der Dispersionselektronen," *Zeitschrift für Physik 4*, 451–471 (1921).

[148] The factor 3 in the numerator of the expression for τ which does not appear in the Kramers-Heisenberg formulas (for linear oscillators) is due to the fact that the average energy of a three-dimensional oscillator in an isotropic radiation field is three times that of a linear oscillator, as shown by F. Reiche, "Zur Quantentheorie der Rotationswärme des Wasserstoffs," *Annalen der Physik 58*, 657–694 (1919) (see p. 693).

energy of each oscillator, satisfying Eq. (1.6), page 11, $U = 3c^3u_{\nu_0}/8\pi\nu_0^2$. Hence

$$J_{cl} = \frac{\pi e^2}{m} N u_{\nu_0} \tag{4.9}$$

Quantum-theoretically, on the other hand, according to Einstein, the energy J_{abs} absorbed per second by transitions from the state i to the state k, where $E_i < E_k$, is given by $J_{abs} = h\nu_{ik}N_iB_k{}^iu_{ik}$, where N_i is the number among the N atoms which are in state i, u_{ik} is the field energy density for the frequency ν_{ik}, and $B_k{}^i = B_i{}^k$ are the Einstein probability coefficients defined on page 112. Under equilibrium conditions the energy J_{qu} emitted per second by the transitions from state k to state i is equal to J_{abs}, so that

$$J_{qu} = h\nu_{ik}N_iB_k{}^iu_{ik} \tag{4.10}$$

Putting $J_{cl} = J_{qu}$ and recalling Eq. (3.7), page 112, Ladenburg concluded that

$$N = N_i \frac{mc^3}{8\pi^2e^2\nu_{ik}{}^2} A_k{}^i \tag{4.11}$$

Since for an atom in state i it has to be assumed[149] that f_i in Eq. (4.8) equals N/N_i, Eq. (4.11) implies that the atom has a varying electric moment of amplitude

$$A = \frac{c^3E}{32\pi^4} \sum_k \frac{A_k{}^i}{\nu_{ik}{}^2(\nu_{ik}{}^2 - \nu^2)} \tag{4.12}$$

where $\nu_{ik} = (E_k - E_i)/h$.

Kramers[150] generalized Ladenburg's result, arguing that if the state i is not the ground state also all states k' have to be taken into account for which $E_{k'} < E_i$ so that the secondary waves can be described as originating from a varying electric moment whose amplitude is given by the formula

$$A = \frac{c^3E}{32\pi^4}\left[\sum_{\substack{k \\ E_k > E_i}} \frac{A_k{}^i}{\nu_{ik}{}^2(\nu_{ik}{}^2 - \nu^2)} - \sum_{\substack{k' \\ E_{k'} < E_i}} \frac{A_i{}^{k'}}{\nu_{k'i}{}^2(\nu_{k'i}{}^2 - \nu^2)} \right] \tag{4.13}$$

where $\nu_{ik} = (E_k - E_i)/h$ and $\nu_{k'i} = E_{k'} - E_i$. In view of Eq. (4.11) with $N/N_i = f_i$ and provided ν_{0k} in Eq. (4.6) for p_k is replaced by ν_{ik} or $\nu_{k'i}$, respectively, Eq. (4.13) can be written $A = \sum f_kp_k - \sum f_{k'}p_{k'}$. Hence, Kramers contended, "the reaction of the atom against the incident

[149] On this point cf. R. Ladenburg and F. Reiche, "Absorption, Zerstreuung und Dispersion in der Bohrschen Atomtheorie," *Die Naturwissenschaften 11*, 584–598 (1923).

[150] H. A. Kramers, "The law of dispersion and Bohr's theory of spectra," *Nature 113*, 673–674 (1924); "The quantum theory of dispersion," *ibid. 114*, 310 (1924).

radiation can thus formally be compared with the action of a set of virtual harmonic oscillators inside the atom, conjugated with the different possible transitions to other stationary states . . . in the final state of the transition the atom acts as a 'positive virtual oscillator' of relative strength $+f$; in the initial state it acts as a negative virtual oscillator of strength $-f$. However unfamiliar this 'negative dispersion' might appear from the point of the classical theory, it may be noted that it exhibits a close analogy with the 'negative absorption' which was introduced by Einstein, in order to account for the law of temperature radiation on the basis of the quantum theory."[151]

In accordance with our principle of using a notation as uniform as possible, our notation of the dispersion formula (4.13) differs from Kramers's original formulation in so far as Kramers denoted the transition frequencies (which we called ν_{ik} and $\nu_{k'i}$) by $\nu_k{}^a$ and $\nu_{k'}{}^e$, where $\nu_k{}^a$ referred to transitions to higher states k (absorption) and $\nu_{k'}{}^e$ to lower states k' (emission), relative to the stationary state under consideration.[152] This difference in notation, irrelevant as it is from the mathematical point of view, seems nevertheless to have some importance from the point of view of our study of the conceptual development of the theory. For, as our notation clearly indicates, Kramers was already in a position to describe the dispersive behavior of the atom by a *two-fold infinite set* of virtual oscillators, an idea of basic importance for the future development of matrix mechanics. But instead of doing so, his attention, guided by his notation, was focused, so to speak, only on the *one-fold infinite set* of virtual oscillators relative to every given stationary state. In compliance with the correspondence principle the dispersion law could not be formulated without recourse to the frequencies ν_{ik} and $\nu_{k'i}$, referring to initial and final stationary states. These are states which potentially can be occupied by the atom, but are not actually occupied. In other words, hypothetical possibilities had to be regarded as physically active efficient causes. In order to overcome this conceptual difficulty it had to be assumed that the virtual radiation field continues to be operative throughout the whole lifetime of a given stationary state, an assumption which, in turn, entailed the statistical interpretation of energy conservation as previously discussed.

Kramers obtained these results in the early spring of 1924. In June, 1924, Born completed his paper "On quantum mechanics,"[153] to which reference has been made at the end of the preceding chapter. In this article Born tried to establish a method by which the classical perturbation theory of multiply periodic nondegenerate systems could be applied to quantum

[151] *Ibid.*, p. 674.

[152] Strictly speaking, Kramers used for positive dispersion $\nu_i{}^a$ and for negative dispersion $\nu_j{}^e$. However, the denotation of a dummy index, over which a summation is carried out, is of course completely irrelevant.

[153] M. Born, "Über Quantenmechanik," *Zeitschrift für Physik 26*, 379–395 (1924).

phenomena which involve external periodic disturbances or internal couplings. So far, Born argued, theories of polyelectronic systems, such as the helium atom, which treated the electronic interactions along classical lines, had to fail since electrons affect each other with frequencies of the order of those of light, but the interaction between matter and light is evidently a "nonmechanical" quantum process; one cannot expect, therefore, that the interactions between electrons can be treated along classical lines either. Born now suggested a solution of the problem by generalizing Kramers's treatment of the interaction between radiation and electrons into a "quantum mechanics" of interactions. In carrying out this program, Born showed that the transition from classical mechanics to what he called "quantum mechanics" can be obtained in accordance with Bohr's correspondence principle if a certain differential is replaced by a corresponding difference. The quantum-theoretic frequency $\nu_{n,n-\tau}$ for a transition from a stationary state characterized by the quantum number n to a stationary state characterized by the quantum number $n' = n - \tau$, it will be recalled, coincides, according to Bohr's correspondence principle, for large n and small τ, with the classical frequency $\nu(n,\tau)$, the τth harmonic (overtone) of the fundamental frequency of the classical motion in state n:

$$\nu_{n,n-\tau} = \nu(n,\tau) = \tau\nu(n,1) \tag{4.14}$$

$\nu(n,1) = \nu$ is the classical fundamental frequency and is equal to the derivative of the Hamiltonian (or total energy) with respect to the action variable,[154] $\nu = dH/dJ$. To prove Eq. (4.14) we note that for H as a function of $J = nh$ Bohr's frequency condition states that

$$\nu_{n,n-\tau} = \frac{H(nh) - H[(n-\tau)h]}{h} = \tau\frac{dH}{dJ} = \tau\nu$$

Now, comparing

$$\nu(n,\tau) = \tau\frac{dH}{dJ} = \frac{\tau}{h}\frac{\partial H}{\partial n} \tag{4.15}$$

with

$$\nu_{n,n-\tau} = \frac{1}{h}[H(n) - H(n-\tau)]$$

Born recognized that the quantum-theoretic frequency $\nu_{n,n-\tau}$ is obtained from the classical frequency $\nu(n,\tau)$ by substituting the difference $H(n) - H(n-\tau)$ for the differential $\tau(\partial H/\partial n)$. Generalizing this relation, Born postulated[155] that for any arbitrary function $\Phi(n)$, defined for stationary states, the differential $\tau[\partial\Phi(n)/\partial n]$ is to be replaced by the difference

[154] See p. 102.

[155] REF. 153, p. 388.

$\Phi(n) - \Phi(n - \tau)$ or symbolically

$$\tau \frac{\partial \Phi(n)}{\partial n} \leftrightarrow \Phi(n) - \Phi(n - \tau) \tag{4.16}$$

Since this recipe for translating classical formulas into their quantum-theoretic analogues will play an important role in the discovery of matrix mechanics, as we shall see in due course, we shall call it briefly "Born's correspondence rule," indicating thereby its intimate relation to Bohr's correspondence principle. For future reference we mention on this occasion that an iteration of Born's correspondence rule shows that

$$\tau \frac{\partial \nu(n,\tau)}{\partial n} \leftrightarrow \nu_{n+\tau,n} - \nu_{n,n-\tau} \tag{4.17}$$

for, by (4.15),

$$\tau \frac{\partial \nu(n,\tau)}{\partial n} = \frac{1}{h} \tau \frac{\partial}{\partial n}\left[\tau \frac{\partial H(n)}{\partial n}\right]$$

$$= \frac{1}{h}\{[H(n + \tau) - H(n)] - [H(n) - H(n - \tau)]\}$$

Werner Heisenberg, who spent the winter semester 1923/24 in Göttingen, assisted Born in this work[156]—and with great profit, as his subsequent comprehensive study of scattering and dispersion, carried out together with Kramers, clearly shows.

Kramers and Heisenberg[157] considered a nondegenerate multiply periodic system whose motion can therefore be described in terms of angle and action variables w, J and whose undisturbed electric moment M can be represented by

$$M = \tfrac{1}{2} \sum C_{\tau_1,\ldots,\tau_f} \exp\left[2\pi i(\tau_1 w_1 + \cdots + \tau_f w_f)\right]$$

where $w_k = \nu_k t + \delta_k$ and where reality requirements impose on the J-dependent Fourier coefficients the conditions $C_{\tau_1,\ldots,\tau_f} = C^*_{\tau_1,\ldots,-\tau_f}$ (the asterisk denotes complex conjugate values). Using the abbreviations $\omega = \tau_1\nu_1 + \cdots + \tau_f\nu_f$, $\omega' = \tau'_1\nu_1 + \cdots + \tau'_f\nu_f$, $\tau_k{}^0 = \tau_k + \tau'_k$, $\omega_0 = \omega + \omega'$, $\delta_0 = \tau_1{}^0\delta_1 + \cdots + \tau_f{}^0\delta_f$, $C_\tau = C_{\tau_1,\ldots,\tau_f}$, $C_{\tau'} = C_{\tau'_1,\ldots,\tau'_f}$, and

$$\frac{\delta}{\delta J} = \sum_1^f \tau_k \frac{\partial}{\partial J_k} \qquad \frac{\delta}{\delta J'} = \sum_1^f \tau'_k \frac{\partial}{\partial J_k}$$

they showed that the classical electric moment M_{cl} of the system, if acted

[156] Born acknowledged Heisenberg's assistance in *ibid.*, p. 380, footnote 1.

[157] H. A. Kramers and W. Heisenberg, "Über die Streuung von Strahlung durch Atome," *Zeitschrift für Physik 31*, 681–707 (1925). This paper was submitted for publication on Jan. 5, 1925, and not, as erroneously printed (p. 681), 1924.

on by a perturbation $E \exp(2\pi i\nu t)$, is given by the expression

$$M_{\text{cl}} = \text{Re}\left[\sum\left(\exp\{2\pi i[(\omega_0 + \nu)t + \delta_0]\}\right.\right.$$

$$\left.\left.\times \sum \frac{1}{4}\left(\frac{\delta C_\tau}{\delta J'}\frac{C_{\tau'}}{\omega' + \nu} - C_\tau\frac{\delta}{\delta J}\frac{C_{\tau'}}{\omega' + \nu}\right)\right)E\right] \quad (4.18)$$

where, for the sake of simplicity, it is again assumed that the incident wave is linearly polarized with its electric vector parallel to the direction of M, and where the first summation extends over all integral indices $\tau_1, \ldots, \tau_f$ from $-\infty$ to $+\infty$, and the second summation similarly over all $\tau_1', \ldots, \tau_f'$. If all τ_k are taken as equal, respectively, to $-\tau_k'$, so that $\omega = -\omega'$ and $\delta/\delta J' = -\delta/\delta J$, the terms obtained in (4.18) represent that part of M_{cl} which vibrates in resonance with the incident radiation of frequency ν.[158] This part, denoted by $M_{\text{cl}}(\nu)$, is consequently given by the equation

$$M_{\text{cl}}(\nu) = \text{Re}\left[\exp(2\pi i\nu t)\sum_{\tau_1,\ldots,\tau_f=-\infty}^{\infty}\frac{1}{4}\frac{\delta}{\delta J}\left(\frac{C_\tau^2}{\omega - \nu}\right)E\right]$$

$$= \text{Re}\left[\exp(2\pi i\nu t)\sum_{\substack{\tau_1,\ldots,\tau_f\\ \omega>0}}\frac{1}{4}\frac{\delta}{\delta J}\left(\frac{C_\tau^2}{\omega - \nu} + \frac{C_\tau^2}{\omega + \nu}\right)E\right]$$

Applying Born's correspondence rule to the classical dipole frequency ω and to the differential operator $\delta/\delta J$, Kramers and Heisenberg obtained the quantum-theoretic electrical moment of the atom in state i,

$$M_{\text{qu}}{}^i(\nu) = \text{Re}\left\{\exp(2\pi i\nu t)\left[\sum_{\nu_{ki}>0}\frac{1}{4h}\left(\frac{C_{ki}{}^2}{\nu_{ki} - \nu} + \frac{C_{ki}{}^2}{\nu_{ki} + \nu}\right)\right.\right.$$

$$\left.\left.- \sum_{\nu_{ik'}>0}\frac{1}{4h}\left(\frac{C_{ik'}^2}{\nu_{ik'} - \nu} + \frac{C_{ik'}^2}{\nu_{ik'} + \nu}\right)\right]E\right\}$$

where $\nu_{ki} = (E_k - E_i)/h$, $\nu_{ki} = -\nu_{ik}$ and where the amplitude C_{ki} of the dipole moment $C_{ki}\exp(2\pi i\nu_{ki}t)$ is connected with the Einstein probability $A_i{}^k$ by the relation[159]

$$A_i{}^k h\nu_{ki} = \frac{(2\pi\nu_{ki})^4}{3c^3}C_{ki}{}^2 \quad (4.19)$$

In virtue of Ladenburg's equation (4.11), now written with the two indices

[158] The nonresonant contributions give rise to the incoherent radiation predicted by A. Smekal in "Zur Quantentheorie der Dispersion," *Die Naturwissenschaften 11*, 873–875 (1923), and experimentally confirmed by C. V. Raman in "A new radiation," *Indian Journal of Physics 2*, 387–398 (1928), and independently by G. Landsberg and L. Mandelstam in "Eine neue Erscheinung bei der Lichtzerstreuung in Krystallen," *Die Naturwissenschaften 16*, 557–558 (1928).

[159] See REF. 102 OF CHAP. 3.

k and i,

$$f_{ki} = \frac{mc^3}{8\pi^2 e^2 \nu_{ki}^2} A_i^k$$

the following equation is obtained,[160]

$$f_{ki} = \frac{2\pi^2 m}{he^2} \nu_{ki} C_{ki}^2 \tag{4.20}$$

and the polarization in state i is given by the expression

$$\alpha_i = \frac{e^2}{4\pi^2 m}\left(\sum_{E_k>E_i} \frac{f_{ki}}{\nu_{ki}^2 - \nu^2} - \sum_{E_{k'}<E_i} \frac{f_{ik'}}{\nu_{ik'}^2 - \nu^2}\right) \tag{4.21}$$

Equation (4.21) is, of course, the quantum-theoretic translation of Eq. (4.8). But contrary to the f_k in Eq. (4.8), which denoted the number of dispersion electrons of kind k and were therefore nonnegative integers, the "f numbers" f_{ki} in Eq. (4.21) are not necessarily integers. They satisfy, however, an important equation, the so-called "f-summation theorem" ("f-Summensatz"), as Thomas[161] and Kuhn[162] have shown. For the sake of simplicity we shall give an outline of the proof of this theorem only for the case of a one-electron system with periodicity degree 1. In this case the dipole moment is given by

$$M = \tfrac{1}{2} \sum_{\tau=-\infty}^{\infty} C_\tau \exp\left[2\pi i\tau(\nu t + \delta)\right] \tag{4.22}$$

in terms of which the average kinetic energy can be calculated as follows:

$$\bar{E}_{\text{kin}} = \frac{m}{2}\frac{\overline{\dot{M}^2}}{e^2} = \frac{m}{2e^2}\nu \int_0^{1/\nu} \tfrac{1}{4}(2\pi i\nu)^2 \sum_\tau \sum_{\tau'} \tau\tau' C_\tau C_{\tau'} \times \exp\left[2\pi i(\tau + \tau')(\nu t + \delta)\right] dt$$

or $\bar{E}_{\text{kin}} = (m\pi^2\nu^2/2e^2) \sum_{\tau=-\infty}^{\infty} \tau^2 C_\tau C_{-\tau}$. Now, the action variable J which[163] equals $2\bar{E}_{\text{kin}}/\nu$ is therefore given by $J = (2\pi^2 m/e^2) \sum_{\tau=0}^{\infty} \tau^2 \nu C_\tau^2$. Differentiation with respect to $J = nh$ yields $1 = (2\pi^2 m/he^2) \sum_{\tau=0}^{\infty} \tau(\partial/\partial n)(\tau\nu C_\tau^2)$ and application of Born's correspondence principle, with due consideration of Eq. (4.20), leads finally to the "f-summation theorem"

$$\sum_k f_{ki} - \sum_{k'} f_{ik'} = 1 \tag{4.23}$$

[160] The factor 3 has now to be omitted. See REF. 148.

[161] W. Thomas, "Über die Zahl der Dispersionselektronen, die einem stationären Zustande zugeordnet sind," *Die Naturwissenschaften 13*, 627 (1925).

[162] W. Kuhn, "Über die Gesamtstärke der von einem Zustande ausgehenden Absorptionslinien," *Zeitschrift für Physik 33*, 408–412 (1925).

[163] See the discussion on p. 100.

5 *The Formation of Quantum Mechanics*

5.1 The Rise of Matrix Mechanics

In spite of its high-sounding name and its successful solutions of numerous problems in atomic physics, quantum theory, and especially the quantum theory of polyelectronic systems, prior to 1925, was, from the methodological point of view, a lamentable hodgepodge of hypotheses, principles, theorems, and computational recipes rather than a logical consistent theory. Every single quantum-theoretic problem had to be solved first in terms of classical physics; its classical solution had then to pass through the mysterious sieve of the quantum conditions or, as it happened in the majority of cases, the classical solution had to be translated into the language of quanta in conformance with the correspondence principle. Usually, the process of finding "the correct translation" was a matter of skillful guessing and intuition rather than of deductive and systematic reasoning. In fact, quantum theory became the subject of a special craftsmanship or even artistic technique which was cultivated to the highest possible degree of perfection in Göttingen and in Copenhagen. In short, quantum theory still lacked two essential characteristics of a full-fledged scientific theory, conceptual autonomy and logical consistency.

The core of the difficulties was, of course, the fact that according to classical physics, which served as the point of departure for quantum-theoretic calculations as long as atomic systems were described in classical terms, the optical frequencies of spectral lines should coincide with the Fourier orbital frequencies of the system's motions, a result not borne out by experiment. The discrepancy was smoothed over by Bohr's heuristically invaluable principle of correspondence. For it made it possible to retain the description of motion in terms of classical kinematics and dynamics but allowed at the same time a certain tailoring of the results so as to fit it to observational data. It was Heisenberg who recognized that this approach is only one alternative. The other alternative, which he chose in his historic paper "On a quantum theoretical interpretation of kinematical and mechanical relations"[1] and which led to the development of matrix mechanics, the earliest formulation of modern quantum mechanics, abandoned Bohr's description of motion in terms of classical physics altogether and replaced it by a description in terms of what Heisenberg regarded as observable magnitudes.

As the following analysis attempts to show, Heisenberg's crucial intervention, if examined with respect to its epistemological tenets, seems to have been made under the influence of those intellectual currents which have been mentioned in our discussion on the philosophical background of nonclassical conceptions. Bohr, it will be recalled,[2] gave a series of lectures in 1922 at the University of Göttingen, whose influence upon Pauli has already been pointed out. These lectures were also attended by Heisenberg. In one of these lectures Bohr spoke on the quadratic Stark effect and expressed his confidence that the theory in its present formulation is on the right track in spite of the as yet unclarified contradictions and the impossibility of calculating exactly the intensities of spectral lines. In particular, Bohr declared, the great success in explaining the linear Stark effect on the basis of the quantum conditions made him feel sure that the proposed interpretation of the quadratic effect must also be correct. Heisenberg, who had already begun his study of the problem of dispersion, differed with Bohr on this issue. Recognizing that the quadratic Stark effect may be considered as a limiting case of dispersion for incident radiation of infinitely low frequency and realizing that a quantum-theoretic treatment of dispersion cannot be worked out along the lines proposed, Heisenberg criticized Bohr's statement. Although Bohr, as Heisenberg recalls,[3] discussed with him privately, after the lecture, this problem and related questions for over three hours, Heisenberg did not retract his challenge. On the contrary, Bohr's frequent remarks that "the experimental situation has to be covered by means of concepts which fit"—in consonance with the Kierkegaard-Høffding insistence that every field of experience required its own conceptions and principles and, interestingly, also in consonance with the philosophies of Comte and Boutroux—seem to have impressed the young Heisenberg strongly and encouraged him in his search

[1] W. Heisenberg, "Über quantentheoretische Umdeutung kinematischer und mechanischer Beziehungen," *Zeitschrift für Physik 33*, 879–893 (1925); *Dokumente der Naturwissenschaft*, vol. 2 (Ernst Battenberg Verlag, Stuttgart, 1962), pp. 31–45.

[2] In this context see REF. 107 OF CHAP. 1.

[3] *Archive for the History of Quantum Physics*, Interview with W. Heisenberg on Nov. 30, 1962.

for an alternative approach. Heisenberg admitted repeatedly that it had been Bohr's influence and thus, ultimately, the philosophical presumptions underlying Bohr's conception of physics which had suggested to him the idea that the prevailing conceptual apparatus was not a categorical necessity or an indubitable catechism.

Thus if Heisenberg's conception of the very possibility of rejecting the description of atomic systems in terms of classical physics may be traced back, via Bohr, to one of the schools of philosophical thought referred to previously, his choice of the particular nature of the new concepts by which he replaced the classical ones goes back to the other philosophical trend mentioned, the positivism or logical empiricism of the early twenties. Interested in philosophy while still a student at the classical "gymnasium"—his first acquaintance with atomic theory was through Plato's *Timaeus*—he was strongly impressed, first by Kant's *Critique of Pure Reason* and later particularly by Wittgenstein's writings. But what most appealed to his mind was Einstein's treatment of the concept of time, his replacement of Lorentz's "local" or "mathematical" time in the so-called Lorentz transformations[4] by the operationally defined and, in this sense observable, relativistic time. Einstein's rejection of the Newtonian, operationally undefinable, conception of simultaneity of spatially separated events and his relativistic reinterpretation of the Lorentz-FitzGerald contraction, previously regarded as an effect unobservable in principle, undoubtedly made a strong impression on Heisenberg. It is known that both in Munich, where Heisenberg had written his thesis under the guidance of Sommerfeld, and in Göttingen, where he was working under Born and where Minkowski lectured, relativity was studied with great fervor.

The preceding remarks will make it clear why Heisenberg insisted on using observable magnitudes as terms for the description of atomic states. From the modern sophisticated point of view a distinction between observable and theoretically inferred magnitudes poses, of course, a highly complicated problem. In fact, when in 1926 Heisenberg confided to Einstein that "the idea of observable quantities was actually taken from his relativity,"[5] Einstein already pointed out that it is the theory which ultimately decides what can be observed and what cannot. One may admit, however, that such a distinction—even if it is, prior to the establishment of a theory, unwarranted—can be adopted as merely a heuristic principle.

Although the explicit formulation of these deliberations was, of course,

[4] The transformation equations as such had already been employed in 1887 by W. Voigt in his study of the Doppler effect for vibrations in an elastic and incompressible medium. See W. Voigt, "Ueber das Dopplersche Princip," *Göttinger Nachrichten 1887*, pp. 41–51. The term "Lorentz transformation" was introduced by H. Poincaré in his paper "Sur la dynamique de l'électron," *Comptes Rendus 140*, 1504–1508 (1905).

[5] *Archive for the History of Quantum Physics*, Interview with W. Heisenberg on Feb. 15, 1963.

of a later date, it was precisely this heuristic approach which Heisenberg adopted, as the following discussion tries to show. Heisenberg rejected the classical notions of position and velocity or momentum of electrons in atoms not only because they were unobservable in the sense that, so far, nobody had measured them directly. For it may always be argued, as Heisenberg explicitly admitted, that future progress in experimental techniques would eventually make it possible to measure these quantities. "This hope," continued Heisenberg—and this is the essential point—"could be regarded as justified if the formal theory [in terms of which these quantities are calculated] can consistently be applied to a distinctly defined domain of quantum-theoretic problems. Experience, however, shows that . . . the reaction of atoms to periodically varying fields cannot be described by these rules nor is their application to polyelectronic systems feasible."[6] Thus, Heisenberg's rejection of these quantities as unobservable was based on two empirical facts, the experimental impossibility of directly measuring them *and* the practical failure of a theory which assumed them to be observable. He replaced these quantities of classical kinematics by the optical quantities of frequency and intensity, or rather dipole amplitude,[7] and investigated whether a theory, assuming these as observable, can be worked out consistently. Basically, Heisenberg's attitude, in this respect, resembled that of Einstein, for whom the concept of Newtonian time had lost its physical significance not only, as he showed in his analysis of the simultaneity of spatially separated events, because of its insusceptibility to operational determination but also because classical physics which assumed this concept as observable conflicted with experience.

A second fundamental innovation in Heisenberg's approach was the way he employed Bohr's correspondence principle. As mentioned before, the translation of classical formulas into the language of quantum theory usually required a great deal of guessing and had only one guideline, the correspondence principle. It had to be used in a separate way for almost every problem, the particular mode of its application depending on the specific data of the problem. While its versatility and fertility made it attractive to the more synthetically minded physicists, such as Bohr, its flexibility and lack of rigidity made it repugnant to the more analytically oriented theoreticians, such as Sommerfeld. Heisenberg, influenced by both Sommerfeld and Bohr, considered now the possibility of "guessing"[8]—in accordance with the correspondence principle—not the solution of a par-

[6] REF. 1, p. 879.

[7] His study of the dispersion problem suggested that not only the squares of the magnitudes of these amplitudes (the intensities) but also their phases are experimentally ascertainable.

[8] "Also war es vielleicht auch möglich, einfach durch geschicktes Erraten eines Tages den Übergang zum vollständigen mathematischen Schema der Quantenmechanik zu vollziehen." W. Heisenberg, "Erinnerungen an die Zeit der Entwicklung der Quantenmechanik," in *Theoretical Physics in the Twentieth Century*, edited by M. Fierz and V. F. Weisskopf (Interscience, New York, 1960), p. 42.

ticular quantum-theoretic problem but the very mathematical scheme for a new theory of mechanics. By integrating in this way the correspondence principle once and for all in the very foundations of the theory, he expected to eliminate the necessity of its recurrent application to every problem individually without jeopardizing its general validity. Through absorbing, so to speak, the correspondence in the foundations of the theory, it would be possible, Heisenberg hoped, to solve quantum-theoretic problems in a mathematically rigorous way without losing the effectiveness of the principle. In the following analysis of what Heisenberg called "the kinematics of quantum theory" we shall see how skillfully he employed Bohr's principle and how he reduced the "guess" to an "almost cogent"[9] procedure.

In classical physics any time-dependent quantity $\xi_n = \xi_n(t)$ can be represented by a Fourier expansion[10]

$$\xi_n = \sum_\tau \xi(n,\tau) = \sum_\tau x(n,\tau) \exp\left[2\pi i\nu(n,\tau)t\right] \tag{5.1}$$

where the τth Fourier component $\xi(n,\tau)$ has the amplitude $x(n,\tau)$ and the frequency $\nu(n,\tau) = \tau\nu(n,1) = \tau\nu$. Now, just as according to Bohr's correspondence principle the quantum-theoretic frequency $\nu_{n,n-\tau}$ corresponds to $\nu(n,\tau)$, Heisenberg assumed that a quantum-theoretic quantity $x_{n,n-\tau}$ corresponds to the Fourier amplitude $x(n,\tau)$ or, symbolically, as

$$\nu(n,\tau) \leftrightarrow \nu_{n,n-\tau} \tag{5.2}$$

so also

$$x(n,\tau) \leftrightarrow x_{n,n-\tau} \tag{5.3}$$

From their definition or from the reality requirement it follows that

$$\nu(n,-\tau) = -\nu(n,\tau) \tag{5.4}$$

$$x(n,-\tau) = x^*(n,\tau) \tag{5.5}$$

$$\nu_{n-\tau,n} = -\nu_{n,n-\tau} \tag{5.6}$$

$$x_{n-\tau,n} = x^*_{n,n-\tau} \tag{5.7}$$

$$\nu(n,\tau) + \nu(n,\tau') = \nu(n,\tau+\tau') \tag{5.8}$$

$$\nu_{n,n-\tau} + \nu_{n-\tau,n-\tau'} = \nu_{n,n-\tau'} \tag{5.9}$$

Heisenberg realized that in spite of the correspondences (5.2) and (5.3) an expression of the kind $\sum_\tau x_{n,n-\tau} \exp(2\pi i\nu_{n,n-\tau}t)$, analogous to (5.1), has no physical significance whatever, if only because of the essentially equal status of the two indices in $\nu_{n,n-\tau}$.

[9] ". . . nahezu zwangläufig." REF. 1, p. 883.
[10] In the sequel, Σ_τ will be used as an abbreviation of $\Sigma_{\tau=-\infty}^{+\infty}$.

Heisenberg, however, assumed that the "set" ("Gesamtheit") $\{x_{n,n-\tau} \exp(2\pi i \nu_{n,n-\tau} t)\}$, or briefly $\{\xi_{n,n-\tau}\}$, may well be chosen as the quantum-theoretic representative of the classical quantity $\xi_n(t)$:

$$\xi_n \leftrightarrow \{x_{n,n-\tau} \exp(2\pi i \nu_{n,n-\tau} t)\} \tag{5.10}$$

In classical physics, now,

$$\xi_n{}^2 = \left\{\sum_{\tau'} x(n,\tau') \exp[2\pi i \nu(n,\tau')t]\right\} \times \left\{\sum_{\tau} x(n, \tau - \tau') \exp[2\pi i \nu(n, \tau - \tau')t]\right\} \tag{5.11}$$

or

$$\xi_n{}^2 = \sum_{\tau} \left\{\sum_{\tau'} x(n,\tau')x(n, \tau - \tau') \exp[2\pi i \nu(n,\tau')t] \exp[2\pi i \nu(n, \tau - \tau')t]\right\} \tag{5.12}$$

so that in virtue of (5.8)

$$\xi_n{}^2 = \sum_{\tau} x^{(2)}(n,\tau) \exp[2\pi i \nu(n,\tau)t] \tag{5.13}$$

where

$$x^{(2)}(n,\tau) = \sum_{\tau'} x(n,\tau')x(n, \tau - \tau') \tag{5.14}$$

Heisenberg now asked: "How has the quantity $\xi_n{}^2$ to be represented in quantum theory?" In other words, if—in agreement with (5.10) –

$$\xi_n{}^2 \leftrightarrow \{x^{(2)}_{n,n-\tau} \exp(2\pi i \nu_{n,n-\tau} t)\} \tag{5.15}$$

what is $x^{(2)}_{n,n-\tau}$?

A straightforward replacement of the classical quantities in (5.12) by the quantum-theoretic quantities in accordance with (5.2) and (5.3) would yield

$$\sum_{\tau'} x_{n,n-\tau'} x_{n,n-(\tau-\tau')} \exp(2\pi i \nu_{n,n-\tau'} t) \exp(2\pi i \nu_{n,n-(\tau-\tau')} t) \tag{5.16}$$

for the τth term in the set representing $\xi_n{}^2$. Since the product of the two phase factors in (5.16) differs from $\exp(2\pi i \nu_{n,n-\tau} t)$, the phase factor required by (5.15), but $\exp(2\pi i \nu_{n,n-\tau'} t) \exp(2\pi i \nu_{n-\tau',n-\tau} t)$, in virtue of (5.9), does yield the required factor, Heisenberg postulated—and this is what he called an "almost cogent" conclusion—that (5.16) has to be replaced by

$$\sum_{\tau'} x_{n,n-\tau'} x_{n-\tau',n-\tau} \exp(2\pi i \nu_{n,n-\tau'} t) \exp(2\pi i \nu_{n-\tau',n-\tau} t) \tag{5.17}$$

which is, of course, equal to

$$\left(\sum_{\tau'} x_{n,n-\tau'} x_{n-\tau',n-\tau}\right) \exp\,(2\pi i \nu_{n,n-\tau} t) \tag{5.18}$$

so that

$$x^{(2)}_{n,n-\tau} = \sum_{\tau'} x_{n,n-\tau'} x_{n-\tau',n-\tau} \tag{5.19}$$

is the required multiplication rule.

He immediately generalized (5.19) for $\xi_n{}^3$ and for the product of two quantities, such as $\xi_n \eta_n$, where $\eta_n \leftrightarrow \{y_{n,n-\tau} \exp\,(2\pi i \nu_{n,n-\tau} t)\}$, and found that, in general, $\xi_n \eta_n \neq \eta_n \xi_n$ since, in general, $\sum_{\tau'} x_{n,n-\tau'} y_{n-\tau',n-\tau} \neq \sum_{\tau'} y_{n,n-\tau'} x_{n-\tau',n-\tau}$. The new kind of multiplication turned out to be non-commutative.

Having completed the "kinematics," Heisenberg proceeded in the second part of his paper to the discussion of the "mechanics" of his new formalism. For the sake of simplicity only linear problems are considered.

In the older theory the solving of a dynamical problem consisted of two steps: the equation of motion $m\ddot{x} + f(x) = 0$ had to be solved and the quantum condition $J = \oint p\,dq = \int m\dot{x}^2\,dt = nh$ had to be satisfied. With the help of (5.2), (5.3), and the new multiplication rule it was not difficult to transform the equation of motion into relations which had to be satisfied by the quantum-theoretic amplitudes $x_{n,n-\tau}$. For the determination of these "transition amplitudes" the quantum-theoretic version of the quantum condition had to be taken into account. To obtain the latter in terms of the new formalism, Heisenberg showed by a straightforward calculation that in virtue of (5.1) and (5.8) the quantum condition is

$$J = \int_0^{1/\nu} m\dot{x}^2\,dt = 4\pi^2 m \sum_\tau \tau\nu(n,\tau)\,|\,x(n,\tau)\,|^2 = nh$$

or, if differentiated with respect to n,

$$h = 4\pi^2 m \sum_\tau \tau \frac{d}{dn}\,(\nu(n,\tau)\,|\,x(n,\tau)\,|^2)$$

Applying the correspondence rule (4.17), page 193, or rather its generalization as used in the Kramers-Heisenberg theory of dispersion,

$$\tau \frac{\partial \Phi(n,\tau)}{\partial n} \leftrightarrow \Phi_{n+\tau,n} - \Phi_{n,n-\tau}$$

Heisenberg obtained

$$h = 4\pi^2 m \sum_\tau \{|\,x_{n+\tau,n}|^2\,\nu_{n+\tau,n} - |\,x_{n,n-\tau}|^2\,\nu_{n,n-\tau}\} \tag{5.20}$$

or

$$h = 8\pi^2 m \sum_{\tau=0}^{\infty} \{ | x_{n+\tau,n}|^2 \nu_{n+\tau,n} - | x_{n,n-\tau}|^2 \nu_{n,n-\tau} \} \tag{5.21}$$

which by Eq. (4.20), page 195, is precisely the Thomas-Kuhn "f-summation theorem" (4.23), page 195.

In the last part of his paper Heisenberg applied his theory to the anharmonic oscillator, classically described by $\ddot{x} + \omega_0^2 x + \lambda x^2 = 0$—the problem of the harmonic oscillator would not have required the application of the new multiplication rule—and found that its energy W, up to the order of magnitude λ^2, is given by the expression

$$W = \frac{h\omega_0}{2\pi} (n + \tfrac{1}{2}) \tag{5.22}$$

Equation (5.22) implied that the energy of the harmonic oscillator ($\lambda = 0$) is likewise given by the same expression in contrast to the result of the older quantum theory $W = nh\nu_0$. Heisenberg applied his new formalism to the rotator, that is, to an electron revolving at constant distance a and with constant angular velocity ω around the nucleus. From Eq. (5.21) and the assumption that at the ground state no radiation is emitted, he obtained

$$\nu_{n,n-1} = \frac{h}{4\pi^2 ma^2} n \tag{5.23}$$

and from Eq. (5.19) for its energy W

$$W = ma^2\pi^2(\nu^2_{n,n-1} + \nu^2_{n+1,n}) = \frac{h^2}{8\pi^2 ma^2} (n^2 + n + \tfrac{1}{2}) \tag{5.24}$$

in agreement with Kratzer's observations.[11] Yet in spite of these promising results and their support by experience, Heisenberg concluded his article with the remarks: "Whether a method of determining quantum-theoretic data in terms of relations between observable magnitudes as at present proposed can be regarded as already satisfactory in principle or whether it is still too crude an approach to the obviously highly complicated problem of a quantum-theoretic mechanics, will become clear only after the presently rather superficially applied procedures will have been subjected to a penetrating mathematical investigation."[12]

The above-mentioned solution of the anharmonic oscillator had been obtained by Heisenberg at the end of May, 1925, while he was recuperating on the island of Heligoland from a heavy attack of hay fever. In June he returned to Göttingen and accepted an invitation to lecture at the Caven-

[11] A. Kratzer, "Störungen und Kombinationsprinzip im System der violetten Cyanbanden," *Münchener Berichte 1922*, pp. 107–118.

[12] REF. 1, p. 893.

dish Laboratory in Cambridge. Still doubtful whether his paper made sense and determined "either to finish it during the few remaining days prior to his departure or to throw it into the flames," he sent it to Pauli, asking him to return it with his comments after two or three days. Pauli's reply was encouraging. Thus, in the middle of July Heisenberg presented the paper to Born, who immediately recognized its importance. Born sent it to the editor of the *Zeitschrift für Physik*, who received it on July 29.

The striking feature in Heisenberg's paper was, of course, its representation of physical quantities by "sets" of time-dependent complex numbers and the peculiar multiplication rule. Even Born was puzzled. "Heisenberg's multiplication rule," Born reminisced in his Nobel Prize Address,[13] "did not give me any rest, and after eight days of concentrated thinking and testing I recalled an algebraic theory which I had learnt from my teacher Professor Rosanes in Breslau." Born recognized that Heisenberg's "sets" were precisely those matrices with which he had become acquainted when in 1903, as a young student at the University of Breslau, he attended lectures on algebra and analytical geometry given by Jacob Rosanes,[14] a pupil of Frobenius.[15]

Heisenberg's strange multiplication rule turned out to be simply the rule "row times column" for the multiplication of matrices, or in symbols, if x denotes the matrix (x_{mn}) and y the matrix (y_{mn}), then the element in the mth row and nth column of the product of these two matrices is given by the formula

$$(xy)_{mn} = \sum_k x_{mk}y_{kn}$$

a rule which had been formulated for the first time, though only for finite square matrices, seventy years earlier by Arthur Cayley.[16]

[13] M. Born, "Die statistische Deutung der Quantenmechanik," *Les Prix Nobel en 1954* (Stockholm, 1955), pp. 79–90; *Nobel Lectures—Physics (1942–1962)* (Elsevier, Amsterdam, London, New York, 1964), pp. 256–267; reprinted in M. Born, *Ausgewählte Abhandlungen* (REF. 116 OF CHAP. 3), vol. 2, pp. 430–441, and in *Dokumente der Naturwissenschaft* (REF. 1), vol. 2, pp. 1–12.

[14] Rosanes's thesis *De polarium reciprocarum theoria observationes* (C. H. Storch, Breslau, 1865) concluded with the words: "Physices leges nisi matheseos sublimioris ope neque exprimi neque perspici possunt," a statement of particular significance for our present discussion.

[15] It will be recalled that G. Frobenius made important contributions to the development of the algebraic theory of matrices, especially in his classic paper "Lineare Substitutionen und bilineare Formen," *Journal für die reine und angewandte Mathematik* (*Crelle*) *84*, 1–63 (1878). In the article "Ueber homogene totale Differentialgleichungen," *ibid. 86*, 1–19 (1879), he introduced the notion of the rank of a matrix.

[16] A. Cayley, "Sept différents mémoires d'analyse," *Journal für die reine und angewandte Mathematik* (*Crelle*) *50*, 272–317 (1855). The third memoir, entitled "Remarques sur la notation des fonctions algébriques" (pp. 282–285), contains Cayley's first discussion of matrices and the multiplication rule: "Il faut faire attention dans la composition des matrices, de combiner les *lignes* de la matrice à gauche avec les colonnes de la matrice à droite, pour former les *lignes* de la matrice composée" (p. 284). It should be noted, however, that the concept of a matrix—as an "arrangement of terms" consisting of lines and columns from which determinants can be formed—as well as its name had already appeared in Sylvester's

In connection with our present discussion of Heisenberg's matrix representation of quantum-mechanical states, the following mathematical digression seems to be of some interest. As is well known, subsequent formalisms of quantum mechanics employed—in addition to matrices and apart from complex-valued functions (Schrödinger wave functions)—also Hilbert space vectors and, after the advent of the quantum mechanics of spinning particles, quaternions as well.[17] The striking point, now, is the fact that precisely these three kinds of mathematical entities, matrices, multidimensional vectors, and quaternions, happened to be equally involved in a historically interesting controversy concerning the priority of the discovery of matrices. Sylvester[18] credited Cayley as the exclusive discoverer of matrices, calling the latter's paper of 1858 "a second birth of algebra, its *avatar* in a new and glorified form." Seven years later Gibbs criticized Sylvester severely for this statement and, referring to Grassmann's *Ausdehnungslehre*,[19] claimed that "the key to the theory of matrices is certainly given in the first 'Ausdehnungslehre'; and if we call the birth of matricular analysis the second birth of algebra, we can give no later date to this event than the memorable year of 1844."[20] Grassmann, as we shall see later on, anticipated a great deal of the theory of linear vector spaces. Now, Gibbs regarded 1844 as a "memorable" year in the annals of mathematics because it was the year of the appearance of Hamilton's first paper

paper "Additions to the articles 'On a new class of theorems . . .' and 'On Pascal's Theorem,'" *Philosophical Magazine 37*, 363–370 (1850), reprinted in *The Collected Mathematical Papers of James Joseph Sylvester* (Cambridge University Press, 1904), vol. 1, pp. 145–151. But its conception as characterizing a linear transformation is due to Cayley, who defined a matrix as "an abbreviated notation for a set of linear equations" and who is to be credited with the first rigorous exposition of an algebraic theory of matrices. Cf. A. Cayley, "A memoir on the theory of matrices," *The Philosophical Transactions of the Royal Society of London 148*, 17–37 (1858); reprinted in *The Collected Mathematical Papers of Arthur Cayley* (Cambridge University Press, 1889), vol. 2, pp. 475–496.

[17] Quaternions had been applied to physics as early as 1867 by P. G. Tait in his *An Elementary Treatise on Quaternions* (Clarendon Press, Oxford, 1867), chaps. 10 and 11 (pp. 248–311). Their use in modern physics was revived by Klein, who showed in 1910, in a lecture before the Göttingen Mathematical Society, that Lorentz transformations, conceived as four-dimensional rotations in Minkowski space, can be conveniently expressed in terms of quaternions; cf. F. Klein, "Über die geometrischen Grundlagen der Lorentzgruppen," *Physikalische Zeitschrift 12*, 17–27 (1911). Their systematic, though never popular, use in quantum mechanics began when it was recognized that the Pauli spin matrices were essentially quaternion basis elements. Recently it has been attempted to describe elementary particles in terms of quaternions; cf. E. J. Schremp, "Isotopic spin and the group-space of the proper Lorentz group," *Physical Review 99*, 1603 (1955). G. R. Allcock, "A space-time model of isospace," *Nuclear Physics 27*, 204–233 (1961). Cf. also O. F. Fischer, *Universal Mechanics and Hamiltons Quaternions* (Axion Institute, Stockholm, 1951).

[18] J. J. Sylvester, "The genesis of an idea, or story of a discovery relating to equations in multiple quantity," *Nature 31*, 35–36 (1884).

[19] H. Grassmann, *Die lineale Ausdehnungslehre* (Wigand, Leipzig, 1878); *Gesammelte mathematische und physikalische Werke* (Teubner, Leipzig, 1894), vol. 1, part 1.

[20] J. W. Gibbs, "Quaternions and the 'Ausdehnungslehre,'" *Nature 44*, 79–82 (1891).

on quaternions.[21] But Hamilton has also been named the discoverer of matrices. The first to do so was Tait. In a letter to Cayley, who to his bad luck had shown that quaternions can be represented as matrices of order 2 with complex elements, Tait, referring to Hamilton's "linear and vector operators," called Cayley's discovery only a modification of Hamilton's ideas. Whereupon Cayley retorted: "I certainly did not get the notion of a matrix in any way through quaternions. . . ."[22] The claim was renewed in 1890 by Taber,[23] who declared: "Hamilton must be regarded as the originator of the theory of matrices, as he was the first to show that the symbol of a linear transformation might be made the subject-matter of a calculus."[24] To give further details on this controversy, which involved W. K. Clifford, A. N. Whitehead, A. Buchheim, and in the twentieth century T. J. I'A. Bromwich, who preferred to be "unable to give any opinion on this matter,"[25] would lead us too far astray from our subject.

Now, matrices, multidimensional vectors, and quaternions are extensions of the concept of real numbers. Beyond the domain of complex numbers, however, extensions are possible only at the expense of Hankel's principle of permanence,[26] according to which generalized entities should satisfy the rules of calculation pertaining to the original mathematical entities from which they have been abstracted. Thus, while associativity and distributivity could be preserved, commutativity had to be sacrificed. It was the price which had to be paid to obtain the appropriate mathematical apparatus for the description of atomic states.

The mathematical process shows a certain analogy to a fundamental epistemological principle which Heisenberg once expressed as follows: "Almost every progress in science has been paid for by a sacrifice, for almost every new intellectual achievement previous positions and conceptions had to be given up. Thus, in a way, the increase of knowledge and insight diminishes continually the scientist's claim on 'understanding' nature."[27]

Prior to 1925 matrices had rarely been used by physicists. Notable exceptions were Mie's nonlinear electrodynamics[28] and, interestingly, Born's

[21] Sir William Rowan Hamilton, "On quaternions; or on a new system of imaginaries in algebra," *Philosophical Magazine 25*, 10–13, 241–246, 489–495 (1844).

[22] Cf. C. G. Knott, *Life and Scientific Work of Peter Guthrie Tait* (Cambridge University Press, 1911), pp. 153, 164.

[23] H. Taber, "On the theory of matrices," *American Journal of Mathematics 12*, 337–395 (1890).

[24] *Ibid.*, p. 337.

[25] T. J. I'A. Bromwich, "Review of Muth's *Elementartheiler*," *Bulletin of the American Mathematical Society 7*, 308–316 (1901).

[26] H. Hankel, *Vorlesungen über die Complexen Zahlen* (Voss, Leipzig, 1867).

[27] W. Heisenberg, "Zur Geschichte der physikalischen Naturerklärung," *Sächsische Berichte 85*, 29–40 (1933).

[28] G. Mie, "Grundlagen einer Theorie der Materie," *Annalen der Physik 37*, 511–534 (1912), *39*, 1–40 (1912), *40*, 1–66 (1913). Cf., for example, p. 525.

work on the lattice theory of crystals.[29] But even in these exceptional cases matrices were not subjected to algebraic operations and, in particular, multiplications of matrices did not appear. On the whole, matrices were regarded as belonging exclusively to the realm of pure mathematics.

In retrospect, it seems almost uncanny how mathematics now prepared itself for its future service to quantum mechanics. Bôcher's *Introduction to Higher Algebra*,[30] the standard textbook on matrix theory, appeared in a German[31] translation in 1910, one year after the publication of Kowalewski's treatise on determinants, the first text to include infinite determinants.[32] Its second and revised edition was completed in May, 1924. At the same time Courant,[33] utilizing Hilbert's lectures, finished in Göttingen the first volume of the well-known *Methods of Mathematical Physics*. Published at the end of 1924, it contained precisely those parts of algebra and analysis on which the later development of quantum mechanics had to be based; its merits for the subsequent rapid growth of our theory can hardly be exaggerated.

One of Courant's assistants in the preparation of this work was Pascual Jordan.[34] Having had to familiarize himself for this purpose with the theory of matrices, Jordan made a thorough study of Bôcher's exposition[35] as well as of other publications on this subject and acquired great skill in the manipulation of matrices.[36]

At the same time Born was looking for a qualified assistant for the project of putting the new matrix mechanics on logically firm foundations. He turned at first to Pauli. Although, as we shall presently see, Pauli was soon to make an important contribution to matrix mechanics, he rejected Born's proposal. It was not easy to find a competent assistant for this project. Physicists either did not know[37] what matrices were or, if they did,

[29] Cf., for example, M. Born, "Über elektrostatische Gitterpotentiale," *Zeitschrift für Physik 7*, 124–140 (1921); *Ausgewählte Abhandlungen* (REF. 116 OF CHAP. 3), vol. 1, pp. 434–450, where reference is made to "the matrix of the coefficients" of a certain algebraic form (p. 129; p. 439).

[30] M. Bôcher, *Introduction to Higher Algebra* (Macmillan, New York, 1907).

[31] M. Bôcher, *Einführung in die höhere Algebra*, translated by H. Beck (Teubner, Leipzig, 1910).

[32] G. Kowalewski, *Einführung in die Determinantentheorie einschließlich der unendlichen und der Fredholmschen Determinanten* (Veit, Leipzig, 1909).

[33] R. Courant and D. Hilbert, *Methoden der mathematischen Physik* (Springer, Berlin, 1924), vol. 1. A second edition appeared in 1931 and an English translation, *Methods of Mathematical Physics* (Interscience, New York, London), in 1953.

[34] See Courant's acknowledgment in the preface of the book, *ibid.*

[35] REF. 30, chaps. 22 to 25.

[36] Cf. *Archive for the History of Quantum Physics*, Interview with P. Jordan on June 17, 1963.

[37] Heisenberg's comment on a letter from Jordan, "Now the learned Göttingen mathematicians talk so much about Hermitian matrices, but I do not even know what a matrix is," illustrates precisely the situation in 1925. It is known that even Born consulted Otto Toeplitz on certain properties of matrices. In this context also the following episode deserves to be mentioned. It will be recounted as described by E. U. Condon, who visited Munich and Göttingen in 1926 as a Fellow

were reluctant to apply them to theoretical problems. This aversion probably explains why Heisenberg's historic paper of 1925 had not been reviewed in the *Physikalische Berichte*, the official *Abstracts* of the German Physical Society, prior to the 1927 issue, and there only in a one-sentence statement written by one of its editors.[38] The report given in 1926 by *Science Abstracts*,[39] issued by the Institute of Electrical Engineers in association with the Physical Society of London and with the American Physical Society, was not much more detailed.

It should be noted, however, that the reluctance with which Heisenberg's matrix mechanics was met has to be attributed not only to the unaccustomed mathematics involved but also, at least in part, to the unfamiliarity of its conceptual background. Thus, for example, Segrè[40] reports that Fermi, having been awarded, in 1923, a fellowship by the Italian Ministry of Education (Ministero dell'Istruzione Pubblica) to study abroad, chose Göttingen to work with Born but left it soon for Leyden. Segrè continues: "Heisenberg's great paper on matrix mechanics of 1925 did not appear sufficiently clear to Fermi, who reached a full understanding of quantum mechanics only later through Schrödinger's wave mechanics. I want to emphasize that this attitude of Fermi was certainly not due to the mathematical difficulties and novelty of matrix algebra; for him, such

---

of the International Education Board. "Hilbert was having a great laugh on Born and Heisenberg and the Göttingen theoretical physicists because when they first discovered matrix mechanics they were having, of course, the same kind of trouble that everybody else had in trying to solve problems and to manipulate and really do things with matrices. So they went to Hilbert for help, and Hilbert said the only times that he had ever anything to do with matrices was when they came up as a sort of by-product of the eigenvalues of the boundary-value problem of a differential equation. So if you look for the differential equation which has these matrices you can probably do more with that. They had thought it was a goofy idea and that Hilbert did not know what he was talking about. So he was having a lot of fun pointing out to them that they could have discovered Schrödinger's wave mechanics six months earlier if they had paid a little more attention to him." Edward U. Condon, "60 Years of Quantum Physics," *Physics Today 15*, 37–49 (Oct. 10, 1962). The quotation is on p. 46.

[38] "Verf. versucht, Grundlagen für eine quantentheoretische Mechanik zu gewinnen, die ausschließlich auf Beziehungen zwischen prinzipiell beobachtbaren Größen basiert ist." *Physikalische Berichte 1927*, p. 1205.

[39] *Science Abstracts 29*, 14 (1926).

[40] E. Segrè, "Biographical Introduction" to *The Collected Papers of Enrico Fermi* (University of Chicago Press, Accademia Nazionale dei Lincei, Roma, 1962), vol. 1, pp. xvii–xlii (quotation on p. xxvii). How discouraged Fermi was after his experience in Göttingen has been well described by Goudsmit (REF. 226 OF CHAP. 3). Ehrenfest, who had early recognized Fermi's genius from his publications, wrote to Uhlenbeck (who at that time was tutoring in Rome the sons of the Dutch Ambassador) and asked him to invite Fermi, who had just returned from Göttingen, to his laboratory in Leiden. Fermi, who after the fiasco in Göttingen was almost ready to give up physics altogether, was thus given a new start. Moreover, as reported by de Latil in his biography of Fermi, it was also through his friendship with Uhlenbeck that in 1930 Fermi visited the University of Michigan, to which Uhlenbeck and Goudsmit had meanwhile moved. The favorable impressions which Fermi gained on this visit to Michigan were, as de Latil pointed out, "a determining factor in Fermi's later decision to emigrate to the United States in his turn." Pierre de Latil, *Enrico Fermi*, translated by L. Ortzen (Souvenir Press, London; Ryerson Press, Toronto, 1965), p. 29.

difficulties were minor obstacles; it was rather the physical ideas underlying these papers which were alien to him." Obviously, the abstract, almost philosophical, nature of Heisenberg's paper and its apparent lack of immediate practical applications made it repugnant to the more experimentally inclined physicists. But precisely the same qualities, viewed as initiating a new era of progress in theoretical physics, appealed to the more philosophically minded theoreticians—as far as they took notice at all. Thus Bohr, at the sixth Scandinavian Mathematical Congress, held in Copenhagen at the end of August, 1925, only a few weeks after the completion of Heisenberg's paper, acclaimed it as an outstanding achievement. To him, of course, there was nothing "alien" in it. On the contrary, he declared: "The whole apparatus of quantum mechanics can be regarded as a precise formulation of the tendencies embodied in the correspondence principle."[41] Concluding his address, he remarked: "It is to be hoped that a new era of mutual stimulation of mechanics and mathematics has commenced."

Now it happened that Born, while traveling by train to Hanover, told a colleague of his from Göttingen about the fast progress in his work but also mentioned the peculiar difficulties involved in the calculations with matrices. It was fortunate and almost an act of providence that Jordan, who shared the same compartment in the train, overheard this piece of conversation. At the station in Hanover Jordan then introduced himself to Born, told him of his experience in handling matrices, and expressed his readiness to assist Born in his work.[42] This, then, was the beginning of a fruitful collaboration which led to the publication of the fundamental Born-Jordan paper "On quantum mechanics,"[43] the first rigorous formulation of matrix mechanics. It was received by the editor of the *Zeitschrift für Physik* on September 27, 1925, exactly sixty days after the Heisenberg paper.

The Born-Jordan paper contains four chapters, the first of which expounds the pertinent theorems of matrix theory. The second chapter develops the foundations of quantum dynamics for nondegenerate systems with one degree of freedom. Representing the classical coordinate q (previously denoted by x) by a matrix $\mathbf{q} = [q_{mn} \exp(2\pi i \nu_{mn} t)]$ and the momentum p by a matrix $\mathbf{p} = [p_{mn} \exp(2\pi i \nu_{mn} t)]$, Born showed that the extremalization of the trace of the matrix

$$\mathbf{L} = \mathbf{p}\dot{\mathbf{q}} - \mathbf{H}(\mathbf{pq})$$

[41] N. Bohr, "Atomic Theory and Mechanics," *Nature 116*, 845–852 (1925) (quotation on p. 852); "Atomtheorie und Mechanik," *Die Naturwissenschaften 14*, 1–10 (1926).

[42] *Archive for the History of Quantum Physics*, Interview with Born in June, 1960 (recorded by P. P. Ewald).

[43] M. Born and P. Jordan, "Zur Quantenmechanik," *Zeitschrift für Physik 34*, 858–888 (1925); *Ausgewählte Abhandlungen* (REF. 116 OF CHAP. 3), vol. 2, pp. 124–154; *Dokumente der Naturwissenschaft* (REF. 1), vol. 2, pp. 46–76.

where $\dot{\mathbf{q}} = [2\pi i\nu_{mn}q_{mn} \exp(2\pi i\nu_{mn}t)]$ and where $\mathbf{H}(\mathbf{pq})$ is the (matrix) Hamiltonian function of the matrices $\mathbf{p}$ and $\mathbf{q}$, leads to the canonical equations

$$\dot{\mathbf{q}} = \partial\mathbf{H}/\partial\mathbf{p} \qquad \dot{\mathbf{p}} = -\partial\mathbf{H}/\partial\mathbf{q}$$

Representing both q and p independently as matrices, and not only q as Heisenberg did, Born established for the first time what was later called the basic *commutation relation* in quantum mechanics. Born's derivation was based on the correspondence principle. Starting from the quantum condition of the older quantum theory,

$$nh = J = \oint p\,dq = \int_0^{1/\nu} p\dot{q}\,dt$$

and expanding p and q into Fourier series,

$$p = \sum_\tau p(n,\tau) \exp(2\pi i\nu(n,\tau)t) \qquad q = \sum_\tau q(n,\tau) \exp(2\pi i\nu(n,\tau)t)$$

Born obtained by simple integration

$$nh = -2\pi i \sum_\tau \tau p(n,\tau)q^*(n,\tau)$$

where use has been made of Eq. (5.5), page 200. From the last equation Born deduced

$$\frac{h}{2\pi i} = -\sum_\tau \tau \frac{\partial}{\partial n}(p(n,\tau)q^*(n,\tau))$$

which he expressed in the language of quantum mechanics in accordance with the translation rule, mentioned on page 202, as

$$\frac{h}{2\pi i} = -\sum_\tau (p(n+\tau, n)q^*(n+\tau, n) - p(n, n-\tau)q^*(n, n-\tau))$$

or by Eq. (5.7), page 200, as

$$\frac{h}{2\pi i} = \sum_\tau (p(n, n-\tau)q(n-\tau, n) - q(n, n+\tau)p(n+\tau, n))$$

Having thus proved the validity of the matrix equation

$$\mathbf{pq} - \mathbf{qp} = \left(\frac{h}{2\pi i}\right)\mathbf{1} \tag{5.25}$$

as far as diagonal elements are concerned ($\mathbf{1}$ is, of course, the unit matrix) Born conjectured that it holds also for the nondiagonal elements. That this is indeed the case was shown by Jordan on the basis of the canonical equations. First it was shown that a matrix with a vanishing time derivative is diagonal. Calculating the time derivative of the matrix $\mathbf{d} = \mathbf{pq} - \mathbf{qp}$, Jordan found in virtue of the canonical equations that $\dot{\mathbf{d}} = 0$, which, as just mentioned, entails that all nondiagonal elements are zero. The fundamental quantization rule (5.25) has thus been rigorously proved. The equations of motion have shown that $\mathbf{pq} - \mathbf{qp}$ is a diagonal matrix. That all diagonal elements are equal to $h/2\pi i$ has been a consequence of the correspondence principle. Moreover, since (5.25), as Born soon realized, is the only fundamental equation in which h appears, the introduction of Planck's constant into quantum mechanics has likewise been a consequence of the correspondence principle. In the present paper Born and Jordan called this equation the "exact quantum condition" and already recognized its axiomatic status within the logical structure of the theory.[44] "I shall never forget the thrill I experienced when I succeeded in condensing Heisenberg's ideas on quantum conditions in the mysterious equation $\mathbf{pq} - \mathbf{qp} = h/2\pi i$," declared Bohr[45] in reviewing his work of 1925.

In the same chapter of their common paper Born and Jordan also proved the energy theorem and Bohr's frequency condition for the case $H = H_1(p) + H_2(q)$ and showed how these conclusions have to be generalized for the case $H = H(p,q)$.

In the third chapter the theory is applied to the harmonic oscillator and Heisenberg's results are retrieved. In contrast to the latter, however, who still needed the correspondence principle to show that transitions occur only between adjacent stationary states,[46] Born and Jordan derived this restriction as a logical conclusion from their theory alone. Following a comprehensive treatment of the anharmonic oscillator, a discussion of the quantization of the electromagnetic field concluded the paper.

Having completed the paper, Born left for a vacation in the Engadine. After his return from Switzerland in September he resumed his work with Jordan and, by correspondence, also with Heisenberg. In the middle of November the famous Born-Heisenberg-Jordan paper "On quantum mechanics II"[47] was completed. Considered as a sequel to the Born-Jordan paper,[48] it generalized the results to systems with an arbitrary finite number of degrees of freedom, introduced canonical transformations, laid the foundations for a quantum-mechanical theory of time-independent as well as

[44] "... die wir die 'verschärfte Quantenbedingung' nennen und auf der alle weiteren Schlüsse beruhen." *Ibid.*, p. 871; p. 137; p. 59.

[45] M. Born, *Physics in My Generation* (Pergamon Press, London, New York, 1956), p. 100. The novelty of Born's result, compared with that of Heisenberg, was frankly admitted by Heisenberg when he declared: "... the fact that xy was not equal to yx was very disagreeable to me. I felt this was the only point of difficulty in the whole scheme, otherwise I would be perfectly happy.... I had written down, as the quantization rule the Thomas-Kuhn sum rule, but I had not realized that this was just $pq - qp$." *Archive for the History of Quantum Physics*, Interview with Heisenberg on Feb. 15, 1963.

[46] Since classically $\tau = 1$, $\Delta n = \pm 1$; cf. p. 115.

[47] M. Born, W. Heisenberg, and P. Jordan, "Zur Quantenmechanik II," *Zeitschrift für Physik 35*, 557–615 (1926), received Nov. 16, 1925; M. Born, *Ausgewählte Abhandlungen* (REF. 116 OF CHAP. 3), vol. 2, pp. 155–213; *Dokumente der Naturwissenschaft* (REF. 1), vol. 2, pp. 77–135.

[48] REF. 43.

time-dependent perturbations, with the inclusion of degenerate cases, and discussed the treatment of angular momenta, intensities, and selection rules from the viewpoint of matrix mechanics. But above all it established a logically consistent general method for solving quantum-mechanical problems and, in addition, uncovered the connection between the mathematical apparatus of matrix mechanics and certain developments in ordinary algebra and analysis. As this short description indicates, the "three-man paper" ("Drei-Männer-Arbeit"), as Born likes to call it, was the first comprehensive exposition of the foundations of modern quantum mechanics in its matrix formulation. From our point of view the last two items in the preceding outline deserve special attention. For the sake of simplicity our analysis of these features will be confined to the discussion of dynamical systems with only one degree of freedom.

Postulating the validity of what is called at first the "fundamental quantum-mechanical relation"[49] and later "the commutation relation"[50]

$$\mathbf{pq} - \mathbf{qp} = \frac{h}{2\pi i}\mathbf{1} \tag{5.26}$$

Born, Heisenberg, and Jordan show that for any function $\mathbf{f}(\mathbf{pq})$ which can be expanded as a power series in $\mathbf{p}$ and $\mathbf{q}$, the following equations hold:

$$\mathbf{fq} - \mathbf{qf} = \frac{h}{2\pi i}\frac{\partial \mathbf{f}}{\partial \mathbf{p}} \tag{5.27a}$$

$$\mathbf{pf} - \mathbf{fp} = \frac{h}{2\pi i}\frac{\partial \mathbf{f}}{\partial \mathbf{q}} \tag{5.27b}$$

since (5.27) hold for $\mathbf{q}$, $\mathbf{p}$, as well as for their sums and products.

By introducing the energy function or Hamiltonian $\mathbf{H}(\mathbf{p},\mathbf{q})$, in analogy to the classical Hamiltonian of the system, and the corresponding canonical equations of motion

$$\dot{\mathbf{p}} = -\frac{\partial \mathbf{H}}{\partial \mathbf{q}} \qquad \dot{\mathbf{q}} = \frac{\partial \mathbf{H}}{\partial \mathbf{p}} \tag{5.28}$$

they derive the energy theorem and Bohr's frequency condition as follows: Since Eqs. (5.27) hold with $\mathbf{f} = \mathbf{H}$, $\dot{\mathbf{p}} = -(2\pi i/h)(\mathbf{pH} - \mathbf{Hp})$, and $\dot{\mathbf{q}} = -(2\pi i/h)(\mathbf{qH} - \mathbf{Hq})$ and hence, by a reasoning similar to that used in the proof of (5.27),

$$\dot{\mathbf{f}} = -\frac{2\pi i}{h}(\mathbf{fH} - \mathbf{Hf}) \tag{5.29}$$

[49] "Fundamentale quantenmechanische Relation." REF. 47, p. 562; p. 160; p. 82.

[50] "Vertauschungsrelation." *Ibid.*, p. 577; p. 175; p. 97. For systems of a number of degrees of freedom these relations are generalized to $\mathbf{p}_n\mathbf{q}_m - \mathbf{q}_m\mathbf{p}_n = (h/2\pi i)\delta_{nm}\mathbf{1}$, $\mathbf{p}_n\mathbf{p}_m - \mathbf{p}_m\mathbf{p}_n = 0$, etc., p. 573; p. 171; p. 93.

If $\mathbf{f} = \mathbf{H}$, Eq. (5.29) yields $\dot{\mathbf{H}} = 0$, which proves the energy theorem and shows that $H_{nm} = W_n\delta_{nm}$, where the time-independent diagonal element W_n is interpreted as the energy of the system in the nth stationary state. Bohr's frequency condition $\nu_{nm} = (W_n - W_m)/h$ follows immediately from (5.29) for $\mathbf{f} = \mathbf{q}$.

As Bohr, Heisenberg, and Jordan continue to explain, the (almost) converse reasoning of the above is also valid, that is, from the assumption $\dot{\mathbf{H}} = 0$ follows (5.28) in the following sense: If two Hermitian matrices $\mathbf{p}^0$ and $\mathbf{q}^0$ (which we assume as being composed of time-independent elements) can be found which satisfy the commutation relation

$$\mathbf{p}^0\mathbf{q}^0 - \mathbf{q}^0\mathbf{p}^0 = \frac{h}{2\pi i}\mathbf{1} \tag{5.30}$$

then

$$\mathbf{p} = (p_{nm}{}^0 \exp{(2\pi i\nu_{nm}t)}) \qquad \mathbf{q} = (q_{nm}{}^0 \exp{(2\pi i\nu_{nm}t)})$$

where $\nu_{nm} = (H_{nn} - H_{mm})/h$, satisfy Eqs. (5.28) and the commutation relation. This can be proved as follows. It is easily verified that for an arbitrary function $\mathbf{f}(\mathbf{pq})$

$$(f_{nm}(\mathbf{pq})) = (f_{nm}(\mathbf{p}^0\mathbf{q}^0) \exp{(2\pi i\nu_{nm}t)}) \tag{5.31}$$

Substituting in (5.31) for $\mathbf{f}(\mathbf{pq})$ the function $\mathbf{pq} - \mathbf{qp}$, one deduces immediately from Eq. (5.30) that $\mathbf{p}$ and $\mathbf{q}$ satisfy Eq. (5.26). Hence (5.27) can be applied and

$$-\left(\frac{\partial\mathbf{H}}{\partial\mathbf{q}}\right)_{nm} = \frac{2\pi i}{h}(\mathbf{Hp} - \mathbf{pH})_{nm} = 2\pi i\nu_{nm}p_{nm} = \dot{p}_{nm}$$

(and similarly for $\dot{q}_{nm}$), which proves the validity of Eq. (5.28).

In consideration of these results the three authors could formulate the process of solving the canonical equations of motion as follows: Find two (time-independent) Hermitian matrices p^0 and q^0 which satisfy the commutation relation and in terms of which the Hamiltonian of the system is a diagonal matrix.

By introducing the concept of canonical transformations, Born, Heisenberg, and Jordan reduced the process of solving the equations of motion and, in particular, of finding the energy values H_{nn} to certain problems in algebra and analysis for which—though under somewhat more restrictive conditions—solutions were available or were the object of contemporary mathematical research. They defined a transformation of the variables $\mathbf{p}$, $\mathbf{q}$ to $\mathbf{P}$, $\mathbf{Q}$ as "canonical" if the commutation relation remains invariant, that is, if with $\mathbf{pq} - \mathbf{qp} = (h/2\pi i)\mathbf{1}$ also $\mathbf{PQ} - \mathbf{QP} = (h/2\pi i)\mathbf{1}$. As shown by Eqs. (5.27), under canonical transformations the canonical equations of motion are invariant. It is also easy to see that with an arbitrary matrix **U**

$$\mathbf{P} = \mathbf{U}^{-1}\mathbf{p}\mathbf{U} \qquad \mathbf{Q} = \mathbf{U}^{-1}\mathbf{q}\mathbf{U} \tag{5.32}$$

is a canonical transformation and that for any function $\mathbf{f}(\mathbf{p},\mathbf{q})$

$$\mathbf{f}(\mathbf{P},\mathbf{Q}) = \mathbf{f}(\mathbf{U}^{-1}\mathbf{p}\mathbf{U},\mathbf{U}^{-1}\mathbf{q}\mathbf{u}) = \mathbf{U}^{-1}\mathbf{f}(\mathbf{p},\mathbf{q})\mathbf{U} \tag{5.33}$$

The last equation made it possible for Born, Heisenberg, and Jordan to reformulate the problem as follows (in view of the preceding considerations all matrices in the following discussion can be assumed to be composed of time-independent elements): If **p** and **q** are given which satisfy the commutation relation, find a canonical transformation such that

$$\mathbf{H}(\mathbf{P},\mathbf{Q}) = \mathbf{H}(\mathbf{U}^{-1}\mathbf{p}\mathbf{U},\mathbf{U}^{-1}\mathbf{q}\mathbf{U}) = \mathbf{U}^{-1}\mathbf{H}(\mathbf{p},\mathbf{q})\mathbf{U} = \mathbf{W} \tag{5.34}$$

is a diagonal matrix.

In the third chapter[51] of their paper Born, Heisenberg, and Jordan succeeded in reducing this process, as reformulated, to a familiar mathematical procedure. For this purpose they associated with every Hermitian matrix $\mathbf{a} = (a_{nm})$ the bilinear form $A(x,z) = \sum_{nm} a_{nm}x_n z_m$ such that $A(x,x^*)$ is real for arbitrary $x_1, x_2, \ldots$. The transformation $x_n = \sum_j u^*_{nj}y_j$ with the matrix $\mathbf{u}^* = (u^*_{nj})$ transforms $A(x,x^*)$ into $B(y,y^*) = \sum_{kj} b_{kj}y_k y^*_j$, where $\mathbf{b} = (b_{kj}) = \tilde{\mathbf{u}}^*\mathbf{a}\mathbf{u}$. The matrix **u** is defined as "orthogonal" if it leaves the Hermitian unit form $E(x,x^*) = \sum x_n x^*_n$ invariant.[52] It is shown that the necessary and sufficient condition for this property is

$$\tilde{\mathbf{u}}^* = \mathbf{u}^{-1} \tag{5.35}$$

For finite matrices, it is now pointed out, there always exists an "orthogonal" transformation "to principal axes" such that $A(x,x^*) = \sum W_n y_n y^*_n$ or, equivalently, there always exists a matrix **u** which satisfies $\tilde{\mathbf{u}}^* = \mathbf{u}^{-1}$ and

$$\mathbf{u}^{-1}\mathbf{a}\mathbf{u} = \mathbf{W} = (W_n\delta_{nm}) \tag{5.36}$$

Born, Heisenberg, and Jordan now refer to the algebraic methods and their elaborations in the *Methoden der mathematischen Physik* by Courant and Hilbert[52a] by means of which the W_n in Eq. (5.36) can be calculated in terms of the a_{nm} without the need of actually performing the transformation. In the same way, they contend, the diagonal elements W_n of the diagonal matrix **W** in (5.34) can be computed in terms of the elements of the (in

[51] *Ibid.*, pp. 581–585; pp. 179–183; pp. 101–105. In the sequel x^* denotes, as previously, the complex conjugate value of x and $\tilde{\mathbf{a}}$ denotes the transpose of the matrix **a**.

[52] Born, Heisenberg, and Jordan seem to have been unaware in 1925 that already in 1902 such transformations had been called "unitary" ("unitaire") by L. Autonne in his paper "Sur l'Hermitien," *Rendiconti del Circolo Matematico di Palermo 16*, 104–128 (1902).

[52a] REF. 33.

general, not diagonal) matrix $\mathbf{H}(\mathbf{p},\mathbf{q})$. The calculation of energy values has thus been reduced to the solution of the eigenvalue problem of Hermitian matrices in linear algebra. Unfortunately, however, as evidenced by Eq. (5.26), quantum-mechanical matrices are in general not of finite order, nor are they even bounded. It had therefore to be assumed that the procedure, which holds for the finite case, remains valid also for infinite matrices if, in addition to the discrete spectrum of eigenvalues (or "point spectrum"), allowance is made for a continuous spectrum of eigenvalues (or "line spectrum"), in accordance with the results obtained by Hilbert and Hellinger.

As Born and his collaborators pointed out, Eq. (5.34) is analogous to the Hamilton-Jacobi differential equation in classical physics and the matrix $\mathbf{U}$ corresponds to the action function. Equation (5.34) or $\mathbf{HU} = \mathbf{UW}$, treated as an equation of finite matrices, implies that for every m

$$\sum_k (H_{mk} - W_n\delta_{mk})U_{kn} = 0 \tag{5.37}$$

which shows that the eigenvalues W_n are the roots of the secular (or characteristic) equation

$$|\mathbf{H} - \lambda \mathbf{1}| = 0 \tag{5.38}$$

Since for each W_n the U_{kn} can be computed, the transformation matrix $\mathbf{U}$ can be determined.

As our following brief discussion will show, the origin and development of the mathematical apparatus under consideration were intimately connected with the techniques of computing planetary motions in astronomy. In fact, just like the mathematics of the older quantum theory and, in particular, the mathematics of its theory of perturbations, the mathematics of matrix mechanics can be traced back to the study of planetary orbits and of the orbits of planetary satellites. Ironically, matrix mechanics, the outcome of Heisenberg's categorical rejection of orbits, had eventually to resort to the mathematics of orbital motions. The very name of Eq. (5.38) already testifies to the truth of this contention. For astronomers called an equation of this type a "secular" equation (from the Latin *saeculum* = generation, *saeculum civile* = a period of 100 years) as it enabled them to determine "secular" (long-period) disturbances of planetary orbits, as regards eccentricities and inclinations about their mean values.

It was with reference to astronomical problems that Lagrange,[53] a few years after Euler's first discussion[54] of this subject, investigated the question

[53] L. de Lagrange, "Recherches sur la méthode de maximis, et minimis," *Miscellanea Philosophico-Mathematica Societatis Privatae Taurinensis* (*Miscellanea Taurinensia*) *1*, 18–32 (1759); *Œuvres de Lagrange*, edited by J. A. Serret (Gauthier-Villars, Paris, 1887), vol. 1, pp. 3–20.

[54] L. Euler, *Introductio in analysin infinitorum* (Lausanne, 1748), vol. 2 (Appendix: De superficiebus secundi ordinis), pp. 379–392, in *Opera Omnia* (Füssli, Teubner, Genf, Leipzig, Berlin, 1945).

of transforming bilinear forms of n variables to sums of squares (transformations to principal axes) and solved it for the case $n = 2$ and $n = 3$. In a paper presented to the Berlin Academy in 1773 Lagrange[55] showed that all roots of the secular equation for a symmetric form of three variables are real. His interest in these problems as well as Laplace's[56] was motivated by the discovery that imaginary roots would have led to functions which, increasing exponentially with time, are incompatible with the assumed stability of the planetary system.[57] Lagrange and Laplace even thought, erroneously as Routh[58] later showed, that the existence of multiple roots is likewise inconsistent with this assumption.[59] The secular equation studied by Lagrange and Laplace was of the sixth degree, only six planets being known at that time. After William Herschel's discovery of Uranus in 1781 numerical methods of solving secular equations of the seventh degree were studied by astronomers and mathematicians, as, for example, by Jacobi[60] in 1846. In the same year Leverrier and Adams discovered Neptune, and secular equations of the eighth degree became the subject of research.[61] Meanwhile, Lagrange's result concerning the reality of eigenvalues[62] of symmetric bilinear forms for $n = 3$ was generalized in 1829, again in connection with astronomical considerations, for an arbitrary finite n by Cauchy[63] and was finally extended to Hermitian forms in 1855 by Charles Hermite.[64] Needless to say, simultaneously with these developments the study of transformations to principal axes engaged the attention also of

[55] L. de Lagrange, "Nouvelle solution du problème du mouvement de rotation d'un corps quelconque," *Mémories de l'Académie des Sciences et Belles-Lettres de Berlin 1773*, p. 85; *Œuvres* (REF. 53), vol. 3, pp. 579–616.

[56] P. S. M. de Laplace, *Traité de Mécanique Céleste* (Paris, 1799–1825), vol. 2, part 1, article 57; *Œuvres Complètes de Laplace* (Gauthier-Villars, Paris, 1878), vol. 1, p. 328.

[57] A similar situation occurs already in elementary problems of linear oscillations where certain parameters are roots of quadratic equations.

[58] E. J. Routh, *A Treatise on the Stability of a Given State of Motion*, Adams Prize Essay for 1877 (Macmillan, London, 1877), pp. 5–9. Cf., on this point, L. de Lagrange, *Mécanique Analytique*, part 2, section 6, article 7.

[59] On early investigations concerning the multiplicity of roots of Eq. (5.38) see H. Seeliger, "Ueber die Gleichungen, von deren Wurzeln die saecularen Veränderungen der Planetenbahnelemente abhängen," *Astronomische Nachrichten 93*, 353–364 (1878).

[60] C. G. J. Jacobi, "Über ein leichtes Verfahren die in der Theorie der Säcularstörungen vorkommenden Gleichungen numerisch aufzulösen," *Journal für die reine und angewandte Mathematik* (*Crelle*) *30*, 51–94 (1846). Jacobi used for his calculations data furnished by Leverrier's paper "Sur les variations séculaires des éléments des orbits," *Additions à la Connaissance pour l'An 1843*.

[61] Cf., for example, F. Tisserand, *Traité de Mécanique Céleste* (Gauthier-Villars, Paris, 1889), vol. 1, pp. 404–430.

[62] The term "eigenvalue" or rather its French equivalent "valeur propre" was, of course, not yet used by Lagrange.

[63] A. Cauchy, "Sur l'équation à l'aide de laquelle on détermine les inégalités séculaires des mouvements des planètes," *Exercices de Mathématique 4* (Anciens Exercices, Paris, 1829), 141–152; *Œuvres Complètes* (Gauthier-Villars, Paris, 1891), vol. 9 (second series), pp. 174–195.

[64] C. Hermite, "Remarque sur un théorème de M. Cauchy," *Comptes Rendus 41*, 181–183 (1855); *Œuvres*, edited by E. Picard (Gauthier-Villars, Paris, 1912), vol. 1, pp. 479–481.

geometers and algebraists.[65] Our insistence, however, on the importance of astronomical calculations of orbits for the development of the mathematics of matrix mechanics gains additional support as soon as we consider the extension to infinite systems of linear equations with infinitely many variables. For, although Kötteritzsch[66] was probably the first to try such an extension, the real development of this process, which proved to be of extreme importance for the later elaboration of matrix mechanics and quantum mechanics in general, had its origin, again, in astronomical computations of orbital motions. It was Hill's paper on the motion of the lunar perigee which initiated the systematic study of infinitely many equations with infinitely many variables. Hill,[67] treating such systems as if they were finite, introduced infinite determinants and applied to them the rules valid for finite determinants. His results agreed with observation, but the legitimacy of his mathematical procedures remained an open question. Recognizing the importance of Hill's paper as well as its analytical deficiencies, Poincaré[68] attempted to put it on logically firm foundations by constructing a theory of determinants of infinite order in a mathematically rigorous way. Poincaré's work, in turn, inspired Helge von Koch[69] to study the use of infinite determinants in the theory of linear differential equations. Soon it became evident that such extensions—and in particular generalizations, to infinitely many variables, of orthogonal transformations of quadratic forms to principal axes—are important not only for the solution of differential equations but also for the theory of integral equations. In fact, it was one of Hilbert's greatest achievements to have recognized that eigenvalues of quadratic forms and eigenvalues occurring in

[65] Cf. J. B. Biot, *Essai de Géométrie Analytique* (Klostermann, Paris, 1810), pp. 254–256. M. Rochat, "Construction des formules qui servent à déterminer directement la grandeur et la situation des diamètres principaux, dans les courbes du second degré," *Annales de Mathématique 2*, 331–335 (1812). J. Plücker, "Allgemeine Methode, eine homogene Funktion beliebig vieler Veränderlicher in eine andere zu verwandeln, welche nur die Quadrate der Veränderlichen enthält," *Journal für die reine und angewandte Mathematik (Crelle) 24*, 287–290 (1842).

[66] E. T. Kötteritzsch, "Ueber die Auflösung eines Systems von unendlich vielen linearen Gleichungen," *Journal für Mathematik und Physik 15*, 1–15, 229–268 (1870).

[67] G. W. Hill, *On the Part of the Motion of the Lunar Perigee Which Is a Function of the Mean Motions of the Sun and Moon* (Wilson, Cambridge, Mass., 1877); *Acta Mathematica 8*, 1–36 (1886). Hill had to solve the equation $\ddot{w} + \theta w = 0$, where

$$\theta = \sum_{-\infty}^{\infty} \theta_k \exp(ikt)$$

Since the integral of the differential equation is

$$w = \sum_{-\infty}^{\infty} b_n \exp[i(n + c)t]$$

b_n and c being constants, he obtained the infinite system of equations

$$\sum_{k=-\infty}^{\infty} \theta_{n-k} b_k - (n + c)^2 b_n = 0$$

for $n = -\infty, \ldots, +\infty$.

[68] H. Poincaré, "Sur les déterminants d'ordre infini," *Bulletin de la Société Mathématique de France 14*, 77–93 (1886).

[69] H. von Koch, "Sur une application des déterminants infinis à la théorie des équations différentielles linéaires," *Acta Mathematica 15*, 53–63 (1891); "Sur les déterminants infinis et les équations différentielles linéaires," *ibid. 16*, 217–295 (1892).

boundary problems of differential and integral equations have a common origin.

In his famous work on the foundations of a general theory of integral equations[70] he considered bounded quadratic forms of infinitely many variables which he defined as follows:[71] the quadratic form $\sum_{k,l=1}^{\infty} a_{kl}x_kx_l$ is bounded if there exists a positive constant C such that for every set of variables $x_1, x_2, \ldots$ satisfying $\sum_{k}^{\infty} |x_k|^2 \leq 1$ and for every n, $|\sum_{k,l=1}^{n} a_{kl}x_kx_l| \leq C$. These forms, now, if subjected to an orthogonal transformation to principal axes, give rise, as he showed, in addition to a series of denumerably many quadratic terms, also to certain integrals; that is to say, their eigenvalue spectrum contains, in addition to discrete values or a "point spectrum," also a continuum of terms or a "line spectrum."[72]

A theory of bounded quadratic forms of infinitely many variables was also constructed by Hellinger[73] independently of Hilbert's theory and without the need of Hilbert's limit transitions from algebra to analysis which involved complicated convergence considerations. Hellinger's result that for such systems the bilinear Hermitian form $\sum H_{mn}x_mx_n^*$ can be transformed to the expression

$$\sum W_ny_ny_n^* + \int W(\varphi)y(\varphi)y^*(\varphi)\,d\varphi \tag{5.39}$$

where the discrete real numbers W_n form the point spectrum and the real-valued functions $W(\varphi)$ of the continuous parameter φ the line spectrum, was assumed by Born and his group to be valid also for systems which are not bounded. It enabled them to derive the continuous spectrum from one and the same theory and to account, as, for example, in the case of the hydrogen spectrum, for four kinds of transitions, transitions from ellipse to ellipse, from ellipse to hyperbola, from hyperbola to ellipse, and from hyperbola to hyperbola, on the basis of a logically consistent approach.

Let us conclude our analysis of the rise of matrix mechanics, the earliest version of quantum mechanics, with some remarks on the relationship between classical mechanics and quantum mechanics. One may right-

[70] D. Hilbert, "Grundzüge einer allgemeinen Theorie der Integralgleichungen," *Göttinger Nachrichten 1904* (1. Mitteilung), pp. 49–91, (2. Mitteilung), pp. 213–259; *1905* (3. Mitteilung), pp. 307–337.

[71] *Ibid. 1906* (4. Mitteilung), pp. 157–227.

[72] Cf. also E. Hellinger, "Die Orthogonalinvarianten quadratischer Formen von unendlich vielen Variablen," *Dissertation* (Göttingen), 1904.

[73] E. Hellinger, "Neue Begründung der Theorie quadratischer Formen von unendlich vielen Variablen," *Journal für die reine und angewandte Mathematik* (*Crelle*) *136*, 210–271 (1909). Cf. also E. Hellinger and O. Toeplitz, "Grundlagen für eine Theorie der unendlichen Matrizen," *Mathematische Annalen 69*, 289–330 (1910).

fully wonder how the formalism or language of quantum mechanics, which, as we shall see in great detail later on, differs from the conceptual point of view most radically from that of classical mechanics, seems nevertheless to include the latter as a special case, as the correspondence principle in its familiar formulation[74] contends. As the present chapter has shown and as will become increasingly clear in the sequel, the evolution of quantum mechanics out of classical physics was *not* a logically continuous process. However inconspicuous the change at the critical point may appear, even if it is only a transposition of indices,[75] it constitutes a logical hiatus of far-reaching consequences. In fact, there is simply no ultimate logical connection between classical and quantum mechanics—"any more than there is one between a sense-datum language and a material object language."[76] How, then, can quantum theory be reduced by a logically continuous inference to classical mechanics as a limiting case, as the well-known extrapolation[77] of the Balmer formula for large quantum numbers seems to exemplify?

To clarify the situation it is advisable to distinguish between *two* principles of correspondence: one referring to the relationship between classical mechanics and the older quantum theory, and one relating classical mechanics to modern quantum mechanics. As to the former—ignoring at present Bohr's farsighted reservation—the popular interpretation of the principle as establishing a logical connection is not wholly unfounded since the older quantum theory was essentially but a modification of classical physics. As to the correspondence principle in the latter significance, the situation is more complicated. As we have seen, the principle, deeply imbedded in the very foundations of the new formalism, played a decisive role in the establishment of matrix mechanics. And yet, Heisenberg's postulated indicial modification, inconspicuous as it was, and the Born-Jordan commutation relations[78] led, so to speak, to a complete logical rupture with the classical scheme of conceptions. As a consequence, the correspondence principle, while leading to numerical agreements between quantum-mechanical and classical deductions, affirmed no longer a conceptual convergence of the results but established merely a formal, symbolic analogy between conclusions derived within the context of two disparate and mutually irreducible theories. It only showed that under certain conditions (for instance, for high quantum numbers or, in classical terms, for great distances from the nucleus) the formal treatments in both theories converge to notationally identical expressions (and numerically

[74] See, however, Bohr's attitude, p. 116 et seq.

[75] See p. 201.

[76] N. R. Hanson, *The Concept of the Positron* (Cambridge University Press, 1963), p. 65.

[77] See, for example, D. Park, *Introduction to the Quantum Theory* (McGraw-Hill, New York, 1964), pp. 12–13.

[78] See p. 210.

equal results)[79] even though the symbols, corresponding to each other, differ strikingly in their conceptual contents.

But one final comment: Poincaré once emphasized that the form of classical mechanics "is due to the influence of celestial mechanics."[80] In the present chapter it has been shown that the mathematics of quantum mechanics is also greatly indebted to the methods that have been used in solving problems of celestial dynamics.

5.2 *Modifications of Matrix Mechanics*

Born's mathematically illegitimate assumption concerning the use of algebraic procedures for unbounded matrices led, as we have seen, in a natural way to full agreement with experience. To put matrix mechanics on sound foundations confronted mathematics with new problems, just as in the case of Hill's work on the lunar perigee a few decades earlier. In fact, "a new era of mutual stimulation of mechanics and mathematics" was ushered in, just as Bohr predicted. In December, 1925, Heisenberg wrote a detailed exposition of the principles of matrix mechanics for the *Mathematische Annalen*,[81] the leading German journal for research in pure mathematics at that time. As a result, many mathematicians became interested in matrix mechanics. Thus, to mention only one example, Aurel Wintner conducted during the winter semester 1927/28 at the University of Leipzig, where Heisenberg had just been appointed professor of theoretical physics, a seminar on the spectral theory of infinite matrices with special consideration of Hermitian unbounded matrices. In spite of numerous important contributions the seminar did not produce an exhaustive solution of the problem and, as Wintner had to admit, "a complete and mathematically satisfactory treatment of quantum-theoretic matrices continues to be a desideratum."[82]

The ushering in of the new era of cooperation between physics and mathematics found its first tangible expression, so to speak, in the collaboration between Max Born and Norbert Wiener at the Massachusetts Institute of Technology. They had met for the first time when Wiener was visiting Göttingen in 1924 and gave a talk on his work on general harmonic analysis, which was very well received. Courant, the head of the department

[79] Just as a gas can be treated by the method of statistical mechanics in the language of discrete particles and also by the method of fluid mechanics in the language of continua with formally identical results.

[80] On this point see, for instance, Emile Meyerson, *Identity and Reality* (George Allen and Unwin, London, 1930; reprinted by Dover, New York, 1965), p. 218.

[81] W. Heisenberg, "Über quantentheoretische Kinematik und Mechanik," *Mathematische Annalen 95*, 694–705 (1926), received Dec. 21, 1925.

[82] "Eine lückenlose, mathematisch befriedigende Theorie der quantentheoretischen Matrizen ist zurzeit noch ein Desideratum." A. Wintner, *Spektraltheorie der unendlichen Matrizen—Einführung in den analytischen Apparat der Quantenmechanik* (Hirzel, Leipzig, 1929), preface, p. vii.

of mathematics—Klein had already retired because of ill health—arranged with Wiener some kind of exchange program according to which Wiener should return to Göttingen in 1926. Meanwhile Born was invited as "Foreign Lecturer" for the fall term 1925/26 at M.I.T. Thus, Born left Göttingen at the end of October, 1925, a few days after the completion of the "Drei-Männer-Arbeit," to join Wiener. Born realized that matrix mechanics, as formulated so far, was incapable of treating aperiodic phenomena such as uniform rectilinear motion, for in such a case no periods are present and the coordinate matrix $\mathbf{q}$ has no off-diagonal elements at all. Born and Wiener, therefore, decided to search for a generalization of matrix mechanics which would embrace periodic as well as aperiodic phenomena.

Now, a few months before Born's arrival in Cambridge, Wiener had written a comprehensive article[83] on the operational calculus in which he discussed the contrast between the rigorous conception of operators, such as Volterra's integral transformations (integral operators) or Pincherle's transformations of one power series into another, and the heuristic approach, "devoid even of the pretense of mathematical rigor," such as Heaviside's method. The main topic of the paper, however, was Wiener's proposed generalization of the Fourier integral which would make it possible to apply integral operators rigorously also to nonanalytic functions.

In his autobiography Wiener gave the following account. "When Professor Born came to the United States he was enormously excited about the new basis Heisenberg had just given for the quantum theory of the atom. Born wanted a theory which would generalize these matrices. . . . The job was a highly technical one, and he counted on me for aid. . . . I had the generalization of matrices already at hand in the form of what is known as operators. Born had a good many qualms about the soundness of my method and kept wondering if Hilbert would approve of my mathematics. Hilbert did, in fact, approve of it, and operators have since remained an essential part of quantum theory."[84]

Let us now give a brief description of how Born and Wiener introduced operators into quantum theory. If $y(t) = \sum_m y_m \exp(2\pi i W_m t/h)$ and, similarly, $x(t) = \sum_n x_n \exp(2\pi i W_n t/h)$, it was easy to see that

$$x_n = \lim_{T=\infty} \frac{1}{2T} \int_{-T}^{T} x(s) \exp \frac{-2\pi i W_n s}{h} \, ds$$

and, hence, that the matrix transformation $y_m = \sum_n q_{mn} x_n$ entails

$$y(t) = \lim_{T=\infty} \frac{1}{2T} \int_{-T}^{T} q(t,s) x(s) \, ds \tag{5.40}$$

[83] N. Wiener, "The Operational Calculus," *Mathematische Annalen 95*, 557–584 (1926), completed Apr. 6, 1925; received Apr. 20, 1925.

[84] N. Wiener, *I Am a Mathematician* (Doubleday, New York, 1956), p. 108.

where $q(t,s) = \sum_{mn} q_{mn} \exp [2\pi i(W_m t - W_n s)/h]$. Equation (5.40) was now interpreted as follows. The "operator"

$$q = \lim_{T=\infty} \frac{1}{2T} \int_{-T}^{T} ds\, q(t,s) \cdots \tag{5.41}$$

transforms $x(t)$ into $y(t)$ or, symbolically,

$$y(t) = qx(t) \tag{5.42}$$

Generalizing Eq. (5.42), Born and Wiener defined operators as follows:[85] "An operator is a rule in accordance with which we may obtain from a function $x(t)$ another function $y(t)$. . . . It is linear if

$$q(x(t) + y(t)) = qx(t) + qy(t)\text{''} \tag{5.43}$$

For the linear operator $D = d/dt$,

$$y(t) = Dq\, x(t) = \lim_{T=\infty} \frac{1}{2T} \int_{-T}^{T} \frac{\partial q(t,s)}{\partial t} x(s)\, ds$$

which shows that Dq conforms to (5.41) if only $q(t,s)$ in (5.41) is replaced by the generating function

$$\frac{\partial q(t,s)}{\partial t} = \frac{2\pi i}{h} \sum_{mn} q_{mn} W_m \exp \left[\frac{2\pi i}{h} (W_m t - W_n s) \right]$$

The matrix associated with Dq, just as (q_{mn}) was with q, is, therefore,

$$(\mathbf{Dq})_{mn} = \left(\frac{2\pi i}{h} W_m q_{mn} \right) = \frac{2\pi i}{h} Wq$$

where $W = (W_n \delta_{mn})$. Similarly, the matrix associated with the operator qD, as a partial integration shows, is

$$(qD)_{mn} = \left(\frac{2\pi i}{h} W_n q_{mn} \right) = \frac{2\pi i}{h} qW$$

The matrix associated with the operator $Dq - qD$ has therefore the elements

$$\frac{2\pi i}{h} (W_m - W_n) q_{mn} = 2\pi i \nu_{mn} q_{mn}$$

since $h\nu_{mn} = W_m - W_n$.

Born and Wiener consequently defined the time derivative $\dot{q}$ of an

[85] M. Born and N. Wiener, "A new formulation of the laws of quantization of periodic and aperiodic phenomena," *Journal of Mathematics and Physics (M.I.T.)* *5*, 84–98 (1925–1926); "Eine neue Formulierung der Quantengesetze für periodische und nicht periodische Vorgänge," *Zeitschrift für Physik 36*, 174–187 (1926), received Jan. 5, 1926.

operator q by the operator equation

$$Dq - qD = \dot{q} \tag{5.44}$$

In addition, the commutation relation

$$pq - qp = \frac{h}{2\pi i} 1 \tag{5.45}$$

as well as the canonical equations

$$\dot{q} = \frac{\partial H(pq)}{\partial p} \qquad \dot{p} = -\frac{\partial H(pq)}{\partial q} \tag{5.46}$$

were interpreted as operator equations for the Hermitian operators p and q, that is, operators whose associated matrices are Hermitian.

On these assumptions they derived the energy theorem and proved that the energy operator H is identical with $(h/2\pi i)D$. Finally, by solving the problems of the harmonic oscillator ($\ddot{q} + \omega_0^2 q = 0$) and of the one-dimensional uniform motion ($\ddot{q} = 0$), they demonstrated the applicability of their new operator mechanics to both periodic and aperiodic phenomena.

It is interesting to note that, had Born and Wiener supplemented their equations (5.44), (5.45), and

$$H = \frac{h}{2\pi i} D \tag{5.47}$$

by the equation

$$p = \frac{h}{2\pi i} \frac{\partial}{\partial q}$$

as a comparison of (5.44) with (5.45) could have suggested, they would have established an operator formulation of wave mechanics. Or as Born declared: "We expressed the energy as d/dt and wrote the commutation law for energy and time as an identity by applying $[t(d/dt) - (d/dt)t]$ to a function of t; it was absolutely the same as for q and p. But we did not see that. And I never will forgive myself, for if we had done this, we would have had the whole wave mechanics from quantum mechanics at once, a few months before Schrödinger."[86]

Yet in spite of this oversight, their idea of representing physical quantities, such as coordinates or momenta, by operators was a remarkable innovation, not only because it made phenomena, which so far had defied matrix mechanics, amenable to analytic treatment, but also because the operator formalism, as we shall see later on, provided the most lucid representation of quantum-theoretic relations and eventually led to a

[86] *Archive for the History of Quantum Physics*, Interview with M. Born on Oct. 17, 1962.

profounder understanding of the theory. Since the notion of operators will gain increasing importance in our subsequent discussions, a brief digression on the development of this conception seems to be appropriate.

Strictly speaking, the operator concept is as old as symbolic algebra, for every symbol in classical algebra denotes a precept either to substitute a number or to perform an arithmetic operation. If we define, however, the concept of an operator in a nontrivial manner as referring to the transformation of a function to another function, as it was defined by Born and Wiener, its origin can be traced back to the discovery of the differential calculus. In fact, it is often claimed that Leibniz[87] was the first to use operators in this sense as, for instance, when he stated that the *modus operandi* of the differential operator $d^n(xy)$ shows a striking analogy to the expansion of $(x + y)^n$. Leibniz's approach was further elaborated by Lagrange,[88] who generalized it to a "new kind of calculus concerning differentiation and integration." In this calculus, which may be regarded as the earliest operator calculus in history, Lagrange employed d/dx systematically as a fictitious quantity, subject to ordinary algebra as, for example, in the "Taylor expansion" $f(x + h) = \exp[h(d/dx)]f(x)$. Lagrange's work was continued by A. M. Lorgna, J. P. Grüson, L. F. A. Arbogast, M. J. F. Francais, S. F. Lacroix, and, especially, Francois-Joseph Servois,[89] who was the first to pay attention to the formal aspects of symbolic operations and introduced the notions of "commutativity" and "associativity" of operators. The earliest detailed study of operator formalisms *in abstracto* was probably Robert Murphy's "First memoir on the theory of analytical operations,"[90] published in 1837. Distinguishing clearly between operand, operator, and result—in Eq. (5.42) these correspond to $x(t)$, q, and $y(t)$, respectively—Murphy stated: "The elements of which every distinct analytical process is composed are three, namely, first the *Subject*, that is, the symbol on which a certain notified operation is to be performed; secondly, the *Operation* itself, represented by its own symbol; thirdly, the *Result*, which may be connected with the former two by the algebraic sign of equality. Thus let a be the *subject* representing, we may suppose, some quantity, b the symbol for multiplication by b, and c the result or product; for greater distinctness let the subject be inclosed in square brackets, the analytical process in this case is $[a]b = c$. Again,

[87] G. W. Leibniz, "Symbolismus memorabilis calculi algebraici et infinitesimalis in comparatione potentiarum et differentiarum, et de lege homogeneorum transcendentali," *Miscellanea Berolinensia 1*, 160 (1710).

[88] L. de Lagrange, "Sur une nouvelle espèce de calcul relatif à la différentiation et à l'intégration des quantités variables," in *Œuvres* (REF. 53), vol. 3, pp. 441–476.

[89] F.-J. Servois, "Essai sur un nouveau mode d'exposition des principes du calcul différentiel," *Annales de Mathématique (Gergonne) 5*, 93–104 (1814).

[90] R. Murphy, "First memoir on the theory of analytical operations," *Philosophical Transactions of the Royal Society of London 127*, 179–210 (1837).

let x^n be the subject, ψ a symbol of operation denoting that x must be changed into $x + h$, and $(x + h)^n$ will evidently be the result, or $[x^n]\psi = (x + h)^n$." Operators may commute or not, or as Murphy, apparently unaware of Servois's terminology, expressed it, "operations are . . . relatively *fixed* or *free;* in the first case a change in the order in which they are to be performed would affect the result, in the second case it would not do so." Thus, to quote Murphy's example, $[x^n]x\psi \neq [x^n]\psi x$, whereas $[x^n]a\psi = [x^n]\psi a$, both being equal to $ax + h^n$.

The use of operators in pure mathematics for the solution of differential equations, a procedure familiar to every student of higher mathematics, characterizes the second stage in the development of the operator conception. This phase was initiated by George Boole's paper "On a general method in analysis,"[91] for which its author was awarded a gold medal by the Royal Society; it was given a detailed exposition in Boole's classic *Treatise on Differential Equations.*[92] Boole's introductory remarks to his method in the *Treatise* are still interesting also from our modern point of view and deserve to be mentioned. "The question of the true value and proper place of symbolic methods," wrote Boole,[93] "is undoubtedly of great importance. Their convenient simplicity—their condensed power—must ever constitute their first claim upon attention. I believe however that, in order to form a just estimate, we must consider them in another aspect, viz. as in some sort the visible manifestation of truths relating to the intimate and vital connexion of language with thought—truths of which it may be presumed that we do not yet see the entire scheme and connexion." Later, under the heading "Forms purely symbolical," he declared:[94] "In any system in which thought is expressed by symbols, the laws of combination of the symbols are determined from the study of the corresponding operations in thought. But it may be that the latter are subject to *conditions of possibility* as well as to laws when possible. And thus it may be that two systems of symbols, differing in interpretation, may agree as to their formal laws whenever they both express operations possible in thought, while at the same time there may exist combinations which really represent thought in the one but do not in the other. For instance, there exist forms of the functional symbol f, for which we can attach a meaning to the expression $f(m)$, but cannot directly attach a meaning to the symbol $f(d/dx)$. And the question arises: Does this difference restrict our freedom in the use of that principle which permits us to treat expressions of the form $f(d/dx)$ as if d/dx were a symbol of quantity? For instance, we can

[91] G. Boole, "On a general method in analysis," *Philosophical Transactions of the Royal Society of London 134,* 225–282 (1844).

[92] G. Boole, *Treatise on Differential Equations* (1st ed. 1859; 2d ed. 1865; 5th ed. 1959; Chelsea Publishing Company, New York).

[93] *Ibid.* (5th ed.), p. vii (preface).

[94] *Ibid* (5th ed.), pp. 398–399.

attach no *direct* meaning to the expression $\exp[h(d/dx)]f(x)$, but if we develop the exponential as if d/dx were quantitative, we have

$$\exp\left(h\frac{d}{dx}\right)f(x) = \left(1 + h\frac{d}{dx} + \frac{1}{1\cdot 2}h^2\frac{d^2}{dx^2} + \text{etc.}\right)f(x) = f(x+h)$$

by Taylor's theorem. Are we then permitted, on the above principle, to make use of symbolic language; always supposing that we can, by the continued application of the same principle, obtain a *final* result of interpretable form? Now all special instances point to the conclusion that this is permissible, and seem to indicate, as a general principle, that the mere processes of symbolical reasoning are independent of the conditions of their interpretation." Boole's principle of admitting symbolic reasoning, even if no immediate interpretation is in sight but has to be searched for, has proved extremely useful in the later development of quantum mechanics. The elaboration of symbolic techniques for the solution of differential equations—as, for example, by Hargreave, Gregory, Brouwin, Carmichael, Forsyth—became in the second half of the last century, especially among British mathematicians, a prolific subject of research.

Of special interest from our point of view is a development which began when Charles James Hargreave presented a paper[95] in June, 1847, to the Royal Society in London on applications of the operator function $\varphi(d/dx)$ to the solution of differential equations, such as, for example, Riccati's equation, in the course of which questions concerning the interchange of symbolic operators began to play an important role.

Hargreave's work was soon generalized by Charles Graves, Erasmus Smith Professor of Mathematics at the University of Dublin from 1843 to 1862 and Bishop of Limerick until his death in 1866. In a series of papers[96] read before the Royal Irish Academy, Graves, a mathematician, theologian, linguist, and archaeologist, developed an extensive and powerful operator formalism which, in one aspect at least, had an influence not only on Heaviside but even on Dirac[97] at an important point of his work.

Graves's paper[98] "On the principles which regulate the interchange of

[95] C. J. Hargreave, "On the solution of linear differential equations," *Philosophical Transactions of the Royal Society of London 138*, 31–54 (1848).

[96] C. Graves, "On the theory of linear differential equations," *Proceedings of the Royal Irish Academy 4*, 88–91 (1847–1850), read on Feb. 28, 1848; "On a generalization of the symbolic statement of Taylor's Theorem," *ibid. 5*, 285–287 (1850–1853), read on Apr. 26, 1852; "On the solution of linear differential equations," *ibid. 6*, 1–3 (1853–1857), read on Nov. 14, 1853; "On a method of solving a large class of linear differential equations by the applications of certain theorems in the calculus of operations," *ibid. 6*, 34–37, read on Jan. 9, 1854; "On the extension of Taylor's Theorem to non-commutative symbols," *ibid. 6*, 302, read on Feb. 11, 1856.

[97] As acknowledged by Dirac himself. Cf. footnote 37, p. 4, in N. R. Hanson, "Copenhagen interpretation of quantum theory," *American Journal of Physics 27*, 1–15 (1959).

[98] *Proceedings of the Royal Irish Academy 6*, 144–152 (1853–1857), read on Dec. 11, 1854.

symbols in certain symbolic equations" began as follows: "Let π and ρ be two distributive symbols of operation, which combine according to the law expressed by the equation

$$\rho\pi = \pi\rho + a$$

a being a constant, or at least a symbol of distributive operation commutative with both π and ρ." Graves then showed that in any symbolic equation $\varphi(\pi,\rho) = 0$, π may be exchanged with ρ and ρ with $-\pi$, yielding the "correlative equation" $\varphi(\rho,-\pi) = 0$. He also proved by induction that $\rho\pi^n = \pi^n\rho + n\pi^{n-1}$ and $\rho\psi\pi = \psi\pi\rho + \psi'\pi$, where $\psi\pi$ denotes any function of integral powers of π. It will have been recognized that Graves's reasoning and formal results recurred in the quantum-mechanical treatment of noncommutative quantities as, for example, in Dirac's q-number algorithm and its handling of the quantum conditions.

The third stage in the development of the operator conception may be characterized by the application of operators to problems in applied mathematics and, in particular, to the analysis of electric networks. The most important—and provocative—contribution was Oliver Heaviside's application of operators to electromagnetic problems. To analyze networks of linear conductors with inductances, capacities, and resistances, Heaviside[99] replaced the differential operator $D = d/dt$ by p and regarded the latter as subject to the ordinary laws of algebra. Applying this method also when partial differential equations were involved, he paid little attention to questions of mathematical rigor. "Mathematics," he declared,[100] "is an experimental science, and definitions do not come first, but later on." The widespread use of the delta function prior to its vindication by the theory of distributions shows that also the development of quantum mechanics was not exempt from Heaviside's antipuristic attitude.

The next and last stage in the development of the operator conception, before its use in quantum mechanics, is characterized by the effort to justify the previous methods by rigorous proofs, either by making use of the theory of functions of a complex variable, as, for example, in the work of T. J. I'A. Bromwich, or, more generally, by the use of integral transforms, on the historical development of which the reader is referred to S. Pincherle's excellent summary in the *Encyklopädie der mathematischen Wissenschaften*.[101] It was in this connection that Wiener became interested in the operator calculus and introduced, together with Born, this powerful mathematical tool into quantum mechanics.

[99] O. Heaviside, "On operators in physical mathematics," *Proceedings of the Royal Society of London* (*A*), *52*, 504–529, *54*, 105–143 (1893); *Electromagnetic Theory* (London, 1899; reprinted by Dover, New York, 1950).

[100] *Ibid. 54*, p. 121.

[101] (Teubner, Leipzig, 1904–1916), vol. 2, part 2, pp. 763–824.

In concluding our brief survey on the early development of operators in mathematical physics, we should like to draw the reader's attention to a rather unknown paper by Ludwik Silberstein,[102] which anticipated to some extent the formal aspects of the operational approach in modern quantum mechanics. The study of symbolic integrals of the equations of the electromagnetic field[103] suggested to Silberstein, who is known mostly for his writings on the theory of relativity, a theory of "physical operators," in terms of which he attempted to give a unified representation of such disparate phenomena as mechanical oscillations, heat conduction, and electrodynamic processes. Defining the "state" ("Zustand") of a physical system by a time-dependent function $\psi(t)$, Silberstein introduced what he called "chrono-operators" $\{{H \atop t}\}$ by means of which the state at time $t = t$ can be determined from the knowledge of the state at time $t = 0$, in accordance with the equation $\psi(t) = \{{H \atop t}\}\psi(0)$. He defined the inverse operator $\{{H \atop t}\}^{-1}$ by $\{{H \atop -t}\}$, showed that $\{{H \atop nt}\} = \{{H \atop t}\}^n$ and that $\{{H \atop t}\} = \{e^{tF}\}$, where F is another operator, connected with H in a definite manner, but independent of t. Silberstein even spoke of the superposition of states and mentioned numerous other details which, including the notation, made their appearance in the operational formalism of quantum mechanics some twenty-five years later. "Every class of physical phenomena," declared Silberstein[104] in 1901, "or at least those amenable to quantitative treatment, will be characterized by corresponding physical operators; and the scientific study of natural phenomena will proceed by detailed examinations of the properties of these operators, based on observation and nurtured by experiment."

Whereas in Cambridge, Massachusetts, matrix mechanics was generalized into an operator calculus by Born and Wiener, in Cambridge, England, it was modified into an algebraic algorithm by Paul Adrien Maurice Dirac. Trained as an electrical engineer—Dirac obtained a B.Sc. in electrical engineering from the University of Bristol in 1921—but unable to find a job in the years of depression, Dirac continued his studies, primarily under the mathematician Peter Fraser, in Bristol until he obtained in 1923 a senior research scholarship for three years at St. John's College in Cambridge. Thus, during the years 1923 to 1926, when he obtained his Ph.D., he worked under R. H. Fowler and published a series of papers on the relativistic dynamics of particles, the Doppler principle in Bohr's theory, the adiabatic principle, and other problems in the older quantum

[102] L. Silberstein, "Versuch einer Theorie der physikalischen Operatoren," *W. Ostwald's Annalen der Naturphilosophie 1903*, vol. 2, pp. 201–254. Cf. also L. Silberstein, "La théorie des opérateurs en physique—la connection des phénomènes dans le temps," *Przegląd Filozoficzny 5*, 425–442 (1902).

[103] L. Silberstein, "Symbolische Integrale der elektromagnetischen Gleichungen aus dem Anfangszustand des Feldes abgeleitet, nebst Andeutungen zu einer allgemeinen Theorie physikalischer Operatoren," *Annalen der Physik 6*, 373–397 (1901).

[104] *Ibid.*, p. 397.

theory.[105] In the course of this work he became thoroughly acquainted with the Hamilton-Jacobi formulation of classical mechanics and the use of uniformizing variables (angle and action variables), primarily through Whittaker's *Analytical Dynamics*, the standard treatise on this subject in England.

Heisenberg, it will be recalled, shortly after his return from Heligoland, had left Göttingen for Cambridge in July, 1925, to lecture at the Cavendish Laboratory. One of his lectures, entitled "On Term-zoology and Zeeman-botany,"[106] delivered on July 28, was sponsored by the Kapitza Club, a weekly colloquium, named after Peter Leonidovich Kapitza, who at that time was the assistant director of magnetic research at the Cavendish. In all his Cambridge lectures, whether in his public lectures at the laboratory or in his colloquium talk at the club, Heisenberg spoke only about the older quantum theory and did not even refer to his new discoveries, probably because he was not yet assured of the correctness of his new approach. Dirac, in fact, became acquainted with the new mechanics not, as it is so often alleged, through Heisenberg's Cambridge lectures at the end of July, but only early in September, several weeks after Heisenberg had left Cambridge. For at that time Fowler obtained from Bohr the proofs of Heisenberg's paper[107] and showed them to Dirac. At first, Dirac "saw nothing useful in it," but after about two weeks he "saw that it provided the key to the problem of quantum mechanics."[108]

Yet, convinced of the indispensability of Hamilton's mechanics for the study of atomic physics, Dirac was not satisfied with Heisenberg's exposition and tried to adapt it to the Hamiltonian formalism. In a few weeks' time he achieved his objective and thus established one of the most profound and useful relations between quantum mechanics and the classical Hamilton-Jacobi formulation of mechanics. He obtained this result by recasting Heisenberg's mechanics into an algebraic algorithm on the basis of which he expected to derive all the formulas of the quantum theory without any explicit use of the Heisenberg products (matrices). Fowler immediately

[105] P. A. M. Dirac, "Dissociation under a temperature gradient," *Proceedings of the Cambridge Philosophical Society 22*, 132–137 (1924); "Note on the relativity dynamics of a particle," *Philosophical Magazine 47*, 1158–1159 (1924); "Note on the Doppler principle and Bohr's frequency condition," *Proceedings of the Cambridge Philosophical Society 22*, 432–433 (1924); "The conditions for statistical equilibrium between atoms, electrons and radiation," *Proceedings of the Royal Society of London (A)*, *106*, 581–596 (1924), 581–596 (1924); "The adiabatic invariance of the quantum integrals," *ibid. 107*, 725–734 (1924); "The effect of Compton scattering by free electrons in a stellar atmosphere," *Monthly Notices of the Royal Astronomical Society 85*, 825–832 (1925); "The adiabatic hypothesis for magnetic fields," *Proceedings of the Cambridge Philosophical Society 23*, 69–72 (1925).

[106] *Archive for the History of Quantum Physics*, Interview with W. Heisenberg on July 5, 1963.

[107] REF. 1.

[108] *Archive for the History of Quantum Physics*, Interview with P. A. M. Dirac on Apr. 1, 1962.

recognized the importance of this work and urged Dirac to publish his results, incomplete as they still were. Thus, Dirac's paper "The fundamental equations of quantum mechanics," completed on November 7, 1925, appeared in the December issue of the *Proceedings of the Royal Society*[109] and marked the beginning of his celebrated contributions to quantum mechanics.

Starting from Heisenberg's representation of quantum-mechanical quantities by means of amplitudes x_{nm} and frequencies ν_{nm}, subject to the multiplication rule

$$(xy)_{nm} = \sum_k x_{nk}y_{km} \tag{5.48}$$

Dirac determined the form of the most general quantum operation d/dv, which satisfies

$$\frac{d}{dv}(x + y) = \frac{d}{dv}x + \frac{d}{dv}y \tag{5.49}$$

$$\frac{d}{dv}(xy) = \left(\frac{d}{dv}x\right)y + x\left(\frac{d}{dv}y\right) \tag{5.50}$$

From (5.50) he concluded that the amplitudes of the components of dx/dv are linear functions of those of x or

$$\left(\frac{d}{dv}x\right)_{nm} = \sum_{n'm'} a_{nm;n'm'}x_{n'm'} \tag{5.51}$$

Substitution of (5.51) in (5.50) yielded

$$\sum_{n'm'k} a_{nm;n'm'}x_{n'k}y_{km'} = \sum_{kn'k'} a_{nk;n'k'}x_{n'k'}y_{km} + \sum_{kk'm'} x_{nk}a_{km;k'm'}y_{k'm'} \tag{5.52}$$

This equation being valid for all values of the amplitudes, Dirac compared the coefficients of $x_{n'k}y_{k'm'}$ on either side and obtained

$$\delta_{kk'}a_{nm;n'm'} = \delta_{mm'}a_{nk';n'k} + \delta_{nn'}a_{km;k'm'} \tag{5.53}$$

Taking $k = k'$, he showed that $a_{nm;n'm'}$ vanishes unless $n = n'$ or $m = m'$ (or both); for $k = k'$, $m = m'$, $n \neq n'$ he obtained $a_{nm;n'm} = a_{nk;n'k}$, which showed that $a_{nm;n'm}$ for $n \neq n'$ is independent of m; similarly, $a_{nm;nm'}$ for $m \neq m'$ was shown to be independent of n. For $k \neq k'$, $m = m'$, $n = n'$ he obtained $a_{nk';nk} + a_{km;k'm} = 0$. Thus, for $k \neq k'$, $a_{nk';nk} = a_{kk'} = -a_{km;k'm}$. Finally, with $k = k'$, $m = m'$, $n = n'$, $a_{nm;nm} = a_{nk;nk} + a_{km;km} = a_{mm} - a_{nn}$

[109] P. A. M. Dirac, "The fundamental equations of quantum mechanics," *Proceedings of the Royal Society of London* (*A*), *109*, 642–653 (1925).

and Eq. (5.51) reduced to

$$\left(\frac{d}{dv}x\right)_{nm} = \sum_k \{x_{nk}a_{km} - a_{nk}x_{km}\}$$

or

$$\frac{dx}{dv} = xa - ax \tag{5.54}$$

which showed that the most general operation on a "quantum variable" x, subject to (5.49) and (5.50), is performed by "taking the difference of its Heisenberg products with some other quantum variable." As the use of the term "Heisenberg product" clearly indicates, Dirac did not yet realize that Heisenberg's theory was a matrix formalism, as Born and Jordan had shown in their paper,[110] completed in September.[111]

Dirac now investigated to what the quantum-mechanical quantity $xy - yx$ corresponds in the classical theory. Classically,[112] $x = x(J,w) = \sum_\tau x_\tau(J) \exp(2\pi i\tau w)$, $y = y(J,w) = \sum_\tau y_\tau(J) \exp(2\pi i\tau w)$, where J and w are action and angle variables, and $x_\tau(J) = x_\tau(nh)$ corresponds, in accordance with Bohr's correspondence principle, to $x_{n,n-\tau}$. Hence

$$x_{n,n-\tau}y_{n-\tau,n-\tau-\sigma} - y_{n,n-\sigma}x_{n-\sigma,n-\tau-\sigma}$$

or, equivalently,

$$(x_{n,n-\tau} - x_{n-\sigma,n-\sigma-\tau})y_{n-\tau,n-\tau-\sigma} - (y_{n,n-\sigma} - y_{n-\tau,n-\tau-\sigma})x_{n-\sigma,n-\tau-\sigma}$$

corresponds to

$$\sigma h \frac{\partial x_\tau(J)}{\partial J} y_\sigma - \tau h \frac{\partial y_\sigma(J)}{\partial J} x_\tau$$

But

$$\sigma y_\sigma \exp(2\pi i\sigma w) = \frac{1}{2\pi i}\frac{\partial}{\partial w}\{y_\sigma \exp(2\pi i\sigma w)\}$$

and similarly for x_τ. Hence, Dirac concluded, the nm component of the quantum-mechanical expression $xy - yx$ corresponds to

$$\frac{h}{2\pi i}\sum_{\tau+\sigma=n-m}\left\{\frac{\partial}{\partial J}[x_\tau \exp(2\pi i\tau w)]\frac{\partial}{\partial w}[y_\sigma \exp(2\pi i\sigma w)] - \frac{\partial}{\partial J}[y_\sigma \exp(2\pi i\sigma w)]\frac{\partial}{\partial w}[x_\tau \exp(2\pi i\tau w)]\right\} \tag{5.55}$$

[110] REF. 43.

[111] Curiously, the term "matrix" did not appear even in Dirac's subsequent paper "Quantum mechanics and a preliminary investigation of the hydrogen atom," *Proceedings of the Royal Society of London* (*A*), *110*, 561–579 (1926), which he wrote in January, 1926, after he had read the Born-Jordan paper. Dirac employed this term for the first time in his third paper, "The elimination of the nodes in quantum mechanics," *ibid. 111*, 281–305 (1926), completed in March, 1926.

[112] See p. 102.

or

$$xy - yx \leftrightarrow \frac{ih}{2\pi}(x,y) \tag{5.56}$$

where

$$(x,y) = \frac{\partial x}{\partial w}\frac{\partial y}{\partial J} - \frac{\partial y}{\partial w}\frac{\partial x}{\partial J} \tag{5.57}$$

is the classical Poisson bracket expression for x and y, which is known to be invariant under a canonical transformation. Since for classical canonical variables p_{cl}, q_{cl} $(q_{\text{cl}},p_{\text{cl}}) = 1$, $(q_{\text{cl}},q_{\text{cl}}) = (p_{\text{cl}},p_{\text{cl}}) = 0$, their corresponding quantum-mechanical variables p, q satisfy

$$pq - qp = \frac{h}{2\pi i} \tag{5.58}$$

which is the Born-Jordan commutation relation (5.25), page 210. In accordance with (5.56) Dirac could take over into quantum mechanics the whole of classical mechanics as far as it could be formulated in terms of Poisson brackets instead of derivatives. In particular, it followed from (5.56) that for any function $f(p,q)$ of the quantum-mechanical variables p and q the equation of motion is

$$\frac{df}{dt} = \frac{2\pi}{ih}(fH - Hf) \tag{5.59}$$

where $H = H(pq)$ is the Hamiltonian of the system.

Dirac could easily generalize Eq. (5.58) to the "quantum conditions" for multiply periodic systems of f degrees of freedom—in fact, in his paper he expounded his theory for this more general case from the very beginning—and obtained for $r, s = 1, 2, \ldots, f$

$$\begin{aligned} q_rq_s - q_sq_r &= 0 \\ p_rp_s - p_sp_r &= 0 \\ p_rq_s - q_sp_r &= \delta_{rs}\frac{h}{2\pi i} \end{aligned} \tag{5.60}$$

Applying (5.56) to the quantum-mechanical coordinate x and the Hamiltonian H, which represents the time-independent energy of the system and has, therefore, according to Heisenberg, nonvanishing amplitudes only for $n = m$, Dirac inferred that

$$x_{nm}H_{mm} - H_{nn}x_{nm} = \frac{ih}{2\pi}\dot{x}_{nm} = -h\nu_{nm}x_{nm}$$

or

$$h\nu_{nm} = H_{nn} - H_{mm}$$

which is "just Bohr's relation connecting the frequencies with the energy differences."[113]

It will have been understood that the decisive step in Dirac's proof of the correlation (5.56) between quantum-mechanical variables and their classical prototypes was the use of Bohr's correspondence principle. By assigning to (5.56) or rather to its equivalent, the quantum conditions (5.60), a postulatory status within the conceptual structure of the theory, Dirac absorbed, so to speak, the correspondence principle as an integral part into the very foundations of his theory and, like Heisenberg, disposed thereby of the necessity of resorting to Bohr's principle each time a problem had to be solved.

The quantization of classical mechanics, if carried out in accordance with (5.56) or, equivalently, (5.60), is more convenient than in Heisenberg's matrix mechanics. Dirac owed this advantage of his theory obviously to the utilization of the Poisson brackets, one of the most powerful analytical tools in classical dynamics. As the name indicates, these brackets had been introduced in mechanics by Siméon Denis Poisson in a paper[114] read at the Institute de France on October 16, 1809. Working on the integration of the equations of motion in the perturbation theory of planetary orbits, Poisson realized the usefulness of these brackets for this purpose. Their remarkable significance, however, for theoretical mechanics in general was fully recognized only in 1842–1843 by Jacobi when he discussed these brackets, and Poisson's theorem in terms of them, in his famous lectures[115] on dynamics at the University of Königsberg.

As is well known, Poisson's theorem, which Jacobi once called "la plus profonde découverte de M. Poisson,"[116] can easily be proved by the use of what is known as the "Jacobi identity"

$$((x,y),z) + ((y,z),x) + ((z,x),y) = 0 \tag{5.61}$$

The proof of this identity in classical mechanics, whether carried out by direct calculation or indirectly,[117] is more laborious than the proof of its quantum-mechanical analogue, as Dirac has pointed out.[118]

In a second paper,[119] dealing primarily with the applicability of the

[113] REF. 109, p. 652.

[114] S. D. Poisson, "Mémoire sur la variation des constantes arbitraires dans les questions de mécanique," *Journal de l'Ecole Polytechnique 8,* 266–344 (1809).

[115] C. G. J. Jacobi, *Vorlesungen über Dynamik,* edited by A. Clebsch (Reimer, Berlin, 1866; 2d ed. 1884).

[116] C. G. J. Jacobi, "Sur un théorème de Poisson," *Comptes Rendus 11,* 529–530 (1841); *Gesammelte Werke,* edited by K. Weierstrass (Reimer, Berlin, 1886), vol. 4, pp. 143–146.

[117] Cf. P. Appell, *Traité de Mécanique Rationelle* (Gauthier-Villars, Paris, 4th ed., 1931), p. 439, where it is called "Identité de Poisson"; H. Goldstein, *Classical Mechanics* (Addison-Wesley, Cambridge, Mass., 1950), p. 252.

[118] REF. 109, p. 650.

[119] P. A. M. Dirac, "Quantum mechanics and a preliminary investigation of the hydrogen atom," *Proceedings of the Royal Society of London* (*A*), *110,* 561–579 (1926), received Jan. 22, 1926.

algebraic formalism to specific problems, Dirac called the quantum-mechanical variables "q-numbers,"[120] in contradistinction to the numbers of classical physics, which satisfy the commutative law and which he called "c-numbers." "At present," Dirac declared, "one can form no picture of what a q-number is like. One cannot say that one q-number is greater or less than another."[121] "In order to be able to get results comparable with experiment from our theory, we must have some way of representing q-numbers by means of c-numbers, so that we can compare these c-numbers with experimental values. The representation must satisfy the condition that one can calculate the c-numbers that represent $x + y$, xy, and yx when one is given the c-numbers that represent x and y. If a q-number is a function of the co-ordinates and momenta of a multiply periodic system, and if it is itself multiply periodic, then it will be shown that the aggregate of all its values for all values of the action variables of the system can be represented by a set of harmonic components of the type $x_{nm} \exp (2\pi i \nu_{nm} t)$, where x_{nm} and ν_{nm} are c-numbers, each associated with two sets of values of the action variables denoted by the labels n and m, and t is the time, also a c-number. This representation was taken as defining a q-number in the previous papers on the new theory [Dirac here referred to the Born-Jordan and the Born-Heisenberg-Jordan papers]. It seems preferable though to take the above algebraic laws and the general conditions [(5.60)] as defining the properties of q-numbers, and to deduce from them that a q-number can be represented by c-numbers in this manner when it has the necessary periodic properties. A q-number thus still has a meaning and can be used in the analysis when it is not multiply periodic, although there is at present no way of representing it by c-numbers."[122]

As this quotation shows, Dirac already envisaged at this stage the idea of a postulatory account of quantum mechanics which leads to a conceptually autonomous and logically consistent system, even if its foundations, underlying the postulates, are deeply rooted in classical dynamics and its ultimate interpretation, connecting the calculus with experience, requires likewise recourse to classical physics. In any case, in the summer of 1926 Dirac completed a postulatory exposition of the algebra of q-numbers which he published in a paper[123] entitled "The quantum algebra."

Dirac also admitted, as we see, that his q-number algorithm—in contrast to the Born-Wiener operator calculus—could not yet account for nonperiodic phenomena. On the other hand, as if to compensate for this deficiency, Dirac derived on his theory the frequencies of the hydrogen atom, a result which Born and Wiener were unable to obtain at that time.

[120] Dirac could not call them "quantum numbers" as this term had a different meaning; cf. REF. 17 OF CHAP. 3.

[121] REF. 119, p. 562.

[122] *Ibid.*, p. 563.

[123] P. A. M. Dirac, "The quantum algebra," *Proceedings of the Cambridge Philosophical Society 23*, 412–418 (1926), received July 17, 1926.

It should be noted, however, that at the same time also Pauli,[124] on the basis of matrix mechanics, arrived at the solution of the hydrogen problem and, in addition, gave a matrix-mechanical derivation of the Stark effect. Finally, Pauli showed in the same paper how the new matrix mechanics could easily solve the problem of crossed electric and magnetic fields (that is, the problem of calculating the energy perturbations in the hydrogen spectrum if the atom is acted on by such fields), a problem which gave rise to insurmountable difficulties if subjected to the treatment of the older quantum theory.[125] As a matter of fact, Pauli completed his matrix-mechanical calculation of the Balmer levels in October, a week or two before Dirac began to work on his second paper. On November 3, Heisenberg wrote to Pauli, after having received from him a draft of his computations: "I need not tell you how delighted I am about the new theory of hydrogen and how pleasantly surprised I am about the speed with which you produced this theory."[126] About Pauli's successful treatment of the problem of crossed fields Heisenberg should have rejoiced even more. For if Pauli's matrix-mechanical derivation of the hydrogen spectrum showed that Heisenberg's theory is not inferior to the older quantum theory, Pauli's solution of the problem of crossed fields proved its superiority over the latter.

It was, however, not the first piece of evidence for this superiority. For it will be recalled that according to Heisenberg's matrix mechanics the energy of the harmonic oscillator in the lowest stationary state was not zero, as inferred by the older quantum theory, but $\frac{1}{2}h\nu_0$. The notion of such zero-point energy, which, as we have seen, was proposed by Planck's so-called "second theory"[127] of 1911 and soon rejected, was nonetheless never completely obliterated. In fact, there were a number of experimental findings, as, for example, those cited by Bennewitz and Simon,[128] which seemed to support the idea. Finally, at the end of 1924, more than half a year before the publication of Heisenberg's historic paper, the existence of the zero-point energy was irrefutably established. In his study of the isotope effect in the vibration band spectrum of boron monoxide Mulliken[129]

[124] W. Pauli, "Über das Wasserstoffspektrum vom Standpunkt der neuen Quantenmechanik," *Zeitschrift für Physik 36*, 336–363 (1926).

[125] O. Klein, "Über die gleichzeitige Wirkung von gekreuzten homogenen elektrischen und magnetischen Feldern auf das Wasserstoffatom, I," *Zeitschrift für Physik 22*, 109–118 (1924). W. Lang, "Über den Bewegungsverlauf und die Quantenzustände der gestörten Keplerbewegung," *ibid. 24*, 197–207 (1924).

[126] W. Heisenberg, "Erinnerungen an die Zeit der Entwicklung der Quantenmechanik," in *Theoretical Physics in the Twentieth Century—A Memorial Volume to Wolfgang Pauli*, edited by M. Fierz and V. F. Weisskopf (Interscience, New York, 1960), p. 43.

[127] See p. 49.

[128] K. Bennewitz and F. Simon, "Zur Frage der Nullpunktsenergie," *Zeitschrift für Physik 16*, 183–199 (1923). Cf. also the historical remarks in C. P. Enz and A. Thellung, "Nullpunktsenergie und Anordnung nicht vertauschbarer Faktoren im Hamiltonoperator," *Helvetica Physica Acta 33*, 839–848 (1960).

[129] R. S. Mulliken, "Isotope effects in the band spectra of boron monoxide and silicon nitride," *Nature 113*, 423–424 (1924); "The isotope effect in band spectra

observed that the shift corresponding to the transition $S_2 = 0$ to $S_1 = 0$ is given by[130] $\frac{1}{2}h[(\nu_2 - \nu_2') - (\nu_1 - \nu_1')]$ a result which could not be accounted for without the assumption of the zero-point energy. Mulliken's observation thus conflicted with the older quantum theory but was in full agreement with the subsequently proposed matrix mechanics. It was the earliest evidence of the superiority of quantum mechanics over the older theory.

5.3 *The Rise of Wave Mechanics*

Simultaneously with the rise of matrix mechanics another, apparently unrelated, conceptual development took place, whose point of departure was not the mechanics of particles but rather the problem of the nature of light, which had challenged man's thirst for knowledge ever since he began to reflect on the physical world. For light, although the principal agent of our knowledge of the external world, never revealed its own identity and, although the intermediary between matter and the finest of man's senses, never removed the veil of its mystery.

Classical physics, as is well known, offered two major hypotheses on the nature of light: Newton's corpuscular doctrine and Huygens's undulatory theory. The first to attempt a reconciliation of these two was Sir William Rowan Hamilton, an ardent admirer of Lagrange. At a time when the wave theory of light, primarily through the work of Fresnel, achieved its most spectacular successes, Hamilton proposed to give to optics, as a formal science, the same "beauty, power and harmony" which Lagrange gave to mechanics. This he tried to achieve by finding one single law of nature which governs both the propagation of light and the motion of particles.[131] With the help of his "characteristic function" he discovered that the actual motion of a point mass in a field of force is governed by the same formal law as the propagation of the rays of light.

In fact, it is often forgotten that Hamilton's famous formulation of mechanics originated in his investigations on optics. Modifying his presentation somewhat for future reference and using modern notation,[132] we may summarize his results as follows.

From Hamilton's variational principle $\delta \int L\, dt = 0$ it is easy to derive, for constant energy E, Maupertuis's principle of least action, the oldest

---

II: The spectrum of boron monoxide," *Physical Review 25*, 259–294 (1925), completed Sept. 12, 1924; revised Dec. 11, 1924.

[130] Primed vibrational frequencies refer to molecules containing the second isotope.

[131] Sir William Rowan Hamilton, "On a general method of expressing the path of light, and of the planets, by the coefficients of a characteristic function," *Dublin University Review 1833*, pp. 795–826.

[132] The notation is the same as in Chap. 3. C denotes a constant.

variational principle of dynamics, $\delta \int 2T\,dt = 0$ or

$$\delta \int [2m(E - U)]^{1/2}\,ds = 0 \qquad (5.62)$$

Comparing (5.62) with Fermat's principle of minimum time, the basis of geometrical optics,

$$\delta \int \frac{n}{c}\,ds = 0 \qquad (5.63)$$

we see that the expression $C[2m(E - U)]^{-1/2}$ plays in mechanics the same role as the phase velocity $u = c/n$ in optics, or symbolically,

$$u = \frac{c}{n} \leftrightarrow C[2m(E - U)]^{-1/2} \qquad (5.64)$$

a relation which may be called "Hamilton's optical-mechanical analogy." This analogy, however, can be extended still further. The action function $S = \int_{t_0}^{t} L\,dt$ defines in configuration space an "action surface" $S(x,y,z,t) =$ const for which, as is well known, $p = \nabla S$ and $\partial S/\partial t = L - pv = -E$. On the other hand, for plane monochromatic waves the wave vector k and the frequency $\omega = 2\pi\nu$ are clearly related to the phase $\varphi = -\omega t + kr$ by the equation $k = \nabla\varphi$ and $\partial\varphi/\partial t = -\omega$. Hence, the surface of constant action of a system of particles is propagated in complete analogy to the surface of constant phase in optics, the wave vector corresponding to the momentum, and the frequency to the energy of the particle.

These results, which Hamilton[133] had published between 1828 and 1837, found little attention for almost a hundred years. Hamilton's mechanics, disjoined from his optics, was, as we know,[134] elaborated and reformulated in a most profound way by Jacobi. Hamilton's optics, disjoined from his mechanics, reappeared in Bruns's theory of the eikonal.[135] But the synthetic point of view, the insistence on a joined formalism of optics and mechanics, which was the chief concern of Hamilton, was almost completely ignored. A notable exception was Felix Klein. He repeatedly stressed the importance of Hamilton's approach, as, for example, in his lecture on

[133] W. R. Hamilton, "Essay on the theory of systems of rays," *Transactions of the Royal Irish Academy 15,* 69–174 (1828), *16,* 1–61 (1830), *16,* 93–125 (1831), *17,* 1–144 (1837); *The Mathematical Papers of Sir William Rowan Hamilton,* edited by A. W. Conway and J. L. Synge (Cambridge University Press, 1931), vol. 1, pp. 1–294.

[134] REF. 115.

[135] H. Bruns, "Das Eikonal," *Leipziger Berichte 21,* 323–436 (1895); also published as a book: *Das Eikonal* (Hirzel, Leipzig, 1895). Bruns's work, although apparently developed independently of Hamilton's theory and not referring to it explicitly, is nevertheless intimately connected with it. In fact, the basic equation, the equation of the eikonal, is essentially Hamilton's differential equation for his characteristic function. The relation between Hamilton's theory and Bruns's work, as well as the importance of the latter for theoretical physics in general—and not only for the construction of optical instruments, for which it was originally intended—was brilliantly exposed by A. Sommerfeld and J. Runge in their paper "Anwendung der Vektorrechnung auf die Grundlagen der geometrischen Optik," *Annalen der Physik 35,* 277–298 (1911).

November 22, 1891, at the Convention of German Scientists in Halle,[136] and deplored the lack of interest in Hamilton's ideas on the part of his contemporaries. Another exception was G. Prange,[137] who dedicated his "Habilitationsrede" in 1921 at the University of Halle to Hamilton's contribution to geometrical optics and their relations to mechanics.

The indifference with which former generations viewed Hamilton's conceptions can easily be explained. For, although the rays, determined by Fermat's principle, do correspond to the trajectories, determined by Maupertuis's principle, and the wave surfaces to those of Hamilton's action, no correspondence could be established between the velocity of the particle and that of the ray. Hamilton's formalism was therefore regarded as only a formal analogy, devoid of any deeper physical significance.

According to Hamilton, Newtonian mechanics corresponds to geometrical optics. But the latter is only an approximation of undulatory optics. Thus, even if it had been realized at that time that the validity of Newtonian mechanics, in view of its correspondence with geometrical optics alone, has limitations and that a more exact mechanics ought to assume undulatory characteristics, the conceptual apparatus of nineteenth-century science would not yet have been ready to accept the conclusion that, just as rays had to be replaced by waves in optics, trajectories of particles had to be superseded by some other conception; for this would have necessarily led to the abrogation of the very notion of the particle itself. For these reasons, not theoreticians but experimentalists were the first to face the wave-particle problem.

From Wilhelm Konrad Röntgen's discovery of x-rays[138] until the famous diffraction experiments performed on Laue's suggestion by Friedrich and Knipping,[139] and even later, the question whether these rays are particles or waves remained unsettled. Their emission from radioactive atoms, together with α and β rays, about whose corpuscular nature there seemed to be no doubt, their power of ionizing[140] a gas through which they pass, and other of their properties suggested a corpuscular nature. W. H. Bragg, even after he had shown that this ionization resulted only indirectly from the ejection of β rays or electrons, declared—anticipating an important

[136] F. Klein, "Ueber neuere englische Arbeiten zur Mechanik," *Jahresbericht der Deutschen Mathematiker Vereinigung 1*, 35–36 (1890–1891); *Gesammelte mathematische Abhandlungen*, edited by R. Fricke and H. Vermeil (Springer, Berlin, 1922), vol. 2, pp. 601–602.

[137] G. Prange, "W. R. Hamiltons Bedeutung für die geometrische Optik," *Jahresbericht der Deutschen Mathematiker Vereinigung 30*, 69–82 (1921).

[138] W. K. Röntgen, "Ueber eine neue Art von Strahlen," *Sitzungsberichte der Physikalisch-Medicinischen Gesellschaft zu Würzburg 137*, 132–141 (1895); *Wiedemannsche Annalen der Physik 64*, 1–11, 12–17 (1898); "On a new kind of rays," *Nature 53*, 274–276 (1896), a translation of the 1895 paper.

[139] W. Friedrich, P. Knipping, and M. Laue, "Interferenzerscheinungen bei Röntgenstrahlen," *Münchener Berichte 1912*, pp. 303–322; *Annalen der Physik 41*, 971–1002 (1913).

[140] The ionizing effect of x-rays was discovered in 1896 by J. J. Thomson. Cf. J. J. Thomson, "The Röntgen rays," *Nature 53*, 391–392 (1896).

feature of the Compton effect:[141] "*One* x-ray provides the energy for *one* beta-ray, and similarly in the x-ray bulb, *one* beta-ray excites *one* x-ray . . . the speed of the secondary beta-ray is independent of the distance that the x-ray has traveled: so the x-ray cannot diffuse its energy as it goes, that is to say, it is a corpuscle."[142] In a paper written conjointly with Porter, Bragg stated: "Energy considerations lead us directly to the supposition that the x- and gamma-rays are corpuscular in nature in so far as each ray is a separate identity moving through space unaltered in form and energy-content, just as an unhindered projectile would do."[143]

These and similar considerations concerning the properties of x-rays led to the paradoxical situation that even after the diffraction experiments had established in an unequivocal manner the undulatory character of these rays, the possibility of these being at the same time also something corpuscular was not wholly dismissed. In fact, reviewing the situation at the end of 1912, W. H. Bragg made this profound and prophetic statement: "The problem becomes, it seems to me, not to decide between two theories of x-rays, but to find, as I have said elsewhere, one theory which possesses the capacity of both."[144]

Bragg's view on the nature of x-rays was shared by a number of experimentalists—among them Maurice de Broglie, an elder brother of Louis de Broglie. He has already been mentioned as the editor of the *Proceedings* of the first Solvay Congress[145] and as a staunch supporter of Compton in the debate with Duane.[146] His laboratory in the rue Byron in Paris specialized in the study of x-rays. "My brother," declared Prince Louis-Victor de Broglie,[147] "considered x-rays as a combination of wave and particle, but, not being a theoretician, he did not have particularly clear ideas on this subject." It was in this laboratory where Maurice initiated his brother to experimental research[148] and where he "insistently drew his attention to the importance and undeniable reality of the dual aspects of particle and wave."[149] In his autobiographical notes Louis de Broglie[150] wrote: "I had long discussions with my brother on the interpretation of his beautiful experiments on the photoelectric effect and corpuscular spectra. . . . These long conversations with my brother about the

[141] See p. 162.

[142] W. H. Bragg, "Corpuscular radiation," *British Association Reports, Portsmouth, 1911*, pp. 340–341. Cf. also his paper "The consequences of the corpuscular hypothesis of the gamma and x-rays and the range of beta-rays," *Philosophical Magazine 20*, 385–416 (1910).

[143] W. H. Bragg and H. L. Porter, "Energy transformations of x-rays," *Proceedings of the Royal Society of London* (*A*), *85*, 349–365 (1911).

[144] W. H. Bragg, "X-rays and crystals," *Nature 90*, 360–361 (1912).

[145] See p. 61.

[146] See p. 162.

[147] *Archive for the History of Quantum Physics*, Interview with L. de Broglie on June 14, 1963.

[148] See REF. 202 OF CHAP. 3.

[149] L. de Broglie, *Savants et Découvertes* (Editions Albin Michel, Paris, 1951), p. 301.

[150] L. de Broglie, "Vue d'ensemble sur mes travaux scientifiques," in *Louis de Broglie—Physicien et Penseur* (Editions Albin Michel, Paris, 1953), p. 459.

properties of x-rays ... led me to profound meditations on the need of always associating the aspect of waves with that of particles."

These speculations, however, were not the only factor which determined Louis de Broglie's direction of thought. It should be recalled that he began his academic studies in preparation for a career in the civil service at the Faculté des Lettres of the University of Paris and in 1910 obtained his L. ès Sc. (Licencié ès Sciences) in history. However, partly as a result of his brother's influence, partly through the impact of Poincaré's *Value of Science* and *Science and Hypothesis*, he became so fascinated with the problems of physics and its philosophy that he decided to change his career. In 1913 he obtained his L. ès Sc. in the physical sciences at the Sorbonne. When he continued his formal studies later at the Faculté des Sciences of the Sorbonne toward a D. ès Sc., he became particularly interested in those branches of physics which dealt with the fundamental problems of time, space, and the structure of matter and light. Thus, for example, Paul Langevin's lectures on relativity and his analysis of the concept of time—Langevin was the first to treat in detail the so-called "clock paradox" in relativity—made a great impression on Louis de Broglie. In fact, his conception of phase waves was intimately related to his reflections on the nature of time. "This difference between the relativistic variations of the frequency of a clock and the frequency of a wave," he wrote,[151] "is fundamental; it had greatly attracted my attention, and thinking over this difference determined the whole trend of my research."

Before we analyze Louis de Broglie's decisive contributions to the conceptual development of quantum mechanics and, in particular, his introduction of what is often called "matter waves,"[152] we shall ask ourselves whether—as it has so often occurred with great innovations in science—his ideas had not been anticipated, at least in part, by speculative thinkers in the past. Because of his optical-mechanical analogy Hamilton is often regarded as a direct precursor of de Broglie's hypothesis. This, however, is only partially true. For Hamilton declared: "Whether we adopt

[151] L. de Broglie, *Une Tentative d'Interprétation Causale et Non-linéaire de la Mécanique Ondulatoire* (Gauthier-Villars, Paris, 1956), p. 4; *Non-linear Wave Mechanics—A Causal Interpretation*, translated by A. J. Knodel and J. C. Miller (Elsevier, Amsterdam, London, New York, Princeton, 1960), p. 5.

[152] In connection with this term it is perhaps appropriate to emphasize that the often repeated story according to which de Broglie took a special interest in spiritualism which allegedly led him to the hypothesis of "matter waves" is completely unfounded. As he told the author, he only once attended a séance and realized that the allegedly ethereal apparition of sparks came from a cigarette lighter. The rumor probably had its origin in a drastic misconception of the scientific terminology, just like the story which associated Einstein with spiritualism. As reported by Seeliger (REF. 180 OF CHAP. 1), a lady who attended a party, given in Einstein's honor by the Oppenheimer family in Frankfurt am Main, confided to Friedrich Dessauer: "Do you know, Herr Professor, what delighted and impressed me most? It was this. Professor Einstein really believes in the fourth dimension."

the Newtonian or the Huyghensian, or any other physical theory, for the explanation of the laws that regulate the lines of luminous or visual communication, we may regard these laws themselves, and the properties and relations of these linear paths of light, as an important separate study, and as constituting a separate science, called often mathematical optics."[153] Hamilton's interest, it is clear, lay wholly in the formal-mathematical aspects and not in the problem of the physical nature of light.

There were in the nineteenth century, however, some unorthodox thinkers who maintained, whether on physical or on philosophical grounds, the ultimate identity between particles and waves. One of the most eloquent proponents of these ideas was Baron Nicolai Dellingshausen. After studying physics at the Universities of Dorpat, Leipzig, and Heidelberg, he returned to his Baltic estate and wrote a number of interesting studies on physics while administering his lands and farms. His major work, an enlarged version of an earlier essay,[154] was published in 1872 under the title *Grundzüge einer Vibrationstheorie der Natur* (*Foundations of a Vibration Theory of Nature*).[155] Basing his ideas on the cartesian conception of extension as the chief characteristic of matter[156] and on the kinetic theory of heat which reduced heat to motion, Dellingshausen identified atoms with standing waves and interpreted motion of particles as a vibrational process. According to his doctrine bodies are merely "extended centers of vibrational motions" and the "nature of matter is extension in motion."[157] A similar theory, though more related to the mathematical problem of the nature of space, was proposed at about the same time by William Kingdon Clifford, the translator of Riemann's works into English. Seeing in Riemann's conception of space the possibility for a fusion of physics with geometry, Clifford interpreted the motion of matter as a manifestation of changes in the curvature of space.[158]

In spite of certain similarities, it would be definitely a mistake to assume that de Broglie had been influenced by such speculations. He did not even know of their existence when he advanced his new ideas. Apart from those more general factors, which we have mentioned previously, there was, however, an additional, more specific, factor which seemed to have determined the direction of his thought. It was a series of papers on what we would today call a hydrodynamic model of a vibrating atom and was written by Marcel Brillouin, whom de Broglie once called "le véritable

[153] REF. 131.

[154] Baron N. Dellingshausen, *Versuch einer speculativen Physik*, 1851.

[155] Published by F. Kluge, Reval, 1872.

[156] Cf. M. Jammer, *Concepts of Space—A History of the Theories of Space in Physics*, with a foreword by Albert Einstein (Harvard University Press, Cambridge, Mass. 1954), p. 97 et seq.; reprinted by Harper and Brothers, New York, 1960.

[157] "Körper sind ausgedehnte Ausgangsorte von Vibrationsbewegungen," REF. 155, p. 373; "Das Wesen Materie ist bewegte Ausdehnung," *ibid.*, p. 402.

[158] REF. 156, chap. 5.

précurseur de la mécanique ondulatoire."[159] In 1919 Brillouin, a professor of physics at the Collège de France, published a paper[160] in which he studied the periodic motion of a vibrating particle in an elastic medium. Assuming that the velocity of the orbiting particle exceeds many times that of the elastic waves in the medium, he showed that with each given position of the particle a finite number n of preceding positions can be associated in such a way that the particle at the given position (at given time) attains precisely those n elastic waves which originated from the particle when it was at the n preceding positions. Considering the case of a constant n as a stationary state, Brillouin attempted to relate this model with Bohr's atom and to account for its optical properties by an appropriate choice of the constants at his disposal.[161] In a second paper[162] he generalized his derivation of discontinuous properties, characterized by integers, and introduced a variable propagation velocity of the elastic waves. He suggested that an appropriate determination of the inverse value of this velocity, which value is a function of the distance r from the nucleus, and which he significantly denoted by $n(r)$, could make it possible to derive, for example, the Balmer terms in Bohr's theory of the hydrogen atom. In a third paper[163] he introduced a particular Lagrangian function for the immediate surroundings of the nucleus where the Hertz-Maxwell equations, as suggested by Bohr's theory, cease to be valid. Referring to the two main suggestions he had made, he concluded this paper with the words: "But perhaps there exists a different, third road on which a young audacious investigator may achieve success."[164] Louis de Broglie undoubtedly was well acquainted with these papers. He even seemed to have received reprints of them from Brillouin himself.[165]

[159] "Il nous paraît juste de souligner que M. Marcel Brillouin a été le véritable précurseur de la mécanique ondulatoire." *Electrons et Photons—Rapports et Discussions du Cinquième Conseil de Physique Tenu à Bruxelles du 24 au 29 Octobre 1927 sous les Auspices de l'Institut International de Physique Solvay* (Gauthier-Villars, Paris, 1928), p. 105. De Broglie's lecture carried the title "La nouvelle dynamique des quanta," pp. 105–132. Similarly, Léon Nicolas Brillouin, Marcel Brillouin's son, declared in *A Review of Scientific Career* (prepared for the American Institute of Physics, 1962), p. 15: "L. de Broglie used some previous suggestions presented by my father from 1919 to 1922, but left by him in a rather incomplete shape."

[160] M. Brillouin, "Actions mécaniques à hérédité discontinue par propagation; essai de théorie dynamique de l'atome à quanta," *Comptes Rendus 168*, 1318–1320 (1919).

[161] "Il semble donc que l'on puisse formuler une hypothèse *dynamique* douée des qualités nécessaires pour représenter les propriétés essentielles de l'atome de Bohr, lorsqu'on saura choisir la loi d'émission comme il convient." *Ibid.*, p. 1319.

[162] M. Brillouin, "Actions à hérédité discontinue et raies spectrales," *Comptes Rendus 171*, 1000–1002 (1920).

[163] M. Brillouin, "Atome de Bohr—Fonction de Lagrange Circumnucléaire," *Journal de Physique 3*, 65–73 (1922).

[164] "Mais peut-être est-ce par une troisième voie toute différente que réussira un jeune chercheur audacieux." *Ibid.*, p. 73.

[165] L. de Broglie declared: "Je connaissais cette note. Je pense même qu'il me l'a envoyé. Oui, elle a certainement joué un certain rôle dans l'ensemble. . . ." *Archive for the History of Quantum Physics*, Interview with L. de Broglie on Jan. 7, 1963.

After these remarks we shall now analyze Louis de Broglie's contributions to the conceptual development of quantum mechanics. In his search for a "synthetic theory of radiation"[166] which combines the aspects of waves and of particles, he naturally turned at first to Einstein's work on light quanta. Following Einstein, he treated black-body radiation as a gas of light quanta and showed in the earliest of his publications on this subject that such a treatment, if subjected to classical statistical mechanics, leads to Wien's distribution law.[167] When he showed this paper to Langevin, he received the answer: "Your ideas are interesting, but your gas has nothing to do with true light." In a second memoir,[168] in which he tried to reconcile Einstein's hypothesis of light quanta with the phenomena of interference and diffraction, he suggested for the first time the idea that it would be necessary to associate with those quanta a certain element of periodicity.[169] Searching for such a periodicity—one should realize that only later, of course, it became clear that if the particle is a photon, the associated frequency, which de Broglie was looking for, is precisely the radiation frequency ν and that the ordinary light wave is precisely the de Broglie wave of the photon—he studied the analogies between the formalism of analytical mechanics and that of undulatory theories, analogies which had "strikingly impressed" his mind from his "early youth."[170] Finally, at the end of the summer of 1923, his conception of phase waves began to take shape. The paper[171] in which he proposed these ideas may rightfully be regarded as initiating the theory of wave mechanics.

De Broglie considered the motion of a particle of rest mass m_0 and velocity $v = \beta c$ with regard to a stationary observer and assumed that the particle is the seat of a periodic internal phenomenon whose frequency ν_0 is given by m_0c^2/h, where h is Planck's constant and m_0c^2 the internal energy of the particle. Now, the stationary observer, de Broglie maintained, ascribes to the particle the energy $m_0c^2/(1 - \beta^2)^{1/2}$ and hence to the periodic phenomenon the frequency $\nu = \nu_0/(1 - \beta^2)^{1/2}$. Looking on the internal phenomenon, however, the observer will see its frequency, owing to the

Moreover, Brillouin's papers are referred to in a footnote to de Broglie's first article on phase waves, "Ondes et Quanta," *Comptes Rendus 177*, 507–510 (1923). The footnote is on p. 507.

[166] REF. 150, p. 460.

[167] L. de Broglie, "Rayonnement noir et quanta de lumière," *Journal de Physique 3*, 422–428 (1922).

[168] L. de Broglie, "Sur les interférences et la théorie des quanta de lumière," *Comptes Rendus 175*, 811–813 (1922).

[169] "L'explication par la théorie des quanta de lumière des phénomènes jusqu'ici interprétés par l'hypothèse des ondulations tels qu'interférence, diffusion, dispersion, etc., paraît fort pénible, et pour la mener à bien, il faudra sans doute faire un compromis entre l'ancienne théorie et la nouvelle en introduisant dans celle-ci la notion de périodicité." *Ibid.*, p. 812.

[170] "Je continuais d'ailleurs à réfléchir à l'analogie entre le formalisme de la mécanique analytique et celui des théories ondulatoires, analogie qui m'avait frappé dès ma première jeunesse." REF. 150, p. 461.

[171] L. de Broglie, "Ondes et quanta," *Comptes Rendus 177*, 507–510 (1923) (Séance du 10 septembre 1923).

relativistic time dilation, being lowered to $\nu_1 = \nu_0(1 - \beta^2)^{1/2}$ and its vibrations varying as $\sin 2\pi\nu_1 t$. It was this discrepancy between ν and ν_1 which so greatly attracted his attention and "determined the whole trend"[172] of his research. To resolve this dilemma, de Broglie introduced "a fictitious wave, associated with the motion of the mobile" ("une onde fictive associée au mouvement du mobile"), which spreads with velocity c/β and frequency ν as defined above. He then could prove that if at the beginning the internal phenomenon of the moving body is in phase with the wave, "this harmony of phase will always persist." Since the body, which at time $t = 0$ coincided with the wave at the origin, is at time $t = t$ at $x = vt$, its internal phenomenon is proportional to $\sin(2\pi\nu_1 x/v)$, whereas the wave at that place is given by

$$\sin\left[2\pi\nu\left(t - \frac{\beta x}{c}\right)\right] = \sin\left[2\pi\nu x\left(\frac{1}{v} - \frac{\beta}{c}\right)\right]$$

But by definition $\nu_1 = \nu(1 - \beta^2)$, which proves the theorem. This result, continued de Broglie, suggests "that any moving body may be accompanied by a wave and that it is impossible to disjoin motion of body and propagation of wave."[173]

De Broglie then applied this result to the motion of an electron on a closed trajectory with period T, assuming that the associated wave describes the same trajectory with the velocity c/β. If at the time $t = 0$ the electron was at the origin O (of the trajectory), he supposed that the wave, starting from O, reaches the electron at the time τ at O', where $OO' = \beta c\tau$. Thus, $\tau = (\beta/c)[\beta c(\tau + T)]$ or $\tau = T\beta^2/(1 - \beta^2)$, and hence the internal phase of the electron changed by

$$2\pi\nu_1\tau = 2\pi\frac{m_0c^2}{h}\frac{T\beta^2}{(1 - \beta^2)^{1/2}} = 2\pi D$$

Claiming that it is almost self-evident that the motion is stable only if "the fictitious wave, passing through O', finds the electron in phase with itself" or that $D = n$ (n is an integer), de Broglie demonstrated that the last equation is just the Sommerfeld quantum condition $J = \int_0^T p\,dq = nh$, since in the present case

$$J = \int_0^T \frac{m_0}{(1 - \beta^2)^{1/2}} v^2\,dt = \frac{m_0\beta^2c^2}{(1 - \beta^2)^{1/2}} T = Dh$$

In a second[174] and a third[175] memoir, published two and four weeks

[172] REF. 151.

[173] L. de Broglie, "A tentative theory of light quanta," *Philosophical Magazine 47*, 446–458 (1924). This article is an English summary of the three 1923 *Comptes Rendus* papers.

[174] L. de Broglie, "Quanta de lumière, diffraction et interférences," *Comptes Rendus 177*, 548–550 (1923) (Séance du 24 septembre 1923).

[175] L. de Broglie, "Les quanta, la théorie cinétique des gaz et le principe de Fermat," *ibid.*, 630–632 (Séance du 8 octobre 1923).

later, de Broglie demonstrated that the velocity $v = \beta c$ of the particle is precisely the group velocity[176] of the phase waves.[177] The statement that the particle follows in every point of its trajectory the ray of its phase wave, that is, the normal to the equiphase surface at that point, was now for him the basic postulate for the dynamics of free particles.

Generalizing this result for the case of curved paths and variable velocities in a field of force and retaining this postulate, de Broglie pointed out that the path can be computed "as in a medium of variable dispersion" by means of Fermat's principle

$$\delta \int \frac{ds}{\lambda} = \delta \int \frac{\nu\, ds}{V} = 0$$

where λ is the wavelength of the phase waves, $V = c/\beta$ their velocity of propagation, and ν their frequency. Fermat's principle can also be expressed by the equation

$$\delta \int \frac{m_0 \beta c}{(1 - \beta^2)^{1/2}}\, ds = \delta \int p\, ds = 0$$

which is precisely Maupertuis's variational principle. This conclusion, de Broglie emphasized, justifies the assumption of the above-mentioned postulate. "The fundamental bond which unites the two great principles of geometrical optics and of dynamics is thus fully brought to light."[178]

In general, the trajectory of the particle will thus be determined by the Fermat-Maupertuis principle. If, however, the particle has to pass through an aperture or slit of dimensions comparable with the wavelength of the phase waves, the particle's path will be curved in accordance with the diffraction of the phase waves. Introducing the additional assumption[179] that the probability of the absorption or scattering of a light quantum by an atom is determined by the geometrical resultant of the phase-wave vectors crossing upon it—a hypothesis which according to de Broglie is

[176] The notion of group velocity may be traced back to Sir George Gabriel Stokes. Cf. his "Notes on hydrodynamics," *Cambridge and Dublin Mathematical Journal 4*, 219–240 (1849); reprinted in *Mathematical and Physical Papers by George Gabriel Stokes* (Cambridge University Press, 1883), vol. 2, pp. 221–242. The formula: group velocity

$$g = \frac{d\nu}{d(\nu/V)}$$

by which de Broglie's contention can easily be proved, was known to Hamilton, as his unpublished papers show. It was published, however, only by Lord Rayleigh in his *Theory of Sound*, vol. 1, section 191 (REF. 74 OF CHAP. 1, 1877 ed., p. 247); cf. REF. 74 OF CHAP. 1. The earliest experimental verification of this formula was given by A. A. Michelson in his famous "Determination of the velocity of light, and of the difference of velocities of red and blue light, in carbon disulphide," *Astronomical Papers Prepared for the Use of the American Ephemeris and Nautical Almanac* (Washington, 1885), vol. 2, pp. 249–258.

[177] The term "phase wave" appears for the first time on p. 549, REF. 174.

[178] "Le lien fondamental qui unit les deux grands principes de l'Optique géométrique et de la Dynamique est mis ainsi en pleine lumière." REF. 175, p. 632.

[179] REF. 174, pp. 549–550.

"analogous to that which is admitted by electromagnetic theory when it links intensity of disclosed light with the intensity of the resultant electric vector"—he reconciled interference and diffraction with the hypothesis of light quanta. He explained, for example, Young's experiment as follows: "Some atoms of light pass through the holes and diffract along the ray of the neighbouring part of their phase waves. In the space behind the wall, their capacity of photoelectric action will vary from point to point according to the interference state of the two phase waves which have crossed the two holes. We shall then see interference fringes, however small may be the number of diffracted quanta, however feeble may be the incident light intensity. The light quanta do cross all the dark and bright fringes; only their ability to act on matter is constantly changing. This kind of explanation, which seems to remove at the same time the objections against light quanta and against the energy propagation through dark fringes, may be generalized for all interference and diffraction phenomena."[180]

These conclusions, de Broglie continued, should apply to electrons as well: "a stream of electrons passing through a sufficiently narrow hole should also exhibit diffraction phenomena."[181] "It is in this direction where one has probably to look for experimental confirmations of our ideas."[182]

This synthesis of waves and particles (or quanta), de Broglie maintained, could be obtained because the phase wave is conceived as "guiding the displacements of the energy."[183] The wave theory, in his view, went too far by denying the discontinuous structure of radiative energy, and not far enough by renouncing its relevancy to mechanics. "The relation between the new dynamics of free particles and the old dynamics is exactly that between wave optics and geometrical optics."[184] He concluded with this statement: "Many of these ideas may be criticized and perhaps reformed, but it seems that now little doubt should remain of the real existence of light quanta. Moreover, if our opinions are received, as they are grounded on the relativity of time, all the enormous experimental evidence of the 'quantum' will turn in favour of Einstein's conceptions."[185]

On November 29, 1924, de Broglie presented his doctoral thesis "Recherches sur la Théorie des Quanta"[186] to the Faculty of Science at the

[180] REF. 173, p. 454.

[181] "Un flot d'électrons traversant une ouverture assez petite présenterait des phénomènes de diffraction." REF. 174, p. 549.

[182] "C'est de ce côté qu'il faudra peut-être chercher des confirmations expérimentales de nos idées." *Ibid.*, p. 549.

[183] "Nous concevons donc l'onde de phase comme guidant les déplacements de l'énergie. . . ." *Ibid.*, p. 549.

[184] "La nouvelle dynamique du point matériel libre est à l'ancienne dynamique ce que l'optique ondulatoire est à l'optique géométrique." *Ibid.*, p. 549.

[185] REF. 173, p. 457.

[186] *Thèses présentées à la Faculté des Sciences de l'Université de Paris pour obtenir le grade de docteur ès sciences physiques par Louis de Broglie, première thèse: Recherches sur la théorie des quanta, deuxième thèse: Propositions données par la Faculté, soutenues le 29 Novembre devant la Commission d'examen: J. Perrin, Président; E. Cartan, Ch. Mauguin, P. Langevin, Examinateurs* (Masson et Cie., Paris, 1924). Cf. also L. de Broglie, "Recherches sur la théorie des quanta," *Annales de Physique 3*, 22–128 (1925).

University of Paris. Essentially a detailed elaboration of his preceding papers, it contained seven chapters: (1) the phase wave, (2) the principles of Maupertuis and of Fermat, (3) the quantum conditions of stability, (4) the quantification of simultaneous motions of two electrical centers, (5) the quanta of light, (6) the diffusion of x- and γ rays, (7) statistical mechanics and the quanta. Besides, it was prefaced by a historical introduction of great interest for the historian of optics. Called by Darwin "truly one of the most important documents in the history of scientific research,"[187] it concluded with these words: "The definitions of the phase wave and of the periodic phenomenon were purposely left somewhat vague . . . so that the present theory may be considered a formal scheme whose physical content is not yet fully determined, rather than a full-fledged definite doctrine."[188] Although the examining committee highly appraised the originality of the work, it did not believe in the physical reality of the newly proposed waves. When asked by Perrin whether these waves could be experimentally verified, de Broglie replied that this should be possible by diffraction experiments of electrons by crystals. In fact, de Broglie seems to have suggested just such an experiment to Alexandre Dauvillier,[189] one of his brother's collaborators in the rue Byron, but Dauvillier was too busy with experimenting on television.[190]

Perrin's question, as we know today, could have been given an affirmative answer on the basis of experimental evidence available at the time, nay, even dating back to 1914, as we shall presently see—if not much earlier.[191]

[187] Sir Charles Darwin, "La découverte scientifique," in REF. 150, p. 192.

[188] "J'ai intentionellement laissé assez vagues les définitions de l'onde de phase et du phénomène périodique. . . . La présente théorie doit donc plutôt être considerée comme une forme dont le contenu physique n'est pas entièrement précisé que comme une doctrine homogène définitivement constituée." REF. 186, p. 110.

[189] See p. 141.

[190] *Archive for the History of Quantum Physics*, Interview with L. de Broglie on Jan. 7. 1963.

[191] Strictly speaking, earlier phenomena, such as radioactive disintegration, may be quoted. For a satisfactory explanation of alpha emission could be given only on the basis of a wave-mechanical treatment in accordance with de Broglie's hypothesis, as was shown by R. W. Gurner and E. U. Condon in their paper "Wave mechanics and radioactive disintegration," *Nature 122*, 439 (1928), and by G. Gamow, "Zur Quantentheorie des Atomkernes," *Zeitschrift für Physik 51*, 204–212 (1928). There is also a widely spread rumor that Phillip Lenard's assistant in Heidelberg, when reexamining some of Lenard's photographs after Lenard's death, recognized traces on them of electron diffraction. Sir George Thomson and Professor E. N. da C. Andrade, who investigated the truth of this rumor, were informed by Professor Otto Haxel of Heidelberg that Mr. Gegusch, the mechanical engineer who developed the photographic plates, covered the outside parts of these plates—for the purpose of reproduction in print–only for purely technical reasons. It has also been established that the focusing of Lenard's cathode-ray beams was not sufficiently homogeneous to exhibit diffraction patterns. (The author is indebted to Sir George Thomson and Professor Andrade for information on this matter; letters of July 7, 1965, and July 19, 1965.) It would nevertheless be interesting to speculate how the course of physics would have been affected had the recognition of the undulatory nature of electrons preceded that of their corpuscular nature. On a field-

In the summer of 1924, Einstein received from Satyandra Nath Bose of Dacca University in India a letter which began with these words: "Respected Sir, I have ventured to send you the accompanying article for your perusal and opinion. I am anxious to know what you think of it. You will see that I have tried to deduce the coefficient $8\pi\nu^2/c^3$ in Planck's law independent of classical electrodynamics, only assuming that the ultimate elementary region in the phase space has the content h^3. I do not know sufficient German to translate the paper. If you think the paper worth publication I shall be grateful if you arrange for its publication in *Zeitschrift für Physik*. Though a complete stranger to you, I do not feel any hesitation in making such a request. Because we are all your pupils though profiting only by your teaching through your writings. . . ."[192]

The article referred to was, of course, Bose's well-known derivation of Planck's radiation law on the basis of what is now known as the Bose-Einstein statistics. Einstein, in fact, immediately recognized the importance of the paper, translated it, and sent it to the editor of the *Zeitschrift für Physik*.[193]

But Einstein did more. Applying Bose's approach to the monatomic gas,[194] he realized that "a far-reaching formal relationship between radiation and gas" ("eine weitgehende formale Verwandschaft zwischen Strahlung und Gas") can be established. For he showed that—if the gas particles are subjected to the new statistics—the mean-square energy fluctuation is given by an expression which, like Eq. (1.19) on page 38, is composed of two additive terms,[195] one corresponding to the Maxwell-Boltzmann statistics of noninteracting molecules and the other to the interference fluctuation associated with undulatory phenomena.

Now, as we know from reliable sources, Langevin, in spite of the lack of any as yet recognized empirical verifications, spoke about de Broglie's work at the fourth Solvay Congress[196] in April, 1924. He also informed

theoretic discussion of the question whether it was purely an accident that electrons first appeared as particles, see R. E. Peierls in "A Survey of Field Theory," *Reports on Progress in Physics 18*, 423–477 (1955), especially pp. 471–473.

[192] The letter, dated June 4, 1924, is found in the *Einstein Estate*, Princeton, N.J.

[193] Bose, "Planck's Gesetz und Lichtquantenhypothese," *Zeitschrift für Physik 26*, 178–181 (1924).

[194] A. Einstein, "Quantentheorie des einatomigen idealen Gases," *Berliner Berichte 1924*, pp. 261–267, issued on Sept. 20, 1924.

[195] A. Einstein, "Quantentheorie des einatomigen idealen Gases," *ibid. 1925*, pp. 3–14, issued on Feb. 9, 1925.

[196] The official proceedings of the Congress—*Conductibilité Électrique des Métaux et Problèmes Connexes, Rapports et Discussions du Quatrième Conseil de Physique Tenu à Bruxelles du 24 au 29 Avril 1924 sous les Auspices de l'Institut International de Physique Solvay* (Gauthier-Villars, Paris, 1927)—contain no record on Langevin's reference to de Broglie's hypothesis. However, A. Ioffe, who attended the Congress, made these interesting remarks: "Je me souviens d'une rencontre avec Paul Langevin en 1924, au congrès Solvay, ou se manifestèrent d'une façon particulièrement nette les contradictions de la physique classique et atomique. Lorentz, créateur de la théorie des électrons, ne put alors que dire son regret de n'être pas mort cinq ans plus tôt, quand l'horizon de

Einstein about it.[197] Einstein, working at that time on the implications of Bose's paper, became naturally very much interested in de Broglie's thesis and asked[198] for a copy, which he received and read in December, 1924. As a result, in his second paper[199] on the quantum theory of the ideal gas, Einstein referred, in the context of his discussion on the interference term in the energy-fluctuation formula, to de Broglie's work and declared: "I shall discuss this interpretation in greater detail because I believe that it involves more than merely an analogy."[200]

After Einstein, now in turn, had drawn Born's attention[201] to de Broglie's thesis, Born discussed it with James Franck, who had headed the department of experimental physics at the University of Göttingen since 1920. Born's student Walter Elsasser, who had come to Göttingen in 1924, also attended the discussion. When Elsasser suggested the performance of a diffraction experiment with free electrons as a possible confirmation of the existence of de Broglie waves, Franck commented that this "would not be necessary since Davisson's experiments had already established the existence of the expected effect."[202]

To understand Franck's remarks, the following developments incident to Davisson's work must be borne in mind. Clinton Joseph Davisson, interested in the secondary electron emission from electrodes in vacuum tubes, had been experimenting since 1919, mostly together with C. H. Kunsman, on the scattering of electrons from metals at the Research Laboratories of the American Telephone and Telegraph Company and Western Electric Company in New York. In their first publication on such experiments Davisson and Kunsman declared: "All the main features of the distribution curves so far observed for the scattering from nickel seem

---

cette théorie semblait serein. Langevin parlait de la thèse de Louis de Broglie comme d'une intéressante tentative de synthèse, d'une issue possible à l'impasse." A. Ioffe, "A la mémoire d'un maître et ami," *La Pensée—Revue du Rationalisme Moderne 1947* (No. 12, May–June), pp. 15–16.

[197] "[Langevin] a évalué à son entière importance la portée des idées de de Broglie sur laquelles Schrödinger fonda les méthodes de la mécanique ondulatoire, et cela avant que les idées de de Broglie se soient condensées en une théorie consistante. Je me rappelle vivement ses explications enthousiastes à ce sujet, mais je me souviens aussi que c'est seulement avec des hésitations et des doutes que je suivais ses développements." A. Einstein, "Paul Langevin," *ibid.*, pp. 13–14.

[198] Langevin told de Broglie: "J'ai parlé de votre thèse à Einstein, ça l'a intéressé; il voudrait avoir un exemplaire." *Archive for the History of Quantum Physics*, Interview with L. de Broglie on June 7, 1963.

[199] REF. 195.

[200] "Ich gehe näher auf diese Deutung ein, weil ich glaube, daß es sich dabei um mehr als um eine bloße Analogie handelt." *Ibid.*, p. 9.

[201] "[Einstein] a attiré mon attention sur la thèse de Louis de Broglie en me disant quelque chose comme ceci: 'Lisez-la, bien qu'elle puisse paraître folle [it looks crazy], elle est tout à fait solide.' " Max Born, "La grande synthèse," in REF. 150, pp. 165–170.

[202] Franck, as Born (*ibid.*) reports, said: "Ce serait très beau, mais non pas nécessaire, car les expériences de Davisson et Davisson et Germer prouvent suffisamment l'existence de l'effet." Since Elsasser corroborated this episode, we do not doubt its occurrence. But Franck's reference was probably to the experiments by Davisson and Kunsman and not to those by Davisson and Germer, which at that time had not yet been carried out.

reasonably accounted for on the supposition that a small fraction of the bombarding electrons actually do penetrate one or more of the shells of electrons which are supposed to constitute the outer structure of the nickel atom and, after executing simple orbits in a discontinuous field, emerge without appreciable loss of energy."[203] Interpreting the angular distribution of the scattered electrons as due to the variation of charge density in the atomic shells, Davisson proposed in 1923 a formula[204] for the confirmation of which he and Kunsman experimented on the scattering of low-speed electrons by platinum and magnesium targets.[205]

Now, when Franck studied the diagrams of the Davisson-Kunsman experiments, he realized that the peaks obtained could be interpreted as due to diffraction phenomena. In fact, Franck and Elsasser soon recognized that the maxima of the angular distributions of the reflected electrons move with increasing velocity of the incident beam toward the direction of the primary beam and thus exhibit features similar to those connected with the well-known phenomena of x-ray diffraction by crystals. They explained therefore the "scattering" as a diffraction process and found that the wavelength involved agreed with the value obtained from de Broglie's formula $\lambda = h/mv$, where m is the mass and v the velocity of the electrons.

Moreover, Elsasser also realized that Ramsauer's strange experimental results[206] could be explained as an interference effect of the de Broglie waves—results which, by the way, were at about the same time independently obtained also by Townsend and Bailey[207] as well as by Chaudhuri,[208] namely, that for slow electrons of energy values below 25 eV the scattering cross section of rare gases, particularly of argon, decreases as if the atom becomes completely transparent to the electron. Strictly speaking, the "Ramsauer effect" or "Ramsauer-Townsend effect," as it was later called, had already been noticed in 1914 by A. Åkesson, a former student of J. R. Rydberg and M. Siegbahn at the University of Lund. When working on his Ph.D. thesis[209] in Lenard's laboratory in Heidelberg,

[203] C. Davisson and C. H. Kunsman, "The scattering of electrons by nickel," *Science 54*, 522–524 (1921). The lobe in figure 1 of this paper (p. 523) at an angle of 75° with the direction of the primary beam, obtained with a nickel target and a bombarding potential of 150 volts, was the first published evidence for the diffraction of de Broglie waves.

[204] C. Davisson, "The scattering of electrons by a positive nucleus of limited field," *Physical Review 21*, 637–649 (1923).

[205] C. Davisson and C. H. Kunsman, "The scattering of low speed electrons by platinum and magnesium," *Physical Review 22*, 242–258 (1923).

[206] C. Ramsauer, "Über den Wirkungsquerschnitt der Gasmoleküle gegenüber langsamen Elektronen," *Annalen der Physik 64*, 513–540 (1921), *66*, 546–558 (1921), *72*, 345–352 (1923).

[207] J. S. Townsend and V. A. Bailey, "The motion of electrons in argon," *Philosophical Magazine 43*, 593–600 (1922).

[208] R. N. Chaudhuri, "The motion of electrons in gases under crossed electric and magnetic field," *Philosophical Magazine 46*, 461–472 (1923).

[209] A. Åkesson, "Über die Geschwindigkeitsverluste bei langsamen Kathodenstrahlen," *Heidelberger Berichte 1914*, 21. Abhandlung. Cf., in particular, footnote 14 on p. 14 of the paper. See also A. Åkesson, "Über die Geschwindigkeitsverluste bei langsamen Kathodenstrahlen und über deren selektive Absorption," *Lunds Universitets Årsskrift 12*, No. 11 (1916).

Åkesson found that the effective scattering cross section of certain gases decreased for slow electrons in a way which was inconsistent with the kinetic theory. A few years later, however, H. F. Mayer claimed in his thesis, also at the University of Heidelberg, that he had disproved Åkesson's results.[210] Eventually, it was Ramsauer who provided an incontestable proof of the existence of the effect by employing what later became known as the "Ramsauer circular method" ("Ramsauer Kreismethode").

Elsasser summarized these conclusions in a paper which he sent to Arnold Berliner, the editor of *Die Naturwissenschaften.* Doubtful whether to publish the paper, Berliner consulted von Laue and Pringsheim, without, however, obtaining any commitment on their part, and also Einstein, who endorsed its publication.[211] Thus Elsasser's note,[212] concluding with an acknowledgment to Franck, was published in the middle of July, 1925.

Interestingly, however, Davisson, whose attention was drawn to this paper, "did not think much of [it] because he did not believe that Elsasser's theory of his (Davisson's) prior results was valid."[213] In fact, Elsasser's note seems to have had little direct influence on Davisson's subsequent work which so brilliantly vindicated de Broglie's hypothesis. What started this famous series of experiments was rather a peculiar accident which occurred in Davisson's laboratory in April, 1925. In the course of an experiment on the angular distribution of electrons scattered by a target of ordinary (polycrystalline) nickel "a liquid-air bottle exploded at a time when the target was at a high temperature; the experimental tube was broken, and the target heavily oxidized by the inrushing air. The oxide was eventually reduced and a layer of the target removed by vaporization, but only after prolonged heating at various high temperatures in hydrogen and in vacuum."[214] When Davisson and Germer resumed their experiments with the nickel target, which, because of recrystallization during the heating, had become a single crystal (or rather a set of a small number of such crystals), they noticed that the angular distribution "completely changed." Thus, when monoenergetic electrons were normally directed against the (111) planes of the target, the pattern showed for a critical electron energy of 54 eV a distinct lobe at a Bragg angle $\varphi = 50°$ and the lobe disappeared when the electron-accelerating voltage was appreciably changed. Since the

[210] H. F. Mayer, "Über das Verhalten von Molekülen gegenüber freien langsamen Elektronen," *Annalen der Physik 64,* 451–480 (1921).

[211] *Archive for the History of Quantum Physics,* Interview with P. Rosbaud on July 8, 1961. Einstein, as reported by Elsasser, said: "Well, darn it, I didn't take my theory of the Bose statistics quite that literally, but I think the man [Elsasser] should be given a chance. It should be published." *Ibid.,* Interview with W. Elsasser on May 29, 1962.

[212] W. Elsasser, "Bemerkungen zur Quantenmechanik freier Elektronen," *Die Naturwissenschaften 13,* 711 (1925).

[213] K. K. Darrow, "The scientific work of C. J. Davisson," *The Bell Journal 30,* 786–797 (1951). Darrow's article contains autobiographical statements by Clinton J. Davisson. The quotation is one of these.

[214] C. J. Davisson and L. H. Germer, "Diffraction of electrons by a crystal of nickel," *Physical Review 30,* 705–740 (1927).

grating constant for the nickel crystal is $a = 2.15$ Å, Bragg's diffraction formula yielded a wavelength $\lambda = 1.65$ Å, which agreed perfectly with its value obtained from de Broglie's formula $\lambda = h/mv \approx (150/V)^{1/2}$ Å for the accelerating potential $V = 54$ volts.

Until these results were obtained, however, it had taken almost another two years after the historic explosion which, as Darrow put it, "blew open the gate to the discovery of electron-waves."[215] "In 1926 Davisson had the good fortune to visit England and attend the meeting of the British Association for the Advancement of Science at Oxford. He took with him some curves relating to the single crystal, and they were surprisingly feeble. He showed them to Born, to Hartree and probably to Blackett. Born called in another Continental physicist (possibly Franck) to view them, and there was much discussion on them. On the whole of the westward transatlantic voyage Davisson spent his time trying to understand Schrödinger's papers, as he then had an inkling (probably derived from the Oxford discussions) that the explanation might reside in them. In the autumn of 1926 Davisson calculated where some of the beams ought to be, looked for them and did not find them. He then laid out a program of thorough search, and on 6 January 1927, got strong beams due to the line-gratings of the surface atoms, as he showed by calculation in the same month."[216]

Meanwhile E. G. Dymond, at the Palmer Laboratory of Princeton University, examined the scattering of electrons by helium at various inclinations to the primary beam. Having passed through two slits after the scattering, the electrons were deflected by a magnetic field into a collector, the field being adjusted so that electrons of only a particular velocity were collected. Dymond concluded that "the occurrence of . . . maxima is strongly suggestive of an interference pattern, as suggested by Elsasser."[217]

In May, 1927, George Paget Thomson, with the assistance of his research student Andrew Reid, completed in Aberdeen the first series of diffraction experiments of electrons by thin films.[218] Thus, while Davisson's experiments were analogous to Laue's method of x-ray diffraction, Thomson's resembled the Debye-Hull-Scherrer technique. Thomson made a beam of monoenergetic cathode rays pass normally through a thin film of celluloid. The rays were then received on a photographic plate which, after being developed, showed a symmetrical pattern similar to those

[215] REF. 213, p. 792.

[216] REF. 213, p. 792.

[217] E. G. Dymond, "Scattering of electrons in helium," *Nature 118*, 336–337 (1926).

[218] G. P. Thomson and A. Reid, "Diffraction of cathode rays by a thin film," *Nature 119*, 890 (1927); G. P. Thomson, "The diffraction of cathode rays by thin films of platinum," *ibid. 120*, 802 (1927); G. P. Thomson, "Experiments on the diffraction of cathode rays," *Proceedings of the Royal Society of London* (*A*), *117*, 600–609 (1928). Cf. Sir George Thomson, "Early work in electron diffraction," *American Journal of Physics 29*, 821–825 (1961).

observed with x-rays. Another collaborator of Thomson's, R. Ironside,[219] performed similar experiments with films of copper, silver, and tin, metals which crystallize in face-centered cubes, while Seishi Kikuchi[220] at the University of Tokyo used mica for this purpose. Summarizing these and his own experiments, Thomson could declare: "Cathode rays behave as waves of wavelength h/mv according to de Broglie's theory of wave mechanics."[221] Likewise, Davisson, at a meeting of the American Optical Society in Washington on November 2, 1928, could announce: "During the last few years we have come to recognize that there are circumstances in which it is convenient, if not indeed necessary, to regard electrons as waves rather than as particles, and we are making more and more frequent use of such terms as diffraction, reflection, refraction and dispersion in describing their behavior."[222] In fact, since 1928 experimental evidence for the correctness of de Broglie's ideas increased enormously. We mention only Rupp's important experiments on electron diffraction by mechanically ruled gratings,[223] the experiments performed by Estermann and Stern[224] and by Estermann, Frisch, and Stern[225] on the diffraction of molecular beams of hydrogen and helium, and the more recent experiments, carried out by Fermi and Marshall[226] and Zinn,[227] within the framework of the Manhattan Project, on the diffraction of neutron beams.

On April 30, 1897, "a date which may fittingly be taken as marking the dawn of modern particle physics,"[228] Joseph John Thomson announced before the Royal Institution of Great Britain his discovery of the electron, for the study of whose corpuscular properties, such as its charge-to-mass ratio, he was awarded the Nobel Prize in 1906. Reviewing his work on cathode rays, he declared in his Nobel Prize Address: ". . . I wish to give

[219] R. Ironside, "The diffraction of cathode rays by thin films of copper, silver and tin," *Proceedings of the Royal Society of London* (*A*), *119*, 668–673 (1928).

[220] S. Kikuchi, "Diffraction of cathode rays by mica," *Proceedings of the Imperial Academy in Tokyo 4*, 271–274, 275–278, 354–356, 471–474 (1928).

[221] REF. 218 (*Nature 120*, p. 802).

[222] C. J. Davisson, "Electrons and quanta," *Journal of the Optical Society of America 18*, 193–201 (1929).

[223] E. Rupp, "Über die Winkelverteilung langsamer Elektronen beim Durchgang durch Metallhäute," *Annalen der Physik 85*, 981–1012 (1928); "Versuche über Elektronenbeugung am optischen Gitter," *Die Naturwissenschaften 16*, 656 (1928); "Versuche zur Elektronenbeugung," *Physikalische Zeitschrift 29*, 837–839 (1928); "Über Elektronenbeugung an einem geritzten Gitter," *Zeitschrift für Physik 52*, 8-15 (1928); "Über Elektronenbeugung an Metallfilmen," *Annalen der Physik 1*, 773–800 (1929); "Das Wesen des Elektrons," *Zeitschrift des Vereins Deutscher Ingenieure 73*, 1109–1114 (1929).

[224] I. Estermann and O. Stern, "Beugung von Molekularstrahlen," *Zeitschrift für Physik 61*, 95–125 (1930).

[225] I. Estermann, R. Frisch, and O. Stern, "Monochromasierung der de Broglie–Wellen von Molekularstrahlen," *Zeitschrift für Physik 73*, 348–365 (1931).

[226] E. Fermi and L. Marshall, "Interference phenomena of slow neutrons," *Physical Review 71*, 666–677 (1947).

[227] W. H. Zinn, "Diffraction of neutrons by a single crystal," *Physical Review 71*, 752–757 (1947). Cf. also E. O. Wollan, E. G. Shull, and M. C. Marney, "Laue photography of neutron diffraction," *ibid. 73*, 527–528 (1948).

[228] N. H. de V. Heathcote, *Nobel Prize Winners in Physics 1901–1950* (Henry Schuman, New York, 1953), p. 50.

an account of some investigations which have led to the conclusion that the carriers of negative electricity are bodies, which I have called corpuscles. . . ."[229]

Forty years later, in 1937, Davisson and George Paget Thomson shared the Nobel Prize "for the experimental discovery of the interference phenomena in crystals irradiated by electrons." If physical reality "is what it is because it does what it does,"[230] then one may feel inclined to say that Thomson, the father, was awarded the Nobel Prize for having shown that the electron is a particle, and Thomson, the son, for having shown that the electron is a wave. But, precisely, what is the physical nature of the de Broglie waves whose reality has been so irrefutably demonstrated?

De Broglie's hypothesis which associated with every particle a wave was soon elaborated into a new theory of mechanics. If there are waves, it was argued, there must be a wave equation. What was now needed was the discovery of this wave equation. It was also assumed[231] that a theory of mechanics, based on such an equation, must include ordinary particle mechanics as a limiting case, just as relativistic mechanics contained Newtonian mechanics in the limit of velocities relative to which the velocity of light may be regarded as infinitely large, a relation symbolically expressed by $c \rightarrow \infty$, or just as undulatory optics contained geometrical optics in the limit of wavelengths which may be regarded as infinitely small if compared with the dimensions of the apparatus, or symbolically for $\lambda \rightarrow 0$. A rigorous proof that in this limit undulatory optics reduces to geometrical optics had been given before 1911, probably for the first time, by Debye,[232] when he showed that the basic equation of undulatory optics, $\Delta u + k^2 u = 0$, can be approximated for large k by the eikonal equation $|\operatorname{grad} W| = n$, the fundamental equation of geometrical optics.[233]

There can be little doubt that Debye's calculation must have been as follows. In the wave equation $\Delta u + n^2 k_0^2 u = 0$ he put $u = A \exp(ik_0 W)$ and treated A and W as slowly varying functions. Since k_0 is by assumption very large, he neglected in the resulting differential equation for W and A all lower powers of k_0 in comparison with the highest power occurring (k_0^2) and obtained $(\nabla W)^2 = n^2$.

[229] *Ibid.*, p. 44.

[230] Cf. M. Jammer, *Concepts of Mass in Classical and Modern Physics* (Harvard University Press, Cambridge, Mass., 1961), p. 153; German translation: *Der Begriff der Masse in der Physik*, translated by H. Hartmann (Wissenschaftliche Buchgesellschaft, Darmstadt, 1964), p. 164.

[231] For a recent critique on this assumption see N. Rosen, "The relation between classical and quantum mechanics," *American Journal of Physics 32*, 597–600 (1964).

[232] This follows from a casual remark made by Sommerfeld and Runge on p. 290 of their paper, REF. 135: "Nach einer gelegentlichen mündlichen Bemerkung von Hrn. Debye läßt sich diese Differentialgleichung [i.e., the eikonal equation] durch einen Grenzübergang aus der Differentialgleichung der Wellenoptik entnehmen."

[233] In these equations u denotes any of the electrical or magnetic field components, depending on time by the factor $\exp(2\pi i\nu t)$, k is the wave number $2\pi/\lambda$, W is the eikonal function, and $n = k/k_0$ the index of refraction.

Now, it will be recalled that the eikonal equation is the optical analogue of Hamilton's differential equation of ordinary mechanics. It should therefore be clear that what was needed for the establishment of the new mechanics was the analogue for mechanics of Debye's limit consideration, but in the reverse direction, that is, an extension of Hamilton's equation into a differential equation which determines waves of the kind envisaged by de Broglie. As we know from reliable sources,[234] Debye was in fact working along these lines. Also E. Madelung,[235] who—like so many others at that time—had serious doubts about the reality of atomic orbits, was trying to devise a "kind of wave theory of the atomic levels." As is well known, however, it was Schrödinger who succeeded in constructing within a few months' time practically the whole of what is now known as nonrelativistic wave mechanics.

Erwin Schrödinger, like Heisenberg, graduated from a "classical gymnasium," where Latin and Greek were major subjects of instruction. His humanistic education greatly influenced his later views on science and life and laid the foundations for a profound interest in the problems of classical and modern philosophy, as his writings eloquently attest.[236] From 1906 until his Ph.D. in 1910 he studied at the University of Vienna, primarily under the guidance of Fritz Hasenöhrl in theoretical physics, Franz Exner in experimental physics, and Wilhelm Wirtinger in mathematics. In 1911 he became Exner's assistant at the Institute. "The old Vienna Institute," he said later,[237] "which had just mourned the tragic loss of Ludwig Boltzmann, the building where Fritz Hasenöhrl and Franz Exner carried on their work and where I saw many others of Boltzmann's students coming and going, gave me a direct insight into the ideas which had been formulated by that great mind. His line of thought may be called my first love in science. No other has ever thus enraptured me or will ever do so again." That Boltzmann had indeed influenced Schrödinger greatly can be shown by a detailed analysis of Schrödinger's early work on the kinetic theory of gases, statistical mechanics, and elasticity.[238] It is particularly interesting

[234] *Archive for the History of Quantum Physics*, Interview with P. S. Epstein on May 26, 1962.

[235] *Ibid.*, Interview with L. Landé on Mar. 6, 1962.

[236] *Science and Humanism* (Cambridge University Press, 1951), *Nature and the Greeks* (Cambridge University Press, 1954), *Mind and Matter* (Cambridge University Press, 1958), *Meine Weltansicht* (Zsolnay, Vienna, 1961), or its English translation by C. Hastings, *My View of the World* (Cambridge University Press, 1964); see also J. Bernstein, "I am this whole world," *The New Yorker*, May 1, 1965, pp. 180–188.

[237] "Antrittsrede des Hrn. Schrödinger," *Berliner Berichte 1929*, pp. c–cii, an address delivered on July 4, 1929, before the Prussian Academy of Science on the occasion of his inauguration to membership in this Academy. Cf. E. Schrödinger, *Science and the Human Temperament*, translated by J. Murphy (George Allen and Unwin, London, 1935; Norton, New York, 1935), p. 13; republished as *Science, Theory and Man* (Dover, New York, 1957), p. xiv.

[238] "Zur kinetischen Theorie des Magnetismus," *Wiener Berichte 121*, 1305–1329 (1912); "Studien über Kinetik der Dielektrika," *ibid.*, 1937–1973; "Dynamik elastisch gekoppelter Punktsysteme," *Annalen der Physik 44*, 916–934 (1914); "Zur

to note in retrospect how, in his papers on the dynamics of elastically bound mass points, he had struggled with the problem of atomicity versus continuity or phenomenology, as the latter approach was still called at that time. This problem, it will be recalled, engaged leading physicists and philosophers of science, such as Boltzmann, Mach, and Volkmann, in heated debates at the turn of the century. Referring to Boltzmann's thought-provoking defense of atomicity,[239] Schrödinger[240] wrote: "Atomicity has the additional task by whose accomplishment alone its superiority over the phenomenological theories can be established. It has to discover and predict conditions under which the differential equations, based on conceptions of continuity, would lead—because of the really atomistic structure of matter—to evidently false conclusions." In the paper to which Schrödinger referred, Boltzmann had asked: "Has not atomistics in its current form great advantages over the presently so fashionable phenomenology? Is it likely that in the foreseeable future phenomenology will produce a theory which has the same potentialities as atomicity?"[241] And Boltzmann continued: "Should it ever happen that a theory will be constructed which is as comprehensive as atomicity and which is based on foundations as intelligible and unassailable as those of Fourier's theory on heat conduction, this would truly be an ideal."[242] It would be interesting to know whether Boltzmann, were he alive in 1926, would have regarded Schrödinger's wave theory of the atom, based on a "diffusion equation" à la Fourier, as such an ideal.

In his early years Schrödinger was working in the physics of continuous media and acquired the mastery of eigenvalue problems which proved so valuable for his future work. He also studied intensively Rayleigh's *Theory of Sound* and, in particular, learned early what important role the notion of group velocity plays in the theory of vibrations, as his paper on the acoustics of the atmosphere[243] clearly shows. In addition, deeply interested in philosophy, he read the writings of Spinoza, Schopenhauer, Mach, Richard Semon, and Richard Avenarius.[244]

In 1920 he left Vienna to assist Max Wien in Jena. Four months later he accepted an associate professorship at the Technische Hochschule in Stuttgart. Finally, after a short professorship in Breslau, he settled in

---

Dynamik der elastischen Punktreihe," *Wiener Berichte 123*, 1679–1697 (1924); "Die Ergebnisse der neueren Forschung über Atom- und Molekularwärmen," *Die Naturwissenschaften 5*, 537–543, 561–566 (1917); "Der Energieinhalt der Festkörper im Lichte der neueren Forschung," *Physikalische Zeitschrift 20*, 420–428, 450–455, 474–480, 497–503, 523–526 (1919).

[239] L. Boltzmann, "Ueber die Unentbehrlichkeit der Atomistik in der Naturwissenschaft," *Wiedemannsche Annalen der Physik 60*, 231–247 (1897), originally published in *Wiener Berichte 105*, 907–922 (1896).

[240] P. 917 of the third paper listed in REF. 238.

[241] REF. 239, p. 232.

[242] *Ibid.*, p. 244.

[243] E. Schrödinger, "Zur Akustik der Atmosphäre," *Physikalische Zeitschrift 18*, 445–454 (1917).

[244] Cf. E. Schrödinger, *Meine Weltansicht*) REF. 236), p. 8.

Zurich in 1921 and stayed there until 1927, when he succeeded Planck in Berlin.

About the external circumstances which made Schrödinger start his important work Debye made the following statement: "Then de Broglie published his paper. At that time Schrödinger was my successor at the University in Zurich, and I was at the Technical University, which is a Federal Institute, and we had a colloquium together. We were talking about de Broglie's theory and agreed that we did not understand it, and that we should really think about his formulations and what they mean. So I called Schrödinger to give us a colloquium. And the preparation of that really got him started. There were only a few months between his talk and his publications."[245]

Debye's choice, it seems, was not accidental. For Schrödinger—and this brings us to a second, more profound incitement—was very eager to study de Broglie's ideas and their possible implications. Interested in statistical mechanics, he had read Einstein's paper[246] on the quantum theory of the ideal gas in which, as we have seen, Einstein expressed his belief that de Broglie's conceptions "involve more than merely an analogy." In fact, it was this remark which induced Schrödinger to study what he referred to at that time as the "de Broglie–Einstein undulatory theory, according to which a moving corpuscle is nothing but the foam on a wave radiation in the basic substratum of the universe."[247] Schrödinger himself acknowledged frankly his intellectual indebtedness to Einstein. "By the way," he wrote to Einstein on April 23, 1926, "the whole thing would not have started at present or at any other time (I mean, as far as I am concerned) had not your second paper on the degenerate gas directed my attention to the importance of de Broglie's ideas."[248]

Schrödinger once confided to Dirac how he began his work. Trying to generalize de Broglie's waves to the case of bound particles, he "finally obtained a neat solution of the problem, leading to the appearance of the energy-levels as eigenvalues of a certain operator. He immediately applied his method to the electron in the hydrogen atom, duly taking into account relativistic mechanics for the motion of the electron, as de Broglie had done. The results were not in agreement with observation. We know now that Schrödinger's method was quite correct, and the discrepancy was due solely to his not having taken the spin of the electron into account. How-

[245] "Peter J. W. Debye—An Interview," with the participation of E. E. Salpeter, D. R. Corson, and S. H. Bauer, *Science* *145*, 554–559 (1964).

[246] REF. 195.

[247] "Die de Broglie–Einsteinsche Undulationstheorie der bewegten Korpuskel, nach welcher dieselbe nichts weiter als eine Art 'Schaumkamm' auf einer den Weltgrund bildenden Wellenstrahlung ist." E. Schrödinger, "Zur Einsteinschen Gastheorie," *Physikalische Zeitschrift 27*, 95–101 (1926).

[248] *Briefe zur Wellenmechanik—Schrödinger, Planck, Einstein, Lorentz,* edited by K. Przibram (Springer, Vienna, 1964), p. 24.

ever, electron spin was unknown at that time, and Schrödinger was very disappointed and concluded that his method was not good and abandoned it. Only after some months did he return to it, and then noticed that if he treated the electron non-relativistically his method gave results in agreement with observation in non-relativistic approximation. He wrote up this work and published it in 1926, and in this delayed manner Schrödinger's wave equation was presented to the world."[249]

The statement "Only after some months did he [Schrödinger] return to it [his new method of treating the hydrogen spectrum as an eigenvalue problem]" referred undoubtedly to the time interval between his two attempts at solving the problem, the first under the impact of Einstein's remarks concerning de Broglie's hypothesis, and the second following Debye's colloquium. And the effect of both these factors upon Schrödinger cannot be doubted.

There are, however, also other accounts for which the present author has been unable to find additional independent evidence. Thus, Victor Henri, a former pupil of the physiologist Albert Dastre of the Sorbonne and since 1924 professor of physical chemistry at the University of Zurich, recalled—according to Edmond H. Bauer[250]—that, while visiting Paris, he received from Langevin a copy of "the very remarkable thesis of de Broglie"; back in Zurich and having not very well understood what it was all about, he gave it to Schrödinger, who after two weeks returned it to him with the words: "That's rubbish." When visiting Langevin again, Henri reported what Schrödinger had said. Whereupon Langevin replied: "I think Schrödinger is wrong; he must look at it again." Henri, having returned to Zurich, told Schrödinger: "You ought to read de Broglie's thesis again; Langevin thinks this is a very good work"; Schrödinger did so and "began his work."

De Broglie's work gave Schrödinger an important clue for his whole approach, as Schrödinger himself admitted. As we have seen on page 244, de Broglie had shown that the Sommerfeld quantum condition can be interpreted as a statement concerning the number of wavelengths which cover the exact orbit of the electron around the nucleus. For Sommerfeld's $\oint p\, dq = nh$ and de Broglie's $p = h/\lambda$ implied that

$$\oint \frac{1}{\lambda}\, dq = n$$

[249] "Professor Erwin Schrödinger—Obituary," by P. A. M. Dirac, *Nature 189*, 355–356 (1961).

[250] *Archive for the History of Quantum Physics*, Interview with Edmond H. Bauer on Jan. 8, 1963. According to a still more doubtful story (told to the author in Zurich) Schrödinger, somewhat à la Archimedes, conceived his "wave mechanics" while bathing in the "Strandbad Zürich" (a bathing place at the lake of Zurich), where he frequently went for a swim during the summer and fall of 1925.

To Schrödinger this equation immediately suggested the idea of a wave-theoretic eigenvalue problem.

In the first part of his monumental paper "Quantization as a problem of proper values"[251] Schrödinger proposed what was later called "the Schrödinger time-independent wave equation." By replacing in the Hamilton equation

$$H\left(q, \frac{\partial S}{\partial q}\right) = E \tag{5.65}$$

the function S, assumed to be separable, by $K \log \psi$, Schrödinger obtained

$$H\left(q, \frac{K}{\psi}\frac{\partial \psi}{\partial q}\right) = E \tag{5.66}$$

where ψ now stands for a product of single-valued functions, each depending on only one of the coordinates q. Equation (5.66) can be written (at least in simple cases) as a quadratic form (in ψ and its first derivatives) equated to zero. Schrödinger then replaced the quantum conditions by the following postulate: ψ has to be a real, single-valued, twice continuously differentiable function for which the integral of the just-mentioned quadratic form over the whole configuration space (q space) is an extremum. The Euler-Lagrange equation, corresponding to this variational integral, is the wave equation.

For the hydrogen atom with potential energy $-e^2/r$, where $r = (x^2 + y^2 + z^2)^{1/2}$, the quadratic form turned out to be

$$F \equiv F\left(x, y, z, \frac{\partial \psi}{\partial x}, \frac{\partial \psi}{\partial y}, \frac{\partial \psi}{\partial z}\right) \equiv (\nabla\psi)^2 - \frac{2m}{K^2}\left(E + \frac{e^2}{r}\right)\psi^2 = 0 \tag{5.67}$$

so that from the variational problem

$$\delta J = \delta \iiint F\,dx\,dy\,dz = 0 \tag{5.68}$$

he obtained in the usual way the wave equation for ψ,

$$\Delta\psi + \frac{2m}{K^2}\left(E + \frac{e^2}{r^2}\right)\psi = 0 \tag{5.69}$$

[251] E. Schrödinger, "Quantisierung als Eigenwertproblem," *Annalen der Physik* *79,* 361–376 (1st communication; 1926), received Jan. 27, 1926; 489–527 (2d communication), received Feb. 23, 1926; *80,* 437–490 (3d communication; 1926), received May 10, 1926; *81,* 109–139 (4th communication; 1926), received June 21, 1926. Reprinted in Schrödinger's *Abhandlungen zur Wellenmechanik* (J. A. Barth, Leipzig, 1st ed. 1926, 2d ed. 1928), pp. 1–55, 85–169; *Dokumente der Naturwissenschaft* (REF. 1, 1963), vol. 4 pp. 9–63, 87–171; *Collected Papers on Wave Mechanics,* translated from the second German edition of the *Abhandlungen* by J. F. Shearer and W. M. Deans (Blackie & Son, London, Glasgow, 1928); *Mémoires sur la Mécanique Ondulatoire,* translated into French by A. Proca (F. Alcan, Paris, 1933), pp. 1–64, 100–196.

and the condition

$$\int df\, \delta\psi \frac{\partial\psi}{\partial n} = 0 \qquad (5.70)$$

where df is an element of the infinite closed surface over which the integral is taken.

To solve Eq. (5.69) Schrödinger introduced spherical polar coordinates, wrote $\psi = \chi(r)u(\theta,\varphi)$, and found that u is a surface harmonic $P_l{}^m(\cos\theta)\cos m\varphi$ or $P_l{}^m(\cos\theta)\sin m\varphi$, where $0 \leq m \leq l$ and $l = 0, 1, 2, \ldots$. The limitation to integral values followed from the postulated single-valuedness of ψ. In order to find the eigenfunctions and eigenvalues[252] of the radial equation for $\chi(r)$, he applied—with the assistance of Hermann Weyl—the Laplace transformation and obtained the following result: Eq. (5.69) or, as it is still called, the Euler equation of the variational problem, has solutions, which are everywhere single-valued, finite, and continuous and tend to zero at infinity as $1/r$, for every positive value of E; for negative values of E, however, such solutions exist only if $me^2/[K(-2mE)^{1/2}]$ is a real integer n. The discrete eigenvalue spectrum thus turned out to be $E = -me^4/2K^2n^2$, $n = 1, 2, \ldots$, which for $K = h/2\pi$ was precisely the Bohr energy spectrum of the hydrogen atom. Schrödinger concluded the first part of his paper with the following remarks on the significance of ψ. "One may, of course, be tempted to associate the function ψ with a *vibrational process* in the atom, a process possibly more real than electronic orbits whose reality is being very much questioned nowadays. Originally, I also intended to lay the foundations for the new formulation of the quantum conditions in this more intuitive manner. But later I preferred to present them in the above neutral mathematical form because it exposes more clearly the essential point, which in my opinion, is the fact that the mysterious 'requirement of integralness' ('Ganzzahligkeitsforderung') no longer enters into the quantization rules but has been traced, so to speak, a step further back by having been shown to result from the finiteness and single-valuedness of a certain space function."[253] Considering the atom as a system of vibrations, Schrödinger thought in a similar vein to have traced back Bohr's frequency condition to the appearance of beats so that Bohr's formula for the frequency of emission, $\nu = (E_1/h) - (E_2/h)$, expresses merely the fact, well-known from acoustics or from the electromagnetic wave theory (heterodyne frequency), that the beat frequency is equal to the difference of two simultaneous characteristic frequencies of

[252] Although the terms "proper value" and "proper function," as used by the translators of Schrödinger's papers, are undoubtedly preferable to the hybridizations "eigenvalue" and "eigenfunction," concession is made to the prevailing use of the latter in current literature. English texts on boundary-value problems, published before the advent of wave mechanics, usually employed the terms "characteristic value" and "characteristic function."

[253] REF. 251 (*Annalen* paper), p. 372.

the emitter. "It is hardly necessary to point out," wrote Schrödinger, "how much more gratifying it would be to conceive a quantum transition as an energy change from one vibrational mode to another than to regard it as a jumping of electrons. The variation of vibrational modes may be treated as a process continuous in space and time and enduring as long as the emission process persists."[254]

The second part of the paper (second communication[255]), which could have preceded the first part from the logical point of view, explained the general ideas which led its author to the formulation of what is here still called "undulatory mechanics" ("undulatorische Mechanik"[256]), the term "wave mechanics" ("Wellenmechanik") being used by Schrödinger for the first time only in the beginning of the third part.[257] Elaborating on Hamilton's optical-mechanical analogy, as outlined above, Schrödinger considered the Hamilton equation

$$\frac{\partial W}{\partial t} + T\left(q, \frac{\partial W}{\partial q}\right) + U(q) = 0 \tag{5.71}$$

where $W = \int (T - U)\, dt$ is the action function or Hamilton's principal function, that is, the time integral of the Lagrangian $L = T - U$, taken along a path of the system and regarded as a function of the end points and the time. Taking, as usual, $W = -Et + S(q)$, where S is Hamilton's characteristic function, he obtained the equations

$$|\nabla W|^2 = 2m(E - U) \tag{5.72}$$

and

$$\frac{\partial W}{\partial t} = -E \tag{5.73}$$

which show[258] that the wave fronts of constant action W travel through space with the phase velocity

$$V = \frac{E}{[2m(E - U)]^{1/2}} \tag{5.74}$$

The particle velocity v, however, is clearly $[2m(E - U)]^{1/2}$. This difference between V and v, Schrödinger maintained, was the reason for the failure of any attempt at establishing, prior to de Broglie's discovery, an undulatory mechanics as an extension of ordinary mechanics in analogy to the generalization of geometrical optics to undulatory optics. "Important conceptions in wave theory, such as amplitude, wavelength, and frequency,

[254] *Ibid.*, p. 375.
[255] *Ibid.*
[256] *Ibid.*, p. 497.
[257] *Ibid.*, p. 438.
[258] This follows immediately from the equation

$$W + dW = W + |\nabla W|\, ds + \frac{\partial W}{\partial t}\, dt = W + (V\, |\nabla W| - E)\, dt$$

or—more generally—the wave *form,* do not at all appear in the analogy, as there exists no mechanical parallel; nor is any reference made to a wave function. Even the definition of W as signifying the phase is not sufficiently clear in the absence of any definition of the wave form. If we regard the whole analogy merely as a convenient means of picturization," Schrödinger continued,[259] "then the defect is not disturbing, and we would consider any attempt at improving upon it as an idle trifling, assuming that the analogy holds precisely with *geometrical* optics, or, if we wish, with a very primitive form of wave optics, and not with the full-fledged undulatory optics. That geometrical optics is only a rough approximation for *light* does not change the situation. To preserve the analogy for the further development of the optics of q space in accordance with a wave theory, we have to be careful not to depart markedly from the limiting case of geometrical optics; that is, we must choose the wavelength sufficiently small, namely, small compared with all the path dimensions. But then the additions do not teach anything new. The picture is only draped with unnecessary ornaments.

"So we might reason to begin with. But even the first attempt to construct an analogy to the wave theory leads to such striking results, that a quite different suspicion arises: for today we do know that our classical mechanics breaks down for very small dimensions and very great curvatures of path. Perhaps this failure corresponds exactly to the failure of geometrical optics, that is, 'the optics of infinitely small wavelengths,' if the obstacles or apertures are no longer great compared with the real, finite wavelength. Our classical mechanics is perhaps the *complete* analogy of geometrical optics and as such is wrong and not in agreement with reality; it fails whenever the radii of curvature and the dimensions of the path are no longer great compared with a certain wavelength, which has a real meaning in q space. Then an undulatory mechanics has to be established, and the most obvious approach to it is the elaboration of the Hamiltonian analogy into a wave theory."

Assuming that for the waves of such an undulatory mechanics the phase velocity is given by Eq. (5.74) and the time dependence by a factor $\exp\,(2\pi i(E/h)t)$, Schrödinger obtained

$$\lambda = \frac{V}{\nu} = \frac{h}{[2m(E - U)]^{1/2}} \tag{5.75}$$

for the wavelength, in accordance with de Broglie's result. From the assumption $\Psi = \psi(q,t) = \psi(q)\exp\,(2\pi i(E/h)t)$, where $\psi(q) \equiv \psi$ is a space function and E/h the frequency ν, and from the general differential

[259] REF. 251 (*Annalen* paper), pp. 496–497.

equation for wave phenomena

$$\Delta\psi(q,t) - \frac{1}{V^2}\ddot{\psi}(q,t) = 0 \tag{5.76}$$

he deduced the fundamental equation

$$\Delta\psi + \frac{8\pi^2 m}{h^2}(E - U)\psi = 0 \tag{5.77}$$

in agreement with his previous result (5.69).

Schrödinger attached great importance to the fact that the particle velocity v is equal to the group velocity g, as the application of the formula $g = d\nu/d(\nu/V)$ immediately shows. For, as he declared, "this fact can be used to establish a much more intimate connection between wave propagation and the motion of the representative point than was possible before. We can attempt to construct a wave group of relatively small dimensions in every direction. Such a wave group will presumably obey the same laws of motion as the image point of the mechanical system. It will then give, so to speak, a *substitute* ('Ersatz') of the image point as far as its dimensions can be ignored, that is, as far as we can neglect any spreading out in comparison with the dimensions of the path of the system."[260] The notion of wave groups or wave packets, as they were later called, had been used in optics as early as 1909 by Debye,[261] and a few years later by von Laue,[262] for an exact analytical representation of cones of rays. It now enabled Schrödinger to link his undulatory mechanics with ordinary particle mechanics. "The true mechanical process," he explained, "will be realized and represented appropriately by the *wave processes* in q space, and not by the motion of *image points* in this space. The study of the motion of image points, which is the object of classical mechanics, is merely an approximation and has, as such, as much justification as geometrical optics or 'ray-optics' has compared with the true optical process. A macroscopic mechanical process will have to be represented as a wave signal of the kind described above, which can approximately enough be regarded as point-like compared with the geometrical structure of the path. We have seen that for such a signal or wave group precisely the same laws of motion are valid as those advanced by classical mechanics for the motion of the image point. This procedure, however, becomes meaningless when the structure of the path is no longer very large compared with the wavelength or indeed is comparable with it. Then we *must* proceed strictly according to the wave theory, that is, we must proceed from the *wave equation*, and

[260] *Ibid.*, p. 499.

[261] P. Debye, "Das Verhalten von Lichtwellen in der Nähe eines Brennpunktes oder einer Brennlinie," *Annalen der Physik 30*, 755–776 (1909).

[262] M. von Laue, "Die Freiheitsgrade von Strahlenbündeln," *Annalen der Physik 44*, 1197–1212 (1914).

not from the fundamental equation of mechanics, in order to include all possible processes. These latter equations are just as useless for the explanation of the microstructure of mechanical processes as geometrical optics is for the explanation of *diffraction*."[263]

In conclusion, Schrödinger applied his theory to the linear harmonic oscillator, to the rigid rotator with a fixed axis, and to that with a free axis, as well as to the vibrational rotator (diatomic molecule). For Planck's oscillator he obtained as eigenfunctions the Hermitian orthogonal functions $\exp(-x^2/2)H_n(x)$, where $H_n(x)$ is the nth Hermitian polynomial, and as its eigenvalues $E_n = (n + \frac{1}{2})h\nu$, $n = 0, 1, 2, \ldots$, in full agreement with Heisenberg's matrix-mechanical result.[264]

For the sake of continuity we shall now turn briefly to the two remaining parts of the paper (the third and fourth communications) and postpone the discussion of another of Schrödinger's papers which chronologically preceded these.

Schrödinger had already indicated at the end of the second part that the applicability of the new theory extended considerably beyond the "directly soluble" problems and that approximate solutions could be obtained for conditions which are sufficiently near to those of soluble problems. The third communication, now, contains a detailed exposition of what became known as the Schrödinger theory of time-independent perturbations. Schrödinger regarded his approach as an extension of a method which Rayleigh introduced in his study of acoustical vibrations. In fact, Rayleigh prefaced his treatment of the vibrations of a stretched string with variable linear density with these words: "The rigorous determination of the periods and types of vibration of a given system is usually a matter of great difficulty, arising from the fact that the functions necessary to express the modes of vibration of most continuous bodies are not as yet recognized in analysis. It is therefore often necessary to fall back on methods of approximation, referring the proposed system to some other of a character more amenable to analysis, and calculating corrections depending on the supposition that the difference between the two systems is small. The problem of approximately simple systems is thus one of great importance, more especially as it is impossible in practice actually to realize the simple forms about which we can most easily reason."[265] This statement, with the appropriate change of interpretation, was fully endorsed by Schrödinger,

[263] REF. 251, p. 506.

[264] The equation

$$\psi'' + \frac{8\pi^2 m}{h^2}(E - \tfrac{1}{2}m\omega^2 x^2) = 0$$

for whose solution Schrödinger referred his readers to p. 261 in Courant-Hilbert, *Methoden der mathematischen Physik*, vol. 1 (REF. 33), had already been solved in 1914 by P. S. Epstein in his dissertation "Ueber die Beugung an einem ebenem Schirm" (Munich, 1914).

[265] John William Strutt, Baron Rayleigh, *The Theory of Sound* (REF. 74 OF CHAP. 1, 2d ed.), vol. 1, pp. 113–114.

who generalized[266] Rayleigh's method in two respects: the coefficients in the differential equations were no longer, as in Rayleigh's treatment, necessarily constant and the procedure was extended to apply also to degenerate cases.

Schrödinger assumed that the Sturm-Liouville eigenvalue problem

$$L\psi + E\psi = 0 \tag{5.78}$$

where L is a self-adjoint linear differential operator of the second order,[267] has for the kth stationary state the normalized eigenfunction solution ψ_k. Taking as the equation for the perturbed case

$$L\psi + E\psi = \lambda r\psi \tag{5.79}$$

where λ is the perturbation parameter and r is a function of the coordinates, characteristic for the perturbation, Schrödinger put

$$E = E_k + \lambda\epsilon_k \qquad \psi = \psi_k + \lambda\varphi_k \tag{5.80}$$

where E_k and ψ_k are the eigenvalue and eigenfunction of the unperturbed problem and ϵ_k and φ_k their "corrections," respectively, for the perturbed case. By substitution, neglecting terms in λ^2, he obtained

$$L\varphi_k + E\varphi_k = (r - \epsilon_k)\psi_k \tag{5.81}$$

Since the nonhomogeneous equation (5.81) has a continuous solution only if its right-hand side is orthogonal to the solution ψ_k of the homogeneous equation (5.78), he concluded that

$$\epsilon_k = \int r\psi_k{}^2\, dx \tag{5.82}$$

and thus obtained the wave-mechanical analogue of the classical theorem that the first-order energy correction is equal to the mean value of the perturbation function averaged over the unperturbed solution.[268] To find φ_k Schrödinger expanded φ_k and the right-hand side of Eq. (5.81), after having substituted in it the value of ϵ_k, in terms of the unperturbed eigenfunctions ψ_i, which he assumed to form a complete orthonormal system. A straightforward calculation yielded

$$\varphi_k = \sum_{\substack{i \\ i\neq k}} \frac{\psi_i \int r\psi_k\psi_i\, dx}{E_k - E_i} \tag{5.83}$$

[266] For this reason the theory is sometimes called, especially by British authors, the "Rayleigh-Schrödinger method"; cf., for example, N. F. Mott and I. N. Sneddon, *Wave Mechanics and Its Applications* (Oxford University Press, 1948; republished by Dover, New York, 1963), p. 73.

[267] To cover the case of using curvilinear coordinates, Schrödinger introduced also a weight function which, for the sake of simplicity, will be ignored in our discussion.

[268] REF. 251, p. 443.

After pointing out that the present procedure could easily be extended to higher orders of approximation, Schrödinger generalized his method for degenerate problems and obtained, in a way well known to every student of quantum mechanics, a secular equation whose roots provide the required energy corrections.

In the remainder of the paper Schrödinger applied his perturbation theory to the Stark effect of the hydrogen atom in two different ways. First, following Schwarzschild and Epstein, he separated the wave equation in parabolic coordinates and then applied his approximation method which led him precisely to the Schwarzschild-Epstein formula, as given on page 108; in the second method, applying his perturbation theory literally as proposed, he expanded the perturbation term, containing the relatively small external field F, in terms of the eigenfunctions of the Kepler problem. He also calculated the intensities of the components of the Stark effect, using the eigenfunctions of the zeroth approximation and the wavelengths of the first-order approximation, and obtained results in fair agreement with observation.[269]

In the fourth communication, the last part of the paper, Schrödinger developed his theory of time-dependent perturbations. Realizing that in Eq. (5.77) the eigenvalue E varies in passing from one stationary state to another and consequently must not occur in the "real wave equation" ("*eigentliche* Wellengleichung") which is to determine the spatial-temporal behavior of the wave function $\Psi \equiv \psi(x,y,z,t)$, Schrödinger eliminated E by means of the equation $\Psi = \psi(q) \exp [2\pi i(E/h)t]$ and obtained

$$-\frac{h^2}{8\pi^2 m} \Delta\Psi + U\Psi = \frac{h}{2\pi i} \frac{\partial \Psi}{\partial t} \tag{5.84}$$

Schrödinger postulated the validity of Eq. (5.84), later called "the Schrödinger time-dependent wave equation," also for time-dependent potential functions U. Recognizing that Eq. (5.84) has the structure of a diffusion equation with an imaginary diffusion coefficient, Schrödinger relaxed his original requirement concerning the reality of ψ and admitted complex-valued functions for what he called the "mechanical field scalar ψ." As an application of the time-dependent wave equation he studied the phenomenon of dispersion, which he treated as a problem involving a periodically varying perturbing potential-energy function U. His well-known first-order approximation result showed that the secondary radiation contains the combination frequencies calculated by Kramers and Heisen-

[269] H. Mark and R. Wierl, "Über die relativen Intensitäten der Starkeffektkomponenten von H_β und H_γ," *Zeitschrift für Physik 53*, 526–541 (1928); "Über die relativen Intensitäten der Starkeffektkomponenten der Balmerlinien H_β und H_γ," *Die Naturwissenschaften 16*, 725–726 (1928); "Weiterer Beitrag zum Intensitätsproblem beim Wasserstoff-Starkeffekt," *Zeitschrift für Physik 55*, 156–163 (1929).

berg[270] and previously predicted by Smekal.[271] Schrödinger concluded his paper with a discussion on the physical significance of ψ. He interpreted $\psi\psi^*$ as a weight function in configuration space which accounts for the electrodynamical fluctuations of the electric space density of the charges. He declared: "The ψ function is to do no more and no less than to offer us a survey and mastery over these fluctuations by a single differential equation. It has repeatedly been pointed out that the ψ function itself cannot and may not in general be interpreted directly in terms of three-dimensional space—however much the one-electron problem seems to suggest such an interpretation—because it is in general a function in configuration space and not in real space."[272]

Schrödinger's brilliant paper was undoubtedly one of the most influential contributions ever made in the history of science. It deepened our understanding of atomic phenomena, served as a convenient foundation for the mathematical solution of problems in atomic physics, solid state physics, and, to some extent, also in nuclear physics, and finally opened new avenues of thought. In fact, the subsequent development of nonrelativistic quantum theory was to no small extent merely an elaboration and application of Schrödinger's work. Interested as we are in the foundations of quantum mechanics, we shall confine our discussion to only those elaborations which had a bearing on the conceptual development of the theory.

One of such basic problems, raised, as we have seen, by Schrödinger at the very beginning of his paper, was the question of exactly what requirements an admissible ψ function must satisfy. The reality requirement, as mentioned, was soon abandoned. In 1930 G. Jaffé,[273] insisting that only physical considerations can serve as a criterion on this issue, contended that ψ must be single-valued, finite, and continuously differentiable throughout configuration space and must possess a second derivative everywhere except at singular points where the potential is discontinuous. That these requirements were too stringent was shown in 1931 by R. M. Langer and N. Rosen.[274] Arguing that the ultimate criterion for the admissibility of ψ is the finiteness of the variational integral J in Eq. (5.68), they declared: "As a rough working rule we may demand of the function that it be integrable in the square and that it be finite and continuous wherever the potential energy is finite. If we introduce singularities (physically non-existent) in the potential energy we must be prepared to put up with singularities in the wave function at those points. When, in the future, we get a deep enough insight into the nature of the physical prob-

[270] REF. 157 OF CHAP. 4.
[271] REF. 158 OF CHAP. 4.
[272] REF. 251 (4th communication), p. 135.
[273] G. Jaffé, "Welchen Forderungen muß die Schrödingersche ψ-Funktion genügen?" *Zeitschrift für Physik 66*, 770–774 (1930).
[274] R. M. Langer and N. Rosen, "What requirements must the Schrödinger ψ-function satisfy?" *Physical Review 37*, 658 (1931).

lems so that we can replace the singular values of the potential energy by their true finite values, the singularities of the wave functions will also be removed. But until then we must be content with results which differ from the facts to the same degree as do the assumptions from which we start." A few weeks later, E. H. Kennard[275] claimed, on the basis of Born's probabilistic interpretation of ψ, which we shall discuss in due turn, that since $\psi\psi^*$ must always be integrable in order that the total probability may be unity, the ψ functions themselves must be quadratically integrable (or, in the continuous-spectrum case, Weyl-normalizable). "The customary requirements of continuity and single-valuedness of ψ or its derivatives are unnecessary as an addition to the fundamental requirement that ψ shall satisfy a certain differential equation. The condition that ψ shall be finite everywhere, which serves so well in atomic theory, is in almost all cases equivalent to the requirement that $\psi\psi^*$ shall be integrable." The basic requirement for ψ, according to Kennard, is that "it shall constitute one of such an orthogonal family in terms of which we can expand any ψ that can occur in Nature."[276]

This brings us to the problem of the completeness of the eigenfunctions, a problem whose basic importance was fully recognized by Schrödinger—especially in connection with the perturbation theory, where it was supposed that the eigenfunctions of the unperturbed case form a complete set. Schrödinger's use of the Laplace transformation for the solution of the radial equation in the Kepler problem had the disadvantage, as he admitted himself,[277] of not showing that the set of the obtained eigenfunctions is complete. In 1927 Eddington[278] showed that the radial equation can be solved in a more elementary way and that its solution is, in fact, implicitly contained in many standard treatises such as that by Whittaker and Watson.[279] In 1928 another short and elementary process for the solution of the radial equation was presented to the American Mathematical Society by T. H. Gronwall.[280] Gronwall also proved the completeness of the eigenfunctions of the hydrogen atom, utilizing the closure property of the spherical harmonics.[281]

Also Schrödinger's postulate concerning the single-valuedness of the wave functions soon became a matter of dispute. Schrödinger had argued[282]

[275] E. H. Kennard, "The conditions on Schrödinger's ψ," *Nature 127*, 892–893 (1931).

[276] *Ibid.*, p. 893.

[277] "... immerhin bestehen noch Lücken. Erstens der Nachweis der Vollständigkeit des *gesamten* nachgewiesenen Systems von Eigenfunktionen." REF. 251, p. 370.

[278] A. S. Eddington, "Eigenvalues and Whittaker's function," *Nature 120*, 117 (1927).

[279] E. T. Whittaker and G. N. Watson, *Modern Analysis* (Cambridge University Press, London, 4th ed., 1927).

[280] T. H. Gronwall, "On the wave equation of the hydrogen atom," *Annals of Mathematics 32*, 47–52 (1931), completed June 30, 1928.

[281] The closure property of the spherical harmonics had been proved by L. Fejér on the basis of the summability of Laplace's series by Cesàro means of the second order; cf. L. Fejér, "Über die Laplacesche Reihe," *Mathematische Annalen 67*, 76–109 (1909).

[282] REF. 251, p. 363.

that l, the wave-mechanical analogue of the azimuthal quantum number in the Bohr-Sommerfeld theory, and m, the analogue of the magnetic quantum number, are necessarily integral since otherwise no uniquely determined dependence of the wave function on the spatial coordinates is assured. The question, expressed in a more modern formulation, was whether the eigenvector representative of the operator L_z, the z component of the orbital angular momentum, is necessarily single-valued. The usual argument is based on Schrödinger's postulate, for it reasons as follows. Since the eigenfunction, say, $\exp(im\varphi)$, of the operator $L_z = (h/2\pi i)(\partial/\partial\varphi)$ has to be single-valued, or $\exp[im(\varphi + 2\pi)] = \exp(im\varphi)$, m must be integral. However, as soon as it was recognized that in the theory of particles with intrinsic spin double-valued functions have to be admitted, Schrödinger's argument seemed specious. To be sure, the requirement of single-valuedness remained valid for observable quantities such as the probability density, but not for probability amplitudes such as the Schrödinger functions.

The first to examine this question more closely was Pauli.[283] As a criterion he proposed to admit only those eigenfunctions which, if subjected to the angular-momentum operator, yield new functions belonging to the same total angular momentum j, i.e., the new functions so obtained should be expressible as linear combinations of the original solutions. This criterion, as Pauli could easily show, excluded the possibility of double-valued orbital angular-momentum eigenfunctions. An appeal to experiment for a decision on this issue was made by David Bohm,[284] who confined the theory to the weaker result, according to which orbital angular eigenvalues are either all integral or all half odd-integral, but added: "Experiment shows that only integral orbital momenta are actually present." Two famous footnotes in Blatt and Weisskopf's treatise on theoretical nuclear physics[285] revived the interest in this problem. Calling Schrödinger's reasoning "fallacious," these authors excluded half odd-integral values in view of certain unpublished results obtained by A. M. Nordsieck. For Nordsieck had shown that the admission of half odd-integral values would lead to the physically unacceptable existence of a probability current density from one pole of a sphere, centered at the origin of the coordinate system, to the other, the two poles acting as source and sink, respectively. Another way to exclude these half odd integers, without referring to the

[283] W. Pauli, "Über ein Kriterium für Ein- oder Zweiwertigkeit der Eigenfunktionen in der Wellenmechanik," *Helvetica Physica Acta 12*, 147–168 (1939). A short discussion of the problem is found already in Pauli's encyclopedia article "Die allgemeinen Prinzipien der Wellenmechanik," *Handbuch der Physik*, vol. 24, edited by H. Geiger and K. Scheel (Springer, Berlin, 2d ed. 1933), p. 126; "Principles of quantum theory," *Encyclopedia of Physics*, vol. 5, part 1, edited by S. Flügge (Springer, Berlin, Göttingen, Heidelberg, 1958), pp. 45–46; reprinted in *Collected Papers* (REF. 154 OF CHAP. 3), vol. 1, pp. 771–938.

[284] D. Bohm, *Quantum Theory* (Prentice-Hall, Englewood Cliffs, N.J., 1951), pp. 389–390.

[285] J. M. Blatt and V. F. Weisskopf, *Theoretical Nuclear Physics* (Wiley, New York, 1952), pp. 783, 787.

Schrödinger representation and his argumentation, was proposed by H. E. Rorschach at the 1962 Southwestern Meeting of the American Physical Society.[286] With the help of dynamical variables which have the character of true vectors (such as **r** and **p**) and which anticommute with the parity operator, he constructed "stepping" operators which lower or raise the l value of state functions and which, by iteration, lead to algebraic equations yielding for the minimum value of l the number zero. In the same year Merzbacher,[287] who regarded the single-valuedness under discussion as resulting from the imposition of boundary conditions, attempted to connect the problem with the Aharonov-Bohm quantum effect[288] in order to enforce a criterion. Buchdahl,[289] on the other hand, resorted to the harmonic-oscillator representation in order to show, "by a straightforward and elementary argument" and without drawing explicitly upon conditions of single-valuedness, that the eigenvalues of the components of the orbital angular momentum are integral. We have purposely gone into these details—for their full comprehension the reader is referred to the original literature—to accentuate the interesting fact that Schrödinger's basic assumptions concerning the mathematical nature of ψ, quite apart from the problem of its physical interpretation, have engaged theoretical research ever since they were published. The recent renewed interest in the problem of the admissibility conditions for wave functions is, of course, also due to the fact that the behavior of field operators in quantum field theory depends on the behavior of the wave functions from which they are formed.

After this digression let us return to 1926, when Schrödinger's derivation of the "integralness" and discreteness of the energy spectrum from the postulated single-valuedness and finiteness of certain space functions surprised the world of physics. To be sure, situations of this kind were not

[286] H. E. Rorschach, "Single-valued wave functions and orbital angular momentum," *Bulletin of the American Physical Society 7*, 121 (1962).

[287] E. Merzbacher, "Single valuedness of wave functions," *American Journal of Physics 30*, 237–247 (1962).

[288] Y. Aharonov and D. Bohm, "Significance of electromagnetic potentials in the quantum theory," *Physical Review 115*, 485–491 (1959).

[289] H. A. Buchdahl, "Remark concerning the eigenvalues of orbital angular momentum," *American Journal of Physics 30*, 829–831 (1962). See also D. Pandres, Jr., "Derivation of admissibility conditions for wave functions from general quantum-mechanical principles," *Journal of Mathematical Physics 3*, 305–308 (1962), and especially M. Kretzschmar, "Must quantal wave functions be single-valued?" *Zeitschrift für Physik 185*, 73–83 (1965). Kretzschmar pointed out that the single-valuedness requirement for angular-momentum eigenfunctions, in contrast to the integralness of their eigenvalues, is not a necessary consequence of basic physical principles, and that the previous derivations of this single-valuedness (Merzbacher, Pandres, et al.) were based on the implicit assumption that $-i\hbar\nabla$ is the only representation of the linear momentum operator. Kretzschmar showed that for other representations the wave functions may be multivalued without leading to inconsistencies, the single-valuedness requirement being merely the consequence of "a pure convention that prescribes the use of only the simplest mathematical representation that is available." Clearly, the argument so often found in textbooks "that eigenvalues must be integer *because* the wave function has to be single-valued" is untenable.

completely unknown at that time, for boundary-value problems, such as that of the stretched string, the tensed membrane or the ball of fluid confined to a spherical shell, had long been solved in mathematical physics. Still, it was striking to see how the enforcement of what seemed completely nonarbitrary conditions led to the same results as the mathematically so unfamiliar and conceptually so sophisticated formalism of matrix mechanics and its variations. After the conceptual cataclysm evoked by the latter it seemed as if Schrödinger's return to quasi-classical conceptions reinstated continuity. Those who in their yearning for continuity hated to renounce the classical maxim *natura non facit saltus* acclaimed Schrödinger as the herald of a new dawn. In fact, within a few brief months Schrödinger's theory "captivated the world of physics" because it seemed to promise "a fulfillment of that long-baffled and insuppressible desire."[290] Einstein was "enthusiastic"[291] about it, Planck reportedly declared, "I am reading it as a child reads a puzzle,"[291] and Sommerfeld was exultant.[291] Göttingen, however, at that time, was rather reserved.[292] Einstein's enthusiasm is not surprising, for he thought he found in Schrödinger's theory the answer to a question which for quite some time had occupied his mind. As early as 1920 he had written to Born: "That one has to solve the quanta by giving up the continuum, I do not believe. . . . I am rather convinced now as before that one has to search instead for an overdetermination by differential equations so that their solutions are no longer of the continuous type. But how?"[293]

It is instructive to compare Schrödinger's wave mechanics with Heisenberg's matrix mechanics. It is hard to find in the history of physics two theories designed to cover the same range of experience, which differ more radically than these two. Heisenberg's was a mathematical calculus, involving noncommutative quantities and computation rules, rarely encountered before, which defied any pictorial interpretation; it was an *algebraic* approach which, proceeding from the observed discreteness of spectral lines, emphasized the element of *discontinuity;*[294] in spite of its renunciation of classical description in space and time it was ultimately a theory whose basic conception was the *corpuscle*. Schrödinger's, in contrast, was based on the familiar apparatus of differential equations, akin to the classical mechanics of fluids and suggestive of an easily visualizable representation; it was an *analytical* approach which, proceeding from a generali-

[290] K. K. Darrow, "Introduction to wave-mechanics," *The Bell System Technical Journal 6*, 653–701 (1927).

[291] *Archive for the History of Quantum Physics*, Interview with Frau Erwin Schrödinger on Apr. 5, 1963. See also REF. 312.

[292] *Ibid.*, but see also REF. 311.

[293] Letter of Jan. 27, 1920, quoted in M. Born, "Erinnerungen an Einstein," *Physikalische Blätter 21*, 297–306 (1965).

[294] It was called "a true discontinuum theory" ("wahre Diskontinuumstheorie") by Born and Jordan, REF. 43, p. 879; p. 145; p. 67.

zation of the classical laws of motion, stressed the element of *continuity;* and, as its name indicates, it was a theory whose basic conception was the *wave.* Arguing that the use of multidimensional (>3) configuration spaces and the computation of the wave velocity from the mutual potential energy of particles is "a loan from the conceptions of the corpuscular theory," Heisenberg[295] criticized Schrödinger's approach as "not leading to a consistent wave theory in de Broglie's sense." In a letter to Pauli he even wrote: "The more I ponder about the physical part of Schrödinger's theory, the more disgusting ('desto abscheulicher') it appears to me." Schrödinger was not less outspoken about Heisenberg's theory when he said: ". . . I was discouraged ('abgeschreckt'), if not repelled ('abgestoßen'), by what appeared to me a rather difficult method of transcendental algebra, defying any visualization ('Anschaulichkeit')."[296]

In spite of these fundamental disparities Schrödinger was always convinced that these two approaches do not clash but rather complement each other.[297] In fact, in the early spring of 1926—prior to the publication of the third communication of his magnum opus—Schrödinger discovered what he called "a formal, mathematical identity" of wave mechanics and matrix mechanics. Schrödinger's proof[298] of the intertranslatability of either formalism into the other proceeded as follows. Recognizing that the Born-Heisenberg matrix relation (5.26)

$$\mathbf{pq} - \mathbf{qp} = \frac{h}{2\pi i}\mathbf{1}$$

corresponds to the wave-mechanical relation

$$\left(\frac{h}{2\pi i}\frac{\partial}{\partial q}\right)q\psi - q\left(\frac{h}{2\pi i}\frac{\partial}{\partial q}\right)\psi = \frac{h}{2\pi i}\psi$$

Schrödinger associated with every physical function $F = F(p,q)$ of the variables p and q the differential operator

$$F\left(\frac{h}{2\pi i}\frac{\partial}{\partial q}, q\right)$$

which he denoted by $[F, \cdot]$. If the functions $u_k = u_k(q)$ form in the con-

[295] W. Heisenberg, "Mehrkörperproblem und Resonanz in der Quantenmechanik," *Zeitschrift für Physik 38,* 411–426 (1926), quotation on p. 412.

[296] REF. 298, p. 735.

[297] REF. 251, p. 513.

[298] E. Schrödinger, "Über das Verhältnis der Heisenberg-Born-Jordanschen Quantenmechanik zu der meinen," *Annalen der Physik 79,* 734–756 (1926); *Abhandlungen zur Wellenmechanik* (REF. 251), pp. 62–84; *Dokumente der Naturwissenschaft* (REF. 1), vol. 3, pp. 64–86. "On the relation between the quantum mechanics of Heisenberg, Born, and Jordan, and that of Schrödinger," *Collected Papers on Wave Mechanics* (REF. 251), pp. 45–61; "Sur les rapports qui existent entre la mécanique de Heisenberg-Born-Jordan et la mienne," *Mémoires sur la Mécanique Ondulatoire* (REF. 251), pp. 71–99.

figuration space of the q a complete orthonormal set, for any wave function $\psi = \psi(q)$ and $\psi' = [F, \psi]$

$$\psi = \sum_k a_k u_k \tag{5.85}$$

and

$$\psi' = \sum_j a'_j u_j \tag{5.86}$$

where

$$a'_j = \int u_j^* [F,\psi] \, dq$$

Hence,

$$a'_j = \sum_k F_{jk} a_k \tag{5.87}$$

where the elements of the matrix (F_{jk}) are given by the equation

$$F_{jk} = \int u_j^* [F,u_k] \, dq \tag{5.88}$$

Equation (5.88) associates—relative to a given basis u_k—with every function $F(p,q)$ a matrix (F_{jk}) and this association, as Schrödinger showed, satisfies the condition that the matrix associated with the sum (product) of two functions is the sum (product) of the matrices associated with the individual terms (factors). Any wave-mechanical equation could therefore consistently be translated into a matrix equation, the operation of F on a wave function ψ corresponding to the application of the matrix (F_{jk}) on the column vector (a_k) whose components are the Fourier coefficients of ψ. In the second part of the paper Schrödinger specified the basic u_i by taking in their place the eigenfunctions ψ_i of the wave equation. The latter could be written as an operator equation

$$[H,\psi] = E\psi \tag{5.89}$$

where $[H, \cdot]$ is the Hamiltonian operator, obtained from the classical prototype $H(p,q)$ just as $[F, \cdot]$ from $F(p,q)$. Its eigenvalues are E_k, satisfying

$$[H,\psi_k] = E_k \psi_k \tag{5.90}$$

Referring now to the matrix-mechanical laws of motion (5.28), Schrödinger demonstrated that with respect to the specified basis the matrix $\mathbf{H}$ is diagonal, since $H_{jk} = \int \psi_j [H,\psi_k] \, dq = E_k \delta_{jk}$, or, in other words, solving (5.89) is equivalent to diagonalizing the matrix $\mathbf{H}$. With the help of the auxiliary theorems

$$\left[\frac{\partial F}{\partial q}, \cdot\right] = \frac{2\pi i}{h}[pF - Fp, \cdot] \tag{5.91}$$

$$\left[\frac{\partial F}{\partial p}, \cdot\right] = \frac{2\pi i}{h}[Fq - qF, \cdot] \tag{5.92}$$

he showed that the matrices constructed in accordance with Eq. (5.88) do in fact satisfy the laws of motion (5.28).

To complete the proof of formal equivalence, Schrödinger had to show not only that the matrices can be constructed from the eigenfunctions but also, conversely, that the functions can be obtained from the numerically given matrices. "Thus the functions do not form, as it were, an arbitrary and particular 'fleshly clothing' ('fleischliche Umkleidung') of a bare matrix skeleton, provided to pander to the need of visualization."[299] To show that such an epistemological superiority of matrix mechanics has no justification, Schrödinger supposed that in the equations

$$q_{jk} = \int u_j u_k \, dq \tag{5.93}$$

the left-hand sides are numerically given and the fu ions u_j are to be found. For this purpose he pointed out that the integr

$$\int P(q) u_j(q) u_k(q) \, dq$$

where $P(q)$ signifies *any* power product of the q's, ca e computed by matrix multiplication. But the totality of these integrals, for fixed j and k, forms what is known as the "moments" of the functions $u_j u_k$. Since it is well known that under very general assumptions the totality of such moments determines uniquely the functions themselves, all the products $u_j u_k$, and hence also all the squares u_j^2, and finally the u_j themselves are ascertainable.

"Today," declared Schrödinger, "there are many physicists who, like Kirchhoff and Mach, regard the task of physical theory as being merely a mathematical description, as economical as possible, of the empirical relations between observable quantities, that is, as a description which represents the relations as far as possible without the intervention of unobservable elements. On this view, mathematical equivalence and physical equivalence have almost the same meaning."[300] In other words, if a physical theory were identical with the set of its mathematical relations, just as Hertz once identified Maxwell's theory with the system of Maxwell's

[299] *Ibid.*, p. 751.

[300] *Ibid.*, p. 751.

equations, wave mechanics and matrix mechanics, according to Schrödinger, would be just one theory.[301]

Schrödinger's article of the formal equivalence of the two theories was received by the editor of the *Annalen* on March 18, 1926. On March 31 Carl Eckart of the California Institute of Technology, whose interest[302] in quantum mechanics was aroused by Born's lecture in Pasadena in the winter of 1925, submitted to the National Academy of Sciences a paper[303] whose declared purpose was "to show, in a purely formal way, that the Schrödinger equation must be the basis for the Born, Jordan and Heisenberg matrix calculus. . . . Lack of space prevents any general discussions," wrote Eckart, "hence, only a very formal solution of the special problem of the single linear oscillator is attempted. All results are capable of immediate generalization, however. The general theory, and an attempt to interpret some of the postulates, will be published in another place."

In fact, on June 7, 1926, Eckart completed his paper "Operator calculus and the solutions of the equations of quantum dynamics,"[304] in which he generalized his former result by constructing an operator calculus for classical dynamics, extending it to quantum dynamics and showing that "the remarkable results of Schrödinger have been included in the same calculus as those of Born and Jordan."[305] This conclusion, Eckart declared, "would seem to be the strongest support which either of the two widely dissimilar theories have thus far received."[306] Eckart's approach, basically a special case of Schrödinger's method, consisted in using the normalized eigenfunctions of the linear harmonic oscillator[307] for the con-

[301] For an interesting discussion on the notions of equivalence and identity of physical theories with special consideration to our subject see N. R. Hanson, "Are wave mechanics and matrix mechanics equivalent theories?" *Czechoslovakian Journal of Physics 11*, 693–708 (1961); *Current Issues in the Philosophy of Science*, edited by H. Feigl and G. Maxwell (Holt, Rinehart and Winston, New York, 1961), pp. 401–428 (with comments by E. L. Hill). Cf. also Hanson, *The Concept of the Positron* (REF. 76), chap. 8 (Equivalence), pp. 113–134, where it is argued that in spite of the formal intertranslatability these theories, *qua physical* theories, cannot be regarded as equivalent prior to Born's probabilistic interpretation of the Schrödinger function. "Born alone deserves the credit for establishing the equivalence as *physical theories* of Wave Mechanics and Matrix Mechanics" (p. 129).

[302] "Born came to Pasadena in the winter of 1925; his lucid lecture aroused my interest. The result was that I spent the spring of 1926 working rather intensively with the operator formulation and was completely familiar with what is now known as the Schrödinger operator (the energy operator) before Schrödinger's papers appeared in Pasadena." *Archive for the History of Quantum Physics*, Interview with C. Eckart on May 31, 1962.

[303] C. Eckart, "The solution of the problem of the single oscillator by a combination of Schrödinger's wave mechanics and Lanczos' field theory," *Proceedings of the National Academy of Sciences 12*, 473–476 (1926).

[304] *Physical Review 28*, 711–726 (1926).

[305] *Ibid.*, p. 726.

[306] *Ibid.*, p. 711.

[307] In his study of the eigenfunctions of the oscillator Eckart was assisted by P. Epstein. In fact, Eckart was explaining his difficulties to Fritz Zwicky when "Paul Epstein came up behind . . . and rather characteristically remarked 'What have you got there?' " Having been informed of the problem, Epstein went to his office and returned "with the whole theory of the Hermite polynomials." *Archive for the History of Quantum Physics*, Interview with C. Eckart on May 31, 1962. See REF. 264.

struction of matrices. His calculation yielded

$$x_{n+1,n} = \sqrt{\frac{(n+1)h}{4\pi m\omega}} \qquad x_{n-1,n} = \sqrt{\frac{nh}{4\pi m\omega}} \tag{5.94}$$

or, as Eckart said, "the matrices of the Born-Jordan theory are really obtained."

As explicitly acknowledged, Eckart was led to the idea of constructing the matrices by integrations of orthonormal eigenfunctions, subjected to differential operators, by an interesting paper by Lanczos.[308] For in December, 1925, Korel Lanczos had shown that the Heisenberg-Born-Jordan theory can be formulated not only in the language of matrices but also, equivalently, in that of integral equations. In fact, Lanczos' reformulation of Heisenberg's mechanics in terms of integral equations preceded Schrödinger's first communication by about four weeks and was therefore, strictly speaking, the earliest continuum-theoretic formalism of quantum mechanics. Lanczos' insistence on the conceptual advantages of such a continuous integral representation, as being more closely related to field conceptions, was little heeded at the time; the use of integral equations with which physicists were—and still are—much less familiar than with differential equations, the absence of any specific example or new result, and certainly also the fact that the publication almost coincided with that of Schrödinger's first communication, explain the relatively cool reception Lanczos' paper was given. As a footnote in Schrödinger's equivalence article indicates,[309] Schrödinger had carefully studied the Lanczos paper and, most probably, he was influenced by it as Eckart was. Schrödinger, however, emphasized that the apparent similarities between his own and Lanczos' paper are, so to speak, only on the surface; in particular, Lanczos' symmetrical kernel $K(s,\sigma)$, whose orthogonal eigenfunctions served Lanczos to construct his matrices, should not be identified with the Green's functions of the wave equation, for the eigenvalues of the former are the reciprocals of those of the latter.

For the sake of historical accuracy it should be noted that according to a statement made by Gregor Wentzel,[310] the mathematical equivalence between matrix and wave mechanics was independently established also by Wolfgang Pauli.

With these remarks we conclude our presentation of Schrödinger's remarkable papers, which were, as Born[311] once said, "of a grandeur un-

[308] K. Lanczos, "Über eine feldmäßige Darstellung der neuen Quantenmechanik," *Zeitschrift für Physik 35*, 812–830 (1926).

[309] REF. 298, p. 754, footnote 1.

[310] G. Wentzel, "Eine Verallgemeinerung der Quantenbedingungen für die Zwecke der Wellenmechanik," *Zeitschrift für Physik 38*, 518–529 (1926), received on June 18, 1926. Cf. footnote 2, p. 522.

[311] M. Born, "Erwin Schrödinger," *Physikalische Blätter 17*, 85–86 (1961). On p. 85 Born said: "Was gibt es Großartigeres in der theoretischen Physik als seine [Schrödinger's] ersten sechs Arbeiten zur Wellenmechanik."

surpassed in theoretical physics." Reviewing Schrödinger's derivation of his wave equation, which according to Planck "plays the same part in modern physics as do the equations established by Newton, Lagrange, and Hamilton in classical mechanics,"[312] we see that Schrödinger's reasoning, as represented in his second communication,[313] was indeed intimately connected with Debye's deduction of ray optics from wave optics. For a comparison of Eq. (5.72), page 261, with the eikonal equation $|\nabla W|^2 = n^2$ showed that $n^2 = 2m(E - U)$, in agreement with Hamilton's "optical-mechanical analogy" (5.64), page 237. As soon as this value of n^2 was introduced into Debye's wave equation $\Delta\psi + n^2k_0^2\psi = 0$, as explained on page 254, and k_0 was identified with $2\pi/h$, Schrödinger's wave equation (5.77), page 263, was obtained.

It seemed strange that Schrödinger's solution of the hydrogen eigenvalue equation resulted precisely in Bohr's energy levels; for it was hard to understand how two methods so dissimilar as Bohr's and Schrödinger's could lead to identical results.

The puzzle was solved by a new wave-mechanical approximation method which was proposed, shortly after the publication of Schrödinger's papers, by Wentzel[314] and, independently, by L. Brillouin[315] and improved by Kramers,[316] the so-called "WKB method."[317] Our discussion of this method will be confined to the one-dimensional case, for which it is most readily applicable, and will deal with only those features which are directly relevant to our present subject.

Writing Schrödinger's equation (5.77) as

$$\psi'' + k_0^2p^2\psi = 0 \tag{5.95}$$

where, as before, $k_0 = 2\pi/h$ and $p^2 = 2m(E - U)$, Wentzel and Brillouin wrote, just as Debye did,

$$\psi = \exp(ik_0W) \tag{5.96}$$

[312] M. Planck, *The Universe in the Light of Modern Physics* (Norton, New York, 1931), p. 30.

[313] REF. 251, pp. 497–510.

[314] REF. 310.

[315] L. Brillouin, "La mécanique ondulatoire de Schrödinger; une méthode générale de résolution par approximations successives," *Comptes Rendus 183*, 24–26 (1926), communicated on July 5, 1926.

[316] H. A. Kramers, "Wellenmechanik und halbzahlige Quantisierung," *Zeitschrift für Physik 39*, 828–840 (1926), received on Sept. 9, 1926.

[317] For chronological as well as logical reasons the term "WBK method" would be more appropriate. The French call it "la méthode BKW"; cf., for example, A. Messiah, *Mécanique Quantique* (Dunod, Paris, 1959), vol. 1, p. 194. English texts sometimes speak of the "JWBK method" or "WKBJ method," as, for example, G. L. Trigg, *Quantum Mechanics* (Van Nostrand, Princeton, N.J., 1964), p. 165, in view of the fact that Harold Jeffreys anticipated part of the mathematical technique of the method in his paper "On certain approximate solutions of linear differential equations of the second order," *Proceedings of the London Mathematical Society 23*, 428–436 (1925).

but specified in addition

$$W = \int_{x_0}^{0} y\,dx \tag{5.97}$$

where y is a function of x and x_0 is a fixed lower limit of integration. From Eqs. (5.95), (5.96), and (5.97) they obtained the Riccati differential equation[318]

$$\frac{h}{2\pi i} y' = p^2 - y^2 \tag{5.98}$$

and by expanding y in powers of h,

$$y = y_0 + \frac{h}{2\pi i} y_1 + \left(\frac{h}{2\pi i}\right)^2 y_2 + \cdots \tag{5.99}$$

and comparing the powers of h on both sides of Eq. (5.98),

$$y_0 = \pm p$$

$$y_1 = -\frac{y_0'}{2y_0}$$

and so on.

The zeroth-order approximation $W = \int y_0\,dx = \int \pm p\,dx$ corresponded, as we see, to the solution of classical mechanics while successive higher-order approximations were expected[319] to approach the rigorous solution of wave mechanics. Treating $(2\pi i/h)y = \psi'/\psi$ as an analytic function over the complex plane and integrating it along a closed contour around all of its poles (with residues $2\pi i$), Wentzel concluded with the help of the theorem of residues that

$$\oint \frac{\psi'}{\psi}\,dx = \frac{2\pi i}{h} \oint y\,dx = 2\pi i n$$

[318] The earliest detailed discussion of this equation was given by the Venetian mathematician Jacobo Riccati in his "Animadversiones in aequationes differentiales secundi gradus," *Acta Eruditorum 1723*, pp. 502–510; *Actorum Eruditorum Supplementa 8*, 66–73 (1724). When, in 1763, D'Alembert solved the boundary-value problem of the inhomogeneous vibrating string and showed that the differential equation $z'' = \lambda\varphi(x)z$ with given $\varphi(x)$ and subject to the condition $z(a) = z(b) = 0$ has solutions for appropriate values of λ, he did so by introducing the new variable $y = z'/z$ and thus obtained what he called—probably for the first time—"Riccati's equation." Cf. *Histoire de l'Académie de Berlin 19*, 244 et seq. (1770).

[319] It was subsequently recognized that the expansion is only semiconvergent and does not lead to exact solutions. Cf. A. Zwaan, "Intensitäten im Ca-Funkenspektrum" (Dissertation, Utrecht), *Archives Néerlandaises 12*, 1–76 (1929), especially pp. 33–54; G. D. Birkhoff, "Quantum mechanics and asymptotic series," *Bulletin of the American Mathematical Society 39*, 681–700 (1933); R. E. Langer, "The asymptotic solutions of ordinary linear differential equations of the second order, with special reference to the Stokes phenomenon," *ibid. 40*, 545–582 (1934).

where n is an integer. He thus obtained the Sommerfeld quantum condition

$$\oint p\,dx = nh$$

However, as Kramers and his collaborators[320] showed, a more exact treatment leads to the result

$$\oint p\,dx = (n + \tfrac{1}{2})h \tag{5.100}$$

These results establish the connection between Bohr's theory of the hydrogen atom of 1913 or its elaborations by Sommerfeld and Schrödinger's wave mechanics. The question why these quantum conditions in spite of their being derived from an approximation method give nevertheless the correct energy values for several important cases, such as the hydrogen atom or the linear oscillator, is a purely mathematical problem whose discussion would lead us too far astray.

In the paper[321] in which Wentzel proposed the new approximation method he had already applied it to the Stark effect of hydrogen. As far as the linear effect was concerned, his result, like that of Schrödinger, was identical with Epstein's formula[322] obtained on the older quantum theory. For the quadratic effect, however, the term containing the square of the field strength F^2 turned out to be[323]

$$\frac{h^6F^2}{2^{10}\pi^6m^3e^6}\,n^4[17n^2 - 3(n_1 - n_2)^2 - 9(n_3 - 1)^2 + 19] \tag{5.101}$$

Wentzel's result (5.101) was also obtained at the same time and independently by I. Waller.[324] It differed from Epstein's expression by the presence of $+19$ and by the appearance of $9(n_3 - 1)^2$ insteady of Epstein's $9n_3^2$.

A deviation from Epstein's result of 1916 concerning the quadratic effect was detected as early as 1921 by Sommerfeld[325] in observations made by Takamine and Kokubu[326] in 1919 at the Mount Wilson Observatory. Their plates—the only material on record on the quadratic effect until

[320] K. F. Niessen, "Über die annähernden komplexen Lösungen der Schrödingerschen Differentialgleichung für den harmonischen Oszillator," *Annalen der Physik 85*, 497–514 (1928); H. A. Kramers and G. P. Ittmann, "Zur Quantelung des symmetrischen Kreisels II," *Zeitschrift für Physik 58*, 217–231 (1929); E. Persico, "Dimostrazione elementare del metodo di Wentzel e Brillouin," *Il Nuovo Cimento 15*, 133–138 (1938).

[321] REF. 310.

[322] See. p. 108.

[323] The same notation is used as on p. 108.

[324] I. Waller, "Der Starkeffekt zweiter Ordnung bei Wasserstoff und die Rydbergkorrektion der Spektra von He und Li$^+$," *Zeitschrift für Physik 38*, 635–646 (1926).

[325] A. Sommerfeld, "Über den Starkeffekt zweiter Ordnung," *Annalen der Physik 65*, 36–40 (1921).

[326] T. Takamine and N. Kokubu, "The effect of an electric field on the spectrum lines of hydrogen," *Memoirs of the College of Science (Kyoto Imperial University) 3*, 271–285 (1919); "Further studies on the Stark effect in helium and hydrogen," *Tokyo Sugaku Buturigakkawi Kizi (Proceedings of the Tokyo Mathematico-Physical Society) 9*, 394–404 (1919).

1925—exhibited a one-sided shift of the whole splitting pattern toward the red, but, as Sommerfeld noticed, not in agreement with Epstein's conclusions. The shift of the middle component of H_γ, corresponding to the transition ($n_1 = 1$, $n_2 = 1$, $n_3 = 3$) → (0,0,2) and (2,2,1) → (0,0,2), was carefully examined in 1925 by Kiuti,[327] who found for a field of 140,000 volts/cm a shift of $\Delta\lambda = 0.6$ Å in contrast to Epstein's theoretical $\Delta\lambda = 0.5$ Å. As Wentzel[328] pointed out, Kiuti's observed value agrees perfectly well with the wave-mechanical result. Later investigations by Rausch von Traubenberg and Gebauer,[329] who applied fields up to 400,000 volts/cm, confirmed this agreement. In fact, von Traubenberg and Gebauer declared at the conclusion of their experiments that "Schrödinger's theory is fully confirmed by our measurements of the red shift as exhibited also by the other components of H_γ."[330]

Just as Mulliken's[331] observation of the shift in the band spectrum of boron monoxide was the earliest evidence for the superiority of Heisenberg's matrix mechanics over the older quantum theory, so was Kiuti's observation of the shift in the line pattern of the hydrogen Stark effect the earliest proof of the superiority of Schrödinger's wave mechanics over the older theory. Both formalisms found support in evidence gathered prior to their establishment.

[327] M. Kiuti, "Further studies of the Stark effect in hydrogen," *Japanese Journal of Physics 4,* 13–18 (1925).

[328] REF. 310, p. 528.

[329] H. Rausch von Traubenberg and R. Gebauer, "Über den Starkeffekt zweiter Ordnung beim Wasserstoff," *Die Naturwissenschaften 16,* 655–656 (1928); "Über den Starkeffekt II. Ordnung bei der Balmerserie des Wasserstoffs," *Zeitschrift für Physik 54,* 307–320 (1929), *56,* 254–258 (1929).

[330] H. Rausch von Traubenberg and R. Gebauer, "Bemerkung zu unserer Arbeit 'Über den Starkeffekt II. Ordnung bei der Balmerserie des Wasserstoffs,'" *Die Naturwissenschaften 17,* 443 (1929).

[331] REF. 129.

6 *Statistical Transformation Theory*

6.1 *The Introduction of Probabilistic Interpretations*

Within a short time after the publication of Schrödinger's papers wave mechanics was successfully applied to a great number of energy-eigenvalue problems. It also soon became clear that the theory could be extended to cope with types of problems not initially envisaged by its originator. Thus, for example, Fock[1] showed as early as June, 1926, that Schrödinger's equation can be generalized to apply also to problems in which the Lagrangian function contains linear velocity-dependent terms. Schrödinger's theory of radiation, based on his time-dependent perturbation theory, made it possible to derive Bohr's frequency condition, which in the older quantum theory had the status of a separate postulate, to calculate the intensities and polarizations of emitted radiations and their selection rules. It provided a logically consistent and mathematically convenient method to solve a great variety of problems, and there could be no doubt that it was a major advance over the older quantum theory. In addition, with the exception of a few cases such as the treatment of the angular momentum it proved to be also more practicable than matrix mechanics.

Schrödinger's physical explanation of his formalism and, in particular, his interpretation of the wave function did not fare so well. In order to account for the fact that a mechanical system can emit electromagnetic waves of a frequency equal to the term difference and to deduce their intensity and polarization, Schrödinger ascribed to ψ an electromagnetic meaning, defining it in the early sections of his paper[2] as a continuous

[1] V. Fock, "Zur Schrödingerschen Wellenmechanik," *Zeitschrift für Physik 38*, 242–250 (1926), received on June 11, 1926.

[2] REF. 251 OF CHAP. 5.

distribution of electricity in actual space. At the end of the fourth communication, where these ideas are further elaborated, he defined $\psi\psi^*$ as the "weight function" of this charge distribution so that $\rho = e\psi\psi^*$ is the electron charge density where e is the charge of the electron. Having derived from his time-dependent wave equation in the well-known manner the continuity equation

$$\frac{\partial\rho}{dt} + \text{div } S = 0 \tag{6.1}$$

connecting the current density

$$S = \frac{eh}{4\pi im} (\psi^* \nabla\psi - \psi \nabla\psi^*) \tag{6.2}$$

with the scalar charge density ρ, he regarded Eq. (6.1) as a convincing confirmation of his electrodynamic interpretation. To explain the empirical fact that the charge of an electron is ordinarily found to be concentrated in a very small region of space, Schrödinger combined his electrodynamic interpretation of the wave function with the idea that the particles of corpuscular physics are essentially only wave groups composed of numerous, strictly speaking, infinitely many, wave functions.

In a paper[3] published in July, 1926, on the relation between microphysics and macrophysics he illustrated his ideas by showing that the phenomenological behavior of the linear harmonic oscillator can be explained in terms of the microphysical conception of wave packets. From the normalized eigenfunctions $(2^n n!)^{-1/2}\psi_n$, where

$$\psi_n = \exp(-\tfrac{1}{2}x^2)H_n(x) \exp(2\pi i\nu_n t) \qquad \text{and} \qquad \nu_n = (n + \tfrac{1}{2})\nu_0$$

he constructed the wave packet $\psi = \sum^{\infty} (A/2)^n\psi_n/n!$, where A is a constant large compared with unity,[4] and showed that the real part of ψ is given by the expression

$$\exp\left[\frac{A^2}{4} - \tfrac{1}{2}(x - A \cos 2\pi\nu_0 t)^2\right]$$
$$\times \cos\left[\pi\nu_0 t + (A \sin 2\pi\nu_0 t)\left(x - \frac{A}{2}\cos 2\pi\nu_0 t\right)\right] \tag{6.3}$$

[3] E. Schrödinger, "Der stetige Übergang von der Mikro- zur Makromechanik," *Die Naturwissenschaften* *14*, 664–666 (1926); *Abhandlungen zur Wellenmechanik* (REF. 251 OF CHAP. 5), pp. 56–61; "On the continuous transition from micro- to macro-mechanics," *Collected Papers on Wave Mechanics* (REF. 251 OF CHAP. 5), pp. 41–44; "Le passage continu de la micro-mécanique à la mécanique macroscopique," *Mémoires sur la Mécanique Ondulatoire* (REF. 251 OF CHAP. 5), pp. 65–70.

[4] Since $z^n/n!$ as a function of n has for large z a single sharp maximum at $n = z$, the important components are those for which $n \approx A^2/2$.

Pointing out that the first factor in (6.3) represents a narrow hump of the shape of a Gaussian error curve, which, at a given moment t, is located in the neighborhood of $x = A \cos 2\pi\nu_0 t$, and that the wave group as a whole does not spread out in space in the course of time, he concluded that the particle picture of the oscillator can be consistently interpreted as resulting from the existence of such a wave packet. "There seems to be no doubt," he continued, "that we can assume that similar wave packets can be constructed which orbit along higher-quantum-number Kepler ellipses and are the wave-mechanical representation of the hydrogen atom."

It was soon recognized, however, that Schrödinger's assumption was erroneous and that wave packets, in general, do spread out in space. In fact, it was shown by Heisenberg[5] that the oscillator is in this respect an exception, the reason being that its successive energy levels happen to differ by equal amounts. Another difficulty was the fact that for poly-electronic systems—strictly speaking, also for the one-electron system if the motion of the nucleus is taken into consideration—ψ is a function in a hypothetical multidimensional space and as such can hardly be regarded as an efficient cause of radiation phenomena as Schrödinger proposed. Other objections to Schrödinger's interpretation were provided by the previously discussed electron-diffraction experiments by crystals and by collision experiments, for it was difficult to understand how in such processes of wave dispersion the stability of a particle could be preserved if it really was but a group of waves.

As a matter of fact, it was precisely in connection with the quantum-mechanical study of such scattering processes that a new interpretation of the wave function was proposed. Born, who in 1954 was awarded the Nobel Prize "for his fundamental work in quantum mechanics and especially for his statistical interpretation of the wave function," as the official decision of the Royal Swedish Academy of November 3, 1954, read, explained the motives of his opposition to Schrödinger's interpretation as follows: "On this point I could not follow him. This was connected with the fact that my Institute and that of James Franck were housed in the same building of the Göttingen University. Every experiment by Franck and his assistants on electron collisions (of the first and second kind) appeared to me as a new proof of the corpuscular nature of the electron."[6]

Searching for a quantum-mechanical explanation of the process of

[5] W. Heisenberg, "Über den anschaulichen Inhalt der quantentheoretischen Kinematik und Mechanik," *Zeitschrift für Physik 43*, 172–198 (1927), received on Mar. 23, 1927; especially pp. 184–185.

[6] M. Born, "Bemerkungen zur statistischen Deutung der Quantenmechanik," in *Werner Heisenberg und die Physik unserer Zeit* (Vieweg, Braunschweig, 1961), pp. 103–108, quotation on p. 103. Cf. also "The statistical interpretation of de Broglie's waves was suggested to me by my knowledge of experiments on atomic collision which I had learned from my experimental colleague James Franck," *Experiment and Theory in Physics* (Cambridge University Press, 1943), p. 23.

collision between a free particle, such as an α particle or an electron, and an atom, Born adopted the formalism of wave mechanics, stating that "among the various forms of the theory only Schrödinger's formalism proved itself appropriate for this purpose; for this reason I am inclined to regard it as the most profound formulation of the quantum laws."[7]

In the paper from which the preceding quotation has been taken Born gave a preliminary report of his quantum-mechanical treatment of collision processes. A more detailed discussion is found in two subsequent articles[8] in which he developed what became known as the "Born approximation." Born's method was essentially an application of the perturbation theory to the scattering of plane waves, the initial and final wave functions being both approximately plane waves far from the scattering center. To the system of an electron of energy $E = h^2/2m\lambda^2$, coming from the $+z$ direction and approaching an atom whose unperturbed eigenfunctions are $\psi_n{}^0(q)$, he ascribed the eigenfunction $\psi_{nE}{}^0(q,z) = \psi_n{}^0(q) \sin(2\pi z/\lambda)$. Taking $V(x,y,z,q)$ as the potential energy of interaction between the electron and the atom and applying the theory of perturbations, he obtained for the scattered wave at great distance from the scattering center the expression

$$\psi_{nE}{}^{(1)}(x,y,z,q) = \sum_m \iint d\omega\, \psi_{nm}{}^{(E)}(\alpha,\beta,\gamma) \sin k_{nm}{}^{(E)}(\alpha x + \beta y + \gamma z + \delta)\psi_m{}^0(q) \qquad (6.4)$$

where $d\omega$ is an element of the solid angle in the direction of the unit vector whose components are α, β, and γ, and where $\psi_{nm}{}^{(E)}(\alpha,\beta,\gamma)$ is a wave function which determines what was subsequently called the differential cross section for the direction (α,β,γ). If this formula, said Born, allows for a corpuscular interpretation, there is only one possibility: $\psi_{nm}{}^{(E)}$, or rather[9] $|\psi_{nm}{}^{(E)}|^2$, measures the probability that the electron which approached the scattering center in the direction of the z axis is found scattered in the direction defined by α, β, γ.[10] Hence, Born continued, wave mechanics does not answer the question: What, precisely, is the state after the collision? but rather the question: What is the possibility of a definite state

[7] M. Born, "Zur Quantenmechanik der Stoßvorgänge," *Zeitschrift für Physik 37*, 863–867 (1926), received on June 25, 1926; *Ausgewählte Abhandlungen* (REF. 116 OF CHAP. 3), vol. 2, pp. 228–232; *Dokumente der Naturwissenschaft* (Ernst Battenberg Verlag, Stuttgart, 1962), vol. 1, pp. 48–52.

[8] M. Born, "Quantenmechanik der Stoßvorgänge," *Zeitschrift für Physik 38*, 803–827 (1926), received on July 21, 1926; *Ausgewählte Abhandlungen* (REF. 116 OF CHAP. 3), vol. 2, pp. 233–257; *Dokumente der Naturwissenschaft* (REF. 7), vol. 1, pp. 53–77. "Zur Wellenmechanik der Stoßvorgänge," *Göttinger Nachrichten 1926*, pp. 146–160; *Ausgewählte Abhandlungen*, vol. 2, pp. 284–298; *Dokumente der Naturwissenschaft*, vol. 1, pp. 78–92.

[9] The statement that $|\psi|^2$ and not ψ itself determines the probability was added by Born when reading the galleys of the preliminary report, REF. 7.

[10] "Will man nun dieses Resultat korpuskular undeuten, so ist nur eine Interpretation möglich." REF. 7, p. 865.

after the collision? In his more detailed paper[11] on collisions he summarized this situation by the often quoted statement: "The motion of particles conforms to the laws of probability, but the probability itself is propagated in accordance with the law of causality."[12]

Born's probabilistic interpretation of the wave function had a threefold root. First of all, as explained, impressed by the corpuscular aspects of collision experiments, Born rejected the undulatory interpretation and tried, instead, to associate the wave function with the presence of particles. In his choice how to carry out this association he was influenced, as he repeatedly admitted,[13] primarily by Einstein's conception of the relation between the field of electromagnetic waves and light quanta. For Einstein, Born asserted, the electromagnetic wave field was a kind of "phantom field" ("Gespensterfeld") whose waves serve to guide the corpuscular light quanta on their path in the sense that the squared wave amplitudes (intensities) determine the probability of the presence of light quanta or, in a statistically equivalent sense, their density. Since the wave function for an ordinary plane light wave of frequency $\nu = E/h$ and wavelength $\lambda = h/p$,

$$u(x,t) = \exp\left[2\pi i\nu\left(t - \frac{x}{c}\right)\right] = \exp\left[\frac{2\pi i}{h}(Et - px)\right]$$

is identical with the de Broglie wave function of a particle of energy E and momentum p, i.e., the eigenfunction of the Schrödinger wave equation

$$\frac{d^2\psi}{dx^2} + \frac{8\pi^2 m}{h^2} E = 0$$

where $E = p^2/2m$, Born found it natural to carry over Einstein's conception of the "phantom field" to particles other than light quanta. Just as the intensity of light waves was a measure of the density of light quanta, Born argued, "it was almost self-understood to regard $|\psi|^2$ as the probability density of particles."[14]

It soon became clear, however, that the notion of probability in Born's interpretation differed from the classical notion as used, for example, in statistical mechanics or in the kinetic theory of gases. For, if a wave field ψ_1 with probability density $P_1 = |\psi_1|^2$ is superposed with a wave field ψ_2 with probability density $P_2 = |\psi_2|^2$, the probability density of the superposed field $\psi = \psi_1 + \psi_2$ is not, as the classical notion of probability would

[11] REF. 8, p. 804.

[12] "Die Bewegung der Partikel folgt Wahrscheinlichkeitsgesetzen, die Wahrscheinlichkeit selbst aber breitet sich im Einklang mit dem Kausalgesetz aus." *Ibid.*, p. 804.

[13] REF. 8. Cf. also M. Born, "Albert Einstein und das Lichtquantum," *Die Naturwissenschaften 11*, 425–431 (1955), especially p. 429. Also *Archive for the History of Quantum Physics*, Interview with M. Born on Oct. 18, 1962.

[14] REF. 6, p. 103.

imply, $P_1 + P_2$, but rather $|\psi_1 + \psi_2|^2 = P_1 + P_2 + \psi_1\psi_2^* + \psi_2^*\psi_1$, where the last two "interference terms" are generally different from zero.

For Einstein the notion of probability, even as he applied it to reconcile his light-quanta hypothesis with Maxwell's theory of electromagnetic waves, was the traditional conception of classical physics, a mathematical objectivization of the human deficiency of complete or exact knowledge but ultimately a creation of the human mind or, as Spinoza expressed it, "a sola imaginatione . . . quod res . . . ut contingentes contemplemur."[15] But Spinoza's statement: "De natura rationis non est, res, ut contingentes, sed, ut necessarias, contemplari,"[16] which dominated scientific thought for centuries with only a few exceptions, such as those referred to in Sec. 4.2, lost its validity for Born and his school. For Born probability, as far as it was related to the wave function, was not merely a mathematical fiction but something endowed with physical reality, for it evolved in time and propagated in space in accordance with Schrödinger's equation. It differed, however, from ordinary physical agents in one fundamental aspect: it did not transmit energy or momentum. Since in classical physics, whether Newtonian mechanics or Maxwellian electrodynamics, only what transfers energy or momentum (or both) is regarded as physically "real," the ontological status of ψ had to be considered as something intermediate. It had—and that brings us to the third root of Born's probabilistic interpretation—precisely that "intermediate kind of reality" which, as Heisenberg[17] had emphasized, transpired in the work of Bohr, Kramers, and Slater in 1924. Now it will be understood that Born's interpretation was indeed influenced by the Bohr-Kramers-Slater conception of the virtual radiation field. Recalling also their treatment of induced emission, etc., we can understand why Heisenberg declared: "Born established in the summer of 1926 his theory of collision processes and interpreted correctly the wave in multidimensional configuration space as a probability wave by developing and elaborating an idea previously expressed by Bohr, Kramers, and Slater."[18]

Laws of nature, as Born and Heisenberg contended from now on, determined not the occurrence of an event, but the probability of the occurrence. For Heisenberg,[19] as he later explained, such probability waves

[15] "It is only through the imagination that we look upon things as contingent with reference to both the past and the future." Benedicti de Spinoza, *Opera quae supersunt omnia* (In Bibliopoli academico, Jena, 1803), (*Ethices* pars 2), vol. 2, p. 217.

[16] "It is the nature of reason to consider things not as contingent but as necessary." (*Ethices* Propositio 44, liber 2), *ibid.*

[17] *Archive for the History of Quantum Physics*, Interview with W. Heisenberg on Feb. 19, 1963.

[18] W. Heisenberg, "Erinnerungen an die Zeit der Entwicklung der Quantenmechanik," pp. 40–47 in *Theoretical Physics in the Twentieth Century* (REF. 218 OF CHAP. 3).

[19] W. Heisenberg, "Planck's discovery and the philosophical problems of atomic physics," pp. 3–20 in *On Modern Physics* (Clarkson N. Potter, New York, 1961; Orion Press, London, 1961).

are "a quantitative formulation of the concept of δύναμις, possibility, or in the later Latin version, *potentia*, in Aristotle's philosophy. The concept that events are not determined in a peremptory manner, but that the possibility or 'tendency' for an event to take place has a kind of reality—a certain intermediate layer of reality, halfway between the massive reality of matter and the intellectual reality of the idea or the image— this concept plays a decisive role in Aristotle's philosophy. In modern quantum theory this concept takes on a new form; it is formulated quantitatively as probability and subjected to mathematically expressible laws of nature."[20]

Having interpreted ψ as a probability wave in the sense just explained but realizing that ψ can be expanded in terms of a complete orthonormal set of eigenfunctions ψ_n of the Schrödinger equation $[H - W, \psi] = 0$,

$$\psi = \sum_n c_n \psi_n \tag{6.5}$$

where in accordance with the completeness relation

$$\int |\psi(q)|^2 \, dq = \sum_n |c_n|^2 \tag{6.6}$$

Born had to ask himself what meaning to ascribe to the c_n. The fact that for a single normalized eigenfunction $\psi(q)$, corresponding to a single particle, the right-hand side of Eq. (6.6) is unity suggested to Born[21] that the integral $\int |\psi(q)|^2 \, dq$ has to be regarded as the number of particles and $|c_n|^2$ as the statistical frequency of the occurrence of the state characterized by the index n. To justify this assumption Born calculated essentially what was later called the expectation value of the energy W for ψ and obtained

$$W = \sum_n |c_n|^2 W_n \tag{6.7}$$

where W_n is the energy eigenvalue of ψ_n.

Born's probabilistic interpretation scored immediate success in the field where it originated and where it could be applied most naturally, in the problems of atomic scattering. Applying Born's approximation method to the scattering of electrically charged particles by a charged scattering center, Wentzel[22] derived Rutherford's classic scattering formula[23] on the basis of wave mechanics. Faxén and Holtsmark[24] and later Bethe[25] and

[20] *Ibid.*, pp. 9–10.
[21] REF. 8, p. 805.
[22] G. Wentzel, "Zwei Bemerkungen über die Zerstreuung korpuskularer Strahlen als Beugungserscheinung," *Zeitschrift für Physik 40*, 590–593 (1926).
[23] REF. 40 OF CHAP. 2.
[24] H. Faxén and J. Holtsmark, "Beitrag zur Theorie des Durchganges langsamer Elektronen durch Gase," *Zeitschrift für Physik 45*, 307–324 (1927).
[25] H. Bethe, "Zur Theorie des Durchgangs schneller Korpuskularstrahlen durch Materie," *Annalen der Physik 5*, 375–40 (1930).

Mott[26] used Born's method for the study of the passage of slow and fast particles through matter in the course of which the Ramsauer-Townsend effect[27] and Dymond's observations[28] found a satisfactory wave-mechanical explanation.

Born himself,[29] and later in collaboration with Fock,[30] attempted to put the probabilistic interpretation on logically stronger foundations by clarifying its relation to classical physics. In August, 1926, at the Oxford meeting of the British Association for the Advancement of Science, Born[31] had declared: "We free forces of their classical duty of determining directly the motion of particles and allow them instead to determine the probability of states. Whereas before it was our purpose to make these two definitions of force equivalent, this problem has now no longer, strictly speaking, any sense. The only question is why the classical definition is so successful for a large class of phenomena. As always in such cases, the answer is: Because the classical theory is a limiting case of the new one. Actually, it is usually the 'adiabatic' case with which we have to do: i.e. the limiting case where the external force (or the reaction of the parts of the system on each other) acts very slowly. In this case, to a very high approximation $c_1^2 = 1$, $c_2^2 = 0$, $c_3^2 = 0, \ldots$, that is, there is no probability for a transition, and the system is in the initial state again after the cessation of the perturbation."

In the mathematical elaboration[32] of these ideas Born considered the general solution

$$\psi(x,t) = \sum_n c_n \psi_n(x) \exp\left(\frac{2\pi i}{h} W_n t\right) \tag{6.8}$$

of the time-dependent Schrödinger equation

$$\Delta\psi - \frac{8\pi^2 m}{h^2} U(x)\psi - \frac{4\pi i m}{h}\frac{\partial\psi}{\partial t} = 0 \tag{6.9}$$

where the $\psi_n(x)$ are assumed as normalized. In contrast to Schrödinger, who interpreted Eq. (6.8), in analogy to acoustics, as asserting the existence of simultaneous eigenvibrations in one and the same atom, Born, adhering

[26] N. F. Mott, "The solution of the wave equation for the scattering of particles by a Coulombian centre of field," *Proceedings of the Royal Society of London* (*A*), *118*, 542–549 (1928); "The quantum theory of electronic scattering by helium," *Proceedings of the Cambridge Philosophical Society 25*, 304–309 (1929); "The scattering of fast electrons by atomic nuclei," *ibid.*, 425–442. "Elastic collisions of electrons with helium," *Nature 123*, 717 (1929).

[27] REFS. 206 and 207 OF CHAP. 5.

[28] REF. 217 OF CHAP. 5.

[29] M. Born, "Das Adiabatenprinzip in der Quantenmechanik," *Zeitschrift für Physik 40*, 167–192 (1926), received on Oct. 16, 1926; *Ausgewählte Abhandlungen* (REF. 116 OF CHAP. 3), vol. 2, pp. 258–283; *Dokumente der Naturwissenschaft* (REF. 7), vol. 1, pp. 93–118.

[30] M. Born and V. Fock, "Beweis des Adiabatensatzes," *Zeitschrift für Physik 51*, 165–180 (1928); *Ausgewählte Abhandlungen* (REF. 116 OF CHAP. 3), vol. 2, pp. 338–353.

[31] M. Born, "Physical aspects of quantum mechanics," *Nature 119*, 354–357 (1926); the paper, read on Aug. 10, 1926, was translated into English by R. Oppenheimer.

[32] REF. 29.

to Bohr's conception that at a given time an atomic system can assume only one stationary state, interpreted, in accordance with the conclusions of his study of scattering,

$$|c_n|^2 = \left| \int \psi(x,t)\psi_n^*(x)\, dx \right|^2$$

as the probability that at the time t the atom is in state n. Studying the effect of an external perturbation on the system, Born compared the approach of classical mechanics with that of quantum mechanics as follows. The basic problem in classical mechanics, he declared, is this: at time $t = 0$ the configuration (positions and velocities) of the system is given; what is the configuration at time $t = T$ if the system has been acted upon by a given force during the interval Δt from $t = 0$ to $t = T$? The basic problem of quantum mechanics, on the other hand, is in his view: until $t = 0$ the probabilities $(|c_n|^2)$ of a configuration are given; what is the probability of a given configuration after the time T, if the system has been acted upon by a given external force during (and only during) the interval Δt? Bohr emphasized that quantum mechanics considers not the individual process, the "quantum transition" or "quantum jump," as causally determined, but rather the a priori probability of its occurrence; this probability is ascertainable by an integration of Schrödinger's differential equation. And Born added, obviously in the spirit of Bohr's philosophy: "Whatever occurs during the transition can hardly be described within the conceptual framework of Bohr's theory, nay, probably in no language which lends itself to visualizability."[33]

Applying Schrödinger's time-dependent perturbation theory, Born solved the basic problem of quantum mechanics as follows. At time $t = 0$ the system is given by

$$\psi(x,0) = \sum_n c_n\psi_n(x) \tag{6.10}$$

as shown by Eq. (6.8). Its evolution is given by Eq. (6.9) with $U(x)$ replaced by $U(x) + F(x,t)$, where $F(x,t)$ vanishes for $t < 0$ and $t > T$. Considering the special case $\psi(x,0) = \psi_n(x)$, Born showed that the integration of the differential equation yields for $t > T$

$$\psi_n(x,t) = \sum_m b_{nm}\psi_m(x) \exp\left(\frac{2\pi i}{h} W_m t\right) \tag{6.11}$$

where the coefficients b_{nm} are uniquely determined by the total behavior of $F(x,t)$ during Δt. Born now declared that $|b_{nm}|^2$ is the probability that the system, which for $t < 0$ was in state n [i.e., $\psi_n(x)$], is for $t > T$ in state m [i.e., $\psi_m(x)$]. In other words, $|b_{nm}|^2$ is the "transition probability"

[33] ". . . überhaupt nicht in einer Sprache, die unserem Anschauungsvermögen Bilder suggeriert." *Ibid.*, p. 172.

("Übergangswahrscheinlichkeit"). Considering, now, the general case where the initial state was given by $\psi(x,0) = \sum_n c_n\psi_n(x)$, and where, consequently, for $t > T$

$$\psi(x,t) = \sum_n c_n\psi_n(x,t) \tag{6.12}$$

with the $\psi_n(x,t)$ given by Eq. (6.11), Born calculated the general transition probability $|C_n|^2$—corresponding to the representation $\psi(x,t) = \sum_n C_n\psi_n(x)$ for $t > T$—and obtained by making use of the orthonormality of the $\psi_n(x)$

$$C_n = \int \psi(x,T)\psi_n^*(x)\,dx = \sum_m b_{mn}c_m \exp\left(\frac{2\pi i}{h} W_n T\right)$$

so that

$$|C_n|^2 = \left|\sum_m c_m b_{mn}\right|^2 \tag{6.13}$$

From Eq. (6.13) Born concluded that the quantum transitions between a state m and a state n cannot be regarded as independent events in the sense of the classical theory of probability; for, were this the case, the right-hand side of Eq. (6.13) should read, in accordance with the theorem of composite probabilities,

$$\sum_m |c_m|^2\,|b_{mn}|^2 \tag{6.14}$$

which generally differs, of course, entirely from $|\sum_m c_m b_{mn}|^2$.

We thus see that in the course of proving the wave-mechanical equivalent of the adiabatic theorem, namely, that for infinitely slow perturbations the transition probabilities are zero, Born advanced two theorems which were destined to play a fundamental role in the further development of quantum theory, its interpretation, and its theory of measurement: (1) the theorem of spectral decomposition according to which there corresponds a possible state of motion to every component ψ_n in the expansion or superposition of ψ; (2) the theorem of the interference of probabilities according to which the phases of the expansion coefficients, and not only their absolute values, are physically significant.

These results were independently obtained at the same time also by Dirac[34] and incorporated into the construction of the transformation theory of quantum mechanics, a generalization of the matrix and wave mechanics as developed so far.

Schrödinger's electrodynamic and Born's probabilistic conceptions

[34] P. A. M. Dirac, "On the theory of quantum mechanics," *Proceedings of the Royal Society of London (A)*, *112*, 661–677 (1926).

were not the only interpretations proposed at that time. At least two other suggestions have to be mentioned which are important not only for historical reasons: they still reverberate in current discussions on the foundations of quantum mechanics.

Realizing that for $\psi = \alpha e^{i\beta}$ with time-dependent α and β Schrödinger's time-dependent wave equation implies

$$\operatorname{div}\,(\alpha^2 \operatorname{grad}\varphi) + \frac{\partial \alpha^2}{\partial t} = 0 \tag{6.15}$$

where $\varphi = -\beta h/2\pi m$, Madelung called attention to the fact that this equation (6.15) has the form of the continuity equation in hydrodynamics if α^2 is taken as the density and φ as the velocity potential of a fluid in motion. Elaborating these ideas, Madelung[35] showed that every eigensolution of the wave equation, although a function of time, can be interpreted as a stationary flow pattern. As the hydrodynamical model represented also other essential features of Schrödinger's theory, Madelung suggested that "there is a chance to treat the quantum theory of atoms on this basis." He had to admit, however, that not all phenomena—and, in particular, those related to absorption processes—could be consistently interpreted as yet by such a hydrodynamical model.

In an alternative interpretation, proposed at about the same time, Louis de Broglie combined Born's probabilistic approach with certain ideas which Einstein had tentatively advanced in 1909, when he regarded light quanta as singularities[36] of a field of waves. In his first paper on this subject de Broglie[37] confined his treatment to light quanta. The solution of the wave equation

$$\Delta u = \frac{1}{c^2}\frac{\partial^2 u}{\partial t^2} \tag{6.16}$$

in classical optics, he argued, is given by a function of the form

$$u = a(x,y,z)\exp\{i\omega[t - \varphi(x,y,z)]\} \tag{6.17}$$

which satisfies the boundary conditions imposed by the presence of screens, apertures, or other obstacles encountered by the waves; in "the new optics of light quanta," on the other hand, he contended that its solution is given by a function of the form

$$u = f(x,y,z,t)\exp\{i\omega[t - \varphi(x,y,z)]\} \tag{6.18}$$

where φ is the same function as before, but $f(x,y,z,t)$ has "mobile singu-

[35] E. Madelung, "Quantentheorie in hydrodynamischer Form," *Zeitschrift für Physik 40*, 322–326 (1926), received on Oct. 25, 1926.

[36] See. p. 38 et seq.

[37] L. de Broglie, "Sur la possibilité de relier les phénomènes d'interférence et de diffraction à la théorie des quanta de lumière," *Comptes Rendus 183*, 447–448 (1926), communicated on Aug. 23, 1926.

larities" ("singularités mobiles") along the curves n normal to the phase fronts φ = const. "These singularities," he declared, "constitute the quanta of radiative energy." Substituting (6.18) in (6.16) and taking only the real part, de Broglie obtained

$$\frac{\partial \varphi}{\partial n}\frac{\partial f}{\partial n} + \tfrac{1}{2}f\Delta\varphi = -\frac{1}{c^2}\frac{\partial f}{\partial t} \tag{6.19}$$

Reasoning that the quotient $f/(\partial f/\partial n)$ vanishes at the position M of the particle, on grounds only partially explained, and knowing that the velocity of the quantum of light when passing through M is given by[38]

$$V = -\left(\frac{\partial f/\partial t}{\partial f/\partial n}\right)_{M,t} \tag{6.20}$$

he concluded that

$$V = c^2\left(\frac{\partial \varphi}{\partial n}\right)_M \tag{6.21}$$

so that φ—just as Madelung's φ—plays the role of a velocity potential. Denoting by ρ the density of the light quanta, de Broglie utilized arguments taken from the mechanics of fluids to show that

$$\rho = \text{const } a^2 \tag{6.22}$$

or, in other words, that the light-quanta density is proportional to the intensity of the radiation.

In a second paper[39] de Broglie carried over these considerations for the interpretation of the Schrödinger wave function and the motion of particles. "In micromechanics as in optics," he declared, "continuous solutions of the wave equation provide merely statistical information; an exact microscopic description undoubtedly necessitates the usage of singularity solutions reflecting the discrete structure of matter and radiation."[40]

In the spring of 1927 de Broglie brought these ideas to maturity and presented them in the form of what he called "the theory of double solution."[41] According to this theory the linear equations of wave mechanics admit two kinds of solution: a continuous ψ function with statistical significance and a "singularity solution" whose singularities constitute the physical particles under discussion. On the relation between the two solutions de Broglie made the following statements: "Particles would then be clearly localized in space, as in the classical picture, but they would be

[38] This is shown as in REF. 258 OF CHAP. 5.

[39] L. de Broglie, "La structure atomique de la matière et du rayonnement et la mecanique ondulatoire," *Comptes Rendus* *184*, 173–274 (1925), communicated on Jan. 31, 1927.

[40] *Ibid.*, p. 274.

[41] L. de Broglie, "La mécanique ondulatoire et la structure atomique de la matière et du rayonnement," *Journal de Physique et du Radium* *8*, 225–241, received on Apr. 1, 1927.

incorporated in an extended wave phenomenon. For this reason, the motion of a particle would not obey the laws of Classical Mechanics according to which the particle is subject only to the action of forces exerted on it in the course of its trajectory, without experiencing any effect from the existence of obstacles that may be situated at some distance outside the trajectory. In my conception, on the contrary, the motion of the singularity was to be dependent on all the obstacles that hindered the free propagation of the wave phenomenon surrounding it, and there would result from this a reaction of the wave phenomenon on the particle—a reaction expressed in my theory by the appearance of a 'quantum potential' entirely different from the potential of ordinary forces. And in this way the appearance of interference and diffraction phenomena would be explained."[42]

6.2 The Transformation Theory

Although the quantum-theoretic transformation theory originated in an attempt to solve a specific problem of rather limited scope, it soon proved itself conducive to such far-reaching generalizations that it led eventually to a unifying synthesis of all the diverse approaches and provided through its abstract formulation of the basic principles a deeper insight into the nature of the theory. A comparison with classical mechanics seems to be instructive. Quantum theory prior to the advent of the transformation theory may be compared with Newtonian mechanics prior to, say, Poisson's introduction of generalized momenta; and just as the development of the canonical formalism in classical dynamics provided in the work of Jacobi, Poincaré, and Appell a profound comprehension of the whole structure of classical mechanics, so the development of the quantum-mechanical theory of transformations culminated with the work of Dirac, Jordan, and von Neumann in the achievement of new vistas which made it possible to view the nonrelativistic quantum mechanics of a finite number of degrees of freedom as a logically consistent, compact, and unified system of thought.

Owing to the expansive trend of this development, a clear-cut or universally accepted definition of the subject matter of the transformation theory is hardly found in literature. To mention only two examples, in the 1960 edition of Sommerfeld's *Atombau und Spektrallinien*[43] the chapter on transformation theory deals almost exclusively with the Fourier transformation of dynamical variables whereas in the 1964 edition of Eisele's *Advanced Quantum Mechanics and Particle Physics*[44] the discussion of the

[42] REF. 151 OF CHAP. 5, p. 86; p. 90.

[43] A. Sommerfeld, *Atombau und Spektrallinien* (Vieweg, Braunschweig, 1960), vol. 2, pp. 201–208.

[44] J. A. Eisele, *Advanced Quantum Mechanics and Particle Physics from an Elementary Approach* (National Book Company, Taipei, Taiwan, 1964), vol. 1, pp. 165–174.

transformation theory focuses exclusively on the relation between the Schrödinger, Heisenberg, and interaction pictures. To cover such extremes it is possible to define the transformation theory as the study of those transformations in quantum theory which leave the results of empirically significant formulas invariant. The various conceptions of the subject matter of the theory then depend on the kind of space with respect to which the performance of transformations is to be considered. In fact, while the early phases of the development were characterized by transformations within the configuration and momentum space, its later elaborations led to the conception of abstract Hilbert spaces as the arena underlying the transformations. In this course the formerly important notion of canonical transformations in quantum mechanics gradually lost its peculiarity and importance.

The early development of the transformation theory from its inception[45] in the spring of 1926 until its elaboration to what is usually called the Dirac-Jordan transformation theory is the subject of our present discussion.

Matrix mechanics, as developed so far, if viewed against the background of classical analytic dynamics, was confined to the treatment of libration coordinates, in contrast to the older quantum theory which worked primarily with angle coordinates. In fact, while for the older quantum theory in its search for canonical transformations which reduce a problem to cyclic motions the *rotator* was the basic prototype, for matrix mechanics it was the *oscillator*.

The specific problem which, as mentioned above, gave rise to the development of the transformation theory even before the advent of wave mechanics was the question: could the Hamilton-Jacobi method of classical dynamics, which, it will be recalled, provided directly the frequencies of multiply periodic systems, be carried over into matrix mechanics? An important step in this direction was made by Dirac in a paper[46] already referred to. In it, incidentally, Dirac introduced a one-letter notation for $h/2\pi$ which, slightly modified,[47] will be used from now on also in our present text. Dirac showed that the determination of the frequencies for multiply periodic systems can be reduced to the problem of finding a system of canonical variables J and w which—similar to their classical prototypes—satisfy the conditions

$$[J_r, J_s] = [w_r, w_s] = 0 \qquad [w_r, J_s] = \delta_{rs}$$

[45] Strictly speaking, rudiments of the theory are already recognizable in the Born-Heisenberg-Jordan paper of November, 1925; see REF. 47 OF CHAP. 5.

[46] REF. 119 OF CHAP. 5.

[47] Dirac's notation "h" for "$(2\pi)^{-1}$ times the usual Planck's constant" as introduced on p. 561, *ibid.*, was the forerunner of $\hbar$. To avoid confusion we shall denote Dirac's "h," in consonance with later usage, always by $\hbar$.

where the quantum-mechanical Poisson brackets $[x,y]$ are defined by

$$xy - yx = ih[x,y]$$

the Hamiltonian H is a function of only the J, and the original p and q describing the system are multiply periodic functions of the w of period 2π. For the hydrogen atom Dirac was able to calculate the J and to retrieve[48] the Balmer formula, but was unable to calculate the intensities. A method for calculating the J was subsequently proposed by Wentzel,[49] who adapted to this end the Sommerfeld method of complex integration to the matrix calculus and applied it successfully in the case of the harmonic oscillator and of the hydrogenic atom.

The introduction of canonical angle and action variables in quantum mechanics was further promoted by Jordan,[50] who gave a rigorous proof of a theorem whose validity had already been assumed six months earlier in the "three-man paper,"[51] namely, that every canonical transformation, that is, a transformation which leaves the commutation relations invariant, can be written as

$$P = SpS^{-1} \qquad Q = SqS^{-1} \tag{6.23}$$

Jordan also proved that every point transformation, in classical mechanics always a canonical transformation, is so also in quantum mechanics. But in spite of Jordan's important results the transformation (6.23) proved of little practical use, primarily because the calculation of reciprocal matrices S^{-1} was a difficult task. Only in the case of infinitesimal canonical transformations, where

$$S = 1 + \lambda S_1 + \lambda^2 S_2 + \cdots \tag{6.24}$$

and consequently

$$S^{-1} = 1 - \lambda S_1 + \lambda^2(S_1^2 - S_2) + \cdots \tag{6.25}$$

could these transformations be carried out to be used in a matrix-perturbation theory for the solution of the principal-axes problem. Matrix mechanics thus was capable of dealing with complicated perturbation problems, once a solution of the unperturbed problem was available, but failed, in general, to solve problems which could not be considered as approximations to solved cases.

A serious difficulty for the introduction of angle and action variables

[48] See p. 234.

[49] G. Wentzel, "Die mehrfach periodischen Systeme in der Quantenmechanik," *Zeitschrift für Physik 37*, 80–94 (1926), received on Mar. 27, 1926.

[50] P. Jordan, "Über kanonische Transformationen in der Quantenmechanik," *Zeitschrift für Physik 37*, 383–386 (1926), received on Apr. 27, 1926; *38*, 513–517 (1926), received on July 6, 1926.

[51] REF. 47 OF CHAP. 5.

in matrix mechanics was the fact that the quantum conditions

$$pq - qp = \frac{\hbar}{i} \tag{6.26}$$

could not be satisfied for constant p, that is, a diagonal matrix p. In May, 1926, London[52] attempted to circumvent this difficulty by generalizing (6.26) to

$$pE(inq) - E(inq)p = nhE(inq) \tag{6.27}$$

where $E(q)$ is defined as the matrix function $\sum_{s=0}^{\infty} q^s/s!$.

For constant $p = (J_{kk})$ London obtained from (6.27)

$$hE_{kl} = (J_{kk} - J_{ll})E_{kl} \tag{6.28}$$

or, since for distinct J_{kk} for each index k only one index l exists with $E_{kl} \neq 0$,

$$J_{kk} - J_{k-1,k-1} = h \tag{6.29}$$

so that in agreement with the older quantum theory

$$J_{kk} = kh + \text{const} \tag{6.30}$$

On the other hand, by taking lim $n \to 0$, he showed that (6.27) reduces to (6.26). Elaborating this approach, London adapted the Hamilton-Jacobi method of introducing action and angle variables to matrix mechanics and concluded that the principal problem for matrix mechanics is the determination of the transformation generator S.

Meanwhile, however, Schrödinger's proof of the formal equivalence between matrix and wave mechanics had appeared and diverted the interest of physicists from the purely matrix-mechanical methods to the wave-mechanical approach. For it was hoped that also these problems could be solved with much less effort by the more advanced analytical apparatus of wave mechanics.

In fact, London[53] himself was the first to carry over, in the fall of 1926, the matrix-mechanical transformation theory, incomplete as it still was, into the conceptual framework of Schrödinger's wave mechanics, which had worked so far only with representations in the configuration space. In contrast to Schrödinger, who had used only point transformations $Q = f(q)$ for solving the eigenvalue equation

$$\left[H\left(q, \frac{\hbar}{i}\frac{\partial}{\partial q}\right), \psi\right] = E\psi \tag{6.31}$$

[52] F. London, "Über die Jacobischen Transformationen der Quantenmechanik," *Zeitschrift für Physik 37*, 915–925 (1926), received on May 22, 1926.

[53] F. London, "Winkelvariable und kanonische Transformationen in der Undulationsmechanik," *Zeitschrift für Physik 40*, 193–210 (1926), received on Sept. 19, 1926.

London, in analogy to the matrix-mechanical use of canonical transformations, considered $S = S(q,P)$ as the generator of the transformation

$$Q = \frac{\partial S[q, (\hbar/i)(\partial/\partial Q)]}{\partial P} \qquad \frac{\hbar}{i}\frac{\partial}{\partial q} = \frac{\partial S[q, (\hbar/i)(\partial/\partial Q)]}{\partial q} \tag{6.32}$$

which, as Jordan[54] had shown, could be written, with $T = T(Q,\partial/\partial Q)$, in the form

$$q = T^{-1}QT \qquad \frac{\partial}{\partial q} = T^{-1}\frac{\partial}{\partial Q}T \tag{6.33}$$

London now showed that the eigenvalues of (6.31) are invariant under the transformation (6.33). For, he said, let

$$\left[\bar{H}\left(Q, \frac{\partial}{\partial Q}\right), \bar{\psi}(Q)\right] = \bar{E}\bar{\psi}(Q) = \bar{E}\bar{\psi} \tag{6.34}$$

be the eigenvalue equation for the transformed function $\bar{\psi}$; here[55]

$$\bar{H} = T^{-1}HT \tag{6.35}$$

By multiplying both sides of Eq. (6.34) by T, he obtained

$$[HT,\bar{\psi}] = \bar{E}[T,\bar{\psi}] \tag{6.36}$$

or, as he tacitly assumed[56] as equivalent,

$$[H,T\bar{\psi}] = \bar{E}[T,\bar{\psi}] \tag{6.37}$$

But (6.37) is identical with (6.31) for

$$\psi = [T,\bar{\psi}] \tag{6.38}$$

and the invariance of the eigenvalues has been established.

Expanding the orthonormal eigenfunction

$$\psi_k(Q) = [T,\bar{\psi}_k(Q)] \tag{6.39}$$

in terms of the orthonormal eigenfunctions $\bar{\psi}_k(Q)$,

$$\psi_k(Q) = \sum_i T_{ik}\bar{\psi}_i(Q) \tag{6.40}$$

London obtained in the usual manner

$$\int \psi_j^*\psi_k\,dQ = \delta_{jk} = \sum_l T_{lj}^*T_{lk} \tag{6.41}$$

or[57]

$$\tilde{T}^*T = 1 \tag{6.42}$$

[54] REF. 50.
[55] See Eq. (5.33), p. 214.
[56] In a footnote, REF. 53, p. 197, London admitted that his proof still lacks rigor.
[57] The tilde ~ denotes, of course, the transpose of a matrix.

Having thus shown that T satisfies the "Hermitian orthogonality condition," that is, in modern terminology, that T is a unitary matrix, London interpreted (6.40) as a rotation, in the infinite-dimensional Hilbert space, of the coordinate system spanned by the orthogonal eigenfunctions. He also showed that the matrix elements F_{kl}, such as those of the electric dipole moment, are invariant under such rotations. For, he contended, if

$$F_{kl} = \int \psi_k^* F \psi_l \, dQ \tag{6.43}$$

$$F_{kl} = \sum_{ij} \int (T_{ik}\bar{\psi}_i)^* T\bar{F}T^{-1}(T_{jl}\bar{\psi}_j) \, dQ = (\tilde{T}^* T\bar{F}T^{-1}T)_{kl} = \bar{F}_{kl} \tag{6.44}$$

Hence, London concluded, under rotations in Hilbert space intensities are rotational invariants whereas energy eigenvalues are affine invariants.

It was only after the completion of his paper that London fully appreciated the intimate connection between his new conceptions and the theory of linear operators in functional spaces or, as it was called at that time, the "theory of distributive functional operations." In a footnote added to the galleys, London referred to a paper[58] by Cazzaniga, to a memoir[59] by Pincherle, and to the latter's résumé article[60] in the *Encyklopädie der mathematischen Wissenschaften* as presentations of the abstract mathematical theory that underlies his new interpretation. When London added this footnote to page 199 of his paper, he could hardly have been aware of its historical importance. It was the first reference to the future language of theoretical physics. For just as the differential and integral calculus was the language of classical dynamics or the tensor calculus that of relativity, the medium of expression for modern quantum mechanics turned out to be the theory of linear spaces and, in particular, of Hilbert spaces, or, more generally, the theory of functional analysis. In view of those circumstances a brief digression on the development of functional analysis prior to London's paper seems to be appropriate.

The earliest conception of a comprehensive theory of linear spaces was probably Grassmann's *Ausdehnungslehre*,[61] whose original version,[62] first published in 1844, was newly formulated and greatly amplified in 1862. In his ambitious attempt to formulate a theory of abstract linear n-dimensional spaces, Grassmann introduced, defined, and employed correctly such fundamental notions as linear independence, bases, inner products, orthogonality, normalization, as well as certain forms of tensorial quantities

[58] T. Cazzaniga, "Intorno ai reciproci dei determinanti normali," *Atti della Reale Accademia delle Scienze di Torino 34*, 495–514 (1898–1899).

[59] S. Pincherle, "Mémoire sur le calcul fonctionnel distributif," *Mathematische Annalen 49*, 325–382 (1897).

[60] REF. 101 OF CHAP. 5.

[61] REF. 19 OF CHAP. 5.

[62] Wigand, Leipzig, 1844; Enslin, Berlin, 1862.

and linear operators. The importance of these innovations for geometry and algebra alike was not recognized until Schlegel[63] drew the attention of mathematicians to these ideas. The first to suggest an extension of Grassmann's approach to spaces of infinitely many dimensions was Peano, whose statement "un sistema lineare può anche avere infinite dimensioni"[64] initiated the theory of linear infinite-dimensional spaces. By combining the algebraization of geometrical concepts with the operator calculus as developed during the early second half of the last century primarily by British mathematicians,[65] mathematicians like Pincherle, Cazzaniga, Calò, Carvallo, Amaldi, and Volterra laid the foundations of functional analysis. For it was soon recognized that certain families of functions display with respect to their algebraic and metric structures properties of the same character as vectors in the finite-dimensional Euclidean spaces of analytical geometry or their newly conceived generalizations to infinitely many dimensions. It was in this connection that Pincherle and Cazzaniga, in the articles referred to by London, defined functional spaces as sets of power series and declared: "Every power series is an element or point of such a space and the set of the coefficients in the series may be regarded as the set of its coordinates."[66] Volterra seems to have been the first to apply the geometry of infinite-dimensional spaces to problems of classical analysis. He was soon followed by Moore, who anticipated a number of important features of modern functional analysis in his so-called "general analysis."[67] Frequently encountered similarities among results obtained in apparently completely disparate branches of mathematics, such as the calculus of variations, the theory of differential and integral equations, problems of oscillations in continuous media, approximations, etc., suggested the establishment of a conceptual unification, or as Moore phrased it: "The existence of analogies between central features of various theories implies the existence of a general theory which underlies the particular theories and unifies them with respect to those central features."[68]

Although intended originally to serve as a maxim of research in pure mathematics, this "general heuristic principle of scientific procedure," as Moore[69] once called it, proved valid also with respect to the various formalisms in quantum physics. In fact, just as functional analysis and its theories of operators in linear spaces both developed from, and led to, a

[63] V. Schlegel, *System der Raumlehre* (Teubner, Leipzig, 1869).

[64] G. Peano, *Calcolo Geometrico secundo l'Ausdehnungslehre di H. Grassmann* (Bocca, Turin, 1888), p. 143.

[65] See pp. 225–227.

[66] REF. 59, p. 331.

[67] E. H. Moore, "On a form of general analysis, with applications to linear differential and integral equations," *Atti di IV. Congresso Internazionale Matematico, Rome*, 1909, vol. 2, pp. 98–114; *Introduction to a Form of General Analysis* (New Haven Mathematical Colloquium, New Haven, Conn., 1910).

[68] E. H. Moore, "On the foundations of the theory of linear integral equations," *Bulletin of the American Mathematical Society 18*, 334–362 (1911–1912), quotation on p. 339.

[69] *Ibid.*, p. 339.

unified point of view with respect to multifarious facts and approaches in mathematics, so also the functional-analytical formalism of quantum mechanics led, as we shall see in due course, to a logical unification of all quantum-theoretic approaches.

With the advent of the modern point-set topology, as founded by Maurice Fréchet and Felix Hausdorff, which made it possible to apply topological notions to algebra and analysis, and with the development of the theory of integral equations through the work of Erik Ivar Fredholm, David Hilbert, Erhard Schmidt, and Frédéric Riesz, functional analysis became one of the most important branches of modern mathematics and, as we shall see later on, the mathematical foundation of quantum mechanics.

Returning now to the paper by London, we note that it started by applying canonical transformations to the wave mechanics of discrete eigenvalue problems and ended up with discrete transformation matrices. A few weeks later Dirac published a paper[70] which began by applying canonical transformations to continuous or discrete matrices in Dirac's matrix mechanics and ended up with the wave mechanics of continuous or discrete eigenvalue problems. Dirac's work thus complemented London's in two respects: it showed, so to speak, the reversibility of the conceptual process under discussion and generalized it to continuous transformations.

Reviewing this crucial phase of his work—for Dirac "the growth of the use of transformation theory . . . is the essence of the new method in theoretical physics," as he stated in the preface to the first edition of his influential textbook *The Principles of Quantum Mechanics*[71]—Dirac commented: "A great deal of my work was just playing with equations and seeing what they give."[72] More specifically, he declared: "After people had established the equivalence between the matrix and the wave theories, I just studied their work and worked on it, and tried to improve it in a way that I had done several times previously. I think the transformation theory came out from that."[73] When studying the work previously done on this subject, Dirac was particularly impressed[74] by Lanczos' paper[75] of 1926 which, as we have seen, reformulated matrix mechanics as a continuum theory in terms of integral equations. Although Dirac, with his characteristic modesty, declared[76] that his transformation theory should be regarded

[70] P. A. M. Dirac, "The physical interpretation of the quantum dynamics," *Proceedings of the Royal Society of London* (*A*), *113*, 621–641 (1926), received on Dec. 2, 1926.

[71] First ed. 1930, 2d ed. 1935, 3d ed. 1947, 4th ed. 1958 (Clarendon Press, Oxford); *Die Prinzipien der Quantenmechanik*, translated by W. Bloch (Hirzel, Leipzig, 1930); *Les Principes de la Mécanique Quantique*, translated by A. Proca and J. Ullmo (Les Presses Universitaires de France, Paris, 1931).

[72] *Archive for the History of Quantum Physics*, Interview with P. A. M. Dirac on May 7, 1963.

[73] *Ibid.*, Interview with P. A. M. Dirac on May 10, 1963.

[74] *Ibid.*, Interview with P. A. M. Dirac on May 14, 1963.

[75] REF. 308 OF CHAP. 5.

[76] REF. 74.

merely as a generalization of Lanczos' paper, he certainly gave his theory a new and independent logical foundation by boldly introducing one of the most useful mathematical devices in quantum theory: the famous δ-function or "Dirac function" as it was subsequently called. His early interest in algebraic operators,[77] his study of Heaviside's operator calculus in electromagnetic theory, his training as an electrical engineer, and his familiarity with the foundations of the modern theory of electric pulses all contributed to this effect. In fact, Dirac declared: "All electrical engineers are familiar with the idea of a pulse, and the δ-function is just a way of expressing a pulse mathematically."[78]

It would be wrong, however, to assume that the earliest appearance of the δ-function was in connection with electrical pulses. It was rather in connection with the use of Green's theorem in the study of Huygens's principle that Gustav Kirchhoff[79] defined the δ-function, which he denoted by $F(\zeta)$, for the first time, in 1882, as follows: "As to the function F we assume that it vanishes for all finite positive and negative values of its argument, but that it is positive for such values when infinitely small and in such a way that

$$\int F(\zeta)\, d\zeta = 1$$

where the integration extends from a finite negative to a finite positive limit." Kirchhoff justified his assumption with the remark that

$$\frac{\mu}{\sqrt{\pi}} \exp\left(-\mu^2\zeta^2\right)$$

approximates F for very large μ, and he emphasized that "$F(\zeta)$ together with its derivatives is for all values of ζ, zero included, a finite and continuous function of its argument."

More than ten years later Heaviside[80] introduced the δ-function or its equivalent into the electromagnetic theory; prior to its use in quantum mechanics it was also employed in statistical mechanics by Paul Hertz[81] in connection with his statistical conception of temperature. That the δ-function is only a convenient mathematical artifice was clearly recognized by Dirac when, in contrast to Kirchhoff, he said: "Strictly, of course, $\delta(x)$ is not a proper function of x, but can be regarded only as a limit of a certain sequence of functions. All the same one can use $\delta(x)$ as though it were a

[77] See REF. 97 OF CHAP. 5.

[78] REF. 74.

[79] G. Kirchhoff, "Zur Theorie der Lichtwellen," *Berliner Berichte 1882*, pp. 641–669, quotation on p. 644; *Vorlesungen über mathematische Optik* (Teubner, Leipzig, 1891), vol. 2, pp. 24–25.

[80] O. Heaviside, "On operators in physical mathematics," *Proceedings of the Royal Society of London* (*A*), *52*, 504–529 (1893), *54*, 105–143 (1893).

[81] P. Hertz, "Statistische Mechanik," in R. H. Weber and R. Ganz, *Repertorium der Physik* (Teubner, Leipzig, Berlin, 1916), vol. 1, part 2, p. 503.

proper function for practically all the purposes of quantum mechanics without getting incorrect results. One can also use the differential coefficients of $\delta(x)$, namely, $\delta'(x)$, $\delta''(x)$ . . ., which are even more discontinuous and less 'proper' than $\delta(x)$ itself."[82]

Thus defining $\delta(x)$, as usually, by the conditions: $\delta(x) = 0$ for all $x \neq 0$ and $\int dx\, \delta(x) = 1$, and showing that

$$\int_{-\infty}^{\infty} f(x)\, \delta^{(n)}(a - x)\, dx = f^{(n)}(a) \tag{6.45}$$

Dirac applied it in order to express the elements of a continuous unit matrix $1(\alpha'\alpha'')$, labeled by the continuous parameters or indices α' and α'',

$$1(\alpha'\alpha'') = \delta(\alpha' - \alpha'')$$

or to express the elements of a general continuous diagonal matrix,

$$f(\alpha'\alpha'') = f(\alpha')\, \delta(\alpha' - \alpha'') \tag{6.46}$$

Considering now the canonical transformation of a dynamical variable g into G,

$$G = bgb^{-1} \tag{6.47}$$

Dirac expressed (6.47) in terms of continuous matrices as follows:

$$g(\xi'\xi'') = \iint (\xi'/\alpha')\, d\alpha' g(\alpha'\alpha'')\, d\alpha''\, (\alpha''/\xi'') \tag{6.48}$$

In (6.48) the primed or multiply primed continuous parameters (c-numbers) number the rows and columns of the matrix elements, (ξ'/α'), (α''/ξ'') denote the "transformation functions" $b(\xi'\alpha')$, $b^{-1}(\alpha'\xi'')$, respectively, and unprimed letters—with the exception of the d, of course—denote dynamical variables (q-numbers). For the equivalent relations

$$Gb = bg \tag{6.49}$$

$$b^{-1}G = gb^{-1} \tag{6.50}$$

Dirac wrote

$$\int g(\xi'\xi'')\, d\xi''\, (\xi''/\alpha') = \int (\xi'/\alpha'')\, d\alpha'' g(\alpha''\alpha') = g(\xi'\alpha') \tag{6.51}$$

$$\int (\alpha'/\xi'')\, d\xi'' g(\xi''\xi') = \int g(\alpha'\alpha'')\, d\alpha''\, (\alpha''/\xi') = g(\alpha'\xi') \tag{6.52}$$

Here the expressions $g(\xi'\alpha')$ and $g(\alpha'\xi')$ "may be regarded as the elements of two matrices that represent the dynamical variable g according to two new more general schemes, in which the rows and the columns of the

[82] REF. 70, p. 625.

matrices refer to different things," and the matrix elements of both G and g are denoted by $g(\xi'\xi'')$ and $g(\alpha''\alpha')$, respectively, since the G is the same function of the transformed variables as g was of the untransformed.[83]

Dirac knew that in a matrix scheme in which the possible values (c-numbers) of the dynamical variable ξ serve to number rows and columns, the matrix representing ξ is diagonal or

$$\xi(\xi'\xi'') = \xi' \, \delta(\xi' - \xi'') \tag{6.53}$$

But, he asked, what, in this same matrix scheme, is the matrix that represents a dynamical variable η which is canonically conjugate to ξ, that is, for which

$$\xi\eta - \eta\xi = i\hbar \tag{6.54}$$

holds? He found that

$$\eta(\xi'\xi'') = -i\hbar \, \delta'(\xi' - \xi'') \tag{6.55}$$

and showed, using integration by parts, that indeed

$$(\xi\eta - \eta\xi)(\xi'\xi'') = i\hbar \, \delta(\xi' - \xi'') \tag{6.56}$$

Considering the transformation between two matrix schemes and applying (6.51) to ξ and η,

$$\xi(\xi'\alpha') = \int \xi(\xi'\xi'') \, d\xi'' \, (\xi''/\alpha') \tag{6.57}$$

$$\eta(\xi'\alpha') = \int \eta(\xi'\xi'') \, d\xi'' \, (\xi''/\alpha') \tag{6.58}$$

Dirac deduced from (6.53) that

$$\xi(\xi'\alpha') = \xi' \, (\xi'/\alpha') \tag{6.59}$$

and from (6.56), using integration by parts, that

$$\eta(\xi'\alpha') = -i\hbar \int \delta'(\xi' - \xi'') \, d\xi'' \, (\xi''/\alpha') = -i\hbar \frac{\partial(\xi'/\alpha')}{\partial\xi'} \tag{6.60}$$

Thus he found that quite generally for any function $f(\xi,\eta)$ of the canonically conjugate dynamical variables ξ and η

$$f(\xi,\eta)(\xi'\alpha') = f\left(\xi', -i\hbar\frac{\partial}{\partial\xi'}\right)(\xi'/\alpha') \tag{6.61}$$

Formula (6.61), Dirac pointed out, makes it possible to obtain the matrix scheme (α) for which a given function $f(\xi,\eta)$ becomes a diagonal matrix

[83] See Eq. (5.33) on p. 214.

so that, in accord with (6.46),

$$f(\alpha'\alpha'') = f(\alpha')\,\delta(\alpha' - \alpha'')$$

For from Eqs. (6.61) and (6.51) it follows that

$$f\left(\xi', -i\hbar\frac{\partial}{\partial\xi'}\right)(\xi'/\alpha') = f(\xi,\eta)\,(\xi'\alpha') = \int (\xi'/\alpha'')\,d\alpha'' f(\alpha''\alpha')$$

$$= f(\alpha')\,(\xi'/\alpha') \qquad (6.62)$$

which is an ordinary differential equation for (ξ'/α'), considered as a function of ξ' and parameterized by α'. In particular, if ξ and η are identified with the position and momentum coordinates q and p, respectively, and $f(\xi,\eta)$ with the Hamiltonian H of the system, and if (ξ'/α') is written[84] as $\psi_E(q)$ with $f(\alpha') = E$, Eq. (6.62) turns out to be

$$H\left(q, \frac{\hbar}{i}\frac{\partial}{\partial q}\right)\psi_E(q) = E\psi_E(q) \qquad (6.63)$$

which is precisely Schrödinger's time-independent wave equation. Dirac thus could declare: "The eigenfunctions of Schrödinger's wave equation are just the transformation functions (or the elements of the transformation matrix previously denoted by b) that enable one to transform from the (q) scheme of matrix representation to a scheme in which the Hamiltonian is a diagonal matrix."[85] The proof as presented has been outlined, for the sake of simplicity, for the simplest conceivable case but can obviously be generalized, as Dirac pointed out, to the general case of a number of degrees of freedom and for eigenvalue spectra which contain both continuous ranges and discrete sets of values.

Having realized that the transformation function (ξ'/α') is a generalization of Schrödinger's wave function ψ just as the differential equation (6.62) is a generalization of Schrödinger's wave equation (6.63), Dirac extended Born's statistical interpretation of ψ to the more general (ξ'/α'). In Born's wave-mechanical treatment of time-dependent perturbations, it will be recalled, the wave function $\psi(q,t)$ of the perturbed system, at time t, was expanded in terms of the unperturbed eigenfunctions $\psi_n(q)$,

$$\psi(q,t) = \sum_n c_n(t)\psi_n(q) \qquad (6.64)$$

and $|c_n(t)|^2$ was interpreted as the probability that a transition to the state characterized by n took place. For a continuous energy range Eq. (6.64) had to be replaced by

$$\psi(q,t) = \int c(E,t)\,dE\,\psi_E(q) \qquad (6.65)$$

[84] This notation is not found in Dirac's paper but added here for additional clarification.

[85] REF. 70, p. 635.

and $|c(E,t)|^2\, dE$ was interpreted as the transition probability to a state whose energy lies between E and $E + dE$. Dirac now reformulated in terms of his transformation matrices the preceding perturbation treatment as well as Born's treatment of the scattering of electrons by atoms and found that full agreement is obtained if it is assumed that "the coefficients that enable one to transform from the one set of matrices to the other are just those that determine the transition probabilities."[86]

This statistical interpretation ascribed a physical meaning to every transformation from one matrix scheme to another—and not only to the transformation from the (q) scheme, in which position matrices are diagonal, to the (E) scheme, in which Hamiltonian or energy matrices are diagonal. Dirac's association of transformation coefficients with probabilities was obtained at about the same time, though formulated only for the discrete case, by Heisenberg[87] on the basis of considerations concerning energy fluctuations. Also Pauli, it seems, independently recognized at that time the possibility of generalizing Born's interpretation. In an important footnote in his paper[88] on Fermi's statistics[89] Pauli formulated Born's interpretation of the Schrödinger function as follows: the probability of finding the position coordinates $q_1, q_2, \ldots, q_f$ of a system of N particles in the volume element $dq_1\, dq_2 \cdots dq_f$ of the configuration space is given by $|\psi(q_1,q_2, \ldots,q_f)|^2\, dq_1\, dq_2 \cdots dq_f$ provided the system is in the state characterized by ψ. In a letter[90] to Heisenberg, Pauli wrote that "Born's interpretation may be viewed as a special case of a more general interpretation; thus, for example, $|\psi(p)|^2\, dp$ may be interpreted as the probability that the particle has a momentum between p and $p + dp$." In general, Pauli argued, there always exists, for every two quantum-mechanical quantities q and β, a function $\varphi(q,\beta)$ which he called "probability amplitude" ("Wahrscheinlichkeitsamplitude") and which has the property that $|\varphi(q_0,\beta_0)|^2\, dq$ is the probability that q has a value between q_0 and $q_0 + dq_0$ if β has the fixed value β_0.

Pauli's remarks furnished the basis on which Jordan[91] proposed at the end of 1926, independently of the work of London and of Dirac, his formulation of the transformation theory. Although ultimately in full agreement with these earlier formulations, as Jordan subsequently recognized, it adopted a totally different point of view. In its formal aspect it was essentially an axiomatic approach and was based on Pauli's probability

[86] *Ibid.*, p. 641.

[87] W. Heisenberg, "Schwankungserscheinungen und Quantenmechanik," *Zeitschrift für Physik 40*, 501–506 (1926), received on Nov. 6, 1926.

[88] W. Pauli, "Über Gasentartung und Paramagnetismus," *Zeitschrift für Physik 41*, 81–102 (1927), received on Dec. 16, 1926. The footnote referred to is on p. 83. *Collected Papers* (REF. 154 OF CHAP. 3), vol. 2, pp. 284–305.

[89] E. Fermi, "Zur Quantelung des idealen einatomigen Gases," *Zeitschrift für Physik 36*, 902–912 (1926).

[90] REF. 8 OF CHAP. 5, p. 44.

[9] P. Jordan, "Über eine neue Begründung der Quantenmechanik," *Zeitschrift für Physik 40*, 809–838 (1927), received on Dec. 18, 1926.

amplitude which was postulated to satisfy three major conditions: (1) $\varphi(q,\beta)$ is independent of the mechanical nature (Hamiltonian function) of the system and depends only on the kinematic relation between q and β; (2) the probability (density) that for a fixed value β_0 of β the quantum-mechanical quantity q has the value q_0 is the same as the probability (density) that, for a fixed q_0 of q, β has the value β_0; (3) probabilities are combined by superposition, that is, if $\varphi(x,y)$ is the probability amplitude for the value x of q at the fixed value y of β, and $\chi(x,y)$ is the probability amplitude for the value x of Q at the fixed value y of q, then the probability amplitude for x of Q at the fixed value y of β is given by

$$\Phi(x,y) = \int \chi(x,z)\varphi(z,y)\,dz$$

In the particular case $Q = \beta$, Jordan's $\Phi(x,y)$ becomes Dirac's $\delta(x - y)$.

According to Jordan p is defined as the momentum canonically conjugate to q if the probability amplitude $\rho(x,y)$, for every possible value x of p at fixed y of q, is given by

$$\rho(x,y) = \exp\left(\frac{xy}{i\hbar}\right) \tag{6.66}$$

Hence, Jordan inferred, for a fixed value of q all possible values of p are equally probable.[92] As he pointed out, $\rho(x,y)$ satisfies the differential equations

$$\left\{x + \frac{\hbar}{i}\frac{\partial}{\partial y}\right\}\rho(x,y) = 0 \tag{6.67}$$

$$\left\{-\frac{\hbar}{i}\frac{\partial}{\partial x} - y\right\}\rho(x,y) = 0 \tag{6.68}$$

Let $\varphi(x,y)$, he continued, be the probability amplitude that Q has the value x at y of q and $\Phi(x,y)$ the amplitude that Q has the value x at $p = y$; then

$$\varphi(x,y) = \int \Phi(x,z)\rho(z,y)\,dz \tag{6.69}$$

Introducing the linear operator

$$T = \int dx\,\Phi(y,x)\cdots \tag{6.70}$$

Jordan could write Eq. (6.69) in the form

$$(x,y) = T\rho(x,y) \tag{6.71}$$

It will have been recognized that Jordan's Eqs. (6.67) and (6.68) are

[92] *Ibid.*, p. 814.

essentially the Schrödinger equation for a free particle, $\rho(x,y)$ being the (conjugate) Schrödinger wave function for a particle of constant momentum. By a skillful application of the operator technique based on the introduction of Hermitian conjugate operators, Hermitian operators, and similar notions which subsequently gained great importance for the modern formulation of quantum mechanics, Jordan proved that his formal theory contains not only Schrödinger's wave mechanics and Heisenberg's matrix mechanics, but also the Born-Wiener operator calculus and Dirac's q-number algorithm as special cases.

From the metascientific or methodological point of view Jordan's transformation theory was an important achievement: it comprised all preceding quantum-mechanical formalisms in one unified formulation. It satisfied a deep-rooted and perennial desire of the human intellect. For the quest for a conceptual synthesis which, as we have seen, lay at the foundation of Moore's "general heuristic principle of scientific procedure"[93] is much older than modern physics or mathematics. In fact, whenever formally diverging theories explained the same phenomena, science had searched for such a unification. Thus, in ancient astronomy, as Theon of Smyrna[94] reported, "Hipparchus deemed it worthy of investigation by mathematicians why on two hypotheses so different from one another, that of eccentric circles and that of concentric circles with epicycles, the same results appear to follow." The analogous problem in the physics of the atom, the question why on four hypotheses so different from one another, those proposed by Heisenberg, Born-Wiener, Schrödinger, and Dirac, "the same results appear to follow," found its answer in Jordan's transformation theory. The only unsatisfactory aspect of Jordan's work was the intricacy of its mathematical exposition, which confined its full appreciation to only a few experts at that time. Kennard's lucidly written résumé[95] of contemporary quantum mechanics, based primarily on Jordan's approach, contributed to a wider dissemination of Jordan's ideas. A full satisfactory and mathematically perspicuous presentation of the transformation theory and its unifying character was made possible, however, only after functional analysis became the language of quantum mechanics.

6.3 *The Statistical Transformation Theory in Hilbert Space*

The transformation theory, both in Dirac's continuous matrix formulation and in Jordan's semiaxiomatic formalism, with its emphasis on the notion

[93] REF. 67.

[94] Theon of Smyrna, *Expositio Rerum Mathematicarum ad Legendum Platonem Utilium*, edited by E. Hill (Teubner, Leipzig, 1878), p. 166.

[95] E. H. Kennard, "Zur Quantenmechanik einfacher Bewegungstypen," *Zeitschrift für Physik 44*, 326–352 (1927), received on July 17, 1927.

of probability amplitudes, was the first indication that it might be possible to dispense with the correspondence principle in the construction of the conceptual edifice of quantum mechanics. For it made it gradually clear that statistical considerations concerning the measuring process might serve instead as the foundation.

For a full appreciation of this important stage in the development of our theory it is worthwhile to review the principal achievements of the transformation theory against the background of the earlier formulations of quantum mechanics. In the latter the energy concept played a dominant role. Schrödinger's eigenvalue problem was that of energy eigenvalues. In Heisenberg's matrix mechanics only those q's and p's were solutions of the problem of motion for which the energy matrix was diagonal. Nondiagonal elements in the solution matrices were associated with the quantum transitions and the diagonal elements with the time averages of the represented magnitude in the stationary states, in accordance with the fact that they correspond with the constant terms in the respective Fourier expansions. As observable were regarded primarily the energy values and the squares of the nondiagonal elements in matrices which—as the electric-dipole matrix—were connected with the emission, absorption, or dispersion of light. This particular choice of what was to be regarded as observable was, of course, conditioned by the historical development of the theory which originated in the study of periodic motions. In fact, the treatment of the free electron, for example, was beyond the conceptual apparatus of this formulation of quantum mechanics. The notion of position had no place in a calculus of matrix elements which were disguised Fourier coefficients.

The transformation theory introduced the experimentally required generalizations by postulating that, in principle, any Hermitian matrix A represents an observable quantity a, on a par with energy, and that the eigenvalues of A are the possible results of measuring a. Through Dirac's introduction of continuous matrices and Born's probabilistic interpretation of Schrödinger's wave function, the notion of position was retrieved. If previously the energy states of an atomic system formed, so to speak, the reference system for theoretical predictions, a more general point of view was now adopted. For, confining our discussion to the discrete nondegenerate case, we may formulate the basic problem of the transformation theory, as envisaged by Dirac and Jordan, as follows: If the measurement of a quantity a resulted in a_m, what is the probability w_{mk} that a subsequent measurement of another quantity b will result in b_k? This was the answer: Let the matrix $A^{(1)} = (a_{mn}) = (\delta_{mn}a_n)$ represent a in the a scheme or coordinate system $\sum^{(1)}$, in which it is diagonal, its diagonal elements $a_{mn} = a_m$ being the possible results of measuring a; let $B^{(2)} = (b_{rs}) = (\delta_{rs}b_s)$ represent b in the b scheme or coordinate system $\sum^{(2)}$, in which it is diagonal, its diagonal elements $b_{rr} = b_r$ being the possible results of measuring b; finally, let U be the unitary matrix which transforms $\sum^{(1)}$ to $\sum^{(2)}$ so that $A^{(2)} = U^{-1}A^{(1)}U = (\sum_k U_{mk}{}^{-1}a_k U_{kn})$ and $B^{(1)} = UB^{(2)}U^{-1} = (\sum_l U_{ml}b_l U_{ln}{}^{-1})$. Then, and this is the answer,

$$w_{mk} = |U_{mk}|^2 \tag{6.72}$$

It was a plausible answer. First of all, since U is unitary,

$$\sum_k w_{mk} = \sum_k |U_{mk}|^2 = 1 \tag{6.73}$$

a condition which such a probability has to satisfy. In addition, if in particular a happens to be the energy, $(B^{(1)})_{nn} = \sum_l b_l |U_{nl}|^2$. Hence, the diagonal term $(B^{(1)})_{nn}$ of $B^{(1)}$ in the energy scheme, which corresponds to the time-independent term in the Fourier expansion of b and signifies therefore the time average of b in the nth energy or stationary state, turns out to be the statistical mean or expectation value.

In the following discussion of the further development of the transformation theory we shall confine ourselves as far as possible to the formal aspects of this development and postpone questions of interpretation and epistemological implications as far as possible to a later stage. It may be argued that greater consistency would have been gained had we deferred also Born's probabilistic interpretation to this later stage. Since, however, the entire development of the transformation theory rested on Born's assumption, its deferment would have made our account rather incomprehensible. On the other hand, we believe it to be possible to postpone the principles of uncertainty and complementarity and related issues—although they preceded chronologically a great part of the subject matter of the following discussion—without seriously impairing the logical coherence of our exposition.

It was most fortunate for the development of our theory that Heisenberg's appeal[96] to pure mathematicians was answered, at the right moment, in Göttingen by Hilbert himself. Since he had been consulted on numerous mathematical aspects of the theory ever since its inception, quantum mechanics was no *terra incognita* to him. In the fall of 1926 he began a systematic study of its mathematical foundations. His assistants were Lothar Wolfgang Nordheim, a former pupil of Born, and the twenty-three-year-old John (Johann) von Neumann, who had just arrived in Göttingen after having obtained a Ph.D. in mathematics from the University of Budapest and a degree in chemical engineering in Zurich. During the winter semester 1926/27 Hilbert also gave a two-hour lecture every Monday and Thursday morning on the mathematical foundations of quantum mechanics

[96] REF. 81 OF CHAP. 5.

("Mathematische Methoden der Quantentheorie"), a résumé of which, edited by Nordheim, was published in the spring of 1927.[97]

Hilbert's approach was, of course, that of a pure mathematician. In his quest for logical rigor he elaborated Jordan's semiaxiomatic theory, based as it was on the notion of probability amplitudes. He first investigated the physical requirements which these amplitudes have to satisfy so as to lead to results in agreement with experience, and then searched for an analytical apparatus in terms of which these requirements could be formulated and satisfied. By a sufficiently stringent formulation of the physical stipulations he attempted to reach a unique determination of the analytical apparatus. What he had in mind was a conceptual system like that of his axiomatization of geometry, where the axioms define the relations between its primitive notions such as "point," "straight line," "plane" in an unequivocal manner and where linear algebra provides an analytical apparatus which satisfies these relations; moreover, the analytical apparatus, conversely, may become the source for new geometrical theorems. In analogy to this procedure, which had proved itself so valuable, Hilbert postulated six relations which amplitudes are supposed to fulfill. In order to establish the analytical apparatus Hilbert associated with every dynamical variable an operator and regarded the corresponding operator calculus, soon to be outlined, as the appropriate analytical apparatus. However, concluding his prefatory remarks, Hilbert had to concede that in physics, contrary to mathematics, the analytical apparatus has often to be surmised before a complete axiomatization of the physical requirements is accomplished and that usually only the interpretation of the formalism leads to an exhaustive postulation of the physical stipulations. "It is difficult to understand such a theory," said Hilbert, "if the formalism and its physical interpretation are not strictly kept apart. Such a separation shall be adhered to even though at the present stage of the development of the theory no complete axiomatization has as yet been achieved. However, what is definite by now is the analytical apparatus which will not admit any alterations in its purely mathematical aspects. What can, and probably will, be modified is its physical interpretation for it allows a certain freedom of choice."[98]

Although Hilbert's contention concerning the finality of the analytical apparatus, as we shall see, was soon to be disproved, his conception of a separate analytical apparatus through the interpretation of which empirical relations can be brought into the form of mathematical statements became a basic element of the methodology of modern theoretical physics. Having for many years been convinced of the importance of integral equations

[97] D. Hilbert, J. von Neumann, and L. Nordheim, "Über die Grundlagen der Quantenmechanik," *Mathematische Annalen 98*, 1–30 (1927), received on Apr. 4, 1927. J. von Neumann, *Collected Works*, edited by A. H. Taub (Pergamon Press, New York, Oxford, London, Paris, 1961), vol. 1, pp. 104–133.

[98] *Ibid.*, p. 3.

also for theoretical physics,[99] Hilbert assumed that the operators to be associated with dynamical variables could be written as integral operators of the type

$$T^{\binom{x}{y}}(f(y)) = \int \varphi(x,y)f(y)\,dy$$

where the kernel of the operator, $\varphi_T = \varphi(x,y)$, provides the probability amplitude. Thus, if T is the operator which transforms canonically the operator q for the position coordinate into the operator F of the variable $F(p,q)$, a function of q and the momentum p, φ_T is the probability amplitude $\varphi(xy;qF)$ relating to q and F, that is, the probability amplitude that for a fixed value y of $F(p,q)$ the coordinate q has a value between x and $x + dx$. If, in particular, $F(p,q)$ is q itself, or p, the corresponding kernels are $x\,\delta(x - y)$ and $(\hbar/i)\,\delta'(x - y)$, respectively. Generalizing these results for any two dynamical variables F_1 and F_2, Hilbert and his collaborators studied the conditions that the relative probability density

$$w(xy;F_1F_2) = \varphi(xy;F_1F_2)\varphi^*(xy;F_1F_2)$$

is real and nonnegative. They concluded that only Hermitian operators are to be admitted. With the help of Dirac's δ-function they derived differential equations from the integral-operator equations and showed, in particular, that the time-independent Schrödinger wave equation

$$\left\{H\left(x, \frac{\hbar}{i}\frac{\partial}{\partial x}\right) - W\right\} \varphi(xW;qH) = 0$$

is the functional equation for the probability amplitude for energy and position coordinate, and that the time-dependent wave equation

$$\left\{H\left(x, \frac{\hbar}{i}\frac{\partial}{\partial x}\right) + \frac{\hbar}{i}\frac{\partial}{\partial t}\right\} \varphi(xt;qT) = 0$$

is obtained if in $\varphi(xy;F_1F_2)$ F_1 is the position operator and F_2 a "time operator," associated with the time $t = t(p,q)$ and conceived as a dynamical variable canonically conjugate to the Hamiltonian $H(q,p)$. $\varphi(xt;qT) = \psi(x,t)$ is consequently the probability amplitude that for a given time t the position coordinate has a value x between x and $x + dx$. Hilbert, von Neumann, and Nordheim pointed out that their result is in full agreement with Born's probabilistic interpretation. For, according to Born, in the general solution of the time-dependent Schrödinger wave equation $\psi(x,t) = \sum_n c_n\psi_n(x) \exp[(i/\hbar)W_nt]$, $|c_n \exp[(i/\hbar)W_nt]|^2$ measures the probability that the atom is found in the nth state and $|\psi_n(x)|^2$ that it is found, while in state n, at x. The probability interference principle consequently shows

[99] See, for example, REF. 7 OF CHAP. 1.

that $\psi(x,t)$ is the amplitude that the atom, irrespective of its state, is found at x.

The Hilbert-Neumann-Nordheim transformation theory, like its predecessors, the theories of Dirac and Jordan, offered a unified formalism in which both matrix and wave mechanics were embedded as special cases. This unification, as von Neumann soon realized, depended critically on the properties of the Dirac δ-functions and was therefore objectionable to at least the same extent as were these functions.

In matrix mechanics, it will be recalled, the basic problem, as Born and Jordan already knew,[100] was the solution of the matrix equation

$$U^{-1}H(pq)U = W \tag{6.74}$$

or

$$UW = HU \tag{6.75}$$

where the diagonal terms of the diagonal matrix W are the energy eigenvalues of the dynamical system under discussion and where U denotes a unitary transformation which diagonalizes H. Using the notation $H = (h_{mn})$, $U = (u_{mn}) = (u_m^{(n)})$, and $W = (w_m\delta_{mn})$, one could write Eq. (6.75) as follows:

$$\sum_k h_{mk}u_k^{(n)} = \lambda u_m^{(n)} \qquad \lambda = w_n \tag{6.76}$$

and regard the elements $u_k^{(n)}$ of the nth column of U as the components of a vector (u_k). The basic problem could consequently be reformulated as that of solving the eigenvalue problem

$$\sum_{k'} h_{mk'}u_{k'} = \lambda u_m \tag{6.77}$$

for given $h_{mk'}$. If it is recalled that in wave mechanics the basic problem was that of solving the eigenvalue problem

$$H\psi(q) = \lambda\psi(q) \tag{6.78}$$

where H denotes the Hamiltonian operator of the system, the similarity between the two equations (6.77) and (6.78) cannot be ignored. It suggests that u_m be regarded as a function of the "discrete variable" m just as $\psi(q)$ is a function of the continuous variable q. Then the right-hand sides of the two equations (6.77) and (6.78) have the same structure.

By the same token, $h_{mk'}$ should correspond to a function of two variables, $h(q,q')$, and the $\sum_{k'}$ to an integral $\int \cdots dq'$. In other words, Eq. (6.77) should be written as

$$\int h(q,q')\psi(q')\,dq' = \lambda\psi(q) \tag{6.79}$$

[100] See p. 214.

Comparison of Eq. (6.79) with Eq. (6.78) yields

$$H\psi(q) = \int h(q,q')\psi(q')\,dq' \tag{6.80}$$

which shows that $h(q,q')$ is an integral kernel of the functional or integral operator H of precisely the type used by Hilbert and his collaborators. Matrix and wave mechanics can therefore be unified if it proves possible to find for every acceptable H a corresponding integral kernel $h(q,q')$ which satisfies Eq. (6.80). Taking for H the identity operator, so that $H\psi(q) = \psi(q)$, one can easily confirm that

$$\psi(q) = \int h(q,q')\psi(q')\,dq' \tag{6.81}$$

can be satisfied only if the integral kernel is Dirac's δ-function $\delta(q - q')$.

Von Neumann, now, contended quite generally that a unification of matrix theory with wave mechanics by setting up a correspondence between the "space" Z of the discrete index m in Eq. (6.77) and the space Ω of the continuous variable q in Eq. (6.78) can be carried out only by transforming a differential operator into an integral operator, a process which necessitates the introduction of "improper" functions such as the Dirac δ-function.

It should now be recalled that in 1927 the δ-function, in spite of its increasing popularity, was mathematically an illegitimate conception. It is so still today if it is conceived as an ordinary function. The justification given by Kirchhoff[101] was faulty simply because there is no limiting function of $(\mu/\sqrt{\pi})\exp(-\mu^2\zeta^2)$ for $\mu \to \infty$. Nor could it be defined as the derivative of the Heaviside function $Y(x)$, for which $Y(x) = 0$ if $x < 0$, $Y(x) = \frac{1}{2}$ if $x = 0$, and $Y(x) = 1$ if $x > 0$, because at $x = 0$ a derivative does not exist. In fact, it was not difficult to see that the requirements $\delta(x) = 0$ for $x \neq 0$ and $\int_{-\infty}^{\infty} \delta(x)\,dx = 1$ are incompatible irrespective of whether the Riemann or the Lebesgue conception of the integral is being used. That the difficulty could be overcome by expressing $\delta(x)$ as a Stieltjes[102] integral or as a distribution became clear only after 1945, when Laurent Schwartz[103] generalized the notion of function, derivative, and Fourier transformation by his theory of distributions. Utilizing results from the theory of linear topological spaces, Schwartz was able to replace the ill-defined δ-function and its derivatives by well-defined linear functionals or distributions which have the immediate advantage of always possessing a derivative which is itself a distribution.

[101] REF. 79.

[102] T. J. Stieltjes, "Recherches sur les fractions continues," *Annales de la Faculté des Sciences de Toulouse 8*, 68–122 (1894); *Œuvres Complètes* (Noordhoff, Groningen, 1914), vol. 2.

[103] L. Schwartz, "Généralisation de la notion de fonction, de dérivation, de transformation de Fourier et applications mathématiques et physiques," *Annales de l'Université de Grenoble 21*, 57–74 (1945); *Théorie des Distributions* (Hermann et Cie, Paris, 1950–1951), 2 vols.

A rigorous definition of the δ-function could also be obtained by a method introduced by Jan G. Mikusiński,[104] who defined generalized functions as the closure of certain ordinary functional spaces with respect to a weak topology. Roughly speaking, generalized functions were defined as certain sequences of ordinary functions, just as real numbers were defined by Cantor as sequences of rational numbers.[105] In contrast to these methods, which also include Marcel Riesz's theory of pseudofunctions,[106] a different approach has recently been proposed toward a rigorous foundation of the theory of δ-functions. By extending, not the concept of function, but that of number—again in a way similar to Cantor's closure process by embedding real numbers in sequences of such—Schmieden and Laugwitz[107] defined infinitely large and infinitely small numbers and showed that on the basis of what they called "continuous continuations" ("stetige Fortsetzungen") of real functions and "normal functions" ("normale Funktionen"), functions can be constructed in a rigorous manner which possess all the properties of Dirac's δ-function. To go into details would lead us too far from our subject. We have gone into this digression to show that the notion of δ-function, as it has been applied by Dirac, Jordan, and Hilbert and his collaborators, can be fully legitimized today. But it could not in 1927.

Arguing that the use of the δ-function leads to "insolvable mathematical difficulties" ("unlösbare mathematische Schwierigkeiten"),[108] von Neumann rejected the proposed fusion of the space Z of the discrete variable m in Eq. (6.77) with the space Ω of the continuous variable q in Eq. (6.78). Instead, elaborating on certain ideas of Hilbert's work on linear integral equations, he developed between 1927 and 1929 a new

[104] J. G. Mikusiński, "Sur la méthode de généralisation de Laurent Schwartz et sur la convergence faible," *Fundamenta Mathematica 35*, 235–239 (1948); "Sur certains espaces abstraits," *ibid. 36*, 125–130 (1948); "Sur les fondaments du calcul opérative," *Studia Mathematica 11*, 41–70 (1949). Certain aspects of Mikusiński's approach are already found in S. Bochner's *Vorlesungen über Fouriersche Integrale* (1932) and in S. L. Soboleff's "Méthode nouvelle à résoudre le problème de Cauchy pour les équations linéaires hyperboliques normales," *Mathematicheskii Sbornik 1*, 39–72 (1936).

[105] Cf. G. Temple, "Theories and applications of generalized functions," *Journal of the London Mathematical Society 28*, 134–148 (1953), and M. J. Lighthill, *Introduction to Fourier Analysis and Generalised Functions* (Cambridge University Press, 1959).

[106] M. Riesz, "L'integral de Riemann-Liouville et le problème de Cauchy," *Acta Mathematica 81*, 1–223 (1949).

[107] C. Schmieden and D. Laugwitz, "Eine Erweiterung der Infinitesimalrechnung," *Mathematische Zeitschrift 69*, 1–39, (1958); D. Laugwitz, "Eine Einführung der δ-Funktionen," *Münchener Berichte 1959*, pp. 41–59.

[108] This characterization is still found, on p. 17, in J. von Neumann, *Mathematische Grundlagen der Quantenmechanik* (Springer, Berlin, 1932; Dover, New York, 1943; Presses Universitaires de France, Paris, 1947; Instituto de Mathematicas "Jorge Juan," Madrid, 1949); translated into English by R. T. Beyer, *Mathematical Foundations of Quantum Mechanics* (Princeton University Press, Princeton, N.J., 1955). The English translation reads: "with great mathematical difficulties," p. 31.

mathematical framework[109] of the theory which subsequently proved to be the most suitable formalism of nonrelativistic quantum mechanics as we use it today, as well as of its extensions, the relativistic quantum mechanics of particles and the quantum theory of fields.

The decisive innovation on which von Neumann based his new approach was his ingenious discovery, early in 1927, that a new formalism of quantum mechanics can be established in view of the fact that not Z and Ω, but rather the sequence space F_Z over Z and the function space F_Ω over Ω are essentially identical. Here F_Z is the set of all sequences $\{u_m\}_1^\infty$ satisfying $\sum | u_m |^2 = 1$ and F_Ω the set of all summable and square-integrable complex-valued functions $\psi(q)$ over Ω, satisfying $\int | \psi(q) |^2 \, dq = 1$. Von Neumann assumed that the indicated normalizations, which, incidentally, exclude trivial solutions, can always be imposed in the treatment of eigenvalue problems of the type envisaged. The sequence space F_Z, it will be recalled, played an important role in Hilbert's theory of integral equations and his related investigations on bounded quadratic forms of infinitely many variables. It was therefore known as the "Hilbert sequence space." Von Neumann's F_Ω, on the other hand, corresponded to what became known as $L_2(\Omega)$ in functional analysis, that is, the space of all summable and square-integrable functions (in the sense of Lebesgue) over the domain Ω. Von Neumann knew that F_Z and F_Ω are isomorphic and isometric, provided the operations of addition, multiplication by a scalar, and the formation of an inner product are appropriately defined. For he recalled that twenty years ago Frédréric (Friedrich) Riesz,[110] like von Neumann a native of Hungary, proved an important theorem which in a slightly modernized version can be formulated as follows: if $\{\varphi_n(x)\}_1^\infty$ is a complete orthonormal system in $L_2(\Omega) = L_2(a,b)$ and $\{a_n\}_1^\infty$ a sequence of (real) constants, then the convergence of $\sum a_n^2$ is a necessary and sufficient condition that there exists in $L_2(a,b)$ an almost everywhere (i.e., up to a range of measure zero) uniquely determined function $f(x)$ such that $a_n = \int_a^b f(x) \varphi_n(x) \, dx$ for all n.

Riesz's paper was presented to the Göttingen Society by Hilbert on March 9, 1907. Four days earlier, at a meeting of the Mathematical Society

[109] J. von Neumann, "Mathematische Begründung der Quantenmechanik," *Göttinger Nachrichten 1927*, pp. 1–57 (presented on May 20, 1927); *Collected Works* (REF. 97), vol. 1, pp. 151–207. "Wahrscheinlichkeitstheoretischer Aufbau der Quatenmechanik," *Göttinger Nachrichten 1927*, pp. 245–272 (presented on Nov. 11, 1927); *Collected Works*, vol. 1, pp. 208–235. "Thermodynamik quantenmechanischer Gesamtheiten," *Göttinger Nachrichten 1927*, 273–291 (presented on Nov. 11, 1927); *Collected Works*, vol. 1, pp. 236–255. "Beweis des Ergodensatzes und des H-Theorems in der neuen Mechanik," *Zeitschrift für Physik 57*, 30–70 (1929); *Collected Works*, vol. 1, pp. 558–598.

[110] F. Riesz, "Ueber orthogonale Funktionensysteme," *Göttinger Nachrichten 1907*, pp. 116–122, presented on Mar. 9, 1907; "Sur les systèmes orthogonaux de fonctions," *Comptes Rendus 144*, 615–619 (1907), read on Mar. 11, 1907.

of Brünn, Ernst Fischer had proved the following theorem: if a sequence of functions $\{f_n\}_1^\infty$, all of which belong to $L_2(a,b)$, converges in the mean [that is, $\lim_{m,n=\infty} \int_a^b (f_m - f_n)^2\, dx = 0$], then there exists in $L_2(a,b)$ a function $f(x)$ to which they converge in the mean [that is, $\lim_{n=\infty} \int_a^b (f - f_n)^2\, dx = 0$]. In a paper[111] presented to the Paris Académie des Sciences on April 29, 1907, in which Fischer published the proof of this theorem, he pointed out that Riesz's conclusion is an immediate consequence of this theorem; for, he said, using Riesz's notation, the convergence of $\sum a_n^2$ entails the convergence in the mean of $\sum a_n\varphi_n$; hence there exists a function $f(x)$ to which $\sum a_n\varphi_n$ converges in the mean; but this implies that $a_n = \int_a^b f\varphi_n\, dx$.

The "Fischer-Riesz theorem," as Riesz's theorem was subsequently called, established therefore a one-to-one correspondence between bounded sequences or vectors $\{u_m\}$ and square-integrable functions $\psi(q)$. As von Neumann realized, this correspondence is linear and isometric, that is, if $\{u_m\} \leftrightarrow \psi(q)$ and $\{v_m\} \leftrightarrow \varphi(q)$, then $\{au_m\} \leftrightarrow a\psi(q)$, $\{u_m + v_m\} \leftrightarrow \psi(q) + \varphi(q)$, and $\sum |u_m|^2 = \int |\psi(q)|^2\, dq$. In view of these results von Neumann reasoned as follows: since F_Z and F_Ω—and not Z and Ω—are the "real analytical substrata" of matrix and wave mechanics and since they are isomorphic, that is, ultimately only different mathematical representations of the same abstract relations, the equivalence between matrix and wave mechanics is a logical consequence of this isomorphism. Moreover, he added, a formulation of quantum mechanics which retains only absolutely essential relations and dispenses with all accidental trappings has to be based on the abstract common structure of F_Z and F_Ω or, as he called it, the "abstract Hilbert space." In von Neumann's formulation of quantum mechanics was found the ultimate fulfillment—as far as the theory of quanta is concerned—of Hipparchus's insistence on the usefulness of investigating "why on two hypotheses so different from one another . . . the same results appear to follow,"[112] which, as we have seen, had already characterized Jordan's line of research.

To this end von Neumann developed an axiomatic theory of the abstract Hilbert space which, defined as an infinite-dimensional complete separable linear vector space with a positive definite metric, contained F_Z and F_Ω as special realizations. After a study of the geometrical properties of this space von Neumann constructed a theory of linear operators whose domains are everywhere dense, abandoning thereby the requirement that operators must be defined everywhere. In the realization F_Ω von Neumann's operators are functionals as studied in functional analysis. Similarly, von

[111] E. Fischer, "Sur la convergence en moyenne," *Comptes Rendus 144*, 1022–1024 (1907).

[112] See REF. 94.

Neumann's defining equation of *adjoint operators* A and $A^\dagger$, namely,[113]

$$(A^\dagger f,g) = (f,Ag)$$

where the parentheses denote the *inner product* of two vectors in the Hilbert space, can be interpreted in the realization F_Ω as the equation

$$\int_\Omega (A^\dagger f)g^*\,d\tau = \int_\Omega f(Ag)^*\,d\tau$$

A is *Hermitian* if $A = A^\dagger$ and A is *continuous* or *bounded* if for a constant C and for all vectors f for which it is defined

$$\|Af\| \leq C\|f\|$$

where the double bars denote the *length* $(f,f)^{1/2}$ of f. A is said to be *definite* if for every f, $(Af,f) \geq 0$; A is *unitary* if $AA^\dagger = A^\dagger A = I$, where I is the unity operator satisfying $If = f$.

Most of these conceptions were not new. Some of them had been used by Fréchet[114] about twenty years earlier when he introduced the language of Euclidean geometry into the theory of functional spaces; in particular, he gave the earliest definition of such notions as separability and completeness of spaces. Other concepts, such as the length $\|f\|$ or the concept of orthogonality, can be traced back to Erhard Schmidt,[115] to Study,[116] and even to Grassmann.[117] Schmidt was the first to recognize the importance of projection operators which, as we know, proved so important for von Neumann's spectral theory. For finite-dimensional spaces such projections had already been introduced by Grassmann, who, in such spaces, had defined also inner products, linear independence, complete orthonormal systems, and linear manifolds. For infinite-dimensional spaces the notion of complete orthonormal systems seems to have been used for the first time by Gram.[118] The inequality $|(f,g)| \leq \|f\|\,\|g\|$, which appears so frequently in von Neumann's theory, had been called, by Poincaré[119] in

[113] We are using von Neumann's notation in which, for complex a, $(af,g) = a\,(f,g)$ and $(f,ag) = a^*(f,g)$.

[114] M. Fréchet, "Sur quelques points du calcul fonctionnel," *Rendiconti del Circolo Matematico di Palermo 22*, 1–74 (1906); "Essai de géométrie analytique à une infinité de coordinées," *Nouvelles Annales de Mathématiques 8*, 97–116, 289–317 (1908). Separability is defined on p. 305.

[115] E. Schmidt, "Über die Auflösung linearer Gleichungen mit unendlich vielen Unbekannten," *Rendiconti del Circolo Matematico 25*, 53–77 (1908), especially chap. 1: "Geometrie in einem Funktionenraum."

[116] E. Study, "Kürzeste Wege im komplexen Gebiet," *Mathematische Annalen 60*, 321–378 (1905).

[117] REF. 19 OF CHAP. 5.

[118] J. P. Gram, "Ueber die Entwicklung reeller Funktionen in Reihen mittels der Methode der kleinsten Quadrate," *Journal für die reine und angewandte Mathematik (Crelle) 94*, 41–73 (1883).

[119] H. Poincaré, "La méthode de Neumann et le problème de Dirichlet," *Acta Mathematica 20*, 59–142 (1896); quotation on p. 73.

1896, "l'inégalité de Schwarz" after its discoverer.[120] Its usefulness became evident after Hellinger and Toeplitz[121] had shown in their study of infinite matrices in how powerful a way it can be applied. Finally, the notion of Hermitian operators goes back, of course, to that of Hermitian forms which had been introduced by Charles Hermite, first[122] for $n = 2$, and one year later[123] for any finite n.

After these historical remarks on some of the most important functional-analytical notions in modern quantum mechanics we shall turn to von Neumann's treatment of the eigenvalue problem. Von Neumann realized that the traditional formulation of this problem $H\psi = \lambda\psi$ and the requirement that its solutions form a complete set, a development which can be traced back to Sturm[124] and Liouville,[125] cannot be carried over into the transformation theory in Hilbert space—the eigenfunction of a free particle is no Hilbert vector; he therefore reformulated the eigenvalue problem by using a method developed by Hilbert in 1906 in his study of integral equations.[126]

To understand von Neumann's reformulation we shall discuss it first for linear vector spaces of N dimensions. Let us assume that the unitary matrix U diagonalizes the Hermitian form $(Ax,x) = \sum_{m,n}^{N} A_{mn}x_m^*x_n$ to

$$\sum_{k=1}^{N} a_k y_k y_k^* = \sum_{r=1}^{M}\left(a^{(r)} \sum_{\rho=1}^{g_r} y_\rho^{(r)} y_\rho^{(r)*}\right)$$

where $a^{(r)}$, with $a^{(1)} < a^{(2)} < \cdots < a^{(M)}$, are the M $(\leq N)$ *different* eigenvalues of A and where $a^{(r)}$, of g_r-fold degeneracy, belongs to the g_r vectors $y^{(r)}$ with components $y_\rho^{(r)}$. Using the projection operator $F^{(r)}$, an N-dimensional matrix whose elements are zero except those corresponding to $a^{(r)}$—which are unity—so that

$$\sum_{\rho=1}^{g_r} y_\rho^{(r)} y_\rho^{(r)*} = (F^{(r)}y,y)$$

[120] H. A. Schwarz, "Ueber ein die Flächen kleinsten Flächeninhalts betreffendes Problem der Variationsrechnung," *Acta Societatis Scientiarum Fennicae 15*, 315–362 (1885); *Gesammelte Mathematische Abhandlungen* (Springer, Berlin, 1890). vol. 1, pp. 251–253.

[121] REF. 73 OF CHAP. 5.

[122] C. Hermite, "Sur la théorie des formes quadratiques," *Journal de Crelle 47*, 343–368 (1854); *Œuvres de Ch. Hermite*, edited by E. Picard (Gauthier-Villars, Paris, 1905), vol. 1, pp. 234–263.

[123] C. Hermite, "Remarque sur un théorème de Cauchy," *Comptes Rendus 41*, 181–183 (1855); *Œuvres* (REF. 122), vol. 1, pp. 479–481.

[124] C. Sturm, "Sur les équations différentielles linéaires du second ordre," *Journal de Mathématique 1*, 106–186 (1836); "Sur une classe d'équations à différences partielles," *ibid.*, 373–444.

[125] J. Liouville, "Sur le développement des fonctions," *ibid.*, 253–265, *2*, 16–35, 418–436 (1837); "D'un théorème dû à M. Sturm," *ibid. 1*, 269–277 (1836).

[126] REF. 71 OF CHAP. 5.

we easily see that

$$(Ax,x) = \sum_r a^{(r)}(F^{(r)}y,y)$$

and hence, since $y = U^{-1}x$,

$$(Ax,x) = \sum_r a^{(r)}(UF^{(r)}U^{-1}x,x)$$

Since also $UF^{(r)}U^{-1}$ is a projector

$$(Ax,x) = \sum_r a^{(r)}(E^{(r)}x,x)$$

or symbolically

$$A = \sum_r a^{(r)}E^{(r)}$$

If, in particular, A is the identity operator I,

$$I = \sum_r E^{(r)}$$

For this reason the range of projectors $\{E^{(r)}\}$ has been called a "resolution of the identity."[127] The result of our discussion for finite-dimensional vector spaces can consequently be formulated as follows: to every Hermitian operator A belongs a (unique) resolution of the identity $\{E^{(r)}\}$ such that $A = \sum_r a^{(r)}E^{(r)}$; that is, $(Ax,x) = \sum_r a^{(r)}(E^{(r)}x,x)$ or, more generally, $(Ax,y) = \sum_r a^{(r)}(E^{(r)}x,y)$ and the eigenvalue problem has been reduced to the determination of the resolution of the identity.

Von Neumann, now, showed that this approach can be generalized in a mathematically rigorous manner to the infinite-dimensional Hilbert space. As a result of this generalization, for whose details the reader is referred to the original literature, von Neumann could formulate the eigenvalue problem as follows: For a given Hermitian operator A a resolution of the identity $\{E(\lambda)\}$ has to be found such that (1) $A = \int_{-\infty}^{\infty} \lambda\, dE(\lambda)$, that is, such that $(Af,g) = \int_{-\infty}^{\infty} \lambda\, d(E(\lambda)f,g)$ holds for all g and all f for which $0 \leq \int \lambda^2\, d(\| E(\lambda)f \|^2) < +\infty$; (2) for $\lambda \leq \lambda'$, $E(\lambda) \leq E(\lambda')$, that is, $E(\lambda') - E(\lambda)$ is a projection operator; (3) for $\lambda \to +\infty$ $E(\lambda)f \to f$; for $\lambda \to -\infty$, $E(\lambda)f \to 0$; (4) $\lambda' \to \lambda$ with $\lambda' > \lambda$, $E(\lambda')f \to E(\lambda)f$; that is, discontinuities occur only to the left.

For bounded (or, equivalently, continuous) operators the problem had been solved by Hilbert, who had shown that to each such operator there belongs one and only one resolution of the identity. Since the important operators in quantum mechanics, however, are in general un-

[127] The German original "Zerlegung der Einheit" is sometimes also rendered as "resolution of the unity."

bounded, von Neumann was faced with the problem of extending Hilbert's spectral theory to the case of unbounded Hermitian operators, a problem which he finally solved[128] in 1929 for all Hermitian operators which are maximal, that is, which possess no proper extensions.[129] For he showed that (1) every Hermitian operator can be extended to a maximal operator and (2) to every maximal operator belongs exactly one resolution of the identity or none.

The new formalism also made it possible for von Neumann to reformulate the statistical foundations of quantum mechanics. Referring to Schrödinger's eigenvalue problem $H\psi_n = \lambda_n\psi_n$, which, for the sake of simplicity, he assumed to possess a pure point spectrum, he noticed that the resolution of the identity for H is given by $E(\lambda)f = \sum_{\lambda_n \leq \lambda} (f,\psi_n)\psi_n$ and for the operator q by

$$F(\lambda)\psi(q) = \begin{cases} \psi(q) & \text{if } q \leq \lambda \\ 0 & \text{if } q > \lambda \end{cases}$$

According to Bohr $\int_J |\psi(q)|^2 \, dq$ is the probability that the position coordinate q of the system in state ψ is found in the interval $J = [q',q'']$ and $\sum_{\lambda_n \epsilon I} |\int \psi(q)\psi_n^*(q) \, dq|^2$ is the probability that the energy of the system in state ψ is found in the interval $I = [\lambda',\lambda'']$. Von Neumann now gave to these two probability statements a unified formulation as follows. If it is only known that the energy is in I, the probability P that q is in J is given by

$$P = \sum_{\lambda_n \epsilon I} \int_{q'}^{q''} |\psi_n(q)|^2 \, dq = \sum_{\lambda_n \epsilon I} | F(q',q'')\psi_n(q)|^2 \, dq$$

$$= \sum_{n=1} \int_{-\infty}^{\infty} | F(q',q'')E(\lambda',\lambda'')\psi_n(q)|^2 \, dq = \sum_{n=1} || F(q',q'')E(\lambda',\lambda'')\psi_n(q)||$$

or symbolically[130]

$$P = [E(I)F(J)]$$

Generalizing this result, von Neumann formulated the basic statistical assumption of quantum mechanics as follows: If $R_1, R_2, \ldots, R_i$ form one set of (commutative) operators representing dynamical variables and $S_1, S_2, \ldots, S_j$ another such set, the former having $E_1(\lambda), \ldots$ as their

[128] J. von Neumann, "Allgemeine Eigenwerttheorie Hermitischer Funktionaloperatoren," *Mathematische Annalen 102*, 49–131 (1929); *Collected Works* (REF. 97), vol. 2, pp. 3–85. Identical results were independently obtained by Marshall Harvey Stone, *Linear Transformations in Hilbert Space* (American Mathematical Society Colloquium Publications, New Haven, Conn., 1932).

[129] B is an extension of A if everywhere where Af is defined Bf is defined and equal to Af.

[130] $F(q',q'') = F(q'') - F(q')$ and similarly for $E(\lambda)$.

resolutions of the identity and the latter $F_1(\lambda), \ldots$, then the probability that the values of the $S_1, \ldots$ lie in the intervals $J_1, \ldots$ if the values of the $R_1, \ldots$ lie in the intervals $I_1, \ldots$ is given by

$$[E(I_1) \cdots E(I_i)F(J_1) \cdots F(J_j)]$$

This statement, von Neumann concluded, contains all the statistical assertions of quantum mechanics which have been made so far. On the basis of these results he offered a mathematically rigorous formulation of the transformation theory which comprised all previous formalisms of quantum mechanics. It will have been observed that this formulation, which von Neumann presented in Göttingen on May 20, 1927, did not postulate any physical or epistemological assumptions other than those on which the transformation theories of Dirac and Jordan were based. In particular, the uncertainty principle which Heisenberg had meanwhile (in March, 1927) proposed did not yet play any part in these considerations. As we shall see, however, this principle not only greatly influenced von Neumann's subsequent elaborations and, in particular, his theory of measurement, but also was of decisive importance for the—at that time generally accepted—interpretation of the whole formalism advanced so far; we shall therefore have to discuss the uncertainty relations and their epistemological implications in greater detail.

But before we conclude the present chapter, we should add a note of caution against an overoptimistic evaluation of von Neumann's great achievement. For there are some serious doubts whether a formulation of quantum mechanics, based on a separable Hilbert space in which, consequently, every orthonormal basis is denumerable, suffices to encompass the solutions of all physically significant problems. In particular, the application of von Neumann's theory to systems with continuous energy states has led to the systematic use of functions not normalizable in the sense of quadratic integrability. Although alternative methods of normalization have been developed (Weyl, Fues, and others[130a]), no canonical set of the resulting wave packets could be established to serve as an orthonormal basis in a space of quadratically integrable functions. To overcome such difficulties, inner-product function spaces of solutions of the Schrödinger equation for free and scattered particles, admitting plane, cylindrical, and spherical waves, have been constructed;[131] but—in contrast to von Neumann's approach and its foundation for the statistical interpretation—in these

[130a] For a recent attempt to unify consistently Dirac's formalism with von Neumann's formulation of quantum mechanics by replacing unnormalizable eigenvectors of operators with continuous spectra by true unit vectors in direct integral decomposition spaces, see A. R. Marlow, "Unified Dirac–von Neumann formulation of quantum mechanics I," *Journal of Mathematical Physics 6*, 919–927 (1965).

[131] E. L. Hill, "Function spaces in quantum-mechanical theory," *Physical Review 104*, 1173–1178 (1956); "State spaces in scattering theory," *ibid. 107*, 877–883 (1957).

spaces an expansion theorem (the possibility of expressing the basic functions of one space in terms of the basic functions of any other) does not generally hold. Furthermore, a mathematically rigorous treatment of the scattering problem in terms of the so-called scattering operator or the S matrix could so far not be worked out in a satisfactory manner within the context of von Neumann's formalism in spite of the importance of the problem, although during the past ten years some promising results along this line have been achieved.[132] Finally, it is well known that for systems described by a quantized field theory von Neumann's separability condition is not satisfied.

[132] J. M. Cook, "Convergence to the Møller wave-matrix," *Journal of Mathematics and Physics* (*M.I.T.*) *36*, 82–87 (1957); J. M. Jauch, "Theory of the scattering operator," *Helvetica Physica Acta 31*, 127–158 (1958); M. N. Hack, "On convergence to the Møller wave operator," *Il Nuovo Cimento 9*, 731–733 (1958); J. M. Jauch and I. I. Zinnes, "The asymptotic condition for simple scattering systems," *ibid. 11*, 553–567 (1959); S. T. Kuroda, "On the existence and the unitary property of the scattering operator," *ibid. 12*, 431–454 (1959).

7 The Copenhagen Interpretation

7.1 The Uncertainty Relations

With the establishment of the statistical transformation theory the formalism of nonrelativistic quantum mechanics was completed in all its essential points. But a formalism, even if complete and logically consistent, is not yet a physical theory. To reach this status, some of its symbols have to be given an operationally meaningful interpretation (or epistemic correlation in the sense of Carnap's "phenomenal-physikalische Zuordnung"[1]). For unless a formalism is linked with certain data of sensory experience in such a way that both the beginning and the end of a chain of theoretical deductions are anchored in experience, it is not verifiable or falsifiable by experiment or observation and consequently not a physical theory.

Apart from Born's probabilistic interpretation of the wave function—the idea of associating expectation values with average results of measurements, though already implicitly referred to by Born, as we have seen, was not yet fully worked out—no epistemic correlations had so far been established between the quantum-mechanical formalism and experience. Even if Bohr's frequency relation was considered as part of the formalism, or rather of its logical conclusions, and spectroscopic confirmation as methodologically unobjectionable, the formalism could not yet be regarded as fully interpreted. For even if certain conclusions of formal deductions could be compared with experience, the early stages of these deductions did not have any clear-cut correlations or associations with sense data. The formalism was, so to speak, interpreted, at best, only at its end. This unusual state of affairs was due, of course, to the peculiar development of the formalism which at first was dominated by the correspondence principle but later detached from it.

[1] Cf. R. Carnap, "Ueber die Aufgabe der Physik," *Kantstudien* *28*, 90–107 (1923).

The situation, however, was still worse. Terms like "place," "velocity," "orbit" continued to play an important part in the representation of the formalism although their very rejection lay at the foundation of the whole development. In any case, it was clear—as the basic commutation relation $pq - qp = h/2\pi i$ incontestably evinced—that these concepts, if at all legitimate, could not simply have their classical significance. Hence, even Born's interpretation, based as it was on the conception of the localizability of a particle or its "place," was, strictly speaking, only an *ignotum per ignotius*. In short, there still remained the open questions: how to interpret the new formalism and, in particular, how to put its symbols or mathematical expressions into correspondence with the concepts of classical physics—for only in this way could operationally meaningful correlations be established.

The urgency of these problems became particularly evident when Schrödinger visited Copenhagen in September, 1926, and lectured on wave mechanics at Bohr's Institute. With his characteristic insistence on an all-pervasive continuity in microphysics he attacked Bohr's views on discontinuity and his basic conception of quantum jumps. In spite of prolonged discussions—some debates began in the morning and ended at night—no agreement between the disputing sides could be reached.[2] Finally, when Bohr referred even to Einstein's paper[3] of 1916 (or 1917) on the transition probabilities as supporting his view that without such discontinuities Planck's radiation law could not be accounted for, Schrödinger reportedly[4] exclaimed: "If one has to stick to this damned quantum jumping, then I regret having ever been involved in this thing"; whereupon Bohr replied: "But we others are very grateful to you that you were, since your work did so much to promote this theory." The Bohr-Schrödinger clash of opinion was highly provocative and stimulated animated discussions in Copenhagen even long after Schrödinger's departure.[5]

Heisenberg realized that the root of this conflict was the lack of a definite interpretation of the quantum-mechanical formalism. Recognizing on grounds similar to those just mentioned that classical notions like "position" or "velocity" cannot be employed in microphysics in the same sense as they were in macrophysics, Heisenberg compared the situation in quantum mechanics with that which would have prevailed in relativity if the formalism of the Lorentz transformations had been combined with a language based on the notions of space and time in their prerelativistic

[2] Cf. E. Schrödinger's later paper "Are there quantum jumps?" *The British Journal for the Philosophy of Science 3*, 109–123, 233–242 (1952).

[3] REF. 100 OF CHAP. 3.

[4] *Archive for the History of Quantum Physics*, Interview with W. Heisenberg on Feb. 25, 1963. Cf. also W. Heisenberg, "Die Entwicklung der Deutung der Quantentheorie," *Physikalische Blätter 12*, 289–304 (1956).

[5] "After Schrödinger's visit we had innumerable discussions about the problem." REF. 4 (Feb. 25, 1963).

meaning. In fact, Einstein's reinterpretation of the concepts of space and time, by which the Lorentz transformations[6] were given a consistent operational meaning, served Heisenberg as an example of how to overcome the present difficulties in quantum mechanics. Just as Einstein reversed the question and—instead of asking how nature can be described by a mathematical scheme—postulated that nature always works so that the mathematical formalism can be applied to it, Heisenberg asked himself: "Well, is it not so that I can only find in nature situations which can be described by quantum mechanics?"[7] "Then I asked," Heisenberg continued, " 'What are those situations which you can define?'. . . Then I found very soon that these are situations in which there was this Uncertainty Relation between p and q. Then I tried to say 'Well, let us assume there is only this possibility of having $\Delta p \; \Delta q \geq h/2\pi$. Does this make a consistent statement? Can I then prove that my experiments never give anything different?' " That Heisenberg's retrospective reflections on the conceptual background of his discovery of the uncertainly principle are correct and are not projections of later ideas, and that his famous thought experiments for the determination of the position of an electron played, indeed, the same methodological role as Einstein's operational definition of the simultaneity of spatially separated events, will be fully corroborated by our analysis of Heisenberg's paper.

For a full comprehension of Heisenberg's approach it is necessary to understand his assessment of the conceptual situation at the end of 1926. The formalism of quantum mechanics, he reasoned, which operates in abstract multidimensional spaces and employs noncommutative quantities, does not admit of ordinary space-time descriptions or causal connections of physical phenomena. And yet, he pointed out, the intuitive space-time descriptions had been carried over from classical physics into quantum mechanics and were being applied indiscriminately. Heisenberg saw in this disparity between formalism and intuitive conceptions the root of serious difficulties and contradictions. Realizing, however, that it would be impossible to construct an independent appropriate conceptual apparatus (in the sense of a descriptive language) which would give an adequate intuitive interpretation of the abstract formalism, since all our conceptualizations are inseparably connected with spatiotemporal ideas, Heisenberg saw no other alternative than to retain the classical intuitive notions but to restrict their applicability.

In a letter to Pauli on October 28, 1926, Heisenberg had already written that it makes no sense to speak of a monochromatic wave at a definite instant or extremely short interval and that, as the basic commu-

[6] See REF. 4 OF CHAP. 5.

[7] *Archive for the History of Quantum Physics*, Interview with W. Heisenberg on Feb. 25, 1963.

tation relation clearly shows, it is "meaningless to speak of the place of a particle with a definite velocity."[8] "But if one does not take it too seriously with the accuracy in using the notions of velocity and position, then it may well make sense."[9] Four months later, after much meditation on these problems, Heisenberg arrived at certain conclusions which he communicated to Pauli in a 14-page letter, dated February 23, 1927. Encouraged by Pauli's enthusiastic reaction,[10] he wrote a comprehensive essay on the contents of this letter, sent it to Pauli for approval, and afterward presented it to Bohr. Bohr, who had just returned from a vacation in Norway, was not quite satisfied with the paper and suggested some modifications. But Heisenberg insisted on his original version and referred to Bohr's suggestions only in a postscript to his paper. Finally, at the end of March, he submitted his paper on the intuitive contents of the quantum-theoretic kinematics and mechanics,[11] in which he advanced his famous uncertainty principle, to the editor of the *Zeitschrift für Physik*.

The uncertainty principle or, as it is also called, the indeterminacy principle had its origin in the Dirac-Jordan transformation theory. In fact, some of its qualitative aspects had already been announced by the proponents of this theory. Thus, in the introduction to his formulation of this theory, Dirac,[12] referring to conjugate variables, declared: "One cannot answer any question on the quantum theory, which refers to numerical values for both the q and the p. One would expect, however, to be able to answer questions in which only the q or only the p are given numerical values, or, more generally, when any set of constants of integration ξ that commute with one another are given numerical values." And Jordan,[13] it will be recalled, arrived at a similar conclusion: "For a given value of q all values of p are equally possible."

Whereas Dirac and Jordan were fully aware, as these statements indicate, that in quantum mechanics, contrary to classical physics, assignments of sharp values both of q and of p are incompatible with each other, Heisenberg investigated the quantitative relation between the theoretically permissible distributions of such values. In other words, he inquired what information can be derived from the transformation theory on the relation between the statistical distribution of q values and that of p values.

To answer this question he assumed that the position coordinate q has a distribution around q' with an uncertainty $\delta q = q_1$, the latter being defined as that distance from q' at which the squared probability amplitude

[8] REF. 8 OF CHAP. 5, p. 44.

[9] "Nimmt man es aber mit Geschwindigkeit und Lage nicht so genau, so hat das sehr wohl einen Sinn." *Ibid.*

[10] Pauli said something like "now it becomes day in quantum mechanics" ("Morgenröte einer Neuzeit"). REF. 7.

[11] W. Heisenberg, "Über den anschaulichen Inhalt der quantentheoretischen Kinematik und Mechanik," *Zeitschrift für Physik 43*, 172–198 (1927).

[12] REF. 70 OF CHAP. 6, p. 623.

[13] REF. 92 OF CHAP. 6.

$|S(\eta,q)|^2$, for a fixed parameter η which need not be further characterized, falls to e^{-1} of its maximum height. Hence, assuming the dependence on q to be that of a Gaussian error curve, Heisenberg could write

$$|S(\eta,q)|^2 = \text{const exp}\left[-\frac{(q-q')^2}{q_1^2}\right]$$

and for $S(\eta,q)$ itself

$$S(\eta,q) = \text{const exp}\left[\frac{-(q-q')^2}{2q_1^2} - \frac{2\pi i p'(q-q')}{h}\right]$$

In

$$S(\eta,p) = \int_{-\infty}^{\infty} S(\eta,q)\,S(q,p)\,dq$$

he replaced, following Jordan,[14] $S(q,p)$ by $\exp(2\pi ipq/h)$ so that

$$S(\eta,p) = \text{const}\int \exp\left[-\frac{(q-q')^2}{2q_1^2} + \frac{2\pi i(p-p')q}{h}\right]dq$$

where $\exp(2\pi ip'q'/h)$ has been absorbed into the new constant. By writing

$$S(\eta,p) = \text{const exp}\left\{-\frac{2\pi^2 q_1^2(p-p')^2}{h^2} + \frac{2\pi i(p-p')q'}{h}\right\}$$

$$\times \int \exp\left\{-\left[\frac{\sqrt{2}\,q_1\pi(p-p')i}{h} - \frac{(q-q')}{\sqrt{2}q_1}\right]^2\right\}dq$$

he obtained after integration

$$S(\eta,p) = \text{const exp}\left[-\frac{2\pi^2 q_1^2(p-p')^2}{h^2} + \frac{2\pi i(p-p')q'}{h}\right]$$

the integral, extended from $-\infty$ to $+\infty$, being a constant. With p_1 defined by the equation

$$q_1 p_1 = \frac{h}{2\pi}$$

he could write

$$|S(\eta,p)|^2 = \text{const exp}\left[\frac{-(p-p')^2}{p_1^2}\right]$$

which by comparison with the first equation of this derivation showed that

[14] $\exp(2\pi ipq/h)$ is, of course the solution of the Schrödinger equation

$$\frac{\hbar}{i}\frac{\partial}{\partial q}S(q,p) = pS(q,p)$$

p_1 has the same significance for p as q_1 had for q. Thus, Heisenberg concluded, the product of the uncertainties δq and δp is $h/2\pi$, or

$$\delta q \, \delta p = \frac{h}{2\pi} \tag{7.1}$$

"The more accurately the position is determined, the less accurately the momentum is known and conversely."[15] In this conclusion Heisenberg saw the "direct intuitive interpretation" ("direkte anschauliche Erläuterung") of the basic commutation relation $pq - qp = h/2\pi i$.

Heisenberg now asked himself whether this reciprocal limitation of precisions is merely a restriction imposed by the mathematical formalism of the transformation theory or whether it is a reflection of a more profound state of affairs. More precisely, he asked "whether it is not possible by a more profound analysis of the kinematical and mechanical conceptions to resolve the contradictions . . . and to obtain an intuitive understanding of the quantum-mechanical relationships."[16]

Heisenberg was hoping that an analysis of the notions of position and velocity, or rather their reinterpretation, would do for quantum mechanics what Einstein's famous analysis of the concept of simultaneity did for the mechanics of high-speed motions; and just as Einstein resolved thereby the contradictions in prerelativistic physics so, Heisenberg believed, his analysis would remove the difficulties of atomic physics. Starting his analysis with the notion of position, Heisenberg declared: "If one wants to clarify what is meant by 'position of an object' ('Ort des Gegenstandes'), for example, of an electron"—and he added "relatively to a given reference system"—"he has to describe an experiment by which the 'position of an electron' can be measured; otherwise this term has no meaning at all."[17] Heisenberg now outlined his well-known thought experiment with the γ-ray microscope, an idea which had occupied his mind as early as 1924, that is, well before he discovered matrix mechanics, as he once confided to one of his pupils.[18] At an early date he had also discussed such a possibility with Borchert Drude, who like his father, Paul Drude, specialized in optics.[19] It is also interesting to note in this context that the problem of the resolving power of the microscope was one of the questions (the others were on the resolving power of the Fabry-Perot plates and on the theory of the storage battery) which Wien had asked but Heisenberg failed to answer in his oral Ph.D. examination at the University of Munich in 1923. It seems that Heisenberg was conscientious enough to study these subjects after the examination. He himself said: "One might even say that in this later work on the γ-ray

[15] "Je genauer der Ort bestimmt ist, desto ungenauer ist der Impuls bekannt und umgekehrt." REF. 11, p. 175.

[16] *Ibid.*, p. 173.

[17] *Ibid.*, p. 174.

[18] M. R. Siddiqi, *Lectures on Quantum Mechanics* (Osmania University Press, Hyderabad-Deccan, 1938), vol. 1, p. 234.

[19] REF. 7.

microscope and the uncertainty relation I used the knowledge I had acquired by this poor examination."[20] Still, in his discussion of the experiment he did not take into account the angular aperture of the objective, an omission which Bohr soon corrected.

In principle, Heisenberg contended, one can determine the position or coordinate q to any desired degree of accuracy. One has only to illuminate the electron with radiation of sufficiently short wavelength and to observe it under the microscope. Owing to the Compton effect, however, the impact of the light quantum upon the electron changes the latter's momentum abruptly, "this change being the greater, the smaller the wavelength of the light used, that is, the more accurately the position is determined."[21] Heisenberg's omission of the angular aperture of the microscope lens made his description of the experiment somewhat vague as to the ultimate reason of the uncertainty in momentum. The reason why on the basis of the theory of the Compton effect the change in momentum cannot be sharply determined is, of course, that the direction of the scattered photon cannot be determined within the bundle of rays entering the microscope.

Analyzing a Stern-Gerlach experiment, Heisenberg showed in a similar vein that the precision in the measurement of energy is smaller, the shorter the time spent by the atom in crossing the deviating field, or

$$\delta E \, \delta t \sim h \tag{7.2}$$

In view of these results Heisenberg could declare: "All the concepts that are used in the classical theory for the description of a mechanical system can also be defined exactly for atomic processes. But the experiments which lead to such definitions carry with them an uncertainty if they involve the simultaneous determination of two canonically conjugate quantities."[22] In fact, it is precisely this uncertainty, as Heisenberg pointed out, which makes the use of classical notions admissible.

Even a notion like "path" or "trajectory" ("Bahn"), which presupposes a sharp knowledge of both position and momentum, can be retained in quantum mechanics, contended Heisenberg. For "the path comes into existence only when we observe it,"[23] he declared. This contention, one of the most provocative statements ever made in physics, was justified by him as follows. If an atom which, say, is in its 1000th excited state is illuminated by light of relatively long waves—this suffices because the dimensions under discussion are large—the electron, owing to the Compton effect, may be found after the interaction somewhere between the 950th and the 1050th state, the particular orbit being indeterminate.

[20] *Archive for the History of Quantum Physics*, Interview with W. Heisenberg on Feb. 11, 1963.

[21] REF. 11, p. 175.

[22] *Ibid.*, p. 179.

[23] "Die 'Bahn' entsteht erst dadurch, daß wir sie beobachten." *Ibid.*, p. 185.

The electron will therefore be represented by a wave packet in configuration space, composed of the eigenfunctions of the states just mentioned, and of a size determined by the precision of the position measurement, that is, by the wavelength of the illuminating light. The packet describes an orbit analogous to that of a classical particle, but spreads with time. Every new observation by which a certain q is selected out of a multitude of possibilities reduces the otherwise spreading wave packet again to the dimensions of the wavelength of the incident light. The constant replacement in each consecutive observation of the wave packet, which has spread out, by a smaller packet gives rise to the orbit as the temporal sequence of the locations where the wave packet has been observed.

From these results Heisenberg drew a conclusion of far-reaching philosophical implications at the end of his paper: "We have not assumed that the quantum theory, unlike classical physics, is essentially a statistical theory in the sense that from exact data only statistical conclusions can be inferred. For such an assumption is refuted, for example, by the well-known experiments by Geiger and Bothe. However, in the strong formulation of the causal law 'If we know exactly the present, we can predict the future' it is not the conclusion but rather the premise which is false. We *cannot* know, as a matter of principle, the present in all its details."[24] And Heisenberg concluded: "In view of the intimate connection between the statistical character of the quantum theory and the imprecision of all perception ('Ungenauigkeit aller Wahrnehmung') it may be suggested that behind the statistical universe of perception there lies hidden a 'real' world ('eine "wirkliche" Welt') ruled by causality. Such speculations seem to us—and this we stress with emphasis—useless and meaningless ('unfruchtbar und sinnlos'). For physics has to confine itself to the formal description of the relations among perceptions."[25]

For a full comprehension of Heisenberg's reasoning the following remarks should be noted. The purely mathematical result $\delta q \, \delta p = h/2\pi$ [or formulas (7.3) and (7.4), which follow], that is, the fact that the product of the errors or standard deviations of the variables is equal to, or exceeds, a certain positive number, does not in itself entail that these variables cannot be simultaneously measured with perfect precision—as examples of two random variables defined on sets of macroscopic objects, such as the height and the weight of the members of a given population, clearly show. Although suggested by two important ideas—namely, (1) the tacit assumption of the perfect similarity (or in a more modern terminology "identity of states of preparation" as in a pure case) of the particles under discussion and (2) the recognition of an unavoidable interaction between the object and the measuring apparatus—Heisenberg's identification of formula (7.1) with an uncertainty relation, as he understood it, had not

[24] *Ibid.*, p. 197.

[25] *Ibid.*, p. 197.

the character of a logically necessary conclusion. Ultimately, like every interpretation of a mathematical relation, it was an act of associating symbols with operations and thus contained a certain element of arbitrariness.

One may be tempted to relate this arbitrariness to the fact that in the years 1925 and 1926 questions concerning the limitations of the sensibility of measuring instruments were much discussed. In 1925 Moll and Burger constructed their thermal relais technique,[26] which proved very useful for the microphotometry in experimental spectroscopy; discussing the sensibility of galvanometers, they declared: "By the application of this technique the sensibility of galvanometers can be increased indefinitely."[27] At the same time, however, the disturbing effects of Brownian fluctuations[28] were already fully recognized. In fact, in the summer of 1926, G. Ising[29] and F. Zernike[30] gave a series of guest lectures on this subject in Göttingen, and these lectures received a considerable amount of attention.[31] We must emphasize, however, that we have found no evidence of any direct or indirect influence of these lectures or discussions upon Heisenberg.

Heisenberg's omission of taking into account the angular aperture of the microscope gave rise to an often repeated misconception. Although his description of the experiment mentioned the deflection of the reflected light quantum by the lens of the microscope, it did not emphasize the indeterminacy of the direction of its reflection but put the stress instead upon the discontinuous change of the momentum of the electron. This was somewhat misleading, for it suggested that this discontinuous momentum change, as an uncontrollable interaction between the object (the electron) and the measuring device (the incident radiation), produces the uncertainty. It should be understood, however, that in the Compton effect, as the Bothe-Geiger and Compton-Simon experiments have shown, the discontinuous change of momentum can be rigorously calculated, provided the direction of the reflected quantum is accurately known. Now, only because this direction cannot be ascertained more accurately than up to the finite angle of the aperture may one speak of an uncontrollable interaction.

[26] W. J. H. Moll and H. C. Burger, "Das Thermorelais," *Zeitschrift für Physik 34*, 109–111 (1925).

[27] W. J. H. Moll and H. C. Burger, "Empfindlichkeit und Leistungsfähigkeit eines Galvanometers," *Zeitschrift für Physik 34*, 112–119 (1925); "Es ist möglich, die Empfindlichkeit eines Galvanometers durch die Anwendung der früher beschriebenen Thermorelais-Methode *beliebig zu vergrößern*." (Italics by the authors), p. 115.

[28] W. Einthoven, W. F. Einthoven, W. van der Horst, and H. Hirschfeld, "Brownsche Bewegingen van een Gespannen snaar," *Physica 5*, 358–360 (1925).

[29] G. Ising, "A natural limit for the sensibility of galvanometers," *Philosophical Magazine 1*, 827–834 (1926).

[30] F. Zernike, "Die natürliche Beobachtungsgrenze der Stromstärke," *Zeitschrift für Physik 40*, 628–636 (1926).

[31] Cf., for example, P. Jordan, "Kausalität und Statistik in der modernen Physik" (Habilitationsvortrag), *Die Naturwissenschaften 15*, 105–110 (1927), especially p. 108.

At the time of the publication of Heisenberg's paper, the philosophical problems which it raised were not immediately apprehended—problems such as whether the uncertainties are of an ontological nature (that is, whether the particle just does not have a determinate position or velocity) or whether they are only of an epistemological nature (that is, whether they are due to the crudeness of the measuring process or, in any case, exist only in our knowledge). From Heisenberg's remarks concerning the notion of "path" or "trajectory" it may be inferred that he would have favored the first alternative. At the present stage philosophical problems of this kind were not yet so much in the limelight of interest as were the methodological questions concerning the definability and measurability of the concepts of quantum mechanics. Heisenberg's basic approach, viewed from this angle, was at that time to regard something as defined if it is measurable. In short, definability was based on, or reduced to, measurability. Furthermore, according to Heisenberg's paper, light was focused not only *in* a measurement, but also *on* measurement as such: it directed the attention of atomic physics to the importance of the concept of measurement which from now on was to assume a central position in all questions relating to the foundations of quantum mechanics. Finally, our assessment of Heisenberg's paper would not be complete if we omitted mentioning that his proposed solution of the problem of causality, his idea that the unascertainability of exact initial values obstructs predictability and, consequently, deprives causality of any operational meaning, was an outstanding contribution to modern philosophy. In fact, modern philosophy, as Schlick[32] later admitted, was taken by surprise since even the mere possibility of such a solution had never been anticipated in spite of the profusion of discussions on this problem for generations.

It took some time until philosophy fully appreciated the implications of Heisenberg's paper for the problem of causality. In fact, even its impact on physics was at first not generally recognized, although the paper, unlike the one on matrix mechanics of July, 1925,[33] was reviewed immediately after its publication in the *Physikalische Berichte*[34] and somewhat later in *Science Abstracts*.[35] Apart from Bohr and Pauli, who, as we know, were intimately acquainted with Heisenberg's ideas, only Kennard, then visiting Copenhagen, fully understood from the beginning the far-reaching sig-

[32] M. Schlick, "Die Kausalität in der gegenwärtigen Physik," *Die Naturwissenschaften 19*, 145–162 (1931). "Soviel auch über Determinismus und Indeterminismus, über Inhalt, Geltung und Prüfung des Kausalprinzips philosophiert wurde—niemand ist gerade auf diejenige Möglichkeit verfallen, welche uns die Quantenphysik als den Schlüssel anbietet, der die Einsicht in die Art der kausalen Ordnung öffnen soll, die in der Wirklichkeit tatsächlich besteht," p. 145.

[33] See p. 208.

[34] *Physikalische Berichte 1927*, p. 1784.

[35] *Science Abstracts 31*, 11 (1928). Both reviews only report that "exact definitions of place, velocity . . . are proposed which are valid in quantum mechanics, and it is shown that canonical conjugated quantities can only be determined simultaneously with a characteristic inexactness. The latter is the particular basis for the appearance of statistical connections in quantum mechanics. The mathematical formulation has been obtained by means of the Dirac-Jordan theory. . . ."

nificance of the uncertainty relations. Already in July, 1927, Kennard, in his previously mentioned review article,[36] called the relations "the core of the new theory" ("der eigentliche Kern der neuen Theorie").[37] He also generalized Heisenberg's derivation to the case of any two canonically conjugate variables and pointed out that the particular choice of the Gaussian probability distribution gives the optimal limit of precision. In addition, he declared that one can no longer speak, as in classical physics, of the true value of a physical quantity since what hitherto was regarded as an "error of observation" has now become an integral part of the theory. In his well-known encyclopedia article[38] Pauli began his exposition of the general principles of quantum theory with the statement of the uncertainty relations. There are indications that it was also due to Pauli that Hermann Weyl's *Theory of Groups and Quantum Mechanics*[39] was the first treatise on quantum mechanics which assigned to the uncertainty relations—which Weyl on a suggestion by Pauli derived on the basis of the Schwarz inequality—an integral part in the logical structure of the whole theory. C. G. Darwin was probably the first to recognize the connection between the uncertainty relations and the inversion of the Fourier integral, as his paper "The electron as a vector wave"[40] discloses.

As may have been expected, the uncertainty principle was regarded in certain quarters as a challenge to experimental ingenuity, and numerous thought experiments were devised to disprove it. Thus, for example, at the Nashville meeting of the American Physical Society on December 30, 1927, Ruark[41] suggested the following idealized experiment. Two slits, a great distance d apart, are equipped with movable shutters. At the time t_1 the first slit is opened for an interval τ and a freely moving particle is made to pass through it. If it is found that the particle moves at such a rate that it passes also through the second slit, which is opened at the time t_2 likewise for an interval τ, then, contended Ruark, since d and $t_2 - t_1$ can be determined with an arbitrary precision, and since τ may be taken sufficiently minute, a simple calculation would show that Heisenberg's uncertainty

[36] REF. 95 OF CHAP. 6.

[37] *Ibid.*, p. 337.

[38] REF. 283 OF CHAP. 5. Pauli's approach was subsequently adopted by numerous authors of textbooks on quantum mechanics. Cf., for example, D. Bohm, *Quantum Theory* (Prentice-Hall, Englewood Cliffs, N.J., 1951); S. Dushman, *The Elements of Quantum Mechanics* (Wiley, New York, 1938); H. A. Kramers, *The Foundations of Quantum Theory* (North-Holland Publishing Co., Amsterdam, 1957); W. Heisenberg, *The Physical Principles of the Quantum Theory*, translated by C. Eckart and F. C. Hoyt (University of Chicago Press, 1930; Dover, New York); L. D. Landau and E. M. Lifshitz, *Quantum Mechanics* (Pergamon Press, London; Addison-Wesley, Reading, Mass., 1958); A. March, *Die Grundlagen der Quantenmechanik* (2d ed., J. A. Barth, Leipzig, 1931); L. Schiff, *Quantum Mechanics* (McGraw-Hill, New York, 1949).

[39] H. Weyl, *Gruppentheorie und Quantenmechanik* (S. Hirzel, Leipzig, 1928; 2d ed. 1931); translated into English by H. P. Robertson as *The Theory of Groups and Quantum Mechanics* (Methuen, London, 1931; Dover, New York, 1950).

[40] *Proceedings of the Royal Society of London (A)*, *117*, 258–293 (1927), received on Oct. 25, 1927.

[41] A. E. Ruark, "Heisenberg's indetermination principle and the motion of free particles," *Bulletin of the American Physical Society 2*, 16 (1927); *Physical Review 31*, 311–312 (1928).

relation is violated. This, however, Ruark declared, is not the case; Heisenberg's "indetermination principle"—he was the first to call it a "principle"—remains valid because "the measuring devices are statistical aggregates of atoms and are themselves subject to fluctuations." A few weeks later, at the New York meeting of the Society on February 24–25, 1928, Ruark[42] revised his argumentation and declared: "The velocity of the particle changes when it passes the first slit, for it may be considered as a group of waves and the frequency of each harmonic train in the group is changed by the modulation due to the shutter. This involves a change of energy, that is, a modified velocity."

Stimulated by Ruark's ideas, Kennard,[43] having returned at the end of 1927 from Copenhagen to Cornell, studied in great detail how what he called the "shutter effect" changes the speed of an electron and concluded in analogy with the Fourier resolution of a finite wave train that this velocity is predictable only in a statistical sense. It is interesting to note that neither Ruark nor Kennard recognized in this context that the uncertainty principle does not refer to the past. In fact, from the idealized experiments which they considered it could easily be inferred that if the position of an electron is precisely measured, after an exact measurement of its velocity, the product $\delta p \, \delta q$ has a value much smaller than h. But this value, as Heisenberg[44] soon pointed out in his Chicago lectures, "can never be used as an initial condition in any calculation of the future progress of the electron and thus cannot be subjected to experimental verification. It is a matter of personal belief," he declared, "whether such a calculation concerning the past history of the electron can be ascribed any physical reality or not."[45] There was, however, one important result to which Ruark was led by his study of these thought experiments. Heisenberg, it will be recalled, declared that a dynamical variable such as the position of a particle can be measured, in principle at least, to any desired degree of accuracy although at the price of the accuracy of the simultaneous measurement of its conjugate variable. Ruark,[46] now, was one of the first to challenge this statement. He pointed out, for example,[47] that in the γ-ray experiment the wavelength of the incident radiation cannot be made smaller than the increase of wavelength due to the Compton effect which, for a scattering angle of 90°, is h/mc. Similar results, though on the basis of relativistic considerations, were obtained also by Flint and Richardson.[48]

[42] A. E. Ruark, "Heisenberg's uncertainty relation and the motion of free particles," *Physical Review 31*, 709 (1928).

[43] E. H. Kennard, "Note on Heisenberg's indetermination principle," *Physical Review 31*, 344–348 (1928).

[44] W. Heisenberg, *The Physical Principles of the Quantum Theory* (REF. 38).

[45] *Ibid.*, p. 20.

[46] A. E. Ruark, "The limits of accuracy in physical measurements," *Proceedings of the National Academy of Sciences 14*, 322–328 (1928).

[47] Today, of course, additional objections concerning such high-energy interactions may be raised.

[48] H. T. Flint and O. W. Richardson, "On the minimum proper time and its

While Ruark and Kennard discussed primarily the experimental aspects of Heisenberg's principle, Condon and Robertson studied its mathematical relation to the formalism of quantum mechanics. In his effort to find a general formulation of the principle which was assumed to hold for any two noncommutative operators, Condon[49] encountered the following difficulties: (1) the noncommutativity of two operators A and B does not imply that the product of their corresponding uncertainties $\Delta A \cdot \Delta B$, which Condon, following Weyl, defined by the variance (or the square of the standard derivation) in accordance with the formula

$$(\Delta A)^2 = \int \psi^* A^2 \psi \, d\tau - \left(\int \psi^* A \psi \, d\tau\right)^2$$

must be greater than, or equal to, some positive lower limit; (2) even if A and B do not commute, certain values of the quantities represented by these operators may both be known precisely; (3) there exist states with regard to which A and B commute although the values of their corresponding quantities cannot be known with unlimited precision. To substantiate (1) Condon considered ΔL_z, the uncertainty of the z component of the angular momentum, for the hydrogen eigenfunction $\psi_{nlm} = R(r) \times \exp(im\varphi) P_l^m(\cos\theta)$; since $L_z = m\hbar$, he concluded that $\Delta L_x \, \Delta L_z = 0$ in spite of the fact that $L_x L_z - L_z L_x \neq 0$. As to (2) he chose the state function ψ_{n00} so that $L_x = L_y = L_z = 0$ and their uncertainties vanish. Finally, as to (3) he pointed out that for ψ_{nl0} (for which $L_z = 0$) $L_x L_y - L_y L_x = i\hbar L_z$ implies that L_x and L_y commute although $\Delta L_x \neq 0$ and $\Delta L_y \neq 0$.

A few weeks later Robertson, a colleague of Condon at Princeton, clarified the situation. He showed[50] that the uncertainty principle can be generally formulated for any two Hermitian operators if in the formulation reference is made to the wave function of the state under consideration. For he proved that for a given normalized function ψ_0 and two Hermitian operators A and B

$$\Delta A \, \Delta B \geq \frac{h|C_0|}{4\pi} \tag{7.3}$$

where $C = (2\pi i/h)(AB - BA)$, $C_0 = \int \psi_0^* C \psi_0 \, d\tau$, $(\Delta A)^2 =$

---

application (1) to the number of chemical elements (2) to some Heisenberg relations," *Proceedings of the Royal Society of London (A)*, *117*, 637–649 (1928).

[49] E. U. Condon, "Remarks on uncertainty principles," *Science 69*, 573–574 (1929), completed on May 10, 1929. In this context it should be noted that P. Jordan's paper "Über eine neue Begründung der Quantenmechanik II," *Zeitschrift für Physik 44*, 1–25 (June, 1927), already implied that $[\varphi, L_z] = i\hbar$, in contrast to $[z, p_z] = i\hbar$, does *not* hold. For an uncertainty relation between ΔL_z and $\Delta\varphi$ see D. Judge, "On the uncertainty relation for angle variables," *Il Nuovo Cimento 31*, 332–340 (1964).

[50] H. P. Robertson, "The uncertainty principle," *Physical Review 34*, 163–164 (1929), completed on June 18, 1929.

$\int \psi_0^*(A - A_0)^2\psi_0\, d\tau$, $A_0 = \int \psi_0^* A\psi_0\, d\tau$, etc., or in a slightly modernized notation[51]

$$(\Delta_\psi A)(\Delta_\psi B) \geq \tfrac{1}{2}\,|\langle D\rangle_\psi| \tag{7.3'}$$

where $iD = AB - BA$. For the special case $A = p$ and $B = q$, Robertson concluded, $C = C_0 = 1$ and

$$\Delta q\,\Delta p \geq \frac{h}{4\pi} \tag{7.4}$$

in full agreement with Weyl's result. Ditchburn[52] soon clarified completely the relation of the inequality (7.4), which he called "Pauli's inequality," with Heisenberg's original "equality" (7.1) by pointing out that in accordance with the general theory of errors $\Delta q = \delta q/\sqrt{2}$ and similarly for p and by proving rigorously (with the assistance of J. L. Synge) that the limiting condition of equality in (7.4) is obtained if and only if the scatter is a Gaussian distribution.

Schrödinger, as we know from his correspondence with Bohr[53] and Einstein,[54] examined very carefully various implications of the uncertainty principle, such as its relation to the discernibility of the discrete energy levels of the atoms of an ideal gas in an enclosure.[55] When in the spring of 1930 he was studying the problem of how to distribute, in an optimal simultaneous measurement of p_0 and q_0 at an instant t_0, the unavoidable uncertainty $h/4\pi$ between the two variables in such a way that at a given later instant t the uncertainty Δq in position will be minimal, Sommerfeld drew his attention to the papers by Condon and Robertson. Schrödinger immediately realized that their result could be improved. For he showed[56] that for any two Hermitian operators A and B in any state ψ

$$(\Delta A)^2(\Delta B)^2 \geq \left(\frac{\overline{AB + BA}}{2} - \bar{A}\bar{B}\right)^2 + \left|\frac{\overline{AB - BA}}{2}\right|^2 \tag{7.5}$$

where $\bar{A} = \int \psi^* A\psi\, d\tau$, etc. Although the first additive squared term on the right-hand side of (7.5) may vanish in particular cases, as in the case of all optimal simultaneous measurements of canonically conjugate variables, Schrödinger's formula (7.5) provided, in general, for the lower

[51] For a very short derivation of (7.3') cf. N. K. Tyagi, "New derivation of the Heisenberg uncertainty principle," *American Journal of Physics 31*, 624 (1963).

[52] R. W. Ditchburn, "The uncertainty principle in quantum mechanics," *Proceedings of the Royal Irish Academy 39*, 73–80 (1930).

[53] *Bohr Archive*, Copenhagen, Letter to Bohr, dated May 13, 1928, and Bohr's answer of May 23, 1928.

[54] *Einstein Estate*, Princeton, N.J., Letter to Einstein, dated May 30, 1928.

[55] Cf. also his criticism of the particle concept on the basis of the uncertainty principle in his paper "What is an elementary particle?" *Endeavour 9*, 109–116 (1950); *Smithsonian Institution, Annual Report* 1950, pp. 183–196; *Science, Theory and Man* (Dover, New York, 1957), pp. 193–223.

[56] E. Schrödinger, "Zum Heisenbergschen Unschärfeprinzip," *Berliner Berichte 1930*, pp. 296–303.

limit of the product of the uncertainties a value greater than previously obtained.

In the early thirties, the far-reaching implications of Heisenberg's principle for physics as well as for epistemology were fully recognized, and the uncertainty principle became an issue of numerous discussions both by philosophically minded physicists and by philosophers interested in the foundations of science.[56a] C. G. Darwin called it a "great achievement,"[57] whereas von Laue[58] refused to accept it as a final barrier in the search for deeper causes; K. R. Popper[59] charged Heisenberg with having tried "to give a causal explanation why causal explanations are impossible." Some even saw in Heisenberg's indeterminacy principle a way to solve the long-standing philosophical conflict between the doctrines of free will and determinism. "If the atom has indeterminacy," argued Eddington,[60] "surely the human mind will have an equal indeterminacy; for we can scarcely accept a theory which makes out the mind to be more mechanistic than the atom." Lucretius' theory of the *exiguum clinamen principiorum*,[61] his doctrine that free will in man is made possible by the "minute swerving of the elements," enjoyed an unexpected revival on the basis of modern quantum mechanics.

To discuss these questions in detail would lead us too far from our subject. But we wish to point out that the uncertainty principle still

[56a] One of the earliest objections raised was that by G. W. Kellner in his papers "Die Kausalität in der Physik," *Zeitschrift für Physik 55*, 44–51 (1929), received Mar. 14, 1929; "Die Kausalität in der Quantenmechanik II,"*ibid. 59*, 820–835 (1930); "Zwei Bemerkungen zu meiner Arbeit 'Die Kausalität in der Physik,'" *ibid. 64*, 147–150 (1930); "Die Kausalität in der Physik," *ibid. 64*, 568–580 (1930). Kellner distinguished between what he called "the technical principle of causality," almost equivalent with Heisenberg's "strong" formulation, and Kant's causality principle which asserts the existence of a uniquely determined structural interconnection between all phenomena ("ein eindeutiger struktureller Zusammenhang zwischen allem"). The validity of the latter, Kellner contended, was not disproved by Heisenberg. H. Margenau, in an article entitled "Causality and modern physics," *The Monist 41*, 1–36 (1931), arrived on similar grounds at the same conclusion that "quantum mechanics does not require the abandonment of the causality principle," a view he later abandoned. A similar opinion was also expressed by L. Brunschvicg in his paper "Science et la prise de conscience," *Scientia 55*, 334–342 (1934). For a related criticism, namely that Heisenberg failed to distinguish between the validity and the applicability of the causality principle, see, for example, the recent paper by K. Hübner, "Zur gegenwärtigen philosophischen Diskussion der Quantenmechanik," *Philosophia Naturalis 9*, 3–21 (1965).

[57] C. G. Darwin, "The uncertainty principle," *Science 73*, 653–660 (1931); "The uncertainty principle in modern physics," *Scientific Monthly 34*, 387–396 (1932).

[58] M. von Laue, "Zu den Erörterungen über Kausalität," *Die Naturwissenschaften 20*, 915–916 (1932); "Über Heisenbergs Ungenauigkeitsbeziehungen und ihre erkenntnistheoretische Bedeutung," *ibid. 22*, 439–441 (1934).

[59] K. R. Popper, *Die Logik der Forschung* (Springer, Wien, 1935), p. 184; *The Logic of Scientific Discovery* (Basic Books, New York, 1959), p. 249. Cf. also Popper's "Zur Kritik der Ungenauigkeitsrelationen," *Die Naturwissenschaften 22*, 807–808 (1934), in which an experiment is suggested to disprove the principle, and von Weizsäcker's critique of this paper, *ibid.*, p. 808.

[60] A. S. Eddington, "The decline of determinism," *Mathematical Gazette 16*, 66–80 (1932).

[61] T. Lucretius Carus, *De rerum natura*, liber II, 292.

continues to play an important role in current research on the foundations of quantum mechanics. Thus, to mention only one example, a rigorous reformulation of quantum mechanics as a general statistical dynamical theory, that is, as a theory in which also the transformation with time of the probability distribution is a purely stochastic process,[62] leads, within the framework of ordinary probability concepts, as Suppes[63] has recently shown, to the conclusion that a genuine joint probability distribution for two variables, such as position and momentum, does not always exist. In other words, "not only are position and momentum not precisely measurable simultaneously, they are not simultaneously measurable at all."[63a]

Having discussed the development of Heisenberg's principle and its implications concerning the notion of causality, as conceived by the physicists at that time, we are now in a position to discuss also its implications with regard to the notions of identity, indistinguishability, and identifiability of physical objects, notions which came to the forefront of scientific interest as soon as the empirical spectra of polyelectronic atoms became the subject of quantum-mechanical investigations.

Our remarks[64] on Pauli's exclusion principle with which our topic is intimately related apply also at present: the subject, bearing upon physics and philosophy alike, seems to have not yet been fully explored from the epistemological and logical point of view. Since the revision of these notions which played an important role in the establishment of quantum statistics led to far-reaching philosophical conclusions, a brief digression on their conceptual development, as far as nonrelativistic quantum mechanics is concerned, seems appropriate.

Let us then start with the notion of "identical" or, as we prefer to call them in this section, "like"[64a] particles, that is, particles exactly equal in all their qualities or physical specifications. The problem whether such like entities really exist in nature has a long history of its own. Most philosophies of nature which, like the Pythagorean doctrine of atoms, based their systems on conceptions of combinatorial discreteness affirmed

[62] Cf., for example, J. E. Moyal, "Quantum mechanics as a statistical theory," *Proceedings of the Cambridge Philosophical Society 45*, 99–124 (1949).

[63] P. Suppes, "Probability concepts in quantum mechanics," *Philosophy of Science 28*, 378–389 (1961).

[63a] *Ibid.*, p. 378.

[64] See end of Chap. 3.

[64a] To avoid misconceptions we restrict the term "identical," unless otherwise qualified, to denote "numerical identity." The often used expression "similar" particles seems not fully adequate since the notion of "similarity"—more than that of "likeness"—admits of a margin of qualitative difference. "Things are *similar* . . . if the qualities they have in common are more numerous than those in which they differ," said Aristotle (*Metaphysics* 1018a 7–17). Nor is the term "indistinguishable" or "indiscernible" appropriate since it refers to subjective apperception and in theory "like" particles may well be "distinguishable," as we shall see. "Like" or "alike" has to be understood in this context as a technical term, not synonymous with "similar." No language, it seems, possesses a special word to specify exact qualitative likeness in contrast to numerical identity. And, after all, the simile reads: "as *like* as two peas."

the idea whereas those schools of thought which espoused the principles of continuity and determinism as their theoretical foundations denied it. Whether Democritus, according to whom "atoms differ in shape and size," taught that atoms of one kind are strictly alike, as so often alleged, seems not to stand the test of documentary evidence. Aristotle's description of Democritean atoms, for example, as having "all of them the same nature just as if each one separately were a piece of gold,[64b] does not decide the issue.

There is no doubt, however, that the Stoics, with their belief in continuity and determinism, categorically denied the existence of like entities. In their view two physical objects, however indistinguishable they may appear, must intrinsically differ from each other, for otherwise no sufficient reason would explain the difference of their positions in space. Thus, for example, Cicero, who became acquainted with the ontological realism of the Stoics through his contact with Diodotus, Antichus of Ascalon, and Posidonius, quoted them as saying that "no hair or no grain of sand is in all respects the same as another hair or grain of sand."[64c] Lucius Annaeus Seneca, the leader of Roman Stoicism, wrote in a similar vein: ". . . amid all this abundance there is no repetition; even seemingly similar things are, on comparison, unlike. God has created all the great number of leaves that we behold: each, however, is stamped with its special pattern. All the many animals: none resembles another in size—always some difference! The Creator has set himself the task of making unlike and unequal things that are different."[64d] Stating that the Academics "confuse things indistinguishable by their attributing one and the same quality to two substances,"[64e] Plutarch introduced the term "indistinguishability" (ἀπαραλλαξία) into philosophical discourse, a term which already appeared in the writings of Philodemus of Gedara.[64f] Plotinus,[64g] the founder of the Neoplatonic school in Rome, Athanasius[64h] in his polemics against

[64b] Aristotle, *De Caelo* 276a 1.

[64c] "Nullum esse pilum omnibus rebus talem, qualis sit pilus alius, nullum granum." M. Tullii Ciceronis *Academicorum Posteriorum Liber II*, chap. 26, section 85; *The Academica of Cicero*, edited by J. S. Reid (Macmillan, London, 1874), p. 66; *The Academic Questions of M. T. Cicero*, translated by C. D. Yonge (George Bell, London, 1887), p. 48.

[64d] *L. Annaei Senecae Ad Lucilium Epistularium Moralium quae supersunt*, edited by O. Hense (Teubner, Leipzig, 1914), p. 542 (Epistula 113); English translation in *Seneca ad Lucilium—Epistulae Morales*, translated by R. M. Gummere (Loeb edition: W. Heinemann, London; Harvard University Press, Cambridge, Mass., 1953), vol. 3, p. 291.

[64e] ". . . ab his omnes res confundi indiscreta ista similitudine, duabus substantiis unam qualitatem inesse contendentibus." *Plutarchi Chaeronensis Scripta Moralia, Graece et Latine* (De communibus notitiis 1077c) (A. F. Didot, Paris, 1841), vol. 2, p. 1318.

[64f] Cf. *Philodemos*: *Peri Semeion kai Semeioseon*, edited by T. Gomperz (Herkulanische Studien, Leipzig, 1865), pp. 6, 37; *Lexicon Philodemeum*, edited by C. J. Voos (J. Muusses, Purmerend, 1934), vol. 1, p. 36.

[64g] *Plotini Opera Omnia*, edited by G. H. Moser and F. Creuzer (Oxonii, 1835), pp. 998–999.

[64h] S. Athanasius, *Orationes adversus Arianos*, Patrologia Graeca, edited by J. P. Migne (Paris, 1857), vol. 26, pp. 186–187.

the Arian doctrine of "homoousia" (coequality of Father and Son), Giordano Bruno,[64i] Nicholas Malebranche,[64j] and many more thinkers of the past concurred with the Stoics on this issue, though on different grounds.

The most pregnant formulation of this viewpoint, based on an argument most akin to that of the Stoa,[64k] was given by Leibniz in his *principium identitatis indiscernibilium:* "no two objects in nature can differ only in number."[64l] In his fourth letter to Samuel Clarke[64m] Leibniz declared: "There is no such thing as two individuals indiscernible from each other." For otherwise, he argued, no sufficient reason could account for the fact that object A_1 is in place a_1 and object A_2 in a_2 rather than A_1 in a_2 and A_2 in a_1.

To discuss in detail the impact of Leibniz's principle of the identity of indiscernibles upon later thought would lead us too far astray. To Kant's criticism of the principle as "spurious"[64n] Salomon Maimon took exception, arguing that "the difference in external temporal-spatial relations has its origin in the difference of internal qualities. . . . Two drops of water could not appear in two places unless they differed in intrinsic qualities; it is only the result of the incompleteness of our conceptions that we do not realize this difference."[64o] For more recent reactions we refer the reader to the literature on this subject.[64p]

[64i] Giordano Bruno, *De Triplici Minimo et Mensura ad Trium Speculativarum Scientiarum et Multarum Activarum Artium Principia Libri V* (Francofurti, 1591).

[64j] "Il est certain que tous les corps naturels, ceux-là même que l'on appelle de même espèce, différent les uns des autres." *Recherche de la Verité,* in *Œuvres de Malebranche* (Charpentier, Paris, 1871), vol. 3, pp. 439–440.

[64k] In his *Grundriss der Geschichte der Philosophie des Altertums* (E. S. Mittler, Berlin, 4th ed., 1903), p. 298, F. Ueberweg already suggested that Leibniz was influenced by the Stoa in this respect. We may add that this conjecture seems to be confirmed in Leibniz's statements "neque . . . duo folia vel gramina in horto perfecte sibi similia reperientur," quoted in *Opuscules et Fragments Inédits de Leibniz,* edited by L. Couturat (F. Alcan, Paris, 1903), p. 519, and "M. d'Alvensleben voulut le refuter par le fait et chercha dans le jardin deux feuilles semblables; il n'en trouva point," written in a letter to Princess Sophie, Oct. 31, 1705.

[64l] "Non dari posse in natura duas res singulares solo numero differentes." *Ibid.,* p. 519. Cf. also Leibniz's *Nouveaux Essais* (II, chap. 27, section 1) and *The Monadology* (1714), in G. M. Duncan, *The Philosophical Works of Leibnitz* (Tuttle, Morehouse and Taylor, New Haven, Conn., 1890), p. 219, where it is argued that, since each monad mirrors the universe from its own point of view, every monad differs from the other.

[64m] *The Leibniz-Clarke Correspondence,* edited by H. G. Alexander (Philosophical Library, New York, 1956), p. 36.

[64n] "Adulterina lex." "Nihil subesse dictitant rationis, cur Deus duabus substantiis diversa assignaverit loca, si per omnia alia perfecte convenirent. Quales ineptiae! Miror gravissimos viros hisce rationum crepundiis delectari," wrote I. Kant in 1755 in *Principiorum Primorum Cognitionis Metaphysicae Nova Dilucidatio* (Proposition XI), *Immanuel Kant's Kleine Logisch-Metaphysische Schriften,* edited by K. Rosenkranz, part I of *Sämtliche Werke* (L. Voss, Leipzig, 1838), p. 34. Cf. also *Kritik der reinen Vernunft* (2d ed., J. F. Hartknoch, Riga, 1787), pp. 319–320.

[64o] S. Maimon, *Versuch einer neuen Logik oder Theorie des Denkens* (E. Felisch, Berlin, 1794; Reuther and Reichard, Berlin, 1912), p. 197; see also p. 134.

[64p] C. S. Peirce, *Collected Papers* (REF. 52 OF CHAP. 4), vol. 4 (1933), p. 251; p. 311. G. E. Moore, *Philosophical Studies* (Trubner, London, 1922), p. 307. F. Waisman, "Über den Begriff der Identität," *Erkenntnis 6,* 56–64 (1936). H. Grelling, "Identitas Indiscernibilium,"

Turning now to those schools of thought which affirmed the existence of like physical entities, we find that this view has only rarely been explicitly professed. This, of course, is not surprising if we recall that for many centuries Aristotelianism dominated scientific thought. In fact, prior to the revival of the atomistic doctrine in the seventeenth century the foremost proponents of this view were the Mutakallimun, the philosophers of the Kalam,[64q] whose atomistic doctrine renounced causal determinism and adopted instead a transcendental principle of constant divine interference. Their conception of atoms, as described by Moses Maimonides, one of the principal sources for our knowledge of the Kalam, was: "All these atoms are perfectly alike; they do not differ from each other in any point."[64r] John Dalton, the originator of modern atomism in the early nineteenth century, used almost the same words when he said that "the ultimate particles of all homogeneous bodies are perfectly alike."[64s]

By ascribing to all atoms of the same chemical element the same qualities, modern classical physics rejected Leibniz's principle as we see. In striking contrast to Maimon's conception, local position was no longer regarded as a clue to qualitative differentiation, an idea which was strongly supported by the theory of relativity. "An atom of oxygen remains the same whether it descends in the brook into the valley or rests in the lake, whether it rises in the vapor or drifts along in the air, whether it is inhaled in the breath or is absorbed in the blood: and in each of these locations it can be replaced by every other atom of its kind," declared Windelband in 1910 in a statement characteristic of the conceptual situation of that time.[64t]

The atomic theory of the nineteenth and early twentieth centuries accepted, as we see, the qualitative identity of particles—but denied their indistinguishability! If with the Stoa and Leibniz determinism, in its logical attire as the principle of sufficient reason, led to the denial of qualitative identity, in modern classical atomism determinism, in its

---

ibid., 252–259. B. Russell, *Human Knowledge: Its Scope and Limits* (George Allen and Unwin, London; Simon and Schuster, New York, 1948), p. 295. L. de Broglie, "L'individualité dans le monde physique," in *Problèmes de Connaissance en Physique Moderne*, Actualités Scientifiques et Industrielles 1066 (Hermann et Cie, Paris, 1949), pp. 61–80. M. Black, "Theid entity of indiscernibles," *Mind 61*, 153–164 (1952). N. L. Wilson, "The identity of individuals and the symmetrical universe," *ibid. 62*, 506–511 (1953). C. B. Martin, "Identity and exact similarity," *Analysis 18*, 83–87 (1958). G. C. Nerlich, "Sameness, difference and continuity," *ibid.* 144–149. J. Bobik, "Matter and individuation," in *The Concept of Matter* (University of Notre Dame Press, Notre Dame, Ind., 1963), pp. 277–288.

[64q] On the Kalam and its physical theories on matter, space, and time see REF. 156 OF CHAP. 5, pp. 60–64.

[64r] M. Maimonides, *The Guide for the Perplexed* (part I, chap. 73, proposition 1), translated by M. Friedländer (George Routledge, London; E. P. Dutton, New York, 2d ed., 1904), p. 120.

[64s] J. Dalton, *A New System of Chemical Philosophy* (R. Bickerstaff, London, 1808–1827; 2d ed., J. Waele, London, 1842), p. 142 (1842).

[64t] W. Windelband, "Über Gleichheit und Identität," *Heidelberger Berichte 1910* (14. Abhandlung), pp. 17–18.

physical attire as the laws of motion, led to the denial of indistinguishability. For it was claimed that two particles, once "told apart," can always be "told apart," for they can always be reidentified, thanks to the uniqueness of the solutions of the equations of motion. If the motion of each particle is represented by its four-dimensional world-line in space-time, a family of *nonintersecting* lines is obtained each of which can be used to characterize or label uniquely the particle whose trajectory it is. It was therefore thought it is always possible, at least in principle, to decide unambiguously whether or not a particle found at a certain time at a certain place is the same as a particle found at another time at the same or another place. In short, particles could be "alike," but were never "indistinguishable." Even in the absence of an *intrinsic* difference of quality the *extrinsic* difference of kinematical behavior was a sufficient criterion for their identifiability; it derived its operational meaning from the assumption of an unrestricted possibility of establishing an unbroken connection between the object at a time t_1 with the object at a time t_2 "by continuous observation (either direct or indirect) through all intermediate time."[64u]

Quantum mechanics, as we now shall see, led to a far-reaching revision on this point. In the light of our preceding remarks it is interesting to note that Heisenberg, one of the first quantum physicists to question the doctrine of determinism, was also the first to recognize the importance of the notion of like particles for the new scheme of conceptions. Already in the spring of 1926 he wrote: "It is a characteristic trait of atomic systems that their constituent parts, the electrons, are equal and subject to equal forces."[64v] To show how to express this "trait" in exact mathematical terms he formulated the Hamiltonian for the system under discussion (a classical system of two coupled oscillators) as a symmetric function of the coordinates and parameters (mass, frequency) of the constituent parts.[64w]

At about the same time Dirac[64x] proposed a new quantum-mechanical approach to the statistical treatment of a system of like particles, though as yet without taking account of the spin, and Fermi[64y] arrived independently at similar results. The Fermi-Dirac statistics, as is well known, allowed only for antisymmetric functions, in contrast to the Bose-Einstein

[64u] P. W. Bridgman, *The Logic of Modern Physics* (Macmillan, New York, 1927), p. 92.

[64v] "Gleich und gleichen Kräften unterworfen," REF. 213 OF CHAP. 3, p. 414.

[64w] Heisenberg's procedure is somewhat reminiscent of Leibniz's definition "Eadem sunt, quorum unum potest substitui alteri salva veritate," *Non inelegans specimen demonstrandi in abstractis*, in *God. Guil. Leibnitii Opera Philosophica*, edited by J. E. Erdmann (Eichler, Berlin, 1840), p. 94, or Christian Wolff's definition "Eadem dicuntur, quae sibi invicem substitui possunt salvo quocunque praedicato," *Philosophia Prima sive Ontologia* (1730), *Gesammelte Werke*, edited by J. Ecole and H. W. Arndt (reprinted by Olm, Hildesheim, 1962; 2d ed. 1736), section 181.

[64x] REF. 34 OF CHAP. 6.

[64y] E. Fermi, "Sulla quantizzazione del gas perfetto monoatomico," *Rendiconti dell'Accademia dei Lincei* *3*, 145–149 (1926); "Zur Quantelung des idealen einatomigen Gases," *Zeitschrift für Physik* *36*, 902–912 (1926).

statistics[64z] for like particles and the classical Boltzmann statistics for distinguishable particles. In November, 1926, Heisenberg[65] was already in a position to apply these ideas to the helium atom with the inclusion of spin. At the same time Fowler[65a] reexamined the whole problem *ab initio.* In the attempt to arrive at "a quite general form of statistical mechanics of which the classical form and Einstein's and Fermi-Dirac's are special cases," Fowler defined like—or, as he called them, "identical"—particles as "having the same set of energy values."[65b] Having thus obtained the status of a fundamental notion in the new scheme of ideas, the conception of like articles was soon fully vindicated by Heisenberg,[65c] Hund,[65d] and Dennison.[65e] Through his correct interpretation of the change in the rotational specific heat of the hydrogen molecule, Dennison could show that protons, just like electrons, are subject to the Fermi-Dirac statistics or the exclusion principle and have the spin $\frac{1}{2}$. Wigner[65f] meanwhile arrived at a successful approach to the general case of N like particles and its relations to group-theoretical considerations, a relation—first suggested to him by von Neumann—which since then became one of the most important conceptual instruments for the study of elementary particles.

In the summer of 1927 the importance of the ideas of exchange degeneracy and exchange energy for our understanding of atomic spectra was already clarified. In June, Heitler and London[65g] published their famous paper on the hydrogen molecule in which they showed the existence of a new kind of saturable, nondynamic forces, the so-called "exchange forces" of attraction or repulsion between like particles, and developed a schematic theory of the homopolar valence[65h] which eventually brought the whole of chemistry under the sovereignty of quantum mechanics.

[64z] REFS. 193, 194, and 195 OF CHAP. 5.

[65] W. Heisenberg, "Schwankungserscheinungen und Quantenmechanik," *Zeitschrift für Physik 40,* 501–506 (1926).

[65a] R. H. Fowler, "General forms of statistical mechanics in the special reference to the requirements of the new quantum mechanics," *Proceedings of the Royal Society of London* (*A*), *113,* 432–449 (1926).

[65b] *Ibid.*, footnote on p. 434.

[65c] W. Heisenberg, "Mehrkörperproblem und Resonanz in der Quantenmechanik II," *Zeitschrift für Physik 41,* 239–267 (1927).

[65d] F. Hund, "Zur Deutung der Molekelspektren II," *ibid. 42,* 93–120 (1927). For an early detailed account of the role of quantum mechanics for the explanation of the structure of atoms and molecules and chemical valency, see F. Hund, "Allgemeine Quantenmechanik des Atom- und Molekelbaues," in *Handbuch der Physik* (edited by H. Geiger and K. Scheel), vol. 24, part 1 (second edition), (Springer Verlag, Berlin, 1933), pp. 561–694; reprinted by Edwards Brothers, Ann Arbor, Michigan, 1943.

[65e] D. M. Dennison, "A note on the specific heat of the hydrogen molecule," *Proceedings of the Royal Society of London* (*A*), *115,* 483–486 (1927).

[65f] E. Wigner, "Über nicht kombinierende Terme in der neueren Quantentheorie II," *Zeitschrift für Physik 40,* 883–892 (1927).

[65g] W. Heitler and F. London, "Wechselwirkung neutraler Atome und homöopolare Bindung nach der Quantenmechanik," *ibid. 44,* 455–472 (1927).

[65h] That chemists had long before recognized that there exist different types of chemical bonds is shown by the fact that the terms "homopolar" ("homöopolar") and "heteropolar" valences had been introduced two decades before the

These results not only lent weight to the concept of like particles; they also showed that like particles may be indistinguishable, that is, may lose their identity, a conclusion which follows from the uncertainty relations or, more precisely, from the impossibility of keeping track of the individual particles in the case of interactions of like particles. For, contrary to classical dynamics, trajectories could no longer be defined as sharp non-intersecting world-lines but had to be conceived as *overlapping* each other. In fact, all papers on exchange phenomena and, in particular, the calculations concerning the ground state of the helium atom, in which the wave functions of the two electrons overlap completely, showed clearly that the classical principle of an unrestricted identifiability of particles had to be abandoned. Moreover, it was possible to show that Heisenberg's uncertainty principle was already contradicted by the idea of an approximately continuous sequence of atomic-configuration measurements, designed to identify electrons in lower-energy states and hence requiring positional uncertainties smaller than average electron distances. In addition, the success of the theory of the scattering of like particles, as first suggested by Oppenheimer[65i] and subsequently developed by Mott[65j] with the well-known interference term, characteristic for the exchange interaction and leading to results at variance with classical calculations, and the agreement with experience of such calculations concerning the effective cross section constituted, so to speak, an experimental verification of the untenability of the principle of unrestricted identifiability.

It must be admitted, however, that these far-reaching conclusions, although implicitly contained in almost all papers on this subject, were only seldom explicitly stated by their authors. It was rather a topic discussed in the writings of the more philosophically interested physicists. Thus, Kellner, in his papers previously referred to,[65k] characterized the new statistics as "precluding the reidentification ('Wiedererkennbarkeit') of a particle if once one has lost track of it"; Langevin discussed this matter in his address to the Paris meeting of the French Society for Physical Chemistry in 1933;[65l] and, in the English-speaking world, C. G. Darwin's writings, especially his at that time widely circulated book on the new conceptions of matter,[65m] attracted the attention of philosophers to these results.

It should also be noted that the renunciation of identifiability or the

---

work of Heitler-London by R. Abegg in his paper "Über die Fähigkeit der Elemente, miteinander Verbindungen zu bilden," *Zeitschrift für anorganische Chemie 50*, 309–314 (1906).

[65i] J. R. Oppenheimer, "On the quantum theory of electronic impacts," *Physical Review 32*, 361–376 (1928).

[65j] N. F. Mott, "The exclusion principle and aperiodic systems," *Proceedings of the Royal Society of London (A)*, *125*, 222–230 (1929).

[65k] REF. 56a.

[65l] See J. Ullmo's report "L'évolution de la notion de corpuscule d'après M. Langevin," *Scientia 55*, 103–117 (1934).

[65m] C. G. Darwin, *The New Conceptions of Matter* (Macmillan, New York, 1931), pp. 196–197.

assumption of total likeness of particles was not always unreservedly accepted. One of the main objections, raised by logicians and physicists alike, was the following argument, recently restated, for example, by Arthur March: "In order to count them (these particles) as two or more, there must be something intrinsic by which they differ from each other."[65n] Number or, more precisely, "being numerable," we may answer, is not necessarily a property of the individuals to be counted; it is rather a property of the system composed by them. In fact, electrons in atoms are never counted, but it is shown that the mathematical consequences of tentatively assigning a number to the system which they compose either agrees or disagrees with experience. It is in this sense that particles, though being alike, indistinguishable, and even unidentifiable, may still be counted collectively. With these remarks we conclude our digression.

7.2 *Complementarity*

The often repeated statement that Bohr derived the notion of complementarity from Heisenberg's uncertainty relations is erroneous from both the historical and the conceptual points of view. Heisenberg's uncertainty relations were for Bohr rather a confirmation of conceptions he had been groping for long before Heisenberg derived his principle from the Dirac-Jordan transformation theory. True, Heisenberg's work prompted Bohr to give his thoughts on complementarity a consistent and final formulation, but these thoughts, as we shall see presently, can be traced back at least to July, 1925.

Although very little is known about the genesis of these ideas in Bohr's mind—from July, 1925, until September, 1927, the most dramatic period in the development of the modern quantum theory, Bohr had published very little on quantum physics and still less on his more intimate philosophical ideas—one thing seems to be certain: Bohr's conception of complementarity originated from his final acceptance of the wave-particle duality. His formerly inflexible opposition to Einstein's light quanta—which found such an eloquent expression in his answer to Einstein's letter shortly after the publication of the Bohr-Kramers-Slater paper,[66] and as a result of which, at least in part, he advocated the highly unconventional conceptions as expressed in this paper—suffered a serious blow through the results of the Bothe-Geiger experiments.[66a]

An early evidence of Bohr's acknowledgment of the wave-particle duality and of his first ideas on how to cope with it is contained in a four-

[65n] "Denn um sie als zwei oder mehr zählen zu können, müßte etwas an ihnen sein, das sie voneinander unterscheidet." A. March, *Das neue Denken der modernen Physik* (Rowohlt, Hamburg, 1957), p. 121.

[66] See p. 187.

[66a] See p. 185.

page postscript ("Nachschrift"), dated July, 1925, to a paper[67] which treated the interaction between atomic systems still along the lines of the Bohr-Kramers-Slater approach. After having written the article (pp. 142–154), Bohr became acquainted with the Bothe-Geiger paper which had meanwhile appeared and found in it a convincing proof that in the Compton scattering the emission of the recoil electron is coupled with the emission of the photoelectron, ejected by the scattered radiation—in blatant contrast to the assumption of the independence of atomic transitions as proposed in the Bohr-Kramers-Slater paper. This assumption, Bohr now declared in the postscript, had to be made because no spatiotemporal mechanism seemed conceivable which could have admitted such a coupling and at the same time would have conformed to the classical electrodynamics which had always proved itself so useful for the description of optical phenomena. "It must, however, be stressed," he declared, "that the question of a coupling or of an independence of the individually observable atomic processes cannot simply be regarded as a differentiation ('Unterscheidung') between two well-defined conceptions concerning the propagation of light in empty space, corresponding perhaps to a corpuscular or undulatory theory of light. It involves rather the problem how far spatiotemporal pictures in terms of which physical phenomena have hitherto been described can be applied to atomic processes. . . . In view of the recent results one should not be surprised if the required extension of classical electrodynamics leads to a far-reaching revolution of the conceptions ("eine durchgreifende Revolution der Begriffe") on which the description of nature has been based so far."[68] Bohr was convinced that the particle picture as suggested by the Bothe-Geiger experiments had to be somehow harmonized with the basic assumption of the Bohr-Kramers-Slater paper, according to which the radiative aspects of atomic interactions have to be described by the wave picture. Facing a similar dilemma in his treatment of the capture of electrons by fast α particles, subject to the conservation of energy, Bohr declared that "we are compelled to acknowledge in this phenomenon a new trait which is not describable in terms of spatiotemporal pictures."[69] And he added that one has to be prepared to envisage processes "which are incompatible with the properties of mechanical models . . . and which defy the use of ordinary space-time models just as the coupling of individual[70] processes in distant atoms defy a wave description of optical phenomena."[71]

[67] N. Bohr, "Über die Wirkung von Atomen bei Stößen," *Zeitschrift für Physik 34*, 142–154 (1925), received on Mar. 30, 1925; "Nachschrift," *ibid.*, 154–157, dated July, 1925.

[68] *Ibid.*, pp. 154, 155.

[69] *Ibid.*, p. 156.

[70] For Bohr's conception of the term "individual" and the increasing importance of the role of this term in Bohr's conceptual scheme, cf. K. M. Meyer-Abich, *Korrespondenz, Individualität und Komplementarität* (REF. 121 OF CHAP. 2), pp. 117–122.

[71] REF. 67, p. 156.

From now on, the wave-particle dilemma seems to have engaged Bohr's attention constantly. In a lecture on "Atomic Theory and Wave Mechanics"[72] delivered before the Copenhagen Academy on December 17, 1926, he discussed this issue at some length. After the lecture he confided to Høffding that he became "more and more convinced of the need of a symbolization if one wants to express the latest results of physics."[73]

In the debates following Schrödinger's visit to Copenhagen and later, especially in his controversy with Heisenberg on the meaning of the uncertainty relations, he rejected the latter's more formal approach and regarded the wave-particle duality as the ultimate point of departure for an interpretation of the theory. He disliked Heisenberg's idea that "nature imitates a mathematical scheme"[74] and his one-sided preference for the intuitive particle conception of classical physics whose applicability should be limited only as far as necessary in order to conform with the quantum-mechanical formalism.

Heisenberg's remark[75] that "Bohr developed his ideas on complementarity" while vacationing in Norway early in 1927 implied that it was during this period that Bohr associated his general ideas on the wave-particle duality with what he subsequently called the quantum postulate according to which an essential discontinuity or not-further-analyzable individuality, a notion completely foreign to classical thought and symbolized by Planck's quantum of action, has to be attributed to every atomic process.

For Bohr, in contrast to Heisenberg, it was not the formalism as such but rather its underlying logic which should be given priority in the attempt to reach an appropriate interpretation of the theory. Thus, with regard to the basic equations

$$E = h\nu \qquad \text{and} \qquad p = hk \tag{7.6}$$

which express the proportionality between energy E and momentum p with the frequency ν and wave number k, respectively, Bohr asked himself how it is possible that these equations relate together characteristics of radiation

[72] Announced in *Nature 119*, 262 (1927).

[73] "Dans sa récente conférence sur 'la doctrine des atomes et le mouvement ondulatoire,' il soutient qu'on ne peut décider si l'électron est un mouvement ondulatoire (dans ce cas on pourrait éviter la discontinuité) ou une particle (avec discontinuité entre les particles). Certaines équations nous conduisent à la première conception, certaines autres à la dernière. Aucune image, aucun mot ne peut répondre à toutes les équations. Dans une conversation qu'il a eue avec moi après la conférence, M. Bohr m'a dit qu'il est toujours plus convaincu de la nécessité de la symbolisation si on veut exprimer les derniers résultats de la Physique." Letter from Høffding to Meyerson, dated Dec. 30, 1926, in *Correspondance entre Harald Høffding et Emile Meyerson* (Munksgaard, Copenhagen, 1939), p. 131.

[74] *Archive for the History of Quantum Physics*, Interview with W. Heisenberg on Feb. 25, 1963.

[75] W. Heisenberg, "Die Entwicklung der Deutung der Quantentheorie," *Physikalische Blätter 12*, 289–304 (1956); "50 Jahre Quantentheorie," *Die Naturwissenschaften 38*, 49–55 (1951).

which, strictly speaking, are contradictory to each other: particle attributes energy and momentum, necessary for the description of the interaction of radiation with matter, as in the photoeffect or in the Compton effect, and frequency and wave number, necessary for the description of the propagation of light, as in the phenomena of interference and diffraction. Planck's constant, Bohr realized, establishes in these equations a relation between two descriptions of radiation that are mutually exclusive, but equally necessary.

It was this mutual exclusion and, at the same time, indispensability of fundamental notions and descriptions which led Bohr to the conclusion that the problem with which quantum physics found itself confronted could not be solved by merely modifying or reinterpreting traditional conceptions. What was needed, he concluded, was a new logical instrument. He called it "complementarity," denoting thereby the logical relation between two descriptions or sets of concepts which, though mutually exclusive, are nevertheless both necessary for an exhaustive description of the situation. In Heisenberg's reciprocal uncertainty relations he saw a mathematical expression which defines the extent to which complementary notions may overlap, that is, may be applied simultaneously but, of course, not rigorously. The uncertainty relations, Bohr contended, tell us the price we have to pay for violating the rigorous exclusion of notions, the price for applying to the description of a physical phenomenon two categories of notions which, strictly speaking, are contradictory to each other. In these relations he also saw a confirmation of his thesis that complementarity would never lead to logical contradiction in spite of the fact that the notions involved are logically exclusive of each other; for, as the uncertainty principle shows, the sharp exhibition of one of such complementary notions necessitates an experimental setup which differs totally from that required for the other, it being the essence of the uncertainty principle that no physical situation can arise which exhibits simultaneously and rigorously (sharply) both complementary aspects of the phenomenon.

The possibility of using mutually contradictory notions for the description of the same physical situation, argued Bohr, arises from the indeterminateness of the concept "observation." For the interaction between the object of observation and the agency of observation—which interaction in accordance with the quantum postulate cannot be neglected as it could in classical physics—makes it impossible to separate sharply the behavior of the atomic system from the effect on the measuring instrument whose behavior must be expressed in classical terms. By combining the atomic system with different classically describable devices one may measure complementary variables, and by expressing the results of these measurements in classical terms one may describe an atomic system in terms of complementary classical pictures.

In view of our discussion on the philosophical background of nonclassical interpretations (Sec. 4.2) it will be understood that Bohr's conception of complementarity was strongly influenced both by Høffding's philosophical teachings and by James's psychological writings. In an article published on the occasion of Høffding's eighty-fifth birthday[76] Bohr acknowledged the impact of Høffding's ideas on his work and, in particular, on his search for an appropriate interpretation of the quantum-mechanical formalism.[77]

Bohr's insistence on the indeterminateness of the concept of observation, in so far as it depends upon which objects are included in the system to be observed and which in the agency that observes, showed a striking similarity to James's analysis of the notion of observation in psychology. The influence of James, whose psychological writings, as we know, Bohr had studied with great interest, may be inferred also from the following fact: Bohr concluded his first paper[78] on the notion of complementarity with the statement that "the idea of complementarity is suited to characterize the situation, which bears a profound analogy to the general difficulty in the formation of human ideas, inherent in the distinction between subject and object"—a statement which clearly referred to one of the main issues dealt with in James's *Principles of Psychology*. Similarly, in a subsequent paper[79] in which Bohr resumed his discussion on the impossibility of a strict separation of the phenomenon from the means of its observation, he reemphasized the analogy between physics and psychology: "Strictly speaking, the conscious analysis of any concept stands in a relation of exclusion to its immediate application,"[80] and he admitted explicitly: "The necessity of taking recourse to a complementary, or reciprocal, mode of description is perhaps familiar to us from psychological problems."[81] Even an allusion to the "Stream of Thought,"[82] a chapter in

[76] N. Bohr, "Ved Harald Høffdings 85-Aars Dag," *Berlingske Tidende*, Mar. 10, 1928.

[77] "M. Bohr déclare qu'il a trouvé dans mes livres des idées qui ont aidé les savants dans 'l'entendement' de leur travail et leur ont été par là d'un réel secours. C'est une grande satisfaction pour moi, qui sens si souvent l'insuffisance de ma préparation spéciale quant aux sciences naturelles." Letter from Høffding to Meyerson, dated Mar. 30, 1928 (REF. 73, p. 149).

[78] N. Bohr, "The quantum postulate and the recent development of atomic theory," *Nature 121*, 580–590 (1928); reprinted in N. Bohr, *Atomic Theory and the Description of Nature* (Cambridge University Press, 1934, 1961), pp. 52–91; "Das Quantenpostulat und die neuere Entwicklung der Atomistik," *Die Naturwissenschaften 16*, 245–257 (1928); reprinted in N. Bohr, *Atomtheorie und Naturbeschreibung* (Springer, Berlin, 1931), pp. 34–59; "Kvantepostulatet og Atomteoriens seneste Udvikling," *Atomteori og Naturbeskrivelse* (Lunos Bogtrykkeri, Copenhagen, 1929), pp. 40–68, also in *Atomteorie og Grundprincipperne for Naturbeskrivelsen* (Schultz, Copenhagen, 1958), pp. 47–76.

[79] N. Bohr, "Wirkungsquantum und Naturbeschreibung," *Die Naturwissenschaften 17*, 483–486 (1929); *Atomtheorie und Naturbeschreibung*, pp. 60–66; "Virkningskvantet og Naturbeskrivelsen," *Atomteori og Naturbeskrivelse*, pp. 69–96; *Atomteorie og Grundprincipperne for Naturbeskrivelsen*, pp. 77–84; "The quantum of action and the description of nature," *Atomic Theory and the Description of Nature*, pp. 92–101. (See REF. 78.)

[80] *Atomic Theory and the Description of Nature* (REF. 78), p. 86.

[81] *Ibid.*

[82] W. James, *The Principles of Psychology*, chap. IX: "The Stream of Thought" (Dover, New York, 1950), vol. 1, pp. 224–290. This chapter was an elaboration of James's paper "On some omissions of introspective psychology," *Mind 9*, 1–26 (1884).

which James attempted to disprove Locke's atomism of ideas and by which, according to von Weizsäcker,[83] Bohr was particularly impressed, may be found at the end of this paper, where Bohr declared: "In particular, the apparent contrast between the continuous onward flow of associative thinking and the preservation of the unity of the personality exhibits a suggestive analogy with the relation between the wave description of the motions of material particles, governed by the superposition principle, and their indestructible individuality."[84]

Before we conclude our discussion on the development of Bohr's complementarity conception, it seems to be worthwhile to make a brief digression on the term "complementary" itself. This may throw some light on the meaning it had when Bohr adopted it as a technical term. Its earliest use in science and also its earliest occurrence in the scholastic curriculum is undoubtedly the geometrical expression "complementary angles." Although, as we know,[85] Bohr was deeply interested in the foundations of geometry and in the relations of mathematics to language, it was probably not this geometrical expression but James's application of the term which had an impact on Bohr's mind. In his discussion of hysteric diseases James[86] described an experiment performed by Pierre Janet,[87] the noted psychologist and neurologist of the Salpêtrière, who like Freud was a pupil of J. M. Charcot and specialized in the field of disordered personality. Studying a patient called Lucie, Janet covered her lap with cards, each bearing a number; he then told her that on waking, after the present state of trance, she should *not see* any card whose number is a multiple of 3. "When she was awakened and asked about the papers on her lap, she counted and said she saw only those whose number was not a multiple of 3. To the 12, 18, 9, etc., she was blind. But the *hand*, when the sub-conscious self was interrogated by the usual method of engrossing the upper self in another conversation, wrote that the only cards in Lucie's lap were those numbered 12, 18, 9, etc., and on being asked to pick up all the cards which were there, picked up these and let the others lie."[88] James quoted this experiment as an illustration of the fact that "in certain persons the total possible consciousness may be split into parts which coexist but mutually ignore each other, and share the objects of knowledge between them. More remarkable still, they are *complementary*."[89] "Few things are more curious," wrote James a few pages later, "than these relations of mutual exclusion. . . ."[90] In James's psychology the sum total of complementary

[83] REF. 121 OF CHAP. 2, p. 125.
[84] REF. 80, p. 99.
[85] A. Petersen, "The philosophy of Niels Bohr," *Bulletin of the Atomic Scientists 1963*, pp. 8–14.
[86] REF. 82, pp. 202–213.
[87] The experiment is described and discussed in P. Janet, *L'Automatisme Psychologique—Essai de Psychologie Expérimentale sur les Formes Inférieures de l'Activité Humaine* (Alcan, Paris, 1889, 10th ed. 1930), pp. 276–277.
[88] REF. 82, pp. 206–207.
[89] *Ibid.*, p. 206.
[90] *Ibid.*, p. 210.

parts, as exemplified by Janet's experiment, makes up the normal totality; in Bohr's physics the sum total of complementary descriptions makes up the description of classical physics.

On September 16, 1927, in the auditorium of the Istituto Carducci in Como, where the International Congress of Physics convened in commemoration of the hundredth anniversary of Allessandro Volta's death (Volta was born and died in Como), Bohr delivered a lecture on "The Quantum Postulate and the Recent Development of Atomic Theory"[91] in which he disclosed for the first time in public his ideas on complementarity. Addressing an audience of leading physicists from all parts of the world,[92] Bohr began his lecture with the words: "I shall try by making use only of simple considerations and without going into any details of technical mathematical character to describe to you a certain general point of view which I believe is suited to give an impression of the general trend of the development of the theory from its very beginning and which I hope will be helpful in order to harmonize the apparently conflicting views taken by different scientists."[93] After stressing the contrast between the classical description, based on the assumption that the phenomenon discussed may be observed without appreciable disturbance, and the description of quantum phenomena, subject to the quantum postulate, according to which to any atomic process an essential discontinuity, or rather individuality, has to be attributed, Bohr declared: "On one hand, the definition of the state of a physical system, as ordinarily understood, claims the elimination of all external disturbances. But in that case, according to the quantum postulate, any observation will be impossible, and, above all, the concepts of space and time lose their immediate sense. On the other hand, if in order to make observation possible we permit certain interactions with suitable agencies of measurement, not belonging to the system, an unambiguous definition of the state of the system is naturally no longer possible, and there can be no question of causality in the ordinary sense of the word. The very nature of the quantum theory thus forces us to regard the space-time coordination and the claim of causality, the union of which characterizes the classical theories, as complementary but exclusive features of the description, symbolizing the idealization of observation and definition respectively."[94] This statement, in which the term "complementary" appeared for the first time and in which spatiotemporal description is referred to as complementary to causal description, contained the essence of what later became known as the "Copenhagen interpretation" of quantum

[91] *Atti del Congresso Internazionale dei Fisici* (Zanichelli, Bologna, 1928), vol. 2, pp. 565–588. The substance of the lecture has been reprinted in the papers mentioned in REF. 78.

[92] The audience included Born, Bose, de Broglie, Brillouin, A. H. Compton, Debye, Duane, Fermi, Franck, Frenkel, Gerlach, Hall, Heisenberg, Kramers, Laue, Lorentz, Millikan, von Neumann, Paschen, Pauli, Planck, Saha, Smekal, Sommerfeld, Stern, Tolman, Wigner, Zeeman.

[93] REF. 91, p. 565.

[94] *Ibid.*, p. 566.

mechanics. It is therefore appropriate to analyze this statement in some detail. The "state of a system," as understood, for example, in ordinary mechanics, is the set of all the (position) coordinates and the momenta of the constituent parts and implies the possibility of using these data for the prediction (or retrodiction) of the structural properties of the system. Prediction (or retrodiction), however, is possible only if the system is closed, that is, unaffected by external disturbances. For an open system, strictly speaking, no "state" can be defined. Now, the evolution of the structural properties of the system is governed by the law of causality; their progress in time is the causal behavior of the system. But according to the quantum postulate, in atomic physics, any observation of the system implies a disturbance. In other words, a system, if observed, is always an open system. A space-time description, however, presupposes observation. Hence, concluded Bohr, the claim of causality excludes spatiotemporal description and vice versa.[95] The usual causal space-time description, that is, the simultaneous use of complementary descriptions, is made possible in classical physics, argued Bohr, merely because of the extremely small value of the quantum of action "as compared to the actions involved in ordinary sense perceptions."

To give a first example of the breakup of the classical causal space-time description into its complementary parts, Bohr contrasted the wave theory of light, an adequate description of the propagation of light in space and time, with the theory of light quanta which accounts for the interaction with matter in terms of energy and momentum. The mutual exclusion of the notions of wave and particle precludes, according to Bohr, a causal space-time description of optical phenomena. "The two views of the nature of light are rather to be considered as different attempts at an interpretation of experimental evidence in which the limitation of the classical concepts is expressed in complementary ways."

After showing that the same situation exists with respect to the description of the constituents of matter (corpuscles vs. de Broglie waves), Bohr pointed out that "the fundamental contrast between the quantum of action and the classical concepts is immediately apparent from the simple formulas which form the common foundation of the theory of light quanta and of the wave theory of material particles." Writing the Einstein–de Broglie relations in the form

$$E\tau = p\lambda = h \tag{7.7}$$

[95] An analogy with this situation is found perhaps also in Spinoza's ideas on the effect of an (introspective) observation of one's own affections on these. Another analogy is, of course, the destructive interference of physical observations on a living organism. Bohr's use, not much later, of the term "wholeness" with respect to physical systems seems to suggest that it is not impossible that, in addition to the above-mentioned associations with psychology, also this biological analogy (to which he later referred) may already have been in the back of his mind at this early stage.

where E and p are the energy and momentum, respectively, as used in the particle picture, and $\tau = 1/\nu$ and $\lambda = 1/\sigma$ the period of vibration and the wavelength, respectively, as used in the wave picture, Bohr showed their relation to the ordinary mode of description by referring to the relativistic equations

$$p = \frac{v}{c^2} E \tag{7.8}$$

and

$$v\, dp = dE \tag{7.9}$$

where v denotes the velocity of the particle and c the velocity of light. Hence, Bohr pointed out, the phase velocity, which is given by ν/σ, turns out to be c^2/v, and the group velocity, which according to de Broglie is defined by $d\nu/d\sigma$, coincides with v. "The possibility of identifying the velocity of the particle with the group velocity indicates the field of application of space-time pictures in the quantum theory. Here the complementary character of the description appears, since the use of wave groups is necessarily accompanied by a lack of sharpness in the definition of period and wavelength, and hence also in the definition of the corresponding energy and momentum as given by relation (7.7)." For in accordance with the well-known Fourier analysis of wave trains

$$\Delta t\, \Delta\nu = \Delta x\, \Delta\sigma = 1 \tag{7.10}$$

where Δt and Δx denote the extension of the wave packet in time and space. But then, by (7.7),

$$\Delta t\, \Delta E = \Delta x\, \Delta p = h \tag{7.11}$$

which clearly shows that any contraction of the wave packet, that is, a sharper definition in the space-time description, is accompanied by an increase in the indefiniteness of the energy-momentum components, that is, a less precise definition in the causal description. This general reciprocal relation between the maximum sharpness of definition of, relativistically speaking, the space-time vector and that of the energy-momentum vector was for Bohr "a simple symbolical expression for the complementary nature of the space-time description and the claims of causality." Bohr then discussed, from the viewpoint of complementarity, Heisenberg's uncertainty relations and the meaning of measurement in quantum theory, and pointed out that an adequate tool for complementary modes of description is offered precisely by the new formalism of quantum mechanics. For this formalism, he contended, is essentially a purely symbolic scheme which permits only predictions, in accordance with the correspondence principle, of results obtainable under conditions specified in terms of classical concepts.

In the discussion,[96] in which Born, Kramers, Heisenberg, Fermi, and Pauli participated, no objections were voiced—but the real issue of the paper was not touched upon. Those who heard Bohr's ideas for the first time found it probably too difficult to comprehend their full meaning. Thus, for example, Léon Rosenfeld, who became one of the most eloquent proponents of the complementarity interpretation[97] and whom Bunge,[98] in his critique of this approach, singled out as the most representative member of this school of thought, admitted,[99] referring to Bohr's lecture: "I did not see, I did not feel, any of the subtlety that was in it." Nor did the Göttingen theorists appreciate its meaning at that time. Wigner, for example, as reported by Rosenfeld,[100] remarked that Bohr's lecture "will not induce any one of us to change his own opinion about quantum mechanics." And according to Wigner,[101] von Neumann had this to say on Bohr's lecture: "Well, there are many things which do not commute and you can easily find three operators which do not commute." What von Neumann alluded to was, of course, Bohr's contention that the complementarity of the wave and particle pictures and the consequent limitation in the application of their corresponding conceptions are adequately reflected in the noncommutativity of certain operators. Von Neumann's criticism may perhaps remind the student of philosophy of a similar question which had been asked in connection with Spinoza's theory of attributes and its subjective interpretation by Erdmann. For, like Bohr's "descriptions of nature," Spinoza's attributes[102] are intellectual modes of perception and, though not satisfying reciprocal relations such as those that are satisfied by Bohr's complementary descriptions, do exclude each other. The epistemological problem: Why does the human mind recognize only two attributes (*res cogitans* and *res extensa*) of the infinite substance which according to Spinoza has an infinitude of attributes? offers an interesting analogue to von Neumann's sally—although it must, of course, be kept in mind that the ontological status of Spinoza's substance differs radically from that which Bohr ascribed to the object of physical description.

It is interesting to note—although it would be rash to draw from it any far-reaching conclusions—that the *Physikalische Berichte* published two separate reviews of the substance of Bohr's lecture, a very short one,

[96] *Discussione sulla comunicazione Bohr* (REF. 91), pp. 589–598.

[97] Cf., for example, L. Rosenfeld, "L'évidence de la complémentarité," in REF. 150 OF CHAP. 5, pp. 43–65; "Strife about complementarity," *Science Progress No. 163*, 393–410 (1953); "Foundations of quantum theory and complementarity," *Nature 190*, 384–388 (1961).

[98] M. Bunge, *Metascientific Queries* (C C Thomas, Springfield, Ill., 1959), pp. 173–209.

[99] *Archive for the History of Quantum Physics*, Interview with L. Rosenfeld on July 1, 1963.

[100] *Ibid.*

[101] *Archive for the History of Quantum Physics*, Interview with E. P. Wigner on Nov. 21, 1963.

[102] "Per attributum intelligo id, quod intellectus de substantia percipit tamquam ejus essentiam constituens." B. de Spinoza, *Ethices*, Definitio IV. See, for example, *Opera* (In Bibliopoli academico, Jena, 1803), vol. 2, p. 55.

written by Elsasser,[103] and an exceptionally long one of almost two pages, written by Smekal,[104] whereas *Science Abstracts* did not mention it at all! Physicists, engaged in the applications of the new formalism to still unsolved problems in atomic physics, were too busy to direct their attention to questions of interpretation, and philosophers were generally still lacking technical knowledge to participate in the debate.

It will have been noted that in his Como lecture Bohr did not explicitly define complementarity. The first to give a precise definition of this notion was Pauli. In his article for the *Handbuch der Physik*[105] Pauli called two classical concepts—and not two modes of description—complementary "if the applicability of the one (e.g., position coordinate) stands in the relation of exclusion to that of the other (e.g., momentum)" in the sense that any experimental setup for measuring the one interferes destructively with any experimental setup for measuring the other.[106] Pauli, as we see, in contrast to Bohr, ascribed complementarity to two notions which belong to the same classical mode of description (e.g., the particle picture) and not to two mutually exclusive descriptions. It was this notion of complementarity which was in his view *the* characteristic feature of quantum mechanics which he therefore proposed to call the "theory of complementarity" ("Komplementaritätstheorie") in analogy to the "theory of relativity." C. F. von Weizsäcker, who like Pauli fully espoused the idea of complementarity—he once called it "the key to the presently best possible understanding of quantum theory"[107]—recognized the subtle divergence between Bohr and Pauli and suggested a distinction between what he called "parallel complementarity" and "circular complementarity." One should speak of the former if, as in Pauli's example—and by his definition—two concepts (as "position coordinate" and "momentum") are complementary in quantum theory without having been so in classical physics or if, as in Hund's example of complementarity,[108] two concepts (as "position" and "wave number") are complementary in quantum mechanics having been so also in classical physics. On the other hand, Bohr's conception of the complementarity between space-time description and causal description should be called "circular complementarity." For, von Weizsäcker contended, it is only via a detour to wave mechanics and the deterministic Schrödinger equation of the wave function that a causal description obtains; but if we wish to arrive at a space-time description, a measurement of classical observables has to be performed

[103] *Physikalische Berichte 9,* 1932 (1928).
[104] *Ibid.*, 1368–1369.
[105] REF. 283 OF CHAP. 5, p. 89.
[106] "Wenn die Benutzbarkeit *eines* klassischen Begriffes in einem ausschließendem Verhältnis zu der eines *anderen* steht, nennen wir diese beiden Begriffe (z.B. Orts- und Impulskoordinaten eines Teichens) mit Bohr *komplementar.*" *Ibid.*
[107] C. F. von Weizsäcker, "Komplementarität und Logik," *Die Naturwissenschaften 42,* 521–529, 545–555 (1955).
[108] F. Hund, *Materie als Feld* (Springer Verlag, Berlin, Göttingen, Heidelberg, 1954), p. 44.

which leads to a statement in terms of space and time; such a reduction of the wave packet, however, destroys the deterministic behavior of the function, as we know. "The complementarity between space-time description and the claim of causality," wrote von Weizsäcker, "is therefore precisely the complementarity between the description of nature in classical notions and in terms of the ψ function."[109] Bohr, as we know from his correspondence with von Weizsäcker[110] on this issue, rejected this interpretation and declared that his expression "claim of causality" referred to the application of the notions of energy and momentum and their conservation principles which determine the course of events.

We have mentioned these divergent conceptions of complementarity as an illustration of the conceptual difficulties which the Bohr interpretation of quantum mechanics had to face. Part of these difficulties originated undoubtedly from a certain vagueness of expression which the Copenhagen interpretation exhibited, at least during its early stages. It has also been claimed, however, that it was precisely due to this vagueness and its concomitant conceptual flexibility that the Copenhagen interpretation managed to survive serious crises. According to Feyerabend, one of the reasons "for the persistence of the creed of complementarity in the face of decisive objections is to be found in the *vagueness* of the main principles of this creed,"[111] this ambiguity being perhaps one of the reasons of its fruitfulness, as Groenewold[112] had pointed out. In this case, it may be said, a characteristic tenet of the Kierkegaard-Høffding philosophy, which, as we have tried to show,[113] had an impact upon Bohr, would have vindicated itself.

Bohr himself, it seems, regarded the first enunciation of his new conceptions more as a program for further elaboration than as a definite statement of an immutable dogma. He was fully aware that both the foundations and the implications of his ideas had still to be subjected to critical investigations. It was therefore for him "a most valuable incentive . . . to reexamine the various aspects of the situation as regards the description of atomic phenomena . . ." and "a welcome stimulus to clarify still further the role played by the measuring instruments"[114] when, a few weeks after the Como Congress, his interpretation became the target

[109] REF. 107, pp. 525–526.

[110] C. F. von Weizsäcker, *Zum Weltbild der Physik* (S. Hirzel, Stuttgart, 8th ed., 1960), p. 330. Cf. also REF. 121 OF CHAP.2, pp. 143–146.

[111] P. K. Feyerabend, "Problems in Microphysics," in *Frontiers of Science and Philosophy* (University of Pittsburgh Press, 1962), p. 193.

[112] *Ibid.*, p. 256.

[113] Sec. 4.2.

[114] N. Bohr, "Discussion with Einstein on epistemological problems in atomic physics," in *Einstein, Philosopher-Scientist*, edited by P. A. Schilpp (The Library of Living Philosophers, Evanston, Ill., 1949), p. 218; reprinted in N. Bohr, *Atomic Physics and Human Knowledge* (Wiley, New York, 1958), pp. 32–66. "Diskussionen mit Einstein über erkenntnistheoretische Probleme in der Atomphysik," in *Albert Einstein als Philosoph und Naturforscher* (W. Kohlhammer, Stuttgart, 1955); reprinted in N. Bohr, *Atomphysik und menschliche Erkenntnis* (Vieweg, Braunschweig, 1958), pp. 32–67.

of severe criticisms at the fifth Physical Conference of the Solvay Institute, which convened in Brussels, under the chairmanship of Lorentz, from October 24 to 29, 1927. Its theme was "Electrons and Photons."[115] But, *de facto*, it provided an excellent opportunity for a fruitful exchange of views on the foundations of quantum mechanics and on its interpretation among the main originators of the theory: Bohr, Born, L. Brillouin, L. de Broglie, A. H. Compton, Debye, Dirac, Ehrenfest, Einstein, R. H. Fowler, Heisenberg, Kramers, Pauli, Planck, and others.

After W. L. Bragg's lecture on the reflection of x-rays[116] and A. H. Compton's talk on the discrepancies between experiment and the theory of electromagnetic radiation,[117] Louis de Broglie delivered a lecture[118] entitled "The new dynamics of quanta," in which he outlined his own early contributions and their elaborations by Schrödinger and Born. Dissatisfied, however, with the latter's purely probabilistic interpretation, which he regarded as contradicting the explanatory aim of theoretical physics, he included in his lecture an exposition of his causal theory. But instead of discussing the theory of double solution as proposed only a few months earlier in his *Journal de Physique* paper[119] and instead of explaining how he conceived the motion of a particle determined by the gradient of the phase common to both solutions of the wave equation, he presented in "an incomplete and diluted"[120] form a simplified version of his original ideas. This "watered-down version" or "pilot-wave theory," as he called it later, postulated an independent existence of the particle but did not regard it as a singularity of a wave field. It was immediately clear that nobody accepted his ideas and that the majority of the participants preferred the purely probabilistic interpretation of Born, Heisenberg, and Bohr. In fact, with the exception of some remarks by Pauli, who justified his objection on the ground of Fermi's treatment of the collision between a particle and a rotator,[121] de Broglie's causal interpretation was not even further discussed at the meeting. Only Einstein once referred to it *en passant*.[122] As de Broglie later admitted, it was partly due to this unfavorable reaction that he abandoned his own ideas and espoused the Copenhagen interpretation from then on. Twenty-five years later, however, Bohm's paper[123] and certain developments in the general theory of relativity to which his assistant, J.-P. Vigier, drew his attention revived his interest in his original conceptions and reconverted him to his causal approach.

[115] *Electrons et Photons—Rapports et Discussions du Cinquième Conseil de Physique Tenu à Bruxelles du 24 au 29 Octobre 1927 sous les Auspices de l'Institut International de Physique Solvay* (Gauthier-Villars, Paris, 1928).

[116] *Ibid.*, pp. 1–43.

[117] *Ibid.*, pp. 55–85.

[118] *Ibid.*, pp. 105–132: "La nouvelle dynamique des quanta."

[119] REF. 41 OF CHAP. 6.

[120] REF. 151 (1960) OF CHAP. 5, p. 90.

[121] REF. 115, p. 280.

[122] *Ibid.*, p. 256.

[123] D. Bohm, "A suggested interpretation of the quantum theory in terms of 'hidden' variables," *Physical Review 85*, 166–193 (1952).

De Broglie's lecture was followed by a detailed exposition by Born and Heisenberg on matrix mechanics, the transformation theory, and its probabilistic interpretation. "Quantum mechanics," they declared, "leads to accurate results concerning average values but gives no information as to the details of an individual event. The determinism which so far has been regarded as the basis of the exact sciences has to be given up. Every additional advance in our understanding of the formulas has shown that a consistent interpretation of the quantum-mechanical formalism is possible only on the assumption of a fundamental indeterminism. . . ."[124] Referring to Heisenberg's uncertainty principle, they stated that "the real meaning of Planck's constant h is this: it constitutes a universal gauge of the indeterminism inherent in the laws of nature owing to the wave-particle duality."[125] And they concluded their lecture with the provocative statement: "We maintain that quantum mechanics is a complete theory; its basic physical and mathematical hypotheses are not further susceptible of modifications."[126]

The last speaker was Schrödinger. He gave a detailed report on the development of wave mechanics with special consideration of problems concerning polyelectronic systems. The climax of the Congress was the general discussion at the end of the meeting. After some brief introductory remarks Lorentz requested Bohr to speak on the epistemological questions which quantum mechanics has to face. Bohr accepted the invitation and presented an exposé which contained essentially the substance of his Como lecture. Einstein, it will have been noted,[127] had not attended this Volta celebration, and it was therefore here in Brussels that he heard for the first time a coherent and comprehensive exposition of the complementarity interpretation. As we know, however—for example, from his correspondence with his intimate friend Paul Ehrenfest—Einstein disliked the Göttingen-Copenhagen approach to atomic physics with its emphasis on discontinuities and acausality ever since it had been formulated. His statement "I look upon quantum mechanics with admiration and suspicion,"[128] made in the summer of 1926, is perhaps the best characterization of his attitude on this issue. It was against his scientific instinct to accept statistical quantum mechanics as a complete description of physical reality. In fact,

[124] REF. 115, p. 160.

[125] *Ibid.*, p. 172.

[126] ". . . nous tenons la *mécanique des quanta* pour une théorie complète, dont les hypothèses fondamentales physiques et mathématiques ne sont plus susceptibles de modification." *Ibid.*, p. 178.

[127] See REF. 92.

[128] "Der Quanten-Mechanik stehe ich bewundernd-misstrauisch gegenüber. . . ." Letter (II–83) from Einstein to P. Ehrenfest, dated Aug. 28, 1926. Cf. also Einstein's letter to Born of Dec. 4, 1926: "Die Quantenmechanik ist sehr achtunggebietend. Aber eine innere Stimme sagt mir, daß das doch nicht der wahre Jakob ist. Die Theorie liefert viel, aber dem Geheimnis des Alten bringt sie uns kaum näher. Jedenfalls bin ich überzeugt, daß *der* nicht würfelt." (*Einstein Estate,* Princeton, N.J.) Cf. also REF. 105 OF CHAP. 3.

as we know from his correspondence with Bohr and Heisenberg,[129] he attempted, having obtained from Bohr a preprint of Heisenberg's paper on the uncertainty principle, to disprove it by showing that it is possible, for instance, to determine the path of a microphysical object with a precision greater than the principle allows. The present Solvay Congress, Einstein thought, was therefore an appropriate opportunity to exchange views on this question and perhaps to reach a final clarification. Thus, both at the sessions and at informal meetings during the evenings Einstein expressed his opposition to the rejection of the principle of deterministic description and challenged Bohr by proposing cleverly devised thought experiments[130] to outwit Heisenberg's principle. Most of these imaginary experiments were designed to show that the interaction between the microphysical object and the measuring instrument is not so inscrutable as Heisenberg and Bohr maintained. Einstein considered the interaction between a photon, for example, and a movable part of a grating and assumed that the momentum change of the latter, due to the interaction, could be accurately measured. He then claimed that on the basis of the conservation principle of momentum the change of momentum of the microphysical object (photon) could be calculated so that both, momentum and position, could be known with an accuracy which the principle prohibits. Bohr's refutation of Einstein's arguments usually was to disclose a fallacy in Einstein's reasoning due to the fact that somewhere in the chain of deductions the uncertainty principle had not been duly taken into account, and that had this been done, no increase in information would have resulted. How ardent these discussions must have been can perhaps be seen from the fact that three years later, at the sixth Solvay Congress[131] in 1930, Einstein and Bohr resumed their debate. In his attempt to disprove Heisenberg's uncertainty relation $\Delta E \, \Delta t \geq h/4\pi$, Einstein considered a box filled with radiation and possessing a shutter in one of its walls. The shutter was assumed to be operated by a clockwork, enclosed in the box, so that a photon could be released during an accurately determinable time interval Δt. Einstein now contended that by weighing the box before and after the emission of the timed radiative pulse the energy of the latter could be computed from the mass-energy relation $E = mc^2$ with an arbitrarily small error ΔE. After a sleepless night over the argument Bohr showed the next morning that, in accordance with Einstein's own general theory of relativity, the effect of the gravitational field on the rate of the

[129] Bohr's letter to Einstein of Apr. 13, 1927, and Heisenberg's letters to Einstein of May 19, 1927, and of June 6, 1927 (*Einstein Estate*, Princeton, N.J.).

[130] For an account of these discussions see REF. 114.

[131] The sixth Solvay Congress convened from October 20 to 25, 1930, and was devoted to the study of the magnetic properties of matter. Cf. *Le Magnétisme—Rapports et Discussions du Sixième Conseil de Physique sous les Auspices de L'Institut International de Physique Solvay* (Gauthier-Villars, Paris, 1932).

clock—which, since rigidly attached to the box, had to be assumed to be moving during the measurement—introduces precisely an uncertainty as required by the Heisenberg relation.[132] Bohr's recourse to Einstein's gravitational theory in order to save the consistency of the complementarity interpretation which was challenged by an argument independent of a theory of non-Newtonian gravitation has later been erroneously criticized by Popper as amounting "to the strange assertion that quantum theory contradicts Newton's gravitational theory, and further to the still stranger assertion that the validity of Einstein's gravitational theory (or at least the characteristic formulas used, which are part of the theory of the gravitational field) can be derived from quantum theory."[133]

Of great importance for the clarification of the meaning of the wave function was Einstein's objection which he had raised at the beginning of the general discussion at the Solvay Congress in 1927. He considered a particle or photon that passes through a slit in a screen covered by a hemispherical photographic film.[134] The wave function associated with the particle or photon will be diffracted and will spread out over the film so that, in accordance with Born's interpretation, practically every point on the film has a nonzero probability of being struck by the microphysical object. As soon, however, as the point of impact is recorded, the probability of finding the object at any other point on the film becomes zero. Einstein now pointed out that to assume with the proponents of the complementarity interpretation that, as long as the object has not been observed, it is virtually present all over the region covered by the wave function, would lead to a contradiction with the theory of relativity or to conclusions which are strange even in the framework of classical conceptions. For the contraction of the wave function to a single point at the moment of the recording could be accounted for only by some kind of action at a distance. The alternative interpretation which ascribes to the microphysical object a distinct trajectory would lead, as Einstein argued, to the conclusion that the description in terms of a wave function is not complete.

As a result of Einstein's objection and similar considerations—as, for example, those which referred to the inevitable use of multidimensional spaces for the representation of the wave function—it became clear that the wave function, as Heisenberg, Pauli, and Dirac had already emphasized during the general discussion, does not in itself represent a course of events in space and time. It rather expresses our knowledge of events. It became clear that what the physicist does is essentially this: from an observation he constructs a wave function which obeys in its progress the laws of quantum mechanics but constitutes at any time merely a catalog of probabilities for the results of subsequent measurements or observations. The

[132] For details see REF. 114, pp. 224–228.
[133] REF. 59, p. 447 (1959).
[134] REF. 115, pp. 253–256.

state of affairs between one observation and the next cannot be described; but how the observation is described depends upon the experimental setup chosen by the experimenter. The results of observations thus can never completely be objectified.

That a consistent interpretation of the accepted formalism of quantum mechanics makes these conclusions inevitable was the result of the general discussion at the Congress. Even Einstein, defeated but not convinced, had to admit that from the logical point of view the theory and its complementarity interpretation form a consistent system of thought.

For the next two and a half decades, the Copenhagen interpretation was the only accepted interpretation of quantum mechanics—and for the majority of physicists it is so even today. It may therefore be said that the search for a general consistent theory of the mechanics of atoms, a search which, as we have seen, was ushered in at the first Solvay Congress of 1911, found its successful completion and finale in the fifth Solvay Congress of 1927.

8 Validation of the Theory

8.1 Some Applications of the Theory

For the time being the epistemological problems of quantum mechanics were generally regarded as of purely academic interest and received relatively little attention. Satisfied that the theory "works," since it provided unambiguous answers whenever invoked, physicists engaged themselves rather in solving problems which so far had defied all previous attempts or which promised to open up new avenues of research. The year 1927 thus not only became the year in which the quantum-mechanical formalism, in all its essential points, received a formal completion and a consistent interpretation; 1927 also witnessed a veritable avalanche of elaborations and applications of the new conceptions and led to new insights in atomic physics to an unprecedented extent. The obvious agreement with previous results or improvement upon them, the mutual consistency of the results obtained, and the variety of phenomena satisfactorily dealt with dispelled all doubts about possible inconsistencies in the new formalism irrespective of its epistemological interpretation.

It would lead us too far to give an exhaustive account of all these achievements. Suffice it to mention only a few typical ones—and primarily those which had a bearing on problems and effects that contributed to the conceptual development of the theory.

How quantum mechanics accounted successfully for the Stark effect has already been recounted.[1] With respect to the Zeeman effect it should be noted that by employing the formal method of matrix mechanics, Heisenberg and Jordan[2] succeeded in deriving Landé's g formula and explaining the anomalous effect for atoms with one valence electron.

[1] See p. 266 and pp. 279–280.

[2] W. Heisenberg and P. Jordan, "Anwendung der Quantenmechanik auf das Problem der anomalen Zeemaneffekte," *Zeitschrift für Physik 37*, 263–277 (1926).

A mathematical scheme to include the Uhlenbeck-Goudsmit spin hypothesis[3] in the wave-mechanical description of a system was proposed by Darwin[4] and by Pauli.[5] Pauli replaced Schrödinger's wave function by a pair of wave functions whose squared absolute values determine the probability density of finding the electron with its spin oriented parallel or antiparallel to an arbitrarily chosen axis of quantization. In analogy to the orbital angular-momentum operators he introduced spin operators, represented by what were subsequently called the "Pauli matrices" which operate on the two-component wave function, a linear superposition of the two wave functions. By studying the transformation of its components under a rotation of the coordinate system and extending the results to systems of more than one electron, Pauli developed a consistent quantum-mechanical spin theory which explained a large number of phenomena but not the origin of the electron's spin. Pauli's theory, like the spinless quantum mechanics on which it was based, failed to satisfy the requirement of relativistic invariance. Dirac's generalization[6] of Schrödinger's equation to a set of four simultaneous partial differential equations of the first order, the so-called "linear Dirac equation," met the requirement of relativity and accounted automatically for the spin properties without abandoning the general physical aspects of the wave-mechanical description. Pauli's formalism turned out to be a low-speed limiting case of Dirac's relativistic theory of the electron.

That for the harmonic oscillator wave mechanics agrees with ordinary mechanics had already been shown by Schrödinger, as we have seen.[7] A more general and direct line of connection between quantum mechanics and Newtonian mechanics was established in 1927 by Ehrenfest,[8] who showed "by a short elementary calculation without approximations" that the expectation value of the time derivative of the momentum is equal to the expectation value of the negative gradient of the potential-energy function.[9] Ehrenfest's affirmation of Newton's second law in the sense of averages taken over the wave packet had a great appeal to many physicists and did much to further the acceptance of the theory. For it made it

[3] See pp. 149–151.

[4] C. G. Darwin, "The Zeeman effect and spherical harmonics," *Proceedings of the Royal Society of London* (*A*), *115*, 1–19 (1927); "The electron as a vector wave," *ibid. 116*, 227–233 (1927).

[5] W. Pauli, "Zur Quantenmechanik des magnetischen Elektrons," *Zeitschrift für Physik 43*, 601–623 (1927); *Collected Papers* (REF. 154 OF CHAP. 3), vol. 2, pp. 306–328.

[6] P. A. M. Dirac, "The quantum theory of the electron," *Proceedings of the Royal Society of London* (*A*) *117*, 610–624 (1928), *118*, 351–361 (1928). For the conceptual relation between Pauli's spin formalism and Dirac's electron theory see E. L. Hill and R. Landshoff, "The Dirac electron theory," *Reviews of Modern Physics 10*, 87–132 (1938).

[7] See pp. 282–283.

[8] P. Ehrenfest, "Bemerkung über die angenäherte Gültigkeit der klassischen Mechanik innerhalb der Quantenmechanik," *Zeitschrift für Physik 45*, 455–457 (1927).

[9] Ehrenfest did not formulate his theorem in terms of expectation values but described the mathematical relation in terms of wave packets.

possible to describe the particle by a localized wave packet which, though eventually spreading out in space, follows the trajectory of the classical motion. As emphasized in a different context elsewhere,[10] Ehrenfest's theorem and its generalizations by Ruark[11] for conservative systems with an arbitrary number of particles do not conceptually reduce quantum dynamics to Newtonian physics. They merely establish an analogy—though a remarkable one in view of the fact that, owing to the absence of a superposition principle in classical mechanics, quantum mechanics and classical dynamics are built on fundamentally different foundations.

That Schrödinger's theory could account also for the photoelectric effect, the main stronghold of the particle view of radiation, was shown in 1926 by Wentzel[12] and by Beck,[13] who gave a wave-mechanical derivation of Einstein's photoelectric equation.[14] Wentzel and Fues[15] explained the Auger effect, the nonradiative rearrangement of atomic electrons in the x-ray region of energy levels, and Gordon[16] and Wentzel[17] provided a wave-mechanical description of the Compton effect. How the problem of like particles and, in particular, that of the helium atom, the great puzzle of the older quantum theory, were successfully solved has already been mentioned at the end of Sec. 7.1. Now it was understood why the older quantum theory with all its perturbation methods could not cope with the problem. Kellner,[18] using Heisenberg's approach, obtained the correct ionization potential (24.47 eV) for helium. By showing that the molecular field, postulated by Weiss[19] twenty years earlier, can be fully accounted for in terms of the quantum-mechanical exchange forces Heisenberg[20] laid the foundations of a quantum theory of ferromagnetism. Furthermore, the adiabatic theorem which through Ehrenfest's work and Burgers' proof concerning the adiabatic invariance of the periodicity moduli of the action integrals played such an important role in the conceptual structure of the older quantum theory, as we saw in Sec. 3.1, could now be reformulated in the language of quantum mechanics: a system which is in a stationary state will stay in this state during an adiabatic process. An elegant proof

[10] REF. 230 OF CHAP. 5, p. 192; p. 207.

[11] A. E. Ruark, "The Zeeman effect and Stark effect of hydrogen in wave mechanics; the force equation and the virial theorem in wave mechanics," *Physical Review 31*, 533–538 (1928).

[12] G. Wentzel, "Zur Theorie des photoelektrischen Effekts," *Zeitschrift für Physik 40*, 574–589 (1926); "Über die Richtungsverteilung der Photoelektronen," *ibid. 41*, 828–832 (1927).

[13] G. Beck, "Zur Theorie des Photoeffekts," *ibid. 41*, 433–452 (1927).

[14] See pp. 35–36.

[15] G. Wentzel and E. Fues, "Über strahlungslose Quantensprünge," *Zeitschrift für Physik 43*, 524–530 (1927).

[16] W. Gordon, "Der Comptoneffekt nach der Schrödingerschen Theorie," *ibid. 40*, 117–133 (1926).

[17] G. Wentzel, "Zur Theorie des Comptoneffekts," *ibid. 43*, 1–8, 779–787 (1927).

[18] G. W. Kellner, "Die Ionisierungsspannung des Heliums nach der Schrödingerschen Theorie," *ibid. 44*, 91–109 (1927).

[19] P. Weiss, "L'hypothèse du champ moléculaire et la propriété ferromagnetique," *Journal de Physique 6*, 661–690 (1907).

[20] W. Heisenberg, "Zur Theorie des Ferromagnetismus," *Zeitschrift für Physik 49*, 619–636 (1928).

of this adiabatic theorem in quantum mechanics was given in 1928 by Born and Fock.[21] Similarly, by a wave-mechanical reformulation of the Born-Heisenberg-Jordan matrix formalism of radiation processes, based on a semiclassical treatment of the electromagnetic field, Klein[22] was able to derive the results obtained by Kramers and by Kramers and Heisenberg in their pre-quantum-mechanical theory of dispersion as described in Sec. 4.3. The insufficiencies of Klein's semiclassical radiation theory and, in particular, the fact that the relation between the emitted radiation and the atomic moment could not rigorously be derived from the general principles of quantum mechanics were soon remedied by Dirac's procedure[23] of quantizing the radiation field, a procedure which marked the beginning of quantum electrodynamics. Finally, it was shown in the fall of 1927 for systems obeying the Bose-Einstein statistics by Jordan and Klein and early in 1928 for systems obeying the Fermi-Dirac statistics by Jordan and Wigner that if the wave function is regarded as a field in ordinary space and time, but treated as a quantum-mechanical operator subject to certain quantum conditions, the usual quantum formalism for particles and the formalism for waves are mathematically equivalent.[24] The Klein-Jordan-Wigner second quantization method was generally considered as a mathematical expression of the wave-particle duality, of the fact that "the particle picture and the wave picture are merely two different aspects of one and the same physical reality." A better agreement between formalism and interpretation was hard to imagine.

Since 1927, the development of quantum mechanics and its applications to molecular physics, to the solid state of matter, to liquids and gases, to statistical mechanics, as well as to nuclear physics, demonstrated the overwhelming generality of its methods and results. In fact, never has a physical theory given a key to the explanation and calculation of such a heterogeneous group of phenomena and reached such a perfect agreement with experience as has quantum mechanics.

[21] M. Born and V. Fock, "Beweis des Adiabatensatzes," *ibid.* *51*, 165–180 (1928). Earlier proofs, given by M. Born, "Das Adiabatenprinzip in der Quantenmechanik," *ibid.* *40*, 167–192 (1926), and by E. Fermi and F. Persico, "Il prinzipio delle adiabatiche e la nozione de forza vivo nella nuova meccanica ondulatoria," *Rendiconti dell' Accademia dei Lincei* *4*, 452–457 (1926), were incomplete in so far as they treated only the nondegenerate case.

[22] O. Klein, "Elektrodynamik und Wellenmechanik vom Standpunkt des Korrespondenzprinzips," *Zeitschrift für Physik* *41*, 407–442 (1927).

[23] P. A. M. Dirac, "The quantum theory of the emission and absorption of radiation," *Proceedings of the Royal Society of London* (*A*), *114*, 243–265 (1927).

[24] P. Jordan and O. Klein, "Zum Mehrkörperproblem der Quantentheorie," *Zeitschrift für Physik* *45*, 751–765 (1927). P. Jordan and E. Wigner, "Über das Paulische Äquivalenzverbot," *ibid.* *47*, 631–651 (1928).

9 Two Fundamental Problems

9.1 Completeness

With the completion of its mathematical formalism, the establishment of its epistemological interpretation, and the accomplishment of its physical applications, quantum mechanics gained wide recognition as a full-fledged theory and gradually became a major subject in the program of the study of physics. The need for a coherent presentation of the theory was widely felt. The formulation most amenable to this end—and therefore also the first to be presented in the form of a textbook—was of course wave mechanics. In fact, the edition[1] of Schrödinger's papers in book form, published at the end of 1926, may be regarded as the earliest text on this subject. It was soon followed by a number of treatises[2] which either confined themselves exclusively to wave mechanics, as, for example, Bigg's work, the earliest textbook on this subject written in English,[3] or presented the topic in a more comprehensive context.[4]

The first complete exposition of the general formalism of quantum mechanics, presented in a logically consistent and axiomatic fashion, based on the notions of "observables" and "states" as primitives, was Dirac's *Principles of Quantum Mechanics*.[5] The impact of the book on the

[1] REF. 251 OF CHAP. 5.

[2] A. E. Haas, *Materiewellen und Quantenmechanik* (Akademische Verlagsgesellschaft, Leipzig, 1928); Louis de Broglie, *La Mécanique Ondulatoire* (Gauthier-Villars, Paris, 1928), *Wellenmechanik*, translated by R. Peierls (Akademische Verlagsgesellschaft, Leipzig, 1929); J. Frenkel, *Einführung in die Wellenmechanik* (Springer, Berlin, 1929); A. Sommerfeld, *Atombau und Spektrallinien, Wellenmechanischer Ergänzungsband* (Vieweg, Braunschweig, 1929).

[3] H. F. Biggs, *Wave Mechanics* (Oxford University Press, 1927).

[4] G. Birtwistle, *The New Quantum Mechanics* (Cambridge University Press, 1928); M. Born and P. Jordan, *Elementare Quantenmechanik* (Springer, Berlin, 1929); A. Landé, "Optik, Mechanik und Wellenmechanik," in *Handbuch der Physik*, edited by H. Geiger and K. Scheel (Springer, Berlin, 1928), vol. 20; W. Pauli, "Allgemeine Prinzipien der Quantenmechanik," *ibid.*, edited by H. Geiger and K. Scheel, vol. 24, part 1 (1933).

[5] REF. 71 OF CHAP. 6.

generation of modern physicists is perhaps best characterized by a remark made by Lennard-Jones when the book appeared: "An eminent European physicist, who is fortunate enough to possess a bound set of reprints of Dr. Dirac's original papers, has been heard to refer to them affectionately as his 'bible.' Those not so fortunate have now at any rate an opportunity of acquiring a copy of the authorized version."[6] The key to Dirac's success in unifying the presentation of the statistical transformation theory was his consistent use of a vector space of states ψ, that is, a space with as many dimensions as there are linearly independent states of the system, each axis corresponding to an independent state and each other state being completely determined by the direction of a vector relative to these axes. In the first half of the book Dirac expounded the general theory and in the second part applied it to a number of important problems such as the electronic structure of atoms, collision problems, and the radiation theory. The second edition (1935), for which he rewrote almost the whole text and which had to undergo only minor modifications in subsequent editions, was for many years the standard text on the subject, hardly "surpassed in brevity and elegance" by later publications in this field.

Regarding mathematics as a handmaid of physics, as we have seen in connection with the introduction of the δ-function, Dirac "tried to keep the physics to the forefront."[7] Not particularly interested in questions concerning mathematical rigor, he advanced his conception of the vector space without realizing at first that it was but a modification of the Hilbert space. Von Neumann's *Mathematische Grundlagen der Quantenmechanik*[8] with its emphasis on mathematical rigor was therefore a timely contribution, supplementary to Dirac's work. "While Dirac presents his reasoning with admirable simplicity and allows himself to be guided at every step by physical intuition—refusing at several places to be burdened by the impediment of mathematical rigor—von Neumann goes at his problem equipped with the nicest of modern mathematical tools and analyzes it to the satisfaction of those whose demands for logical completeness are most exacting."[9]

The first half of von Neumann's monumental book is essentially an elaboration of his papers published in 1927, whereas the second half, inspired by the ideas of Bohr and Heisenberg, concerns itself with the problems of causal description and measurement. Dealing with ensembles of systems, von Neumann distinguished between "homogeneous ensembles" ("einheitliche Gesamtheiten") or, as Weyl used to call them, "pure cases" ("reine Fälle")[10] and "mixtures" ("Gemische"). The former were defined

[6] *Mathematical Gazette 15*, 505 (1931).
[7] Preface to the first edition.
[8] REF. 108 OF CHAP. 6.
[9] H. Margenau, Review in *Mathematical Gazette 17*, 493 (1933).
[10] H. Weyl conceived this notion at the same time, independently of von Neumann. Cf. H. Weyl, "Quantenmechanik und Gruppentheorie," *Zeitschrift für Physik 46*, 1–46 (1927).

as represented by a single normalized state function ψ, expressible in terms of a complete orthonormal system $\psi = \sum a_n\psi_n$; in it every physical quantity has the same expectation value, whether measured in the whole ensemble or in any of its subensembles; it thus represents maximal knowledge. Mixtures were defined as corresponding to physical conditions for whose specification a set of ψ's was required: $\psi^{(1)}, \psi^{(2)}, \ldots, \psi^{(k)}, \ldots$, each $\psi^{(k)}$ being present with a probability $p^{(k)}$; a mixture $\sum p^{(k)}\psi^{(k)}$ thus describes a system in which knowledge is not maximal.[11] Following a suggestion by Landau,[12] von Neumann[13] introduced, for the unique characterization of the statistics of an ensemble, the "statistical operator" U which eventually, under the name of "ρ matrix," became a major tool in quantum statistics. Von Neumann showed that quite generally $\langle A \rangle = \mathrm{Tr}\,(UA)$ and that $U^2 = U$ is a criterion for the existence of a pure case.

In the fourth chapter of the book von Neumann applied these notions to his famous analysis of the problem whether the statistical theory of quantum mechanics as established constitutes a logically closed theory or whether it could be reformulated as an entirely deterministic theory by the introduction of hidden parameters, that is, additional variables which, unlike ordinary observables, are inaccessible to measurements and hence not subject to the restrictions of uncertainty relations. He came to the conclusion that "the present system of quantum mechanics would have to be objectively false, in order that another description of the elementary processes than the statistical one may be possible."[14] To prove this theorem he first showed that dispersion-free ensembles do not exist. For in such ensembles, he argued, $\langle A^2 \rangle = \langle A \rangle^2$ for all operators A or equivalently $\mathrm{Tr}\,(UA^2) = (\mathrm{Tr}\,(UA))^2$; substituting for A the operator $P_{[\psi]}$, he obtained[15] $U = 0$ or $U = 1$; but $U = 0$, leading to $\langle A \rangle = 0$ and furnishing no information at all, had to be ruled out and $U = 1$, implying in a space of infinitely many dimensions $\langle 1 \rangle = \infty$, is not normalizable and could correspond only to a state which is not dispersion-free. Having thus demonstrated that ensembles are never dispersion-free, von Neumann showed that homogeneous ensembles (pure cases) do exist by proving that any ensemble whose statistical operator is a projection operator is homogeneous. Finally, he pointed out, the statistical nature of homogeneous

[11] The expectation value for an operator A is in the pure case $\langle A \rangle_\psi = \Sigma a_i^* A_{ik} a_k$ and in the mixture $\Sigma p^{(k)} \langle A \rangle_{\psi^{(k)}}$.

[12] L. Landau, "Das Dämpfungsproblem in der Quantenmechanik," *Zeitschrift für Physik 45*, 430–441 (1927).

[13] The first explicit reference to the statistical operator is found in von Neumann's paper "Wahrscheinlichkeitstheoretischer Aufbau der Quantenmechanik," p. 253. See REF. 109 OF CHAP. 6.

[14] REF. 108 OF CHAP. 6, German ed. p. 171, English ed. p. 325.

[15] $P_{[\psi]}$ is the projection operator $|\psi\rangle\langle\psi|$ in Dirac's notation. For an objection against the deduction of $U = 1$ or 0 from $\langle U \rangle = \langle U \rangle^2$ see J. Tharrats, "Sur le théorème de von Neumann concernant l'indéterminisme essentiel de la Mécanique quantique," *Comptes Rendus 250*, 3786–3788 (1960), and for a rebuttal of this objection see J. Albertson, "Von Neumann's hidden-parameter proof," *American Journal of Physics 29*, 478–484 (1961).

states cannot be removed by conceiving them as mixtures of substates, each of them associated with a definite set of values of hidden parameters, and by assuming that the dispersion of the homogeneous state results from an averaging over such "actual" states; because then, he declared, a homogeneous ensemble could be represented as a mixture of two different ensembles, contrary to its definition, and because dispersion-free ensembles "which would have to correspond to the 'actual' states do not exist," as previously demonstrated.

Von Neumann thus succeeded in bringing the foremost methodological problem of quantum mechanics down from the realm of speculation into the reach of empirical decision. For the essence of his demonstration was this: as long as observation and experiment enforce upon us the present formalism of quantum mechanics, it is logically impossible to complete this formalism to a deterministic description of physical processes.[16]

It is interesting to note that at the same time and independently of von Neumann's work, Solomon,[17] utilizing the technique of Stieltjes's treatment of the "problem of moments," proved that the introduction of hidden parameters leads to inconsistencies with the accepted formalism. In contrast to von Neumann's demonstration, which, as may have been noticed, was independent of the existence of noncommuting operators, Solomon's proof presumed the uncertainty principle.

In view of the decisive importance of von Neumann's hidden-parameter theorem for the foundations of quantum mechanics various attempts were subsequently made to prove the theorem by different methods. We mention only Destouches-Février's proof,[18] which was based on the existence of

[16] For (mostly critical) reformulations of von Neumann's proof see P. K. Feyerabend, "Eine Bemerkung zum Neumannschen Beweis," *Zeitschrift für Physik 145*, 421–423 (1956); I. I. Zinnes, "Hidden variables in quantum mechanics," *American Journal of Physics 26*, 1–4 (1958); G. Schulz, "Kritik des v. Neumannschen Beweises gegen die Kausalität in der Quantenmechanik," *Annalen der Physik 3*, 94–104 (1959); J. Albertson, see REF. 15; J. Albertson, "The statistical nature of quantum mechanics," *British Journal for the Philosophy of Science 13*, 229–233 (1962); D. Bohm, "Hidden Variables in the Quantum Theory," in *Quantum Theory*, edited by D. R. Bates (Academic Press, New York, London, 1962), vol. 3, pp. 345–387, especially pp. 350–351; A. Komar, "Indeterminate character of the reduction of the wave packet in quantum theory," *Physical Review 126*, 365–369 (1962).

For objections raised against von Neumann's proof see D. Bohm, "A suggested interpretation of the quantum theory in terms of 'hidden variables,' " *Physical Review 85*, 180–193 (1952); I. Fenyes, "Eine wahrscheinlichkeitstheoretische Begründung und Interpretation der Quantenmechanik," *Zeitschrift für Physik 132*, 81–106 (1952); Louis de Broglie, *La Physique Quantique, restera-t-elle indéterministe?* (Gauthier-Villars, Paris, 1953); Louis de Broglie, "Wird die Quantenphysik indeterministisch bleiben?" *Physikalische Blätter 9*, 488–497, 541–548 (1953); W. Weizel, "Ableitung der Quantentheorie aus einem klassischen, kausal determinierten Modell," *Zeitschrift für Physik 134*, 264–285 (1953); A. F. Nicholson, "On a theory due to I. Fenyes," *Australian Journal of Physics 7*, 14–21 (1954).

[17] J. Solomon, "Sur l'indéterminisme de la Mécanique quantique," *Journal de Physique 4*, 34–37 (1933).

[18] P. Destouches-Février, "Une nouvelle preuve du caractère essentiel de l'indéterminisme quantique," *Comptes Rendus 220*, 553–555 (1945).

noncommuting operators, and a recent demonstration by Jauch and Piron[19] which is independent of the assumption, tacitly made by von Neumann but untenable in systems with superselection rules,[20] namely, that every projection operator is an observable, and also independent of the assumption that the expectation values of noncompatible observables are additive, a property of states which was rigorously established[21] only in 1957. In addition, these authors showed that "the hidden variable interpretation is only possible if the theory is observably wrong"; that is, they showed that any introduction of hidden variables to "complete" the theory modifies it to such an extent that it leads to empirically refutable results, a point not raised in von Neumann's analysis.

9.2 Observation and Measurement

Another aspect of quantum mechanics which von Neumann regarded as so far insufficiently dealt with was the problem of observation and measurement, that is, the relation between elements of the physical theory and human experience, a subject to which he devoted the last two chapters of his book.[22] For he realized that the accessibility of quantum-mechanical results to the observer raises a serious problem for the following reason. In classical physics the interaction between object and observer, if not regarded as wholly negligible, was in principle eliminable since the measurement process was fully analyzable in terms of the equations of motion alone. In fact, measurement, in classical physics, constituted a chapter of applied physics: physical theory and the analysis of measurement employed the same category of concepts. In quantum mechanics, however, a fundamental disparity exists between the language of the theory and the language of ordinary experience, which latter, as Bohr repeatedly stressed, is the indispensable medium for the ultimate formulation of results of measurement.

Inspired by Bohr's paper[23] of 1929 on the quantum of action and the description of nature, von Neumann espoused the idea that every quantum-

[19] J. M. Jauch and G. Piron, "Can hidden variables be excluded in quantum mechanics?" *Helvetica Physica Acta 36*, 827–837 (1963).

[20] G. C. Wick, A. S. Wightman, and E. P. Wigner, "The intrinsic parity of elementary particles," *Physical Review 88*, 101–105 (1952).

[21] A. M. Gleason, "Measures on the closed subspaces of a Hilbert space," *Journal of Mathematics and Mechanics 6*, 885–893 (1957).

[22] REF. 108 OF CHAP. 6, chaps. 5 and 6. A somewhat simplified account of von Neumann's theory of measurement has been given by F. London and E. Bauer, *La Théorie de l'Observation en Mécanique Quantique*, Actualités Scientifiques et Industrielles No. 775 (Hermann et Cie, Paris, 1939), pp. 22–47. Cf. also Louis de Broglie, *La Théorie de la Mesure en Mécanique Ondulatoire* (Gauthier-Villars, Paris, 1957), pp. 15–41, and G. Ludwig, *Die Grundlagen der Quantenmechanik* (J. Springer, Berlin, Göttingen, Heidelberg, 1954), chap. 5, pp. 122–165.

[23] REF. 79 OF CHAP. 7.

mechanical measuring process involves an unanalyzable element. He postulated that, in addition to the continuous causal propagation of the wave function governed by the Schrödinger equation, the function undergoes, in the case of a measurement, due to an intervention of the observer on the object, a discontinuous, noncausal, and instantaneous change. Thus, in a mixture, characterized by the statistical operator U, a measurement of the observable R, supposed to have a complete orthonormal set of eigenfunctions $\varphi_1, \varphi_2, \ldots$ corresponding to the (discrete, nondegenerate) eigenvalues $\lambda_1, \lambda_2, \ldots$, changes U to

$$U' = \sum (\varphi_n, U\varphi_n) P_{[\varphi_n]}$$

an irreversible process, whereas the causal, continuous, and reversible change is given by the unitary transformation

$$U_t = \exp\left(-\frac{2\pi itH}{h}\right) U \exp\left(\frac{2\pi itH}{h}\right)$$

where H is the energy operator (Hamiltonian) of the system. In the particular case of a pure state $\psi = \sum a_n\psi_n$, expanded in the eigenfunctions of the observable R to be measured, the measurement of R "reduces" ψ to ψ_k, the eigenstate whose eigenvalue λ_k is the result of the measurement. According to von Neumann this "reduction of state" or, more generally, "reduction of wave packet" is an unavoidable element in every measurement process and assures that the same result will be obtained for an immediately following measurement of the same observable ("projection postulate").

Von Neumann explained the insufficiency of continuous transformations $U \to U_t$ by pointing out that in a measurement not the system S itself but rather S combined with a measuring apparatus M (and ultimately with the observer) is the object of observation. "The theory of the measurement is a statement concerning $S + M$, and should describe how the state of S is related to certain properties of the state of M (namely, the positions of a certain pointer, since the observer reads these). Moreover, it is rather arbitrary whether or not one includes the observer in M, and replaces the relation between the state S and the pointer positions in M by the relations of this state and the chemical changes in the observer's eye or even in his brain."[24] More specifically, the situation as conceived by von Neumann may be described as follows. A physical system S is supposed to be in the state $\psi = \sum a_n\psi_n(x)$, where $\psi_n(x)$ is the eigenfunction belonging to the eigenvalue λ_n of the observable R to be measured. To measure R, S has to be coupled with a measuring apparatus M whose pointer readings $g_0, g_1, \ldots$ (supposed to form a discrete set) correspond to the eigenfunctions

[24] REF. 108 OF CHAP. 6, p. 352 (English ed.).

$\varphi_0(y), \varphi_1(y), \ldots$ of an observable G, $\varphi_0(y)$ representing the zero (initial) state of M with g_0. Before the coupling the state function of the system $S + M$ is

$$\Psi_1(x,y) = \varphi_0(y) \sum a_n \psi_n(x)$$

and after the coupling

$$\Psi_2(x,y) = \sum a_n \psi_n(x) \varphi_n(y)$$

that is, a superposition of different states $\psi_n \varphi_n$ with probabilities $|a_n|^2$. The indefiniteness of G (and, correspondingly, of the pointer readings g_n) is resolved by the reduction $\Psi_2 \rightarrow \psi_k(x)\varphi_k(y)$ which occurs at the moment when the observer becomes aware that the pointer reading is g_k. If, instead, a second apparatus M' had been applied to measure G, then in exact analogy the final state of $S + M + M'$ before the reading would have been $\Psi_2' = \sum a_n \psi_n \varphi_n \varphi_n'$ and so on. Von Neumann showed that the insertion of such additional instrumental (or "objective") stages does not affect the outcome of the measurement, namely, the value λ_k of R, a conclusion by which he justified the adherence to the principle of the psychophysical parallelism according to which, as Bohr contended, it "must be possible so to describe the extraphysical process of the subjective perception as if it were in reality in the physical world."[25]

Von Neumann's truncation of an otherwise infinite regress $S + M + M' + \cdots$ by resorting to the subjective perception of the observer or, as London and Bauer[26] formulated it, to the observer's "faculty of introspection," a process which finds its physical expression in the "reduction of the wave function," introduced the factor of human consciousness as an integral part into the formulation of quantum mechanics. As von Neumann admitted himself, he was greatly influenced in this respect by Leo Szilard, who, at that time Privatdocent in Berlin, studied the problem to what extent the intervention of an intelligent being on a thermodynamic system may produce a *perpetuum mobile* of the second kind. Szilard showed that owing to the very measurement performed by such a being—in fact, it suffices to ascribe to the measuring agency merely a memory—the system's behavior changes distinctly and may lead to a decrease of the entropy, thus violating the second law of thermodynamics, unless the measuring

[25] *Ibid.*, p. 419.

[26] "L'observateur a un tout autre point de vue: pour lui c'est seulement l'objet x et l'appareil y qui appartiennent au monde extérieur, à ce qu'il appelle 'objectif.' Par contre il a *avec lui-même* des relations d'un caractère tout particulier: il dispose d'une faculté caractéristique et bien familière, que nous pouvons appeler la 'faculté d'introspection': il peut se rendre compte de manière immédiate de son propre état. C'est en vertu de cette 'connaissance immanente' qu'il s'attribue le droit de se créer sa propre objectivité, c'est-à-dire de couper la chaine de coordinations statistiques exprimées par $\Sigma a_n \psi_n(x) \varphi_n(y) \chi_n(z)$ en constatant: 'Je suis dans l'état χ_k' ou plus simplement: 'je vois $G = g_k$' ou même directement: '$R = \lambda_k$.' " REF. 22, p. 42.

process itself implies an entropy increase.[27] Szilard, in turn, was led to these thought-provoking ideas—somewhat reminiscent of Maxwell's "demon"—through his study of Smoluchowski's paper[28] on the limitations of the second law which Smoluchowski had read in Göttingen, on the invitation of the Wolfskehl committee,[29] shortly before his appointment to the directorship of the Institute of Physics at the University of Cracow in 1913. Smoluchowski's conception of an intellect that is constantly cognizant of the instantaneous state of a dynamical system and thus able to invalidate the second law of thermodynamics without performing work was probably the earliest logically unassailable speculation about a physical intervention of mind on matter. It paved the way, as we have seen, toward von Neumann's far-reaching conclusion that it is not possible to formulate the laws of quantum mechanics in a complete and consistent way without a reference to human consciousness. Von Neumann fully realized that this conceptual procedure led to the inexorable result that "experience only makes statements of this type: an observer has made a certain (subjective) observation; and never any like this: a physical quantity has a certain value,"[30] or, as Heisenberg[31] once said, "the laws of nature which we formulate mathematically in quantum theory deal no longer with the particles themselves but with our knowledge of the elementary particles."

Although unobjectionable from the purely logical point of view, von Neumann's theory of measurement soon became the target of severe criticisms. Thus, to mention a recent example, his theory was censured as being "founded on a radically subjectivistic (solipsistic) philosophy."[32] It was also pointed out[33] that the transformation of a pure state into a mixture, as envisaged by von Neumann, could not easily be reconciled with certain conservation principles. Margenau and his school,[34] insisting on

[27] L. Szilard, "Über die Entropieverminderung in einem thermodynamischen System bei Eingriffen intelligenter Wesen," *Zeitschrift für Physik 53*, 840–856 (1929). Cf. also his former paper "Über die Ausdehnung der phänomenologischen Thermodynamik auf die Schwankungserscheinungen," *ibid. 32*, 753–788 (1925).

[28] Marie (Ritter von) Smoluchowski, "Gültigkeitsgrenzen des zweiten Hauptsatzes der Wärmethorie," in *Vorträge über die kinetische Theorie der Materie und Elektrizität* (Teubner, Leipzig, Berlin, 1914), pp. 89–121; *Pisma Marjana Smoluchowskiego—Œuvres de Marie Smoluchowski*, edited by L. Natanson (Imprimerie de l'Université Jaguellonne, Cracow; Béranger, Paris, 1927), pp. 361–398.

[29] See p. 25.

[30] REF. 24, p. 420.

[31] W. Heisenberg, "The representation of nature in contemporary physics," *Daedalus 87*(3), 99 (1958).

[32] A. Daneri, A. Loinger, and G. M. Prosperi, "Quantum theory of measurement and ergodicity conditions," *Nuclear Physics 33*, 297–319 (1962).

[33] E. P. Wigner, "Die Messung quantenmechanischer Operatoren," *Zeitschrift für Physik 133*, 101–108 (1952).

[34] H. Margenau, "Critical points in modern physical theory," *Philosophy of Science 4*, 337–370 (1937); "Philosophical problems concerning the meaning of measurement in physics," *ibid. 25*, 23–33 (1958), reprinted in *Measurement—Definitions and Theories* (Wiley, New York, 1959), pp. 163–176; "Measurements and quantum states," *ibid. 30*, 1–16, 138–157 (1963). J. L. McKnight, "The quantum theoretical concept of measurement," *ibid. 24*, 321–330 (1957), reprinted in *Measurement*, pp. 192–203; "Measurement in quantum mechanical systems, an investigation of foundations," Dissertation

the importance of a sharp distinction between "state-preparation" and "measurement," rejected the projection postulate on the ground that a measurement often destroys the state corresponding to the eigenvalue obtained as, for instance, in the measurement of the energy of an atom by observing its emission spectrum where it is precisely the decay of the state which renders possible the measurement. Viewing the measuring process as an interaction between a micro object and a macro system, in the course of which the latter evolves from a thermodynamically metastable state to a stable state, depending on the state of the micro object, Ludwig and his school[35] tried to find the answer to the problem in the ergodic behavior of macro observables. Numerous other attempts have been made to modify von Neumann's theory of measurement or to replace it by what seemed to these authors[36] a more acceptable solution of the problem.

While most of these attempts suggested modifications of only certain parts of von Neumann's formulation of quantum mechanics or even of only the interpretation as given by him, the basic algebraic structure of the formalism as a whole also soon became the object of important generalizations. In the search for a completely rigorous mathematical foundation of the theory and a solution of the difficulties mentioned at the end of

(Yale University, 1957); "An extended latency interpretation of quantum mechanical measurement," *Philosophy of Science 25*, 209–222 (1958).

[35] G. Ludwig, "Der Messprozeß," *Zeitschrift für Physik 135*, 483–511 (1953); "Zur Deutung der Beobachtung in der Quantenmechanik," *Physikalische Blätter 11*, 489–494 (1955); "Zum Ergodensatz und zum Begriff der makroskopischen Observablen," *Zeitschrift für Physik 150*, 346–374 (1958), *152*, 98–115 (1958). A. Daneri, A. Loinger, and G. M. Prosperi (REF. 32).

[36] P. Jordan, "On the process of measurement in quantum mechanics," *Philosophy of Science 16*, 269–278 (1949). D. Bohm, *Quantum Theory* (Prentice-Hall, Englewood Cliffs, N.J., 1951), pp. 583–623. G. Lüders, "Über die Zustandsänderung durch den Messprozeß," *Annalen der Physik 8*, 322–328 (1951). H. J. Groenewold, "Information in quantum measurements," *Proceedings of the Amsterdam Academy (B)*, *55*, 219–227 (1952). E. Schrödinger, "The philosophy of experiment," *Il Nuovo Cimento 1*, 5–15 (1955). H. Everett, " 'Relative State' formulation in quantum mechanics," *Reviews of Modern Physics 29*, 454–465 (1957). P. K. Feyerabend, "Zur Quantentheorie der Messung," *Zeitschrift für Physik 148*, 551–559 (1957); "On the quantum theory of measurement," in *Observation and Interpretation: A Symposium of Philosophers and Physicists*, Proceedings of the Ninth Symposium of the Colston Research Society, University of Bristol, April 1–April 4, 1957 (Butterworth, London, 1957; reprinted by Dover, New York, 1962), pp. 121–130. G. Süssmann, "An analysis of measurement," *ibid.* (Colston Papers), pp. 131–136; "Über den Messvorgang," *Münchener Berichte* (Heft 88, 1958). H. S. Green, "Observation in quantum mechanics," *Il Nuovo Cimento 9*, 880–889 (1958). A. Landé, "Zur Quantentheorie der Messung," *Zeitschrift für Physik 153*, 389–393 (1959). L. Durand III, "On the theory of measurement in quantum mechanical systems," *Philosophy of Science 27*, 115–133 (1960). H. Wakita, "Measurement in quantum mechanics," *Progress of Theoretical Physics 23*, 32–40 (1960), *27*, 139–144, 1156–1164 (1962). H. Araki and M. M. Yanase, "Measurement of quantum mechanical operators," *Physical Review 120*, 622–626 (1960). R. Mould, "Quantum theory of measurement," *Annals of Physics 17*, 404–417 (1962). J. Albertson, "Quantum-mechanical measurement operator," *Physical Review 129*, 940–943 (1962). A. Shimony, "Role of the observer in quantum theory," *American Journal of Physics 31*, 755–773 (1963). E. P. Wigner, "The problem of measurement," *ibid.*, 6–15 (1963). D. I. Blokhintsev, *Principles of Quantum Mechanics* (Allyn and Bacon, Boston, 1964), pp. 65–70.

Chap. 6, two main lines of attack have been adopted. The first of these, aiming at an explicit algebraic generalization of the operator calculus and the structure of the Hilbert space, began when Jordan[37] introduced into quantum mechanics the so-called quasi-multiplicative[38] commutative algebra, that is, an algebra which differs from the noncommutative, but associative, matrix algebra in so far as the associative product AB (of two matrices) is replaced by the commutative expression $\frac{1}{2}(AB + BA)$. Jordan justified this proposal by pointing out that "the whole statistics of measurements on quantum-mechanical systems involves relations of only quasi multiplication and not of complete multiplication between measurable quantities."[39] Using the abbreviation

$$\{A,B,C\} = (AB)C - A(BC)$$

Jordan showed[40] that any two quantum-mechanical observables A, B satisfy $\{A,B,A^2\} = 0$ and that the last equation holds whenever $[A,B,A^2] = 0$, where the square brackets are defined, in terms of ordinary multiplication, by $[A,B,C] = (AB)C - A(BC)$. Jordan therefore suggested that hypercomplex algebras for which $[A,B,A^2] = 0$, which he called "r-number algebras" ("r-Zahl-Algebren"), are the appropriate generalizations of the conventional algebra as used in von Neumann's formalism. A few months later, however, Jordan, von Neumann, and Wigner,[41] having made a penetrating analysis of these algebras, established, with the help of a theorem of Albert,[42] the equivalence between almost all (real[43]) r-number algebras and algebras whose elements are ordinary (real) matrices, with products defined by quasi multiplication, the only exception being the algebra of all three-rowed Hermitian matrices with elements in the real nonassociative algebra of Cayley numbers.[44]

[37] P. Jordan, "Über die Multiplikation quantenmechanischer Größen," *Zeitschrift für Physik 80*, 285–291 (1933).

[38] The term "quasi multiplication" for replacement of AB by $\frac{1}{2}(AB + BA)$ was introduced by Jordan, *ibid.*, p. 286.

[39] *Ibid.*, p. 286.

[40] P. Jordan, "Über eine Klasse nichtassoziativer hyperkomplexer Algebren," *Göttinger Nachrichten 1932*, pp. 569–575; "Über Verallgemeinerungsmöglichkeiten des Formalismus der Quantenmechanik," *ibid. 1933*, pp. 209–217.

[41] P. Jordan, J. von Neumann, and E. Wigner, "On an algebraic generalization of the quantum mechanical formalism," *Annals of Mathematics 35*, 29–64 (1934).

[42] A. A. Albert, "On a certain algebra of quantum mechanics," *Annals of Mathematics 35*, 65–73 (1934).

[43] The field of coefficients was taken as real to meet "an old objection to the quantum mechanical formalism, namely, that it does not confine itself to the real domain." REF. 41, p. 29.

[44] Cayley, "On quaternions," *Philosophical Magazine 26*, 210–211 (1845); *Collected Mathematical Papers* (see REF. 16 OF CHAP. 5), vol. 1, p. 127. On Cayley numbers see L. E. Dickson, *Algebras and Their Arithmetics* (University of Chicago Press, Chicago, 1923; Stechert and Co., New York, 1938; reprinted by Dover, New York, 1960). The structure of Cayley-number algebras was studied by L. E. Dickson, "On quaternions and their generalizations and the history of the eight square theorem," *Annals of Mathematics 20*, 155–171 (1918); H. Freudenthal, "Oktaven, Ausnahmegruppen und Oktavengeometrie," *Mathematisch Instituut der Rijksuniversiteit te Utrecht 1951*; M. Zorn, "Theorie der alternativen Ringe," *Abhandlungen des Mathematischen Seminars der Hamburger Universität 8*, 123–147 (1930).

In 1959 Finkelstein, Jauch, and Speiser[45] showed that only three types of Hilbert spaces are possible in quantum mechanics: the real Hilbert space, the complex Hilbert space, and the quaternion Hilbert space. The first type, that is, a Hilbert space whose scalar multipliers are real numbers, became the subject of extensive investigations by Stueckelberg and his collaborators,[46] while Finkelstein, Jauch, Schiminovitch, and Speiser,[47] as well as Dyson[48] and Emch,[49] studied the structure of a quaternion quantum mechanics,[50] that is, an operator algebra in a Hilbert space whose scalar multipliers are quaternions. Reference to this possibility has been made, in a different context, at the beginning of Chap. 5.[51] Recently, Goldstine and Horwitz[52] investigated the properties of a Hilbert space with Cayley numbers as scalar multipliers.

The second line of development began with the by now classical work of Birkhoff and von Neumann,[53] who showed that the calculus of quantum-mechanical propositions, concerning results of measurements, is formally indistinguishable from the calculus of linear subspaces of the infinite-dimensional Hilbert space—pure states corresponding to one-dimensional subspaces (rays)—with respect to set products, linear sums, and orthogonal complements, and that the relations between measurements of different observables are reflected by the (ortho-complemented) lattice structure of such subspaces, while the distributive law $a \cap (b \cup c) = (a \cap b) \cup (a \cap c)$, characteristic for the propositional calculus of classical systems, no longer holds. Thus, the lattice structure of subspaces—reflecting the fact that measurements of different observables are apt to interfere with each

[45] D. Finkelstein, J. M. Jauch, and D. Speiser, "Notes on quaternion quantum mechanics," *Cern Reports* (Theoretical Study Division) *59–7*, *59–9*, *59–17* (1959); "Zur Frage der Ladungsquantisierung," *Helvetica Physica Acta 32*, 258–260 (1959).

[46] E. C. G. Stueckelberg, "Quantum theory in real Hilbert space," *Helvetica Physica Acta 33*, 727–752 (1960); E. C. G. Stueckelberg and M. Guenin (part II), *ibid. 34*, 621–628 (1961); E. C. G. Stueckelberg, C. Piron, and H. Ruegg (part III), *ibid. 34*, 675–698 (1961); E. C. G. Stueckelberg and M. Guenin (part IV), *ibid. 35*, 673–695 (1962).

[47] D. Finkelstein, J. M. Jauch, S. Schiminovitch, and D. Speiser, "Foundations of quaternion quantum mechanics," *Journal of Mathematical Physics 3*, 207–220 (1962); D. Finkelstein, J. M. Jauch, and D. Speiser, "Quaternionic representations of compact groups," *ibid. 4*, 136–140 (1963).

[48] F. J. Dyson, "The threefold way, algebraic structure of symmetry groups and ensembles in quantum mechanics," *ibid. 3*, 1199–1215 (1962).

[49] G. Emch, "Mécanique quantique quaternionienne et relativité restreinte," *Helvetica Physica Acta 36*, 739–769, 770–788 (1963).

[50] The first to conceive the idea of a quaternion Hilbert space was probably H. Wachs; the first to suggest the possibility of its use in quantum mechanics was P. Jordan in his paper "Über die Multiplikation quantenmechanischer Grössen II," *Zeitschrift für Physik 87*, 505–512 (1934). Its operator algebra was first studied by O. Teichmüller, "Operatoren im Wachsschen Raum," *Journal für die reine und angewandte Mathematik 174*, 73–124 (1936).

[51] See p. 205.

[52] H. H. Goldstine and L. P. Horwitz, "On a Hilbert space with non-associative scalars," *Proceedings of the National Academy of Sciences 48*, 1134–1142 (1962); "Hilbert space with non-associative scalars I," *Mathematische Annalen 154*, 1–27 (1964).

[53] G. Birkhoff and J. von Neumann, "The logic of quantum mechanics," *Annals of Mathematics 37*, 823–843 (1936); *Collected Works* (REF. 97 OF CHAP. 6), vol. 4, pp. 105–125.

other—was found to provide a rigorous logicomathematical formulation of the uncertainty relations and complementarity properties. In 1952 Wick, Wightman, and Wigner[54] showed in their important work on superselection rules that there are physical systems for which the sets of operators representing observables do not form an irreducible system in the space of the state vectors, as previously assumed, that not every self-adjoint operator is an observable, and that certain subspaces in a Hilbert space of state vectors cannot be connected by observables, the state being subject to selection rules. In 1963 Jauch and Piron[55] made it clear that this situation can be incorporated without major difficulties into the lattice structure of the quantum-mechanical propositional calculus. A recent study by Horwitz and Biedenharn[56] on the structure of a quantum theory which works with a Hilbert space over an arbitrary finite algebra with unity quantity promises to establish important connections between these two main lines of development.

The preceding brief description of recent elaborations of von Neumann's formalism is intended to show that today von Neumann's work still forms the main foundation for theoretical research in quest of a mathematically rigorous formulation of quantum mechanics.

[54] REF. 20. Cf. also REF. 33.
[55] REF. 19.
[56] L. P. Horwitz and L. C. Biedenharn, "Intrinsic superselection rules of algebraic Hilbert space," *Helvetica Physica Acta 38*, 385–408 (1965).

Concluding Remarks

Any further discussion of the problems of completeness (hidden variables) and measurement (observation), which, as the preceding references have shown, lead us to current research on the foundations of quantum mechanics, would be beyond the scope of the present study.

There is one question, however, which deserves to be discussed in conclusion of our exposition, namely, whether the conceptual scheme which quantum mechanics has established and which forms the foundation of current research in atomic physics constitutes, if viewed from a broad historical vantage, a complete break with the past in the sense of being an isolated chapter in the history of human thought or whether it is merely another stage in an uninterrupted course of scientific thought and another link in a chain of intellectual tendencies.

It was one of the major objectives of our study to investigate to what extent precisely the conceptual foundations of quantum mechanics evolved from earlier conceptions and what factors played a decisive role in this process of formation. As we have seen, science never worked in a vacuum and every innovation, however radical it was, had its roots in the past. Hence, from the psychological, heuristic, and methodological point of view the answer is clear.

But our present question goes deeper. It asks whether the resulting conceptual scheme, apart from its formative aspects in the mind of its originators, falls in line with the historical development of physics through the ages. In one respect, at least, this seems to be the case.

Reviewing the historical development of physical thought, we may say that Aristotelian physics—and most of its medieval modifications—was a physics of an essentially unlimited number of qualities or properties which were assumed to inhere in the object under investigation. Atomism, as already conceived by the Greeks, attempted to reduce the immense diversity of physical qualities to a limited number of fundamental properties possessed by the atom. Guided by the logical principle that an observed property of a physical object cannot be explained by merely ascribing the property to the constituent parts of the object, atomists were soon led to distinguish two categories of observable properties: those which were conceived as belonging to the substances, independent of the observer—and these were the geometrical and dynamical properties—and those which, like color, heat, sound, or taste, were conceived as not being "in the object," but merely "phantasms" in the "sentients."[1]

Classical physics, with mechanics as its foundation, was but a quantitative attempt to implement this program of "reduction." When Galileo differentiated between "primary" and "secondary" qualities and described physical objects "as bounded and as possessing this or that shape, . . . as in this or that place during this or that time, as in motion or at rest, as in contact or not with some other body, as being one, many, or few . . ."[2] and regarded these properties alone as real, he codified the program of classical physics.

It is commonly claimed that classical physics, in contrast to modern physics, provided a visualizable model of physical reality. Strictly speaking, this is not true. For as soon as the atom was regarded as endowed with only shape, position, and motion and divested, for example, of color, the methodological gain in intellectual unification was paid for by the loss of visualizability or picturability. In fact, how could a colorless object be "pictured"?[3] If we speak of the "picturability" of Newtonian physics, we use this term in a more "pallid" and abstract sense: we mean that those properties which presumably "do really exist in the bodies themselves"[4] can be presented by geometrical-kinematical models whose colors (and other secondary qualities) are completely irrelevant to the purpose for which they were designed, as, for example, to serve as didactic aids in the instruction of mathematics.

In classical physics the primary qualities of shape, position, and motion (later also mass and charge) were regarded as objective features of physical reality or as inseparable attributes of matter, independent of, and irreducible to, observation. Apart from Berkeley's well-known philosophical rejection of any distinction between primary and secondary properties,[5] the first breach in this classical conception, from a logical point of view, was probably Herbart's analysis in which he attempted to

[1] Thomas Hobbes, *Elements of Philosophy—The First Section, Concerning Body*, in *English Works* (Bohn, London, 1839), vol. 1, pp. 389–391.

[2] Galileo Galilei, *Il Saggitore* (1623).

[3] The etymological root of the word "picture" is the Latin *pingere* (= to paint, to cover with color or "pigment").

[4] John Locke, *An Essay concerning Human Understanding* (Otridge and Son, London, 1812), book 2, chap. 8.

[5] Cf. also Gottfried Wilhelm Leibniz, "De modo distinguendi phaenomena realia ab imaginariis," *Philosophische Schriften*, edited by C. J. Gerhardt (Olms, Hildesheim, 1961), vol. 7, p. 322: "De corporibus demonstrare possum non tantum lucem, calorem, colorem et similes qualitates esse apparentes, sed et motum et figuram et extensionem."

show that the notion of *one* thing with *many* properties is self-contradictory. Herbart[6] resolved this paradox by reducing *properties*, including primary qualities, to *relations* of one thing to others. In a similar vein, though more from the standpoint of the epistemology of physics, Stallo pointed out that objects are known only through their relations to other objects. He declared: "They have, and can have, no properties, and their concepts can include no attributes, save these relations, or rather, our mental representation of them. Indeed, an object can not be known or conceived otherwise than as a complex of such relations. In mathematical phrase: things and their properties are known only as functions of other things. . . ."[7] The idea that properties of objects are effects ("Wirkungen") exerted by them either on our senses or on other natural objects was stressed also by Helmholtz when he said that ". . . each quality or property of a thing is, in reality, nothing else but its capability of exercising certain effects upon other things . . . it can never depend upon the nature of one agent alone, but exists only in relation to, and dependent on, the nature of some second object, which is acted upon."[8] Also Høffding, whose philosophy, as we have seen, influenced Bohr so much, declared: "The 'qualities' of a thing are indeed nothing more than the different forms and ways in which this thing influences that thing or is influenced by it. They are a thing's capabilities of doing and suffering."[9]

These statements, if interpreted as including also what was regarded as primary qualities, found their full corroboration in modern physics. At first, the theory of relativity revealed that the geometric-kinematic properties of position, time, and velocity (as well as length, size, duration, and mass), previously regarded as objective features, depend upon the frame of reference. Subsequently, quantum mechanics showed in addition that these properties are relative also to the means of observation.[10] Even the notion of "individuality" or "identity" of elementary particles which, because of its ontological implications, enjoyed a status above and beyond that of "properties" or "qualities," loses in quantum mechanics, and particularly in the context of exchange phenomena, as we have seen,[11] its

[6] Johann Friedrich Herbart, *Schriften zur Metaphysik* (Allgemeine Metaphysik, 1828), *Sämtliche Werke*, edited by G. Hartenstein (Voss, Leipzig, 1851), vol. 3, p. 19.

[7] J. B. Stallo, *The Concepts and Theories of Modern Physics* (Appleton and Co., New York, 1881; Harvard University Press, Cambridge, Mass., 1960), p. 156.

[8] Hermann von Helmholtz, *Handbuch der Physiologischen Optik* (2d ed., Voss, Hamburg, Leipzig, 1896), pp. 588–589; *Treatise on Physiological Optics*, edited by J. P. C. Southall (The Optical Society of America, 1925; reprinted by Dover, New York, 1962), vol. 3, pp. 20–21; *Popular Lectures on Scientific Subjects*, translated by E. Atkinson (Longmans, Green & Co., London, 1873), pp. 260–261.

[9] H. Høffding, *Religionsphilosophie*, translated by F. Bendixen (Reisland, Leipzig, 1901), p. 31; *The Philosophy of Religion*, translated by B. E. Meyer (Macmillan, London, New York, 1906), p. 34.

[10] Cf. Victor F. Lenzen, "The concept of reality in physical theory," *Proceedings and Addresses of the American Philosophical Association 18*, 321–344 (1945)

[11] See end of Sec. 7.1

universal applicability which it possessed in classical physics, where particles could unambiguously be associated with sharply defined trajectories continuous in space and time.

Contrary to the Aristotelian physics of qualities and in contrast to the Newtonian physics of primary properties, the language of quantum mechanics is a language of *interactions* and not of *attributes: processes*, and not *properties*, are the elements of its syntax. Every attempt, for example, to describe in terms of attributes (or their mixtures) the spins in an unpolarized beam of electrons is doomed to failure.

But the human mind does not readily resign to such a renunciation, to such a replacement of properties by interactions (or "particles" by "occupation numbers"). Even Pauli, when defining the spin as "a peculiar, classically not describable two-valuedness,"[12] added "in the quantum-theoretic properties of the optical electron." And although the revocation of primary qualities invalidates the classical distinction between kinematics and dynamics, with statics as a special case of the latter,[13] and thus certainly denies any logical priority of the former, even Heisenberg regarded the uncertainty relations as a means to reinstate that distinction and "to obtain thereby an intuitive understanding of the quantum-mechanical relations."[14] Much of the present dispute on the foundations of quantum mechanics seems to derive from the reluctance to renounce the attribution of primary qualities to elementary particles.

That the classical belief in the reality of physical properties, even if only in the reality of primary qualities such as position or momentum, is inconsistent with the conceptual scheme of quantum mechanics was strikingly illustrated by the widely discussed Einstein-Podolsky-Rosen paradox.[15] Defining a physical theory as *complete* only if "every element of the physical reality" has "a counterpart in the physical theory," Einstein, Podolsky, and Rosen based their argument on the following criterion of physical reality: "If, without in any way disturbing a system, we can predict with certainty the value of a physical quantity, then there exists an element of physical reality corresponding to this physical quantity."[16] Thus, in their view, the possibility of an exact evaluation of a dynamical quantity establishes this quantity as an element of reality, entitled to have a numerical counterpart in the theory, if the theory is complete.

With these epistemological assumptions in mind, the authors considered a combination of two partial systems I and II, which had interacted with each other only during a certain interval of time $0 \leq t \leq T$, and

[12] *See* p. 138.
[13] Cf. M. Jammer, "Statics," in *The Encyclopedia of Physics* (Reinhold, New York, 1966), p. 679.
[14] REF. 11 OF CHAP. 7.
[15] A. Einstein, B. Podolsky, and N. Rosen, "Can quantum-mechanical description of physical reality be considered complete?" *Physical Review 47*, 777–780 (1935).
[16] *Ibid.*, p. 777.

showed that the formalism of quantum mechanics makes it possible, without in any way disturbing II, to deduce the precise value of a variable a_{II} of II from the measurement, after T, of a variable a_I in I. From their epistemological criterion of reality they thus concluded that there exists an element of physical reality—let us call it e_{II}—which corresponds to the exactly calculated value of a_{II}. In fact, e_{II} must have existed prior to the measurement of a_I because, in view of the absence of any interaction after T, the measurement itself could not have affected II in any way.

If instead of a_I another variable a'_I had been measured in I, the exact value of another variable a'_{II} in II could have been predicted, again without in any way disturbing II, and another element of reality e'_{II} would have been established (which likewise must have existed prior to the measurement of a'_I).

The authors then showed that a_I and a'_I can be chosen in such a way that a_{II} and a'_{II} belong to noncommuting operators A_{II} and A'_{II}, respectively.[17] Since a wave function can precisely specify at most only one eigenvalue of two noncommuting operators at a time, either e_{II} or e'_{II}—but never both of these two elements of reality—can have a counterpart in the theory. Quantum mechanics, they therefore concluded, does not provide a complete description of physical reality.

The challenge was soon answered, at least from the viewpoint of the complementarity interpretation of the theory, by Bohr's insistence[18] on the essential influence of the procedure of measurement on the conditions underlying the very definition of physical quantities. Considering these conditions as an inherent element of any phenomenon to which physical reality can be attributed, Bohr pointed out that a mechanical system, even though having ceased to interact dynamically with other systems, does not constitute an independent seat of "real" attributes. Bohr's rejection of the possibility of associating quantities with physical systems in a possessive manner, which rejection invalidated the epistemological premise of the paradox, was clearly but an expression of the fact that, within the conceptual framework of the Born-Bohr-Heisenberg interpretation, quantum mechanics is ultimately a physics of processes and not of properties, a physics of interactions and not of attributes, even not of primary qualities of matter.

From this point of view quantum mechanics may rightfully be regarded as falling in line with the general development of theoretical physics.

[17] For details see Appendix B.

[18] N. Bohr, "Quantum mechanics and physical reality," *Nature 136*, 65 (1935); "Can quantum-mechanical description of physical reality be considered complete?" *Physical Review 48*, 696–702 (1935).

Appendix A

Derivation of the Equation $u_\nu = \frac{8\pi\nu^2}{c^3} U$ *

Planck considered a linear oscillator whose dipole moment $f(t)$ vibrates in the z direction and whose total energy $U_0 = \frac{1}{2}Kf^2 + \frac{1}{2}L\dot{f}^2$. For $U_0 = \text{const}$,

$$Kf + L\ddot{f} = 0 \tag{A.1}$$

its proper frequency is

$$\nu = \frac{1}{2\pi}\sqrt{\frac{K}{L}} \tag{A.2}$$

and

$$(f_{\max})^2 = \frac{2U_0}{K} \tag{A.3}$$

According to Hertz,[1] such an oscillator emits per period of vibration the energy

$$\frac{32\pi^4\nu^3 U_0}{3c^3 K} \tag{A.4}$$

For small damping, (A.1) has to be replaced by

$$Kf + L\ddot{f} + \frac{\sigma}{\pi}\sqrt{KL}\,\dot{f} = 0 \tag{A.5}$$

where σ is the logarithmic decrement (for amplitudes). Hence, by (A.4),

* Equation (1.6), p. 11.

[1] H. Hertz, "Die Kräfte electrischer Schwingungen, behandelt nach der Maxwell'schen Theorie," *Wiedemannsche Annalen der Physik 36*, 1–22 (1889).

the logarithmic decrement for the energy is $2\sigma = 32\pi^4\nu^3/3c^3K$ and

$$\sigma = \frac{16\pi^4\nu^3}{3c^3K} \tag{A.6}$$

In an external field of intensity Z, (A.5) becomes

$$Kf + L\ddot{f} + \frac{\sigma}{\pi}\sqrt{KL}\,\dot{f} = Z$$

or, by (A.2) and (A.6),

$$\ddot{f} + 2\sigma\nu\dot{f} + 4\pi^2\nu^2 f = \frac{3c^3\sigma}{4\pi^2\nu} Z \tag{A.7}$$

If Z is represented by a Fourier series,

$$Z = \sum_{n=1}^{\infty} C_n \cos\left(\frac{2\pi nt}{T} - \theta_n\right)$$

then for large T

$$f(t) = \frac{3c^3}{16\pi^3\nu^2} \sum \frac{C_n T}{n} \sin\gamma_n \cos\left(\frac{2\pi nt}{T} - \theta_n - \gamma_n\right) \tag{A.8}$$

where

$$\cot\gamma_n = \frac{\pi(\nu^2 - n^2/T^2)}{\sigma\nu n/T} \tag{A.9}$$

Since σ is small, $\sin\gamma_n \approx 0$ unless $n \approx \nu T$. Hence

$$f(t) = \frac{3c^3}{16\pi^3\nu^3} \sum C_n \sin\gamma_n \cos\left(\frac{2\pi nt}{T} - \theta_n - \gamma_n\right) \tag{A.8$'$}$$

where

$$\cot\gamma_n = \frac{2\pi(\nu - n/T)}{\sigma\nu} \tag{A.9$'$}$$

The average energy $U = U_\nu$ of the oscillator, given by $U = K\overline{f^2} = (16\pi^4\nu^3/3c^3\sigma)\overline{f^2}$, is therefore

$$U = \frac{3c^3}{32\pi^2\sigma\nu^3} \sum C_n^2 \sin^2\gamma_n \tag{A.10}$$

On the other hand, the intensity J of the exciting radiation, obtained by averaging Z^2 over the time interval from $t = 0$ to $t = T$, is

$$J = \overline{Z^2} = \tfrac{1}{2}\sum C_n^2 \tag{A.11}$$

and can be written as

$$J = \int_0^\infty J_\mu \, d\mu \tag{A.12}$$

where J_μ is the energy absorbed by a linear oscillator of proper frequency μ and negligible damping. Hence

$$J_\mu = \kappa_\mu U_\mu \tag{A.13}$$

where κ_μ is a function of μ. Comparison of (A.11) with (A.12) shows that

$$\tfrac{1}{2} \sum C_n^2 = \frac{3c^3}{32\pi^2} \int_0^\infty \left(\frac{\kappa_\mu}{\sigma \mu^3} \sum C_n^2 \sin^2 \gamma_n \right) d\mu$$

or after interchange of integration and summation,

$$1 = \frac{3c^3}{16\pi^2} \int_0^\infty \frac{\kappa_\mu \sin^2 \gamma_n}{\sigma \mu^3} \, d\mu \tag{A.14}$$

Since only those μ need to be considered for which $\mu \approx \nu$ ($\approx n/T$),

$$1 = \frac{3c^3 \kappa_\nu}{16\pi^2 \sigma \nu^3} \int_0^\infty \left(1 + \frac{4\pi^2(\mu - \nu)^2}{\sigma^2 \nu^2}\right)^{-1} d\mu$$

Since the last integral is $\sigma\nu/2$, $\kappa_\nu = 32\pi^2\nu^2/3c^3$ and, by (A.13),

$$J_\nu = \frac{32\pi^2\nu^2}{3c^3} U_\nu \tag{A.15}$$

Finally, as the density of the field energy is given by

$$u = \frac{1}{8\pi}(E^2 + H^2) = \frac{1}{8\pi} 6\overline{Z^2} = \frac{3}{4\pi} J$$

where E and H are the electrostatic and magnetic-field vectors, it follows that

$$u_\nu = \frac{3}{4\pi} J_\nu$$

or

$$u_\nu = \frac{8\pi\nu^2}{c^3} U_\nu$$

which is Eq. (1.6) on page 11. For details of this derivation the reader is referred to Planck's writings.[2]

[2] M. Planck, "Über irreversible Strahlungsvorgänge," *Annalen der Physik 1*, 69–122 (1900) (see REF. 42 OF CHAP. 1); "Vereinfachte Ableitung der Schwingungsgesetze eines linearen Resonators im stationären Felde," *Physikalische Zeitschrift 2*, 530–534 (1901); *Physikalische Abhandlungen und Vorträge* (REF. 42 OF CHAP. 1), vol. 1, pp. 758–762.

Appendix B

The Einstein-Podolsky-Rosen Paradox

Let x_1 denote the variables used to describe system I and x_2 those for II. The wave function Ψ of the combined system I + II can then be expressed, for $t > T$, by the biorthogonal expansion

$$\Psi(x_1,x_2) = \sum_{n=1}^{\infty} \psi_n(x_2) u_n(x_1) \tag{B.1}$$

where $u_n(x_1)$ are the eigenfunctions of A_1 (the operator representing the observable a_{I} in I) and $\psi_k(x_2)$ describes the state of II if a measurement of a_{I} yielded a_k, the eigenvalue of A_{I}, belonging to $u_k(x_1)$. In the continuous case (B.1) has to be replaced by

$$\Psi(x_1,x_2) = \int_{-\infty}^{\infty} \psi_\alpha(x_2) u_\alpha(x_1)\, d\alpha \tag{B.2}$$

where α now denotes the continuous eigenvalues of A_{I}. If the combined system I + II is supposed to be composed of two particles whose position coordinates are x_1 and x_2, respectively, and if

$$\Psi(x_1,x_2) = \int_{-\infty}^{\infty} \exp\left[\frac{2\pi i}{h}(x_1 - x_2 + x_0)p\right] dp \tag{B.3}$$

where x_0 is an arbitrary constant, for A_{I} the momentum of particle I may be chosen. Since

$$u_p(x_1) = \exp\left(\frac{2\pi i}{h} p x_1\right) \tag{B.4}$$

is then an eigenfunction of A_{I} corresponding to the eigenvalue p, (B.2) shows that

$$\psi_p(x_2) = \exp\left[-\frac{2\pi i}{h}(x_2 - x_0)p\right] \tag{B.5}$$

describes the state of II, if a measurement of A_{I} in I yielded the value p. But (B.5) is an eigenfunction of the operator $A_{\text{II}} = P = (h/2\pi i)\,\partial/\partial x_2$, corresponding to the eigenvalue $-p$ of the momentum of II.

If, however, the position coordinate of I is chosen as A'_{I}, its eigenfunction is $\delta(x_1 - x)$, corresponding to the eigenvalue x. Since (B.2) can now be written

$$\Psi(x_1,x_2) = \int_{-\infty}^{\infty}\int_{-\infty}^{\infty}\left\{\exp\left[\frac{2\pi i}{h}(x - x_2 + x_0)p\right]\delta(x_1 - x)\,dx\right\}dp \tag{B.6}$$

it is clear that

$$\int_{-\infty}^{\infty}\exp\left[\frac{2\pi i}{h}(x - x_2 + x_0)p\right]dp = h\,\delta(x - x_2 + x_0) \tag{B.7}$$

describes the state of II, if a measurement of a'_{I} in I yielded the value x. But (B.7) is an eigenfunction of the operator $A'_{\text{II}} = Q = x_2$, corresponding to the eigenvalue $x + x_0$ of the position coordinate of II. The remark that A_{II} and A'_{II} are noncommutative completes the formal presentation of the paradox. For further details and attempts at resolution the reader is referred to the literature on this subject.[1]

[1] E. Schrödinger, "Discussion of probability relations between separated systems," *Proceedings of the Cambridge Philosophical Society 31*, 555–563 (1935). A. E. Ruark, "Is the quantum-mechanical description of physical reality complete?" *Physical Review 48*, 446–447 (1935); H. Margenau, "Quantum-mechanical description," *ibid. 49*, 240–242 (1936); A. Einstein, "Physics and reality," *Journal of the Franklin Institute 221*, 349–382 (1936); W. H. Furry, "Note on the quantum-mechanical theory of measurement," *Physical Review 49*, 393–399 (1936); "Remarks on measurements in quantum theory," *ibid.*, 476; H. Margenau, "Critical points in modern physical theory," *Philosophy of Science 4*, 337–370 (1937); "Einstein's conception of reality," in *Albert Einstein: Philosopher-Scientist*, edited by P. A. Schilpp (The Library of Living Philosophers, Evanston, Ill., 1949), pp. 243–268; D. Bohm, *Quantum Theory* (Prentice-Hall, Englewood Cliffs, N.J., 1951), pp. 611–622; D. Bohm and Y. Aharonov, "Discussion of experimental proof for the paradox of Einstein, Rosen, and Podolsky," *Physical Review 108*, 1070–1076 (1957); A. Peres and P. Singer, "On possible experimental tests for the paradox of Einstein, Podolsky, and Rosen," *Il Nuovo Cimento 15*, 907–915 (1960); D. R. Inglis, "Completeness of quantum mechanics and charge-conjugation correlations of theta particles," *Reviews of Modern Physics 33*, 1–7 (1961); D. H. Sharp, "The Einstein-Podolsky-Rosen paradox re-examined," *Philosophy of Science 28*, 225–233 (1961); H. Putnam, "Comments on the paper of David Sharp," *ibid.*, 234–237; J. S. Bell, "On the Einstein Podolsky Rosen paradox," *Physics 1*, 195–200 (1964); E. Breitenberger, "On the so-called paradox of Einstein, Podolsky, and Rosen," *Il Nuovo Cimento 38*, 356–360 (1965).

Name Index

Name Index

A

B

C

D

E

F

G

H

I

J

K

L

M

N

O

P

R

S

T